D0653929
DUCHY-STOKE CLIMSLAND
172741

Principles of Crop Improvement

Principles of Crop Improvement

N.W. Simmonds
ScD, AICTA, FRSE, FIBiol
Honorary Professor of Agriculture, University of Edinburgh

J. Smartt
PhD, DSc, DTA (Trin), Dip Agric Sci
Reader, School of Biological Sciences, University of Southampton

with chapter 9 contributed by

S. Millam
HND, BSc, PhD
Scottish Crop Research Institute, Dundee
and
W. Spoor
BSc, PhD
SAC, University of Edinburgh

Blackwell
Science

Blackwell Science Ltd
Editorial Offices:
Osney Mead, Oxford OX2 0EL
25 John Street, London WC1N 2BL
23 Ainslie Place, Edinburgh EH3 6AJ
350 Main Street, Malden
MA 02148 5018, USA
54 University Street, Carlton
Victoria 3053, Australia
10, rue Casimir Delavigne
75006 Paris, France

Other Editorial Offices:

Blackwell Wissenschafts-Verlag GmbH
Kurfürstendamm 57
10707 Berlin, Germany

Blackwell Science KK
MG Kodenmacho Building
7–10 Kodenmacho Nihombashi
Chuo-ku, Tokyo 104, Japan

First published by Longman Group Ltd 1979
Second edition published by Blackwell Science Ltd, 1999

Set in 10/12.5pt Times
by DP Photosetting, Aylesbury, Bucks
Printed and bound in Great Britain by
MPG Books Ltd, Bodmin, Cornwall

DISTRIBUTORS

Marston Book Services Ltd
PO Box 269
Abingdon
Oxon OX14 4YN
(*Orders:* Tel: 01235 465500
Fax: 01235 465555)

USA
Blackwell Science, Inc.
Commerce Place
350 Main Street
Malden, MA 02148 5018
(*Orders:* Tel: 800 759 6102
781 388 8250
Fax: 781 388 8255)

Canada
Login Brothers Book Company
324 Saulteaux Crescent
Winnipeg, Manitoba R3J 3T2
(*Orders:* Tel: 204 837-2987
Fax: 204 837-3116)

Australia
Blackwell Science Pty Ltd
54 University Street
Carlton, Victoria 3053
(*Orders:* Tel: 03 9347 0300
Fax: 03 9347 5001)

A catalogue record for this title is available from the British Library

ISBN 0-632-04191-9

Library of Congress
Cataloging-in-Publication Data is available

For further information on
Blackwell Science, visit our website
www.blackwell-science.com

Contents

Preface

This is the second edition of a work that first appeared in 1979 under the authorship of one of us (NWS); this edition was largely the work of the other (JS) but we are happy to present it jointly and hope that the product will prove as useful as the first edition seems to have been. This preface is not greatly changed from the original foreword, because the subject has not fundamentally changed in 20 years. However, there has been some pruning, several new examples are added and a new chapter on biotechnology (by Drs Millam and Spoor) has been inserted. We hope that the revised product will be both used and useful.

Objective

The objective of this book is to provide a broad, general introduction to the principles of plant breeding. It is not about the practice (which would inevitably be a vast crop-by-crop compilation far beyond our competence); nor is it about the science to which plant breeding naturally appeals, namely genetics, in its more applied aspects. There is, it is true, a good deal of cytogenetics in this book but only as much as has seemed necessary to illuminate the plant breeding ideas at issue. The more interesting problems in plant breeding mostly lie beyond the current reach of genetical analysis but the reader will, we hope, come to agree that genetical ideas offer the only useful way of thinking about them other than at a purely empirical level.

General plan

The book starts with a sketch of crop evolution because of our opinion (and authors must surely be allowed some opinions) that plant breeding is simply, so to speak, the continuation of crop evolution by other means. There is, we believe, no study which so illuminates the breeding of a crop as an understanding of its history. Furthermore, different crops have a way of illuminating each other's problems: the potato breeder does well to think about cassava, the sorghum breeder about cotton and the wheat breeder about tobacco. So the examples in this book are deliberately drawn from a wide range of crops and places and the reader may find what seemed at first to be some strange juxtapositions; they are deliberate, springing from the conviction that plant breeding principles are universal even though practical applications vary greatly in detail. There is, perhaps inevitably, some emphasis upon the crops with which we are most familiar, however, and we confess to a general bias towards field crops at the expense of fruits and vegetables. Choices have to be made, however, and, certainly, any wider treatment would have been longer.

After crop evolution in Chapter 1, the second chapter provides a short general sketch of plant breeding, in the hope that the reader will thereby find his way the more readily through the later, inevitably somewhat more complex, materials. In introducing the subject thus and in providing chapter summaries, we are but following the excellent advice sometimes given to authors (we regret we cannot recall the source): say what you are going to say, say it, then say you've said it. So there are deliberate repetitions in this book but we hope it cannot be described as repetitious.

The rest of the book, Chapters 3–11, follows a more or less logical course, from objectives and genetical principles, through breeding plans (inevitably the heaviest chapter), trials and multiplication, diseases, special techniques and genetic conservation to the social context. The last will no doubt be regarded by some as something of an intrusion in what ought to be a soberly technical work. However, we live in an age in which 'relevance' is demanded of applied science; plant breeding is certainly relevant to the economic success of agriculture and, less directly, to wider questions of social change as well.

Readership

This book is aimed at the honours degree–early postgraduate level of student. It assumes familiarity with basic botany and cytogenetics, a modest knowledge of statistics and some understanding of agriculture. This is a good deal to ask and we expect that some students may find the going difficult in places. But it could hardly have been otherwise: any serious study of an applied science has a way of demanding of the student a rather wide knowledge of diverse fields. Plant breeding is no exception to this rule. The breeder can hardly know too much about all aspects of his crop; furthermore, knowledge of different, even very different, crops is never amiss, for illuminating parallels have a way of turning up in unexpected places. So this will not be found, we think, to be an easy book but it is as easy as we could make it without compromising either the essential diversity of the subject matter or the need for brevity and economy of space.

Figures and tables

We have made the figures and tables in this book, together with the often extensive notes appended to them, carry a considerable load of information,

little of it repeated in the text. The object was saving of space and ease of reference. A number each of the tables and figures are also used to introduce numerical examples for the student to work. Many of the figures are of the flow-diagram kind which lends itself very well to describing diverse processes in which any one step depends upon one or more preceding steps. The following conventions are adopted in such diagrams:
(1) 'White boxes' with black lettering contain material things (usually plants, populations etc.), the subjects of the operations in hand, while 'black boxes' with white lettering contain processes or operations;
(2) functional and historical relations among things and processes are indicated by arrows, vertical or horizontal to the page; when a relation is classificatory rather than causal, the arrowhead is omitted; (3) relative importance is indicated, where necessary, by bold lines or boxes and appropriate weight of type-face, relative unimportance by dashed arrows or lines.

Literature

This is a textbook, not a scholarly monograph, so we have deliberately minimized references in the text because they are, we think, obstructive of a reasonably smooth exposition. On the other hand, we assume that readers will be serious students so have included a fairly extensive bibliography, with references roughly classified by major topics, at the end of each chapter. This is a compromise but, we hope, a workable one. As to the contents of the bibliography, we obviously had to be selective (there already exist, after all, some sixty-seven volumes of *Plant Breeding Abstracts*). We have included some historically important older works but have otherwise concentrated upon the literature of, roughly, the last few decades, favouring, where possible, reviews and books which would provide wider reference lists.

A word of advice to the student may be helpful here. In reading up subjects (as distinct from specific crops) the literature lists to the relevant chapters will give him a start; for deeper search, the indexes of *Plant Breeding Abstracts* will take him further. For individual crops, a short outline will be available in *The Evolution of Crop Plants* (Smartt and Simmonds, 1995) and a number of monographic treatments are listed in Chapter 2; *Plant Breeding Abstracts*, again, will yield comprehensive reference lists.

Acknowledgements

It is a pleasure to acknowledge the help of the following colleagues who read parts of the draft and made helpful criticisms: the late Dr C. Auerbach, Professor A.H. Bunting, Professor D.S. Falconer, Dr J.H. Lennard, Mr R.F. Lord, Dr H.D. Patterson, Professor N.F. Robertson, Dr P.M. Smith, Dr W. Spoor, the late Mr W.S. Wise, Professor M.M. Yeoman. Any errors or faults of treatment which remain are ours alone.

We thank also the following colleagues who provided information, duly acknowledged in the appropriate place: Professor R.W. Allard, Dr F.J. England, Dr B.M. Gerard, Dr P.C. Harper, Dr W.A. Russell, Dr D.L. Shaver, Dr M. Talbot, Dr C.E. Taylor and Dr J.T. Walker.

We have also benefited from discussion with several other colleagues which encouraged us to produce this revised edition, most notably Professors C. N. Law, M. D. Gale, C. A. St. Pierre and Dr P. K. Bretting; for this we are grateful.

Thanks are also due to the authors and publishers for permission to base tables on copyright material as follows: Dr R. W. Allard and Messrs Wiley (Table 4.5); the late Dr A. M. Evans and Messrs Wiley (Table 4.9); *Biological Reviews of the Cambridge Philosophical Society* and The Cambridge University Press (Tables 4.10, 5.16 and Figure 10.5); the Editor of *Theoretical and Applied Genetics* and the Crop Science Society of America (Table 5.4); (Dr A. Marani and the Editor of *Euphytica* (Table 7.2); the Editor of *Heredity* (Table 7.3); The Crop Science Society of America (Table 7.4); the Editor of *Phytopathology* (Table 7.5).

Envoi

We hope this book will be useful and that it will help the student: to comprehend the scientific nature of plant breeding as a specialized, applied aspect of evolutionary and cytogenetic studies; to understand something of the generality of its principles, diversity of practice notwithstanding; to appreciate the diversity of scientific disciplines to which it makes some secondary appeal; and to understand that plant breeding is an intensely practical art having widely beneficial social consequences. It goes without saying that we shall be pleased to have suggestions for the improvement of any later editions that may be called for.

N.W. Simmonds
J. Smartt
February 1999

Chapter 1
The Evolution of Crops

1.1 General

Plant breeding deals with materials which have already been subjected to evolutionary alteration in the past, sometimes to very profound alteration over long periods of time. In this chapter we shall sketch the main features of the evolution of crop plants, concentrating upon those that seem most important from the plant breeding viewpoint. The subject is a huge one and there is no comprehensive survey of the whole field in the English language. Treatments of individual crops will be found in Simmonds (1976a) and Smartt and Simmonds (1995), upon which this chapter is essentially based. Table 1.1 summarizes some of the leading features and provides examples additional to those cited in this chapter.

1.2 Time scales

Wild plants taken into cultivation become genetically altered in the process. At some ill-defined stage along the line, the altered populations may be said to have been domesticated; definitions of the word vary but it is probably generally acceptable to say that a plant population has been domesticated when it has been substantially altered from the wild state and certainly when it has been so altered as to be unable to survive in the wild. Our knowledge of old domestications is always inferential and exact dates are never known.

Information takes the form of archaeological or historical data which show that a certain crop had been domesticated by such and such a time. All early dates are therefore merely earliest available records, subject to revision by later research. The accuracy of dating early plant remains has improved enormously in recent decades, with the general adoption of radio-carbon methods; but well-preserved archaeological materials are scarce and it is as well to remember that reliable early dates are yet available for rather few crops. It is notable that most of those that are available come from sites in arid places such as the eastern Mediterranean and, in the New World, Mexico and Peru. Such sites favour

Table 1.1 Some leading features of crop evolution (1).

Family and genus	Crop	Place	Age	Use	Reproduction	Ploidy
ACTINIDIACEAE						
Actinidia deliciosa	kiwifruit	S. China	R	fr.	ob/sd	6*x*
AGAVACEAE						
Agave spp.	sisal, etc.	C. America	?	fb	ob/cl	2*x*–5*x*
AMARANTHACEAE						
Amaranthus spp.	amaranths	C. America	E	ce	obib/sd	2*x*
A. caudatus	amaranth	Andes	E	ce	obib/sd	2*x*
ANACARDIACEAE						
Mangifera	mango	N. India	E	fr	ob/sdcl	2*x*
Anacardium Occidentale	cashew	S. America	L	nt	ob/sd	2*x*–4*x*
Pistacia vera	pistachio	C. Asia	E	nt	ob/sd	2*x*
ARACEAE						
Alocasia	–	S. India	?	tb	?ob/cl	2*x*
Colocasia	taro	India	E	tb	?ob/cl	2*x*, 3*x*
Cyrtosperma	–	?Indonesia	?	tb	?ob/cl	?
Xanthosoma	tanier	Trop. America	?	tb	?ob/cl	2*x*
BOMBACACEAE						
Ceiba pentandra	kapok	C., S. America	R	in	ob/sd	many *x*
BROMELIACEAE						
Ananas	pineapple	Central S. America	?L	fr	ob/cl	2*x*
CAMELLIACEAE						
Camellia	tea	S.E. Asia	E	bv	ob/sdcl	2x

CANNABINACEAE						
Cannabis	hemp	C. Asia	?E	ib/os/dr	ob/sd	2*x*
Humulus	hops	Europe	L	fl	ob/cl	2*x* 3*x*
CARICACEAE						
Carica	papaya	C. America	?	fr	obib/sd	2*x*
CHENOPODIACEAE						
Beta	sugar beet	Europe	R	in	ob/sd	2*x*
Chenopodium	quinoa	Andes	E	ce	obib/sd	2*x*
COMPOSITAE						
Carthamus	safflower	Near East	E	os	ib/sd	2*x*
Chrysanthemum	pyrethrum	S. Europe	R	in	ob/sd	2*x*
Helianthus annuus	sunflower	N. America	E	os	ob/sd	2*x*
H. tuberosus	Jer. artichoke	N. America	?L	tb	ob/cl	6*x*
Lactuca	lettuce	Mediterranean	E	vg	ib/sd	2*x*
CONVOLVULACEAE						
Ipomoea	sweet potato	Western S. America	?A	tb	ob/cl	auto 6*x*
CRUCIFERAE						
Brassica campestris	turnip, etc.	Med., Near East	E	fd/vg/os	ob/sd	2*x*
B. oleracea	kale, etc.	Med., Near East	E	fd/vg/os	ob/sd	2*x*
B. napus	swede,rape	Europe	L	fd/vg/os	ibob/sd	allo 4*x*
B. nigra	black mustard	Near East	?E	fl/os	ob/sd	2*x*
B. juncea	brown mustard	N. India	?E	fl/os	ib/sd	allo 4*x*
B. carinata	Ethiopian M.	E. Africa	?E	fl/os	?ob/sd	allo 4*x*
Sinapis	white mustard	Mediterranean	?	fl/os	ibob/sd	2*x*
Raphanus	radish	Near East	E	fd/vg/os	ob/sd	2*x*

Contd

Table 1.1 *Contd.*

Family and genus	Crop	Place	Age	Use	Reproduction	Ploidy
CURCURBITACEAE						
Cucumis sativus	cucumber	India	?E	vg	obib/sd	$2x$
C. melo	muskmelon	Africa	?L	fr	obib/sd	$2x$
Citrullus	watermelon	Africa	E	fr	obib/sd	$2x$
Cucurbita spp.	squashes	C. America	A/E	vg	obib/sd	$2x$
DIOSCOREACEAE						
Dioscorea alata	yam	S.E. Asia	E/A	tb	ob/cl	$3x$–$8x$
D. esculenta	yam	S.E. Asia	A	tb	ob/cl	$4x$
D. rotundata	yam	Trop. Africa	A	tb	ob/cl	$4x$
D. cayenensis	yam	Trop. Africa	A	tb	ob/cl	$4x$–$14x$
D. trifida	yam	S. America	?	tb	ob/cl	$6x$–$9x$
ERICACEAE						
Vaccinium	blueberry, cranberry, bilberry, lingonberry	N., S. America, Europe, Asia	R	fr.	ob/sd	$2x$ $4x$ $6x$
EUPHORBIACEAE						
Aleurites fordii	tung	C. China	E	os	ibob/sd	$2x$
A. montana	tung	S. China	E	os	ob/sdcl	$2x$
Hevea	rubber	Amazon (S.E. Asia)	R	in	ob/sdcl	$2x$
Manihot	cassava	S. America	?E	tb	ob/cl	$2x$
Ricinus	castor	E. Africa	?E	os	ob/sd	$2x$
GRAMINEAE						
Avena strigosa-brevis	oat	Iberia	?E	fd	ib/sd	$2x$
A. abyssinica	oat	Ethiopia	?E	ce	ib/sd	auto $4x$
A. sativa-byzantina	oat	Europe	E	ce	ib/sd	allo $6x$
A. nuda	oat	?C. Asia	?E	ce	ib/sd	allo $6x$
Eleusine	finger millet	E. Africa	E	ce	ib/sd	$4x$

Hordeum	barley	E. Med.	A	ce	ib/sd	2*x*
Oryza sativa	rice	Indo-China	E	ce	ib/sd	2*x*
O. glaberrima	rice	W. Africa	E	ce	ib/sd	2*x*
Pennisetum	bullrush millet	W. Africa	E	ce	ob/sd	2*x*
Saccharum	sugarcane	N. Guinea	E	in	ob/cl	many-*x*
Secale	rye	Near East	E	ce	ob/sd	2*x*
Setaria italica	foxtail millet	Eurasia	E	ce	ibob/sd	2*x*
Secalotriticum	triticale	Mexico, Canada	R	ce	ib/sd	allo 6*x*
Sorghum	sorghum	Trop. Africa	E	ce	ibob/sd	2*x*
Triticum monococcum	einkorn	Near East	A	ce	ib/sd	2*x*
T. timopheevii	wheat	Near East	A	ce	ib/sd	allo 4*x*
T. turgidum	emmer	Near East	A	ce	ib/sd	allo 4*x*
T. aestivum	bread wheat	Near East	A	ce	ib/sd	allo 6*x*
Zea	maize	C. America	E/?A	ce	ob/sd	2*x*
GROSSULARIACEAE						
Ribes nigrum	blackcurrants	Europe	L	fr	ob/cl	2*x*
R. sativum	redcurrants	Europe	L	fr	ob/cl	2*x*
LAURACEAE						
Persea	avocado	C. America	E	fr	ob/clsd	2*x*
LEGUMINOSAE						
Arachis	groundnut	Central S. America	E	nt	ib/sd	allo 4*x*
Cajanus	pigeon pea	India	?E	pu	ibob/sd	2*x*
Cicer	chickpea	Near East	E/A	pu	ib/sd	2*x*
Glycine	soybean	China	E	pu	ib/sd	2*x*
Lathyrus sativus	grasspea	India, S. Europe, Asia, N. Africa	E	pu	ibob/sd	2*x*
Lens	lentil	Near East	A	pu	ib/sd	2*x*
Lupinus	lupins	N. East, Americas	A	pu	ibob/sd	many *x*
Medicago	alfalfa	Near East	E	fd	ob/sd	auto 4*x*
Phaseolus vulgaris	common bean	C. or S. America	E/A	pu	ib/sd	2*x*
Ph. coccineus	runner bean	C. America	E/L	pu	ibob/sd	2*x*

Contd

Table 1.1 *Contd.*

Family and genus	Crop	Place	Age	Use	Reproduction	Ploidy
LEGUMINOSAE (Contd)						
Ph. acutifolius	tepary bean	C. America	E	pu	ib/sd	2*x*
Ph. lunatus	Lima bean	C. or S. America	E/L	pu	ib/sd	2*x*
Pisum	pea	Near East	A	pu	ib/sd	2*x*
Psophocarpus tetragonolobus	winged bean	?Africa	R	pu	ibob/sd	2*x*
Vicia	bean	Near East	E	pu	ob/sd	2*x*
Vigna	Asiatic grams	India, Far East	E	pu	ibob/sd	2*x*
Vigna unguiculata	cowpea	Trop. Africa	E	pu/fd	ib/sd	2*x*
LILIACEAE						
Allium cepa	onion	C. Asia	E	vg	ob/sd	2*x*
A. fistulosum	Japanese b. onion	E. Asia	E	vg	ob/sd	2*x*
A. schoenoprasum	chive	Europe	L	vg	ob/sdcl	2*x*–4*x*
A. chinense	rakkyo	E. Asia	?E	vg	ob/cl	2*x*–4*x*
A. sativum	garlic	C. Asia	E	vg	ob/cl	2*x*–4*x*
A. ampeloprasum	leek	Near East	E	vg	ob/sdcl	4*x*, 6*x*
A. tuberosum	Chinese chive	E. Asia	?E	vg	ob/sd	2*x*, 4*x*
LINACEAE						
Linum	flax, linseed	Near East	A	fb/os	ib/sd	2*x*
MALVACEAE						
Abelmoschus	okra	?Africa, ?India	?	vg	ib/sd	many-*x*
Gossypium herbaceum	cotton	Near East	E	fb	ibob/sd	2*x*
G. arboreum	cotton	India	E	fb	ibob/sd	2*x*
G. hirsutum	upland cotton	C. America	E	fb	ibob/sd	allo 4*x*
G. barbadense	sea island cotton	Western S. America	E	fb	ibob/sd	allo 4*x*

MORACEAE						
Artocarpus	bread-fruit	S.E. Asia–W. Pacific	?E	vg	ob/cl	2*x*, 3*x*
Ficus	fig	Near East	E/?A	fr	ob/cl	2*x*
MUSACEAE						
Musa	bananas	S.E. Asia	E	fr	ob/cl	2*x*–4*x* (auto, allo)
MYRTACEAE						
Eugenia	clove	S.E. Asia	E	fl	ibob/sd	2*x*
OLEACEAE						
Olea	olive	Near East	E	os	ob/sdcl	2*x*
PALMAE						
Bactris gasipaes	peach palm	C., S. America	?E	fr	ob/sd	2*x*
Cocos	coconut	S.E. Asia	?E	os	ob/sd	2*x*
Elaeis	oil palm	W. Africa (S.E. Asia)	R	os	ob/sd	2*x*
Phoenix	date	S.W. Asia	E/?A	fr	ob/cl	2*x*
PEDALIACEAE						
Sesamum	sesame	?E, Africa, ?India	E	os	ibob/sd	2*x*
PIPERACEAE						
Piper	pepper	India	?	fl	ob/cl	4*x*–8*x*
POLYGONACEAE						
Fagopyrum	buckwheat	E. Asia	L	ce	ob/sd	2*x*

Contd

Table 1.1 *Contd.*

Family and genus	Crop	Place	Age	Use	Reproduction	Ploidy
ROSACEAE						
Fragaria	strawberry	Europe	R	fr	ob/cl	allo 8*x*
Prunus spp.	cherries	America and W. Asia	E	fr	ob/cl	2*x*, 4*x*
Prunus spp.	plums	W. Asia, America and Europe	E	fr	ob/cl	2*x*–6*x*
P. armeniaca	apricot	W. Asia	?E/L	fr	ibob/cl	2*x*
P. persica	peach	China	?E/L	fr	ibob/cl	2*x*
P. amygdalus	almond	W. Asia	?E/L	nt	ob/cl	2*x*
Malus	apple	E. Asia	E	fr	ob/cl	2*x*, 3*x*
Pyrus	pear	E. Asia	E	fr	ob/cl	2*x*, 3*x*
Rubus	red raspberries	Europe	L	fr	ob/cl	2*x*
Rubus	blackberries	Europe and N. America	R	fr	ob/cl	4*x*, 8*x*
RUBIACEAE						
Cinchona	quinine	Andes (Indonesia)	R	dr	ob/sdcl	2*x*
Coffea canephora	coffee	W. Africa	L	bv	ob/sdcl	2*x*
C. arabica	coffee	E. Africa	L	bv	ib/sdcl	allo 4*x*
RUTACEAE						
Citrus	citrus	S.E. Asia	?E	fr	ob/cl	2*x*
SOLANACEAE						
Capsicum	peppers	C. & S. America	A	fl/vg	ib/sd	2*x*
Lycopersicon	tomato	Western S. America	?E	vg	ib/sd	2*x*
Nicotiana	tobacco	Andes	?E	dr	ib/sd	allo 4*x*
Solanum melongena	eggplant	India	E	vg	ib/sd	2*x*
S. tuberosum	potato	Andes	E	vg	ob/cl	2*x*–5*x* (auto 4*x*)

STERCULIACEAE						
Cola	kola	W. Africa	R	dr	ob/sdcl	4*x*
Theobroma	cacao	N.W.S. America	?L	bv	ob/sdcl	2*x*
TILIACEAE						
Corchorus spp.	jutes	India	R	fb	ib/sd	2*x*
UMBELLIFERAE						
Daucus	carrot	S.W. Asia	?E/L	vg	ob/sd	2*x*
VITACEAE						
Vitis	grape	S.W. Asia	E	fr	ob/cl	2*x*

This table is based upon Simmonds (1976a), and Smartt and Simmonds (1995) q.v. for fuller information. Families are listed alphabetically and genera are (usually) alphabetical within families. Species names are given only where necessary. The treatment is, inevitably, highly condensed and several cases of closely allied crops are treated jointly. The column headed Place is the region in which the crop is thought, or known, to have originated, *either* having made the transition from wild to domesticated there *or* having risen there *de novo* from previously cultivated ancestors (e.g. strawberry); domestication remote from wild ancestors is indicated by brackets (e.g. rubber, oil palm, quinine). Age (where known) is indicated by: A (ancient), before 5000 BC; E (early), 0–5000 BC; L (late), AD 0–1700; R (recent), after AD 1700; there is a good deal of informed guesswork here.

Uses are indicated as follows: ce, cereal; pu, pulse; tb, tuber; vg, vegetable; os, oilseed; fr, fruit; nt, nut; fl, flavour; dr, drug; bv, beverage; fb, fibre; in, industrial; fd, fodder. The column headed Reproduction indicates breeding and propagation systems: ob (outbred, habitually crossed, suffers inbreeding depression); ib (inbred, usually/always selfed, tolerant of inbreeding); ibob (in-out-bred, usually nearer ib than ob); sd (seed propagated); cl (clonal); sdcl (mixed system of propagation). The last column indicates Ploidy; when known, the nature of polyploidy is indicated by allo, auto-.

preservation of human artefacts (including plant remains) and have, naturally, been attractive to archaeologists. Plant remains would have little chance of prolonged survival in moist, warm, climates so the sample we have is biassed. Nevertheless it probably gives us reasonable terminal dates for some crops; for the rest, the majority, we must rely on reasoned guesswork (i.e. biological and historical inference).

The oldest records we have of undoubtedly domesticated materials come from many sites in the Near East (modern Turkey, Syria, Israel, Iran, Iraq) dated around 6000–7000 BC. Einkorn and emmer wheats, flax, pea, vetch and lentil all occur. So domestication of these crops can be inferred to have started at least eight or nine thousand years ago. In the New World, comparable dates have been recorded for *Phaseolus vulgaris* in Peru and Mexico and *Cucurbita* domesticates are probably of similar age; for maize, curiously, the evidence suggests a rather later date (about 5000 BC).

With better knowledge, other crops may well join the list of very early domesticates. On present evidence, a great many evolved from the wild state during the nine millennia or so intervening between the earliest domestications and the present. Thus rice, sorghum, soybeans and sugarcane were probably all domesticated a few millennia BC while others came into cultivation only since the time of Christ. At the later extreme, several crops, three of them, at least, important ones, have very recent origins indeed. Thus sugar beet developed as a crop in eighteenth-century Europe, while rubber and oil palm were domesticated around the end of the nineteenth century, all three as the result of deliberate policy decisions to the effect that new crops were wanted to fulfil certain functions. The forage grasses and clovers are also quite recent crops and it is arguable that many have yet to make the transition to domestication.

Origins have been accompanied by extinctions. No doubt, some crops, perhaps even many, have disappeared without trace. Three recent extinctions (or incipient extinctions) of crops that had at least local significance are sainfoin (*Onobrychis*), indigo (*Indigofera*) and cañahua (*Chenopodium pallidicaule*).

Viewed broadly, we certainly can not say that domestication of crops has been limited to specific historical periods; domestication (and, no doubt, extinction) are continuously spread over time, a reflection of the ever-changing demands of human societies for new and improved agricultural products.

The essential facts about age, drawn together from Table 1.1, are summarized in Table 1.2. Allowing for a good deal of guesswork and uncertainty, it looks as though rate of domestication (per 1000 years, say) has not changed very greatly over time.

1.3 Geography of crop evolution

Every crop originated somewhere and the wide diffusion of places of origin around the world was recognized at least a century ago by De Candolle (1886)

Table 1.2 Places and times of domestication.

Type	Age	Near East and Europe	Central and East Asia	Africa	America	Totals
Cereals and pulses	A	7	0	0	0	7
	E	4	4	6	9	23
	L	0	1	0	0	1
	R	0	0	0	1	1
	Totals	11	5	6	10	32
Tubers, vegetables and fruits	A	0	1	2	2	5
	E	10	17	2	7	36
	L	6	3	1	3	13
	R	2	0	0	1	3
	Totals	18	21	5	13	57
Other crops	A	1	0	1	0	2
	E	7	10	3	4	24
	L	2	0	2	1	5
	R	2	2	2	1	7
	Totals	12	12	8	6	38
Totals		41	38	19	29	127
Totals		A	E	L	R	
		14	83	19	11	

Notes: A, ancient, 7000–5000 BC; E, early 0–5000 BC; L, late AD 0–1750; R, recent, after AD 1750.
Rates (domestications per 1000 years): A – 7, E – 17, L and R – 15 (*rough* estimates).

working with botanical, historical and linguistic evidence (but with more or less Lamarckian assumptions). In this century, Vavilov and his colleagues, working with the great plant collections assembled by the Institute of Plant Industry in St. Petersburg, developed the idea systematically (Vavilov, 1951). He concluded that each crop had a characteristic primary centre of diversity which was also its centre of origin. He recognized twelve areas in which, he argued, all our major crops had been domesticated. These were: Abyssinia, Mediterranean, Persia, Afghanistan, Indo-Burma, Siam–Malaya–Java, China, Mexico, Peru, Chile, Brazil–Paraguay, USA.

Vavilov's views were very influential and indeed still form the basis of our present understanding. But they now need considerable modification, for three reasons. First, it is now apparent (Harlan, 1975a) that only rarely can we speak of a centre of origin in the sense of a relatively small, well-defined patch of country; more often the initial phases of evolution seem to have been spread out over large, even very large, rather ill-defined areas, crops travelling with man and evolving *en route*; thus diffuse, regional (even multiple) origins replace the idea of unique local ones. Second, the associated idea of twelve or so areas of outstanding evolutionary activity disappears with the idea of localized origins: we now think of a geographical continuum of domestication and crop evolution wherever agriculture is practised, in Old and New Worlds. And, third, there now seems to be little connection between source of the wild ancestors, area of domestication and area of evolutionary diversification. Sometimes two or more of these coincide, sometimes they do not.

Our knowledge of the geography of crop evolution has certainly improved very greatly in the sixty years that have elapsed since Vavilov's time. Thus, tropical West Africa did not even appear in Vavilov's map, but we now know it to have been an important area for the evolution of at least six major crops. However, our knowledge, though much improved, is yet far from perfect and many outstanding problems and uncertainties remain. The now generally accepted notion of a geographical continuum is unlikely to be shaken, however; the only rule seems to be that the geographical history of every crop is unique.

Geographical sources, by very broad regions, are summarized in Table 1.2, condensed from Table 1.1. The very wide diffusion of domestication through all the continents, except Australasia, comes out very well; until very recently, Australia lacked the basic requirement, namely settled agriculture, a necessary, even sufficient, condition for active domestication. If the list were longer, incidentally, Australia would appear as the recent source of the *Macadamia* nut, now a domesticate; no doubt other potential crops lie waiting there?

1.4 People

To the question: Who was responsible for domestication and crop evolution? the answer must be: all sorts of people from peasants to PhDs.

Until about 200 years ago (and then only in a few technically advanced, temperate countries), crop evolution was in the hands of farmers, the users of the products; and it still is today for many crops in poorer countries. Probably, the total genetic change achieved by farmers over the millennia was far greater than that achieved by the last hundred or two years of more systematic science-based effort.

Nothing is directly known about how farmers in the ancient past achieved results but something may be inferred from recent observation and historical records. First, peasants tend to be keen and competent botanists who are well aware of differences between closely related plants in taxonomic and economic characters. Selection of preferred planting material is widely, perhaps rather generally, practised. Furthermore, some people even have an understanding of the maintenance of pure seed stocks; for example, certain American Indians who even knew how to maintain maize populations in isolation. And, second, underlying conscious selection, there is always natural selection which, as we shall see below, is both universal and effective; the art of cultivation is perhaps the peasant's most potent contribution.

The peasant however is nearly everywhere declining and his home-grown stocks are being replaced by the products of plant breeding multiplied by seedsmen of one kind or another, private or state-supported. So the current phase of crop evolution is rapidly passing into the hands of professional plant breeders, a trend which has been at least locally apparent for 200 years. At the level of domestication the same is true: sugar beet, rubber and pyrethrum all owed their origins as crops to state-supported research workers. Rubber as a crop was, to a remarkable extent, the achievement of one man, H.N. Ridley, long Director of the Singapore Botanical Garden, who lived long enough to see the crop he virtually created grow into an industrial giant. And the US Department of Agriculture maintains a New Crops Research Branch that is devoted, in part, to new domestication.

There have been many speculations about the social circumstances and motives of the men who first domesticated crops in ancient times. They include the following, though we should add that they are plausible rather than secure. Thus, for some people, agriculture may have been an escape from ecological circumstances in which food was scarce because edible wild plants and catch-able game were scarce or where survival depended upon tedious seasonal migrations; perhaps some people just got tired of living in tents and always being hungry so looked, ingeniously, for an alternative? This sort of transition would no doubt have been encouraged by the occurrence, both of fertile, well-watered patches of land on which plants obviously grew well and of favoured wild food plants as subjects for improvement. Again, the idea of agriculture may have come, as Sauer has argued, especially readily to fisher-folk, who are necessarily fairly sedentary and tend to be technologically inclined; a man who can make a net or a fish-hook can make a hole in the ground. Nor should we forget that man is a highly differentiated species and that human groups differ from each other. Thus there could well have been a human-genetic element in crop-domestication, as Darlington has suggested; perhaps some people recog-nized the idea of agriculture but just didn't fancy it? The possible speculations, in short, are endless and not very profitable. I doubt whether any generalization is possible beyond the not very helpful one that ecological opportunity, human skill and interest and a large random element must have been involved.

1.5 Plant sources and products

The 140–150 crops listed in Table 1.1 (see also Table 1.2) are assigned to about 40 families and 96 genera, plus or minus a few each to allow for nomenclatural uncertainty. If the minor crops covered in the Appendix to Simmonds (1976a) were included, the number of crops would rise to about 230 and of families and genera to 64 and 180. Clearly, the sources of our crops are very diverse systematically but still represent only a small fraction of the total 300 families and 3000 genera roughly estimated for the angiosperms as a whole.

This diversity reflects the diversity of human needs for the products: 13 broad categories of use are listed in Table 1.1. It also reflects the diversity of places in which crops were domesticated; every agriculture needs cereals, tubers, vegetables, fruits and fibres and Tables 1.1 and 1.2 show that they were developed locally from whatever plant sources were available. So, important cereals and tubers originated from independent sources in widely separated places and, not unnaturally, showed parallel features in doing so; the tuberous aroids and yams provide striking examples of widely separated independent domestications of related materials.

Families vary widely in overall economic importance from those few that have produced several important crops (Gramineae – cereals and fodders; Leguminosae – pulses and fodders; Solanaceae – diverse products; Cruciferae – fodders, vegetables and oil seeds) to others, the majority, that have produced but few.

Multiple uses for a single crop species are not uncommon and there must have been many changes of evolutionary direction. The *Brassica* crops, flax/linseed and hemp are perhaps the most striking examples in Table 1.1. Safflower (shown only as an oilseed in Table 1.1) probably started its domestic life as a dye plant, a use which is now virtually extinct.

1.6 Evolutionary processes

Neo-Darwinian evolution

The early years of this century saw the joining of two major strands of biological thought. Darwin's ideas on the differential breeding of better adapted individuals were coupled with Mendel's analysis of hereditary difference. These notions supplemented by Weissman's idea of the continuity of the germ-plasm and Johannsen's demonstration of the phenotype–genotype relationship laid the foundations for the coherent theory of evolutionary processes that was soon to emerge. This theory is well described as Neo-Darwinian because it is founded on the selection theory of Darwin and on the genetics of his successors. A satisfactory broad picture of evolution up to the species level emerged by the time of the Second World War and was surveyed in two key works published about that time: Huxley (1940) and Dobzhansky (1941, 2nd edn, 1949). Later studies have amplified the picture without changing it (e.g. Stebbins, 1950; Grant, 1971).

The leading features of the theory (as applied to diploids) are as follows. Plant species are commonly geographically differentiated (into subspecies, ecotypes, clines, etc.) as a result of natural selection operating upon genetic variability. Variability is maintained by heterozygosity supplemented by gene-flow between populations; and heterozygosity itself is adaptively adjusted, upwards and downwards, by various cytological and genetical mechanisms. Reproductive isolation between populations leads to speciation and, since it generally develops gradually, speciation is a continuous rather than a discontinuous process. At any one time, therefore, a group of related plants is likely to comprise biologically distinct species, which hybridize not at all or only with difficulty, and a variety – often a great variety – of intra-specific groups between which various levels of genetic exchange are possible and do, in fact, occur. All this may be summarized by saying that adaptation is procured by successive gene-substitutions in evolving populations, leading to local differentiation and, ultimately, to speciation.

Plants, but very few animals, have another evolutionary resource, namely polyploidy. The customary distinction between autopolyploidy (structurally identical genomes, unrestricted recombination) and allopolyploidy (genomes so differentiated as to restrict pairing and recombination to homologous chromosomes, functionally diploid) is useful and valid but not infrequently obscured by intermediate situations (segmental allopolyploidy). Polyploidy may sometimes be selectively advantageous *per se*. More generally its significance lies in: (1) facilitating recombination that would be restricted at the diploid level; (2) in permitting adjustment of the mating system (towards inbreeding); (3) in offering (in allopolyploids) an opportunity for permanent interspecific hybridity; and (4) in offering (again in allopolyploids) an opportunity for long-term diploid differentiation by way of adaptive adjustment of duplicate loci.

The essential feature of Neo-Darwinian micro-evolution therefore emerges as gene substitution; polyploidy is important (as shown by its frequent occurrence in wild plants) but only as a preliminary to gene-substitution processes which would not otherwise have been possible. We shall see below that cultivated plants obey the same evolutionary rules as wild ones.

Selection, natural and human

General

The popular notion of selection is of 'nature red in tooth and claw', with the implication of zygote mortality as the principal element. Differential reproduction as between genotypes, however, is pervasive and probably far more important.

In the evolution of crop plants we can usefully distinguish (in principle though not always in practice) between natural and human selection. The former is selection (i.e. differential reproduction which may, but which will commonly not, be expressed as zygote mortality) inherent in the cultivation of a genetically heterogeneous population at a particular place and time. The latter is the result of conscious decision by the farmer, or plant breeder, to keep the progeny of this or that parent in preference to others. In either case, there will (tend to) be evolutionary change in the form of improved adaptation, that is, enhanced reproductive fitness in the joint circumstances of cultivation and the breeder's desires. The relative importance of natural and human selection has surely changed over the millennia, in favour of the latter but that there was some conscious selection in the earliest phases and that natural selection in the modern plant breeder's plots is still significant seem equally certain.

We shall now examine some examples of these two kinds of selection. The first section (A) of Table 1.3 refers to plant characters which we can either infer were significantly changed during the earlier phases of crop evolution or know have been the recent object of plant breeders' attentions.

Table 1.3 Some leading features of crop evolution.

A. Morphological and chemical features

1. Reduced plant size, determinate growth, dwarfing, associated usually with favourable partition but not conspicuously with shorter life cycle:
   Sunflower, barley, rice wheats, soybean, pea, beans (*Phaseolus*), cowpea (and probably other cereals and pulses), some hemp, bananas, coconut, apple and pear (rootstocks), peppers, tomatoc, potatoes, kola.
2. Reduced plant size and woodiness associated with shorter life cycle (trend to annuality):
   Brassica crops, radish, castor, cassava, rye, pigeon pea, *Phaseolus* beans, flax/linseed, cottons, buckwheat.
3. Taller, less branched plants, leading to fewer, larger inflorescences or fibrous stems:
   Grain amaranths, sunflower, maize, flax, hemp, jute.
4. Altered photoperiod/vernalization requirements, associated with latitudinal/climatic adaptation:
   Grain amaranths, beets, lettuce, *Brassica* crops, radish, rice, sugarcane, rye, sorghum, wheats, maize, pigeon pea, pea, soybean, *Phaseolus* beans, cowpea, onion, hemp, buckwheat, potato, jute, carrot.
5. Reduced spininess:
   Sisal, pineapple, lettuce, yams, castor, okra, blackberry, eggplant.
6. Reduction of toxic constituents:
   Mango (resins), quinoa (saponins), safflower (polyphenolics), *Brassica* crops (various S-compounds), cucurbits (cucurbitacins), yams (alkaloids), cassava (CN-glycosides), pulses generally (trypsin inhibitors, haemagglutinins, amino acids, CN-glycosides, cotton (gossypol), eggplant (bitter principle), potatoes (steroidal alkaloids).
7. Development of attractive colours/patterns:
   Grain amaranths, sugarcanes, maize, various pulse seeds (e.g. *Phaseolus, Vicia*), *Capsicum* peppers, potatoes, carrot, tomato.
8. Non-shattering infructescences/fruits:
   Grain amaranths, quinoa, safflower, lettuce, *Brassica* crops, castor, grass-cereals and pulses generally, flax/linseed, hemp, buckwheat, tobacco.
9. Reduced seed (or tuber) dormancy:
   Quinoa, yams (tuber), oats, rice, rye, wheats, pulses, buckwheat, potatoes (tuber).
10. Multiple uses, changes of evolutionary direction:
    Beets, safflower, *Brassica* crops, radish, flax/linseed, hemp, *Capsicum* peppers, cucurbits.

Table 1.3 *Contd.*

B. Cytogenetic features

1. Autopolyploidy, sometimes very recent, usually associated with non-seed crops in which fertility is of little importance or positively disfavoured:
   Sugarbeet (3x), pyrethrum (3x, 4x), oats (4x, *abyssinica*), alfalfa (4x but perhaps with segmental allo-4x features), cloves (4x), bread-fruit (3x), hop (3x), bananas (3x, 4x, also allo-3x-4x), potatoes (Andigena–Tuberosum Groups, 4x).
2. Allopolyploidy, mostly long-established, in highly seed-fertile crops:
   Brassica napus, *carinata* and *juncea* (all 4x), oats (*sativa–byzantina–nuda* complex, 6x), triticale (6x), wheats (*timopheevii* and *turgidum*, 4x; *aestivum*, 6x), groundnut (4x), cottons (*hirsutum*, *barbadense*, 4x), strawberry (8x, probably mixed auto and allo), coffee (*arabica*, 4x), tobacco (*rustica* and *tabacum*, both 4x).
3. Ill-defined polyploidy:
   Jerusalem artichoke (6x), yams (3x–14x), finger millet (4x, probably allo-), sugarcane (many-x), *Allium* spp. (3x–6x), okra (many-x), black pepper (4x–8x), *Prunus* fruits (4x–6x), apples and pears (3x), blackberries (4x–8x), kola (4x).
4. Clonal propagation for products other than seed, associated with various derangements/reductions of flowering and sexual reproduction and varying degrees of seed-sterility:
   Sisal and relatives, tuberous aroids, pineapple, sweet potato, yams, cassava, sugarcane, several *Allium* spp., bread-fruit, fig, bananas, black pepper, strawberry, blackberries, citrus, potatoes, grape.
5. Enhanced inbreeding, in widely varying degree:
   Papaya, *Brassica* crops, cucurbits, rice, alfalfa, beans, (*Vicia*), flax/linseed, hemp, clove, coconut, oil-palm, strawberry, some *Prunus* fruits, raspberries, *Capsicum* peppers, tomato, cacao, grape.
6. Extensive hybridization, after primary domestication, with wild or weedy relatives, leading to recombination/introgression, some of it quite recent; hybridization leading to allopolyploidy (see B2) excluded:
   Sisal and relatives, grain amaranths, safflower, lettuce, *Brassica* crops, yams, cassava, rice, sugarcane, sorghum, maize, black and red currants, alfalfa, cowpeas, cottons, hemp, hop, strawberry, *Prunus* fruits, apple, pear, raspberries and blackberries, potatoes, grape.
7. Incipient speciation within cultivated species suggested (but may represent earlier diversity of hybridization):
   Safflower, barley, rice, groundnut.
8. Examples of important major-gene effects in crop evolution:
   Safflower (non-shattering infructescence), cucurbits (sex-expression, parthenocarpy), castor (sex expression, waxiness), barley (non-shattering ear, six-row character, free-threshing), wheats (non-shattering ear, free-threshing, non-homoeologous pairing in 6x), maize (cob characters), peas (seed characters), fig (persistence of unpollinated syconia), bananas (vegetative parthenocarpy), coconut (dwarfness), oil palm (fruit characters), blackberries (non-spininess), tobacco (non-demethylation of nicotine), peppers (pungency of fruit), grape (hermaphroditism).

Natural selection

Examples of predominantly natural selection are provided by sections A4, 8 and 9. The significance of photoperiod/vernalization adaptation in crop evolution has never, we think, been well explored and we are still largely ignorant of the genetic control of variation in this respect. The list in A 4 (incomplete as it is) makes it clear that adaptation of this kind has been, and still is, very important.

Predominantly natural selection seems clear: the ill-adapted plant simply does not flower (or tuber) in season, so leaves no progeny. Tropical maizes are useless in temperate latitudes; many a potato seedling fails to tuber in long days; contrariwise, day-neutral wheats and rices are widely adapted as to latitude and have been a feature of the 'green revolution' for this very reason (Section 10.4).

Seeds or fruits shed before due harvest are generally lost and A8 lists (certainly incompletely) cases in which we have evidence of predominantly natural selection for inhibition of the normal dehiscence/dispersal mechanisms. A few (B8 – safflower, wheat and barley) are known to be due to major gene mutants (or their complex equivalents). In some crops at least this character, coupled with loss of dormancy, must have been one of the decisive steps in the transition from the wild state to irreversible domestication.

Wild plants living in strongly seasonal climates generally show seed dormancies which tend both to defer germination to a favourable time of year and to spread it over years. The farmer who reaps dry seed and keeps it dry, however, has no need of deep dormancy which will commonly be, not merely neutral, but positively dysgenic. As with photoperiod, our genetic knowledge of the change is regrettably sparse but that reduction of dormancy to a low level, even to zero, has been frequent is certain (A9); it has probably been even more common and important than is yet appreciated. Selection for the character has presumably been largely neutral: the seed which does not grow at the right time does not yield seed in season. On the other hand a low, quickly transient dormancy may be favourable in minimizing the risk of sprouting in a wet harvest so reduction rather than elimination may have been the dominant feature. Incidentally, close relatives of cultivars, of various origins, but with dormant seeds, are liable to become weeds, even serious ones: wild oats in Europe, wild rices in Asia, and the buckwheat species, *Fagopyrum tataricum*, in Canada are three examples. Weeds and crops have, sometimes, much in common, as we shall see later.

In tuber crops the same principles apply, except that here the evolutionary pressure has probably been on achieving just the right level of dormancy to fit the needs both of consumption and planting. Thus the dormancy of some Andean potatoes is nicely adjusted to place: low dormancy where the crop can be grown and consumed continuously, higher dormancy where cultivation must be seasonal.

A very important area of natural selection which could not, we found, be usefully listed in Table 1.3 is disease resistance. We shall explore this subject more fully later (Chapter 7). Here it will suffice to note that diseases, whether caused by fungi, bacteria, viruses or animals, which impair yield or which destroy the harvested products of the crop must be unceasing agents of natural selection for resistance. Primitive cultivations do have plant diseases, plenty of them; but I think it is a fair generalization that they rarely have devastating epidemics (locusts and birds excepted). For this, the heterogeneity of primitive cultivations is often held responsible; heterogeneity, that is, both between and within crops so that disease incidence on susceptibles is buffered by the intervening immune species and by resistant individuals of the crop itself. The situation is so complex that it seems impossible to assess the extent and importance of selection for disease resistance in primitive mixed cultivations. In agricultures which have moved towards the growing of extensive more or less pure stands of a crop, however, the effect of natural selection is clear: it is to tend to produce the race-non-specific resistance (HR) discussed later in this book (Chapter 7). As a generalization, this statement admittedly goes a little beyond the rather sparse facts that are yet available but it has the merits of fitting those we know, of being genetically reasonable and of offering a testable hypothesis.

Human selection

Examples in which predominantly human selection may be inferred are provided by Table 1.3 (A3, 5, 6, 7). At least, it is hard to imagine natural selection as having a significant role in these cases, whereas strong human selection is plain for: few, big, easily harvested heads; long, easily extracted fibres; non-prickly plants; palatable products; and pretty colours and patterns. Some of these changes affect yield of product, some bear on the convenience and safety of the cultivator or consumer and some could be called cosmetic or aesthetic.

As a complement to the negative selection against toxic, or at least unpalatable, constituents listed in A6 there must have been positive human selection for succulence, sweetness and low fibre contents in virtually all the fruit crops. Reduced seed content has also been a feature of the evolution of some, perhaps many, fruits, at least in those in which infertility could be compensated by clonal reproduction (Table 1.3, B1) and adequate fruit growth ensured by physiological stimuli other than seed development (e.g. parthenocarpy or its equivalent in pineapples, figs, bananas and bread-fruit; stenospermocarpy in some grapes). Human selection of the appropriate mutant forms when they occurred must be presumed.

Combined natural and human selection seems likely for the development of yet other characters, particularly those (A1, 2) connected with plant size and habit. To the extent that smaller plants offer possibilities of improved partition towards product with less waste (Chapter 3), there must have been natural selection both for it (more seed) but also against it (reduced competitiveness against large neighbours). But smaller, and sometimes shorter-lived, plants certainly emerged in many crops and it is hard to believe that it was always a perception, by the cultivator, of greater yield or of convenience of harvest which ensured their preservation.

Cytogenetic features

Local adaptation

We saw above that the essential feature of the evolution of wild plants was adaptive change by gene-substitution; mutation and hybridization provide the genetic materials and opportunities for recombination are widened by polyploidy. The same is true of cultivated plants: the fundamental process is gene-substitution, whether at the diploid or polyploid level.

Cultivated plants of wide distribution have become extensively differentiated into locally adapted races. Scores of examples could be given but a few will have to suffice: the numerous races of local maize in tropical America; the five main races of sorghum in Africa; the three races of rice in Southeast Asia; the diverse forms of cowpea in Africa and India; the two races of pigeon pea in India; the three major groups of tetraploid potatoes in Chile, the high Andes and north temperate countries. Some degree of geographical differentiation is indeed universal and normally turns out to reflect both environmental and cultural adaptation. Thus the wheats and barleys, which seem to show rather little of the more or less orderly racial differentiation instanced above, are quite sharply contrasted as between Europe and North America: the long, moist, cool summers of the former have called forth quite different adaptation in the crops from the short, dry hot ones of the latter; and social demands have produced quite different technological adaptations in respect of bread-making and brewing. Varieties are simply not interchangeable between the two areas. Our crops, then, are very generally differentiated geographically, by past selection for environmental adaptation and for end-use, often, but not always, showing recognizable racial differentiation in the process.

The degree of differentiation clearly varies but rarely even approaches the level of reproductive isolation which would imply incipient speciation. Four possible exceptions are given in Table 1.3 (B7) but even these are open to the interpretation that the near-isolation observed reflects biological differentiation *before* domestication (as in the tetraploid cottons). Speciation *in* cultivation would indeed be surprising, having regard to man's habit of carrying his crops with him and to the certainty (see below) that gene-flow between local races and between crops and weeds has long been a feature of crop evolution.

Geographical spread has, then, been accompanied by genetic differentiation into locally adapted populations among which more or less distinctive geographical races can not infrequently be recognized. Adaptation, however, is not the sole cause of geographical differentiation; there is also genetic drift or the 'founder effect'. For some crops (e.g. *arabica* coffee, oil palm and rubber) we know that huge populations have been founded in distant places from very small numbers of ancestors; for others (e.g. coconuts) the same may reasonably be inferred. More generally, it seems very likely that the early movement of crops was often on a narrow genetic base, successive movements (and adaptations) narrowing it still further. Jointly, the effects of small founder populations *plus* local adaptive pressures must tend to produce both geographical differentiation and tolerance of a degree of inbreeding: the probability of identity of alleles by descent rises. This effect must, I think, have been all-pervasive in crop evolution (stock evolution too, perhaps?) even though it can be directly inferred in relatively few instances (Table 1.3, B5).

We shall return to the question of breeding systems later. Meanwhile it is enough to note the existence of trends towards enhanced inbreeding but coupled with a contrary process: episodes of outcrossing. For this there is abundant evidence (Table 1.3, B6): between hitherto separate elements of the crop itself (as in the origin of US corn belt maize); and between cultivated forms and wild or weedy relatives (as in sugarcane, rice, sorghum, alfalfa and strawberry); in early times and in recent plant breeding (many disease resistances have been deliberately introgressed from wild forms). These outcrossing episodes have contributed not only to the crops but also, often, to the wild/weed plants with which they sometimes hybridize: crops and weeds sometimes evolve together (Harlan, 1975a).

The broad picture that emerges, therefore (Fig. 1.1), is of a recurrent pattern of geographical spread followed by local adaptation and enhanced inbreeding, alternating with outcrossing episodes. Outcrossing would widen the genetic base and no doubt usually depress short-term adaptation but enhance the possibilities of later advance.

Polyploidy

It will be evident from Table 1.1 that most crop evolution has depended upon gene-substitution at the diploid level and that polyploidy has had no place in the history of many important crops (e.g. rice, barley, sorghum, most pulses). But it has had a place in some, as listed in Table 1.3 (B1, B2, B3) (see also Fig. 1.1). The following points emerge. Autopolyploidy, sometimes old (alfalfa, bananas and potatoes), sometimes the product of recent breeding (sugar beet, hops, clover and pyrethrum), is a feature of crops grown for vegetative products in which seed fertility is not of first importance or may even be positively disfavoured (as in the bananas). Evidently there has been selection for polyploidy *per se* related to vegetative vigour; this is clear for some of the crops listed and may reasonably be presumed for others. Many of the allopolyploids listed in Table 1.3 (B2) are, by contrast, crops in which seed production is critically important; regular homologous pairing has thus widened the genetic possibilities without incurring impaired fertility. In some, the polyploidy preceded domestication (tetraploid wheats, groundnut, cottons and *arabica* coffee); in others, it was concurrent with or followed domestication (*Brassica* crops, oats, hexaploid wheats, strawberries and tobacco). The latest example is triticale, a product of recent plant breeding which will be discussed later (Section 8.2). Though these allopolyploids contain several highly successful and seed-fertile crops it must not be presumed that new allotetraploid and -hexaploid strains are automatically fertile from the start; they are often not and we shall see later that considerable genetic adjustment is necessary. But allopolyploids at least seem to have a readier capacity for such adjustment than autotetraploids.

The allopolyploid bananas listed in Table 1.3 (B1) are quite distinct; they were positively selected for seed-sterility. The crops listed in Table 1.3 (B3) are a mixed lot about which we know too little to make confident generalizations. In

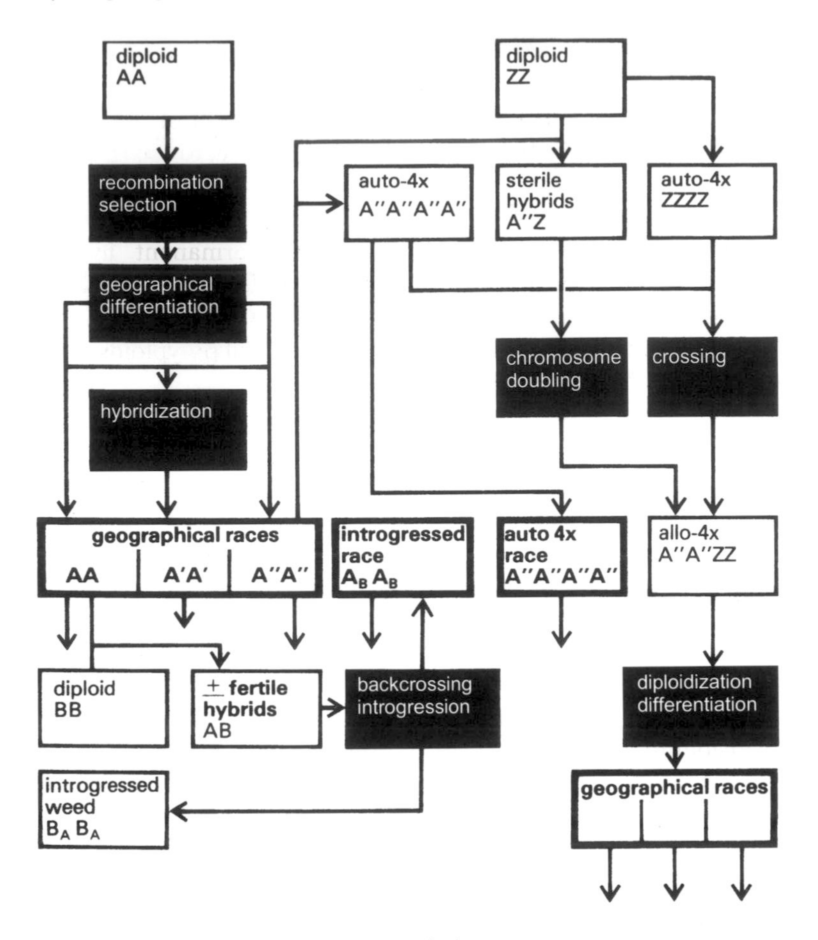

Fig. 1.1 A generalized diagram of crop evolution.
Notes: On the left is shown differentiation at the diploid (AA) level and formation of geographical races or distinctive locally adapted populations; also the two-way introgression between weed and crop that is often found (as in rices and sorghums). On the right is shown the effects of polyploidy, auto- and allo-, as the case may be. In crops grown for seed, polyploidy must be succeeded by diploidization and differentiation at the polyploid level. Uneven polyploidy (3x-, 5x-), found in several seed-infertile clonal crops, is, for simplicity, excluded.

some (e.g. yams, sugarcanes, *Allium* crops) seed fertility is irrelevant and might even have been selected against as a negative correlated response to vegetative yield; in others (the fruits) there could have been selection for just the right level of infertility, enough to ensure fruit set but no more.

Finally, there is an aspect of polyploidy which has been little explored but which may be more important than it now seems: polyploids of outbred diploids frequently display some derangement of the mating system whether by suppression of self-incompatibility (as in the potatoes and some *Prunus* fruits) or disruption of sexual differentiation (as in hemp). Self-pollination is thus facilitated in plants which, by reason of polyploidy, are better able to tolerate the consequences than the diploids from which they come. Polyploidy may therefore sometimes have been a means of escape from outbreeding and therefore, also, in the light of the discussion above, an aid to adaptation in small, isolated populations.

This discussion shows clearly that the significance of polyploidy is certainly varied and complex. Vegetative vigour, permanent hybridity, enhanced inbreeding and seed sterility have all been elements in its history. But let us remind ourselves that it is only a supplement to gene-substitution as an evolutionary mechanism, not a substitute. Successful polyploids themselves have to adapt to the polyploid level.

Mating systems

The generalizations offered earlier in this chapter about the broad features of crop evolution are independent of the mating systems of the crops concerned, the subject to which we now turn. Darwin made the essential distinction between outbreeders (species with various mechanisms promoting cross-pollination and showing inbreeding depression) and inbreeders (self-pollinators, vigorous over generations of inbreeding). Darwin's distinction stands, though amplified by later researches.

Most plants (and virtually all animals) are outbreeders, promoting crossing by a considerable variety of genetical and morphogenetic mechanisms (Table 1.5). Populations are variable and carry a load of the deleterious recessive genes which are responsible for the inbreeding depression, immediately and dramatically evident on one generation of close inbreeding in, for example, maize or diploid potatoes. Individuals in outbred populations are highly heterozygous. The population carries a segregational load due to the occurrence of unfit individuals, the products of chance concurrence of deleterious recessive alleles; but it gains as a whole from a high recombination rate, evolutionary flexibility and the opportunity to exploit favourable recombinants.

Table 1.4 Association between breeding system and life cycle in crop plants.

System	Annual	Perennial
inbred	32	1
in-out-bred	15	5
outbred	23	47

Inbreeders, by contrast, have achieved quite a different balance. Deleterious recessives are rare, the segregational load is trivial, prolonged inbreeding is tolerated, individuals tend to homozygosity, and the population tends to consist of a mixture of inbred lines. Local adaptation is high but at the expense of long-term flexibility of response to change. In practice, both in wild plants and crops, even what seem to be rigorous inbreeding is rarely perfect and the difference, a few per cent of cross-pollination, may be very important indeed in sustaining recombination, heterozygosity and adaptability (see Section 4.3). So inbreeders may not be quite so inbred as we tend to think and there are some crops (Tables 1.1, 1.4, 1.5) which, though basically inbred in the sense that they tolerate selfing, are sufficiently outcrossed to be worth distinguishing as in-outbreeders in some contexts.

In nature, outbreeding is clearly the basic state and inbreeding has evolved from it on many separate occasions, especially in short-lived herbs and in

Table 1.5 Breeding systems in crop plants.

Key	Examples
1. Monoecious	
2. Flowers monoclinous	
3. Self-pollinated	Safflower, lettuce, oats, barley sorghum*, rices, triticale, wheats, groundnut, soybean, lentil, pea, cowpea, flax, cottons*, sesame, *arabica* coffee, *Capsicum* pepper, tomato, tobacco, egg plant, jutes.
3. Cross pollinated	
4. Self-compatible	
5. Wind-pollinated	amaranths, sugarcane, sorghum*, olive,
5. Insect-pollinated	sisal, avocado, alfalfa, *Vicia*, onion, cottons*, carrot.
4. Self-incompatible	
5. Homomorphic	
6. Sporophytic	pyrethrum, 2x *Brassica* crops, *Sinapis*, radish.
6. Gametophytic	pineapple, rye (2-locus), ryegrasses (2-locus), 2x clovers, most 2x *Prunus*, *Pyrus* and *Malus*, 2x *Rubus*, 2x coffee (?), 2x potatoes, *Beta* (multilocus).
6. Other system	kola, cacao.
5. Heteromorphic	sweet potato, buckwheat, quinine.
2. Flowers diclinous	mango, cucurbits, yams, rubber, cassava, castor, maize, banana, coconut, oil palm.
1. Dioecious	papaya (m, f, h), *montana* tung (m, f), fig, (f, h), hop (m, f), hemp (m, f. (h)), grape (m, f (h)).

Note: In-out-bred plants that appear under two headings are marked with an asterisk*.

populations at the edges of geographical range where inbreeding may be regarded as an aid to close adaptation in small isolated groups. Among cultivated plants, most inbreeders derived from inbred ancestors, but there has been a general tendency, as we saw above (and see Table 1.3, B5), towards enhancement of inbreeding; two crops may even have made the entire transition, though with the aid of allopolyploidy, from outbreeding to inbreeding (*Brassica napus* and tobacco, both developed in cultivation from self-incompatible ancestors).

As to the relation between life cycle and breeding system, this comes out very clearly among our crops, as shown in Tables 1.4 and 1.5. The great majority of perennials are outbreeders whereas the annuals are mostly inbreeders.

1.7 Summary

In this chapter we have asked and attempted to answer five questions about domestication and crop evolution, as follows:

1. *When* did it happen? Domestication has been taking place continuously for about 9000 years at a roughly constant rate; evolutionary change following domestication has also been continuous.
2. *Where* did it happen? Domestication and crop evolution have occurred and are occurring wherever agriculture was or is practised; we perceive a continuum of evolutionary activity rather than discrete centres.
3. *Who* was responsible? All sorts of people from peasants to professional plant breeders have been responsible both for domestication and subsequent evolution; nowadays the professionals are becoming dominant.
4. *What* plants were involved? A systematically very diverse group of families and genera provided our crop plants; the diversity reflects the diverse ecological situations in which our crops have evolved and the diversity of human needs for the products.
5. *How* did it happen? The mechanisms of crop evolution are the same as those of natural (Neo-Darwinian) evolution. They involve geographical–ecological differentiation of populations under selection for reproductive fitness; gene-substitution processes coupled with recurrent hybridization are fundamental, supplemented in plants, both wild and cultivated, by polyploidy. Selection in crops is a complex blend of natural and human components.

To these five questions we may usefully add another: *Whither?* The answer to this must lie with the plant breeders, on whom the responsibility for future crop evolution now largely (and increasingly) rests. There is no evidence that the rates of domestication and crop evolution are decelerating, rather the reverse. So new crops will emerge and old ones will change still further. Plant breeding *is* the current phase and in it lies the future of crop evolution. There can be no better basis for a view of the future of a crop than a thorough understanding of its past.

1.8 Literature

Evolution general

Dobzhansky (1949); Dobzhansky *et al.* (1977); Grant (1963, 1971, 1977); Huxley (1940, 1942); Stebbins (1950, 1971a, b).

Evolution, cultivated plants

Anderson (1952); Crane and Lawrence (1938); Darlington (1973); Darlington and Wylie (1956); Darwin (1876, 1882); De Candolle (1886); Galinat (1965); Harlan (1956, 1971, 1972, 1975a, b, 1976a, b); Harlan and De Wet (1975); Harlan, De Wet and Price (1973); Harris (1972); Hawkes (1983); Heiser (1973); Hutchinson (1965, 1971a, 1974); Hutchinson *et al.* (1976); Li (1970); Pickersgill (1977); Purseglove (1968, 1972); Sauer (1952); Schwanitz (1955, 1966); Simmonds (1976a); Smartt (1990), Smartt and Simmonds (1995); Smith (1968, 1969); Ucko and Dimbleby (1969); Vavilov (1951); Zohary and Hopf (1988); Zhukovsky (1962).

Chapter 2
Basic Features of Plant Breeding

2.1 Introduction

Plant breeding is the current phase of crop evolution and it proceeds by the same mechanisms that are responsible alike for the evolution of wild populations of plants and of cultivated ones in earlier times. The fundamental mechanism is adaptational change by gene-substitution under selection, followed by a degree of isolation of the differentiated products: hybridization and polyploidy widen the scope for selection without changing the basic mechanism. The features distinguishing plant breeding from earlier phases of crop evolution are that selection is predominantly artificial rather than artificial and natural combined and that objectives and strategies are defined on the basis of more or less secure scientific knowledge of: (*a*) the agriculture and the society that are to be served; and (*b*) of the essential genetic features of the crop. So the plant breeder is an applied evolutionist working towards defined objectives by tolerably well understood methods. The rest of this book is, in effect, an exploration of that statement.

It would be nice to be able to set out the subject in a simple, linear fashion, each topic following logically from its predecessor. Unfortunately, this is impossible. The ideas are reticulately rather than linearly related. So the purpose of this chapter is to try to provide a bird's-eye view of the whole subject in the hope that the reader will thereby find his way the more readily through the rest of this book. Figure 2.1 presents a summary of much of this chapter.

2.2 Objectives and decisions

Every plant breeding programme must have well defined objectives which are both economically and biologically reasonable. Economic criteria are important, even if not stated in strict monetary terms, because the breeder must be assured that he is trying to produce varieties that farmers and end-users will actually want. This subject is discussed further in Section 2.5 below and Chapter 11 (see also Fig. 2.2). The biological objectives are determined by scientific knowledge and general 'feel' for the crop; they are discussed in Chapter 3 where we shall see that yield and quality (fitness for purpose) are invariable compo-

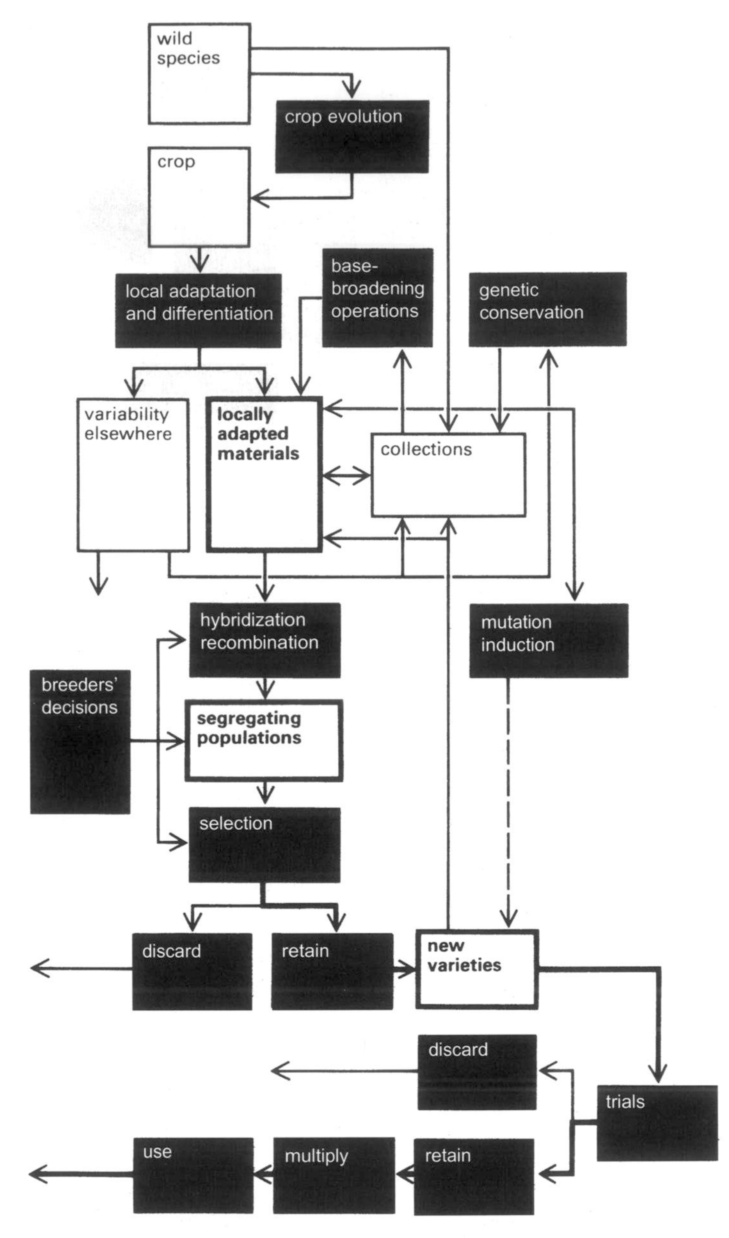

Fig. 2.1 The general pattern of 'mainstream' plant-breeding.
Notes: The diagram brings out the fact that the breeder works with a pool of locally adapted parental materials, intermittently supplemented from elsewhere to maintain an adequate genetic base. The breeder's decisions include: what parents to use, how to combine them, what to select for, how to select, what to discard and what to keep. They are further explored in Fig. 2.2. New varieties submitted by the breeder for practical exploitation are subject to further decisions on the basis of official trials; those that pass will be exposed to farmers' decisions as to whether or not they will adopt them on the basis of economic merit.

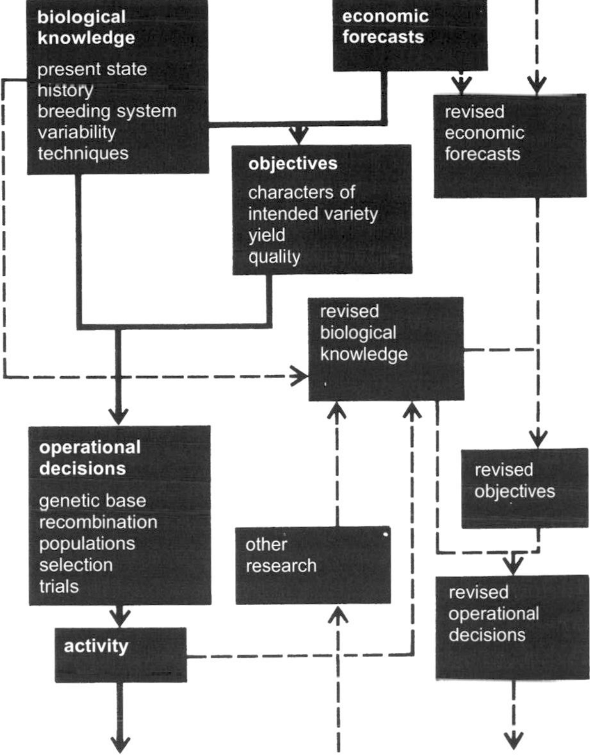

Fig. 2.2 The plant breeder's decisions on objectives and methods. Biological knowledge and economic forecasts are always imperfect so decisions are open to cyclical revision.

nents. Yield is in a sense dominant and is biologically very complex indeed; broadly, we can distinguish between efforts to increase the surviving biomass of the crop and efforts to improve its partition towards the crop fraction and away from waste.

Objectives, then, are the first of the plant breeder's decisions. Others follow from the nature of the crop, from the parental material in hand and from the objectives themselves. Thus the breeder will have to decide (*a*) what parents to include and why (will he have to turn to yet unknown foreign sources?); (*b*) what pattern of crossing and passage through the generations to adopt (which will be largely determined by the mating and propagation system of the crop); (*c*) what methods of selection to use (which will depend largely on the genetics of yield and quality and on the technology of the latter); and (*d*) how he will decide upon the ultimate release or discard of the products. The recurrent nature of these decisions is illustrated in Figs 2.1 and 2.2. The latter brings out the fact that changing knowledge may enforce revision; objectives are sometimes overtaken by external events.

2.3 The pattern of plant breeding

Local adaptation

Crops, as we have seen, are geographically differentiated into locally adapted races, ecotypes, varieties, etc. The breeder is concerned with performance in one environment or a limited group of similar environments because no one variety ever excels over the entire range of any widespread crop. Local adaptation is always evident: hence the facts of crop evolution and, indeed, of plant breeding. Since any one breeder can only work effectively in one relatively small area, it follows that the individual breeder can do no more than seek to promote local adaptation in one or a few similar environments but that the totality of breeders working on a crop will, as a body, promote the evolutionary differentiation of the crop as a whole. Here (see Fig. 2.1) we shall confine our attention to the single breeder, while noting that he is but one agent of local adaptation among many.

The genetic base

The genetic variability available in a crop always greatly exceeds what one breeder can effectively handle. In general, he must work with parental material which is more or less locally adapted. This will include (see Fig. 2.1): (*a*) locally successful varieties (including his own earlier products); (*b*) strains which, from performance in collections, appear to offer local adaptation or, at least, specific desired characters; (*c*) the products of operations specifically designed to widen the genetic base; (*d*) the products of mutation-induction experiments; and (*e*) the products of genetic engineering and biotechnology. The relative importance of these sources will vary rather widely with the breeding history of the crop; they are explored more fully in Section 5.9, mutation work in Section 8.5, and biotechnology in Chapter 9.

Recombination

Enhanced adaptation follows from the selection and isolation of new genotypes, the products of genetic recombination, which are better adapted than their parents. Recombination, both between and within chromosomes, is therefore a crucial phase of any plant breeding programme. The breeder sets up crosses in order to generate recombinants, some of which, he hopes, will be transgressive with respect to the better parent. The amount of recombination in one breeding cycle will depend upon several genetic factors (including relationship of the parents, breeding system, propagation system, population size; see Chapters 4 and 5); generally, it is in the interests of progress that it should be maximized though, in practice, it often seems to be rather restricted.

Selection

Selection is differential reproduction of the products of recombination (Sections 4.6, 5.8). In the breeder's plots, both natural selection (unavoidably imposed by the local environment) and human (or artificial) selection both operate (Section 1.6); the breeder chooses what to keep and what to throw away, but he does so in populations which have already suffered the impress of the local environment. In Section 4.5 we shall meet the idea of heritability as a major factor in artificial selection. Highly heritable characters are but little affected by the environment and are easily and efficiently selected; characters of low heritability, by contrast, are much more difficult and demanding. The distinction dominates most breeding plans.

Besides this dominant genetical aspect of selection there is also an important operational one. In any one year, the breeder will be handling thousands of distinct entities (individuals, populations, lines, clones). Any observations or measurements he makes on them in the interests of selection for yield or quality characters must therefore be as quick and simple as possible. Generally, selection in the early years of a cross, when numbers are large, must perforce be on a visual basis. Subsequently, when numbers have been reduced, actual measurements take over: yields are measured in small plots, specific disease resistance tests are run and quality characters (Section 3.3) estimated from appropriate physical or chemical measurements. A great deal of effort and ingenuity has gone into this aspect of plant breeding.

Selection for disease resistance (which is mainly to be regarded as a determinant of biomass and therefore of yield – Section 3.2) occupies a conspicuous, sometimes a dominant, place in plant breeding programmes, a place which is further explored in Chapter 7. We shall see there that methods vary with the kind of disease and that plant breeders have had very varied success in their efforts: sometimes, excellent or good control has been achieved; on other occasions there have been recurrent failures due to adaptation by the pathogen under natural selection. The underlying biological problems far transcend in importance and interest the immediate practical questions of plant breeding methodology.

Populations

We saw in Chapter 1 that crop plants may be grouped broadly into outbreeders and inbreeders; the former have various morphogenetic mechanisms that favour crossing (Table 1.5), tend to carry deleterious recessives in the heterozygous state and show inbreeding depression; the latter are (at least predominantly) self-pollinated, have been purged of deleterious recessives and are tolerant of prolonged selfing. There is a correlation between breeding system and life cycle such that nearly all perennial crops are outbreeders and most inbreeders are annuals (Table 1.4). Perennials *may* be propagable as clones which are, for good genetical and agricultural reasons, favoured, if feasible; they permit the fixation, in one step, of well adapted recombinants. By contrast, some perennials and all annuals must be seed-propagated. Among them, three basic kinds of population are possible: (1) the inbred pure line in inbreeders; (2) the open-pollinated population (including the synthetic and similar variants) in outbreeders; it is somewhat variable and heterozygosity is high; and (3) the hybrid variety, constructed by crossing more or less inbred lines, typically in outbreeders; it is little variable and heterozygosity is very high.

So there are four basic plant breeding populations (Section 5.2) determined by mating and propagation systems: inbred lines (IBL), open-pollinated populations (OPP), hybrids (HYB) and clones (CLO). The population type adopted is the dominant feature of any plant breeding programme. However, we shall see that there is perhaps room for some scepticism both about the restriction to these four and about the choice of plan in any particular crop; adaptation, adaptability and economic considerations sometimes seem to be in conflict.

Trials and multiplication

The fundamental assumption of plant breeding is that selection practised in the breeder's plots is eugenic, in the sense of promoting adaptation of new varieties to the prevailing agricultural environment. The undoubted successes of plant breeding show that this assumption is, at least sometimes, correct. The decision system whereby a new variety goes or does not go forward from the breeder's plots to agricultural use is essentially a hierarchy of trials intended to predict agricultural performance (Section 6.2). The extent to which trials do, in fact, predict is as we shall see uncertain but, at all events, some sort of decision system is essential: there are always far more potential varieties than successful ones.

The decision to exploit a new variety agriculturally is accompanied, necessarily, by a decision to multiply it so that seed or planting material shall be commercially available. The principles of multiplication (but not the practice) are outlined in Section 6.4: the key points are purity, cleanliness and health of stocks.

Genetic conservation

Long-term progress in plant breeding depends upon provision of an adequate store of genetic variability in the form of diverse parents for inclusion in the genetic base of breeding programmes (Fig. 2.1). But successful plant breeding replaces old varieties by new ones and, the more successful it is, the fewer are the new varieties (which, in addition, are often inter-related). So local variability in crops tends to decline, a trend which has surely accelerated since the end of the Second World War. One outcome is a locally narrow genetic base, apparent in slow breeding progress (Section 5.9), or pathological crises (Chapter 7). Another is a world-wide decline in the variability available to breeders. This has only come to general attention in the past thirty or so years. It is now universally accepted that the decay of variability in actual agriculture is so rapid that the future genetic base must be safeguarded by a great international programme of genetic conservation. In carrying this out, there is no feasible alternative to building and maintaining collections of crop varieties and wild relatives as stored seed stocks or as living plants, according to biological circumstances. We shall see (Section 10.3) that, though good progress has been made in assembling the necessary collections, there is a long way to go.

Genetic conservation, as currently conceived, relates primarily to existing crops. What, then, of crops not yet domesticated? As we shall see (Section 10.2) new domestications are occurring now and will continue into the indefinite future. They imply the need for an even wider view of genetic conservation.

2.4 Supplementary techniques – Biotechnology

General

So far we have been discussing what one might call 'mainstream' plant breeding: the process of recombining locally adapted stocks to produce enhanced local adaptation. It must be emphasized that the vast majority of plant breeding operations are of this nature. There are, however, a number of techniques, all of a genetical character, which supplement mainstream breeding by providing materials which could not otherwise have been available or which offer possible economies in time, space or effort. These are outlined here and discussed in some detail in Chapter 8 and in Chapter 9 under the heading of 'Biotechnology'.

Wide hybridization and polyploidy

Most wide hybridizations used in plant breeding involve locally adapted parents on one side and ill-adapted or wild parents on the other. The purpose is usually the introduction of a desired character (commonly, but not always, a disease resistance) from the foreign parent, and the breeding plan adopted – usually backcrossing to adapted stocks – is designed to restrict the ultimate genetic contribution of the foreign parent to, as nearly as possible, the desired gene and to it alone. There must have been thousands of such gene transfers made in the present century. Some have been very successful indeed; others have failed because the gene transferred turned out to be ineffective against a new race of the pathogen evolved in response to it (Chapter 7). Backcrossing in of foreign genes has, then, been a substantial element of plant breeding, and it has also been the scene of some very elegant cytogenetics (Section 8.3). We shall see that there are exceptions to the usual rule of rigorous backcrossing; occasionally, a genetic contribution from the foreign parent much larger than one gene (and its closely linked neighbours) turns out to be favourable.

Polyploidy, as we have seen in Chapter 1, is already well established in our crops, sometimes carried over from the wild, sometimes a product of selection at or after domestication. From the plant breeder's viewpoint, induced polyploidy has three main aspects, two of them concerned with the exploitation of wide crosses. First, autopolyploids of diploid crops may occasionally be useful as such, though much less often than was hoped forty to fifty years ago; their *gigas* character is sometimes favourable in crops produced for vegetative products but the sterility consequent on erratic meiosis has so far forbidden their exploitation in seed-producing crops. However, second, they are sometimes useful in making difficult allopolyploids; and third, new allopolyploids themselves occasionally offer exciting possibilities for imitating nature in the creation of new crop species. One new allopolyploid crop (triticale) can be said to be established and a few others are in prospect. But we shall see in Section 8.2 that new polyploids, whether auto-or allo-, are never immediately successful: they must always first be worked up into a state of local adaptation, including adequate seed fertility, before they can enter into the normal mainstream of plant breeding.

Mutation induction

Knowledge that ionizing radiations damage chromosomes and cause point mutations goes back to Muller's work on *Drosophila* in the 1920s. Knowledge to the effect that a variety of chemicals could give the same effects came later and it was not until the late 1940s that the idea that induced mutations might be useful in plant breeding became common currency. It was partly stimulated by the sudden wide availability of active isotope sources, a spin-off from wartime nuclear weaponry. Now, fifty years later, chemical agents are probably as often used as physical ones and the technique has a secure, but minor place, in the battery of methods available to the plant breeder (Section 8.5). Earlier over-optimism, to the effect that induced mutations were about to revolutionize plant breeding, has given place to a more sober appreciation of the technique as a valuable supplement to more conventional techniques in certain, rather restricted, circumstances. Those circumstances are: (1) that it should be genetically plausible to assume that a certain specific mutant should be possible – even if yet unknown; (2) that efficient screening methods should be available; the expected mutant must be either readily visible or otherwise easily detected; and (3) that the proposed programme is compatible with the breeding and propagation system of the crop; thus, methods of extracting the mutants in usable form and of cleaning up the cytogenetic background effects of treatment should be apparent.

These requirements are fairly restrictive and very many programmes failed, especially in the early days of over-optimism, to produce anything useful because they were not fulfilled. Nowadays we see mutation-induction simply as one technique which is occasionally useful in enlarging the genetic base of a programme in a limited and highly specific fashion.

In vitro methods and biotechnology

Knowledge that many animal cells are capable of prolonged growth in sterile culture goes back to the work of Alexis Carrell in the 1890s. The techniques have long been highly sophisticated and have made an enormous contribution to the biochemistry and cell biology of animal materials, both within and outwith the medical context. Comparable studies of plant cells lagged far behind. Though embryo cultures had begun and a few plant tissue cultures were in hand in the 1930s, no significant advances came until after the late war, since when progress has been rapid.

The field, in relation to plant breeding (Section 9.2), is now a complex one, complex enough, at least, to defy simple classification. It includes: (1) the manipulation of vegetative meristems and embryos as a sort of *in vitro* horticulture; these have uses in freeing clones from systemic virus diseases, in multiplying clones of difficult subjects, in genetic conservation and in making certain difficult crosses; (2) the manipulation of tissue or free-cell cultures so as to produce genetically novel meristems which can be grown up to whole plants and find uses in plant breeding; these include the products of natural cytological accidents, of mutation induction work, of meiotic segregation (haploids) and of cell fusion *in vitro* and somaclonal variation.

DNA technology in the context of crop improvement has a two-fold application: first the production of useful transgenic or transformed plants; second in the development of sophisticated selection procedures using molecular markers and the production of genetic maps, especially in species which have not been mapped in the traditional way. This technology will be considered in more detail in Chapter 9.

In general, the approaches listed under (1) have already been of very considerable value and hold much promise for the future. Those listed under (2) are of uncertain value; as in the early days of induced polyploids and induced mutations, there has been a certain bandwagon effect and a tendency to overestimate the value of novelty. *In vitro* techniques are better regarded simply as yet another supplement to mainstream plant breeding: they provide, according to circumstances, either special methods of propagation or recombinants not otherwise readily available which, however, still have to be worked up by conventional methods.

The same can be said of the products of transformation technology. The potential is enormous but any rewards from its application are likely to be very hard won indeed. The cost in time, effort and resources will be vast while the products will still have to be utilized through traditional breeding programmes; they will supplement, not supplant nor supersede them.

2.5 The social context

Successful new varieties, as we saw above, are rare; the great majority of potential ones fail, somewhere along the line from the breeder's plots to trials to the market place. Farmers and end-users are the ultimate arbiters of success, and their criteria are economic. They pose the question: is this or that new variety more profitable than the old one? An economic analysis of the place of plant breeding in general agricultural improvement is attempted in Chapter 11. We shall see there that plant breeding is simply one component of change, sometimes more, sometimes less important, in comparison with changes in husbandry. The data available are not good (the subject has had remarkably little critical study), but they seem to indicate that plant breeding has had rather less effect on yields than husbandry changes in advanced temperate agricultures in the past forty to fifty years.

This is not the whole story, however, because profitability depends on costs as well as returns. Our analysis of the economic effects of new varieties will show that both reduced costs and enhanced returns per unit area or per unit of production are apparent. So innovating farmers, those who first grow the new variety, make better profits than their more conservative neighbours. But, as the new variety spreads, the market mechanism adjusts, producer profits fall and consumers benefit from cheaper supply. Agriculture is said by economists to be in a state of near-perfect competition, so the process of market adjustment is rapid. The outcome is a fair profit to farmers (no more), the economic benefits being widely diffused through society as a whole, as consumers.

This leads on to the idea of cost-benefit analysis. The question of which consumers benefit and by how much poses some difficult economic problems; these we shall not explore but it is at least clear that plant breeding, in broad social terms, does indeed generate substantial benefits and is remarkably free of unfavourable side-effects (the economists' 'externalities'). On the cost side of the cost-benefit relation, we shall see that plant breeding is a slow process that has to pay for its many failures as well as its successes.

The foregoing is true of all plant breeding, no matter who does it: the commercial breeder or the worker in the state-supported institute. If successful, they both generate social benefits; they differ in that the costs of the former must be met out of the normal commercial cash-flow, while those of the latter derive from society as a whole in the form of taxation. This difference, as we shall see, leads to some differentiation of objectives between the two.

The reader will note here that, with this mention of objectives, the general survey set out in this chapter becomes circular. The plant breeder is always working in a socio-economic context in which the expected (or possible or hoped-for) end-point helps to determine his early decisions, made many years before. Explicit economic calculations very rarely form a part of the breeder's decisions and, perhaps, in view of the uncertainties of economic forecasting, they should not. But a lively appreciation of economic circumstances is important, if only to identify those situations in which, through ill-judgement or ill-chance, economically unreasonable objectives have been chosen. In that case, revision of objectives and methods may be necessary (Fig. 2.2).

2.6 Summary

In this chapter we have taken a very broad view of plant breeding as a whole. The breeder's objectives ar founded on biological knowledge of the crop (giving a sense of what is *possible*), plus some economic understanding of what the market will want many years ahead. His operational decisions are determined by the objectives *plus*, once again, biological knowledge. They include: choice of parents, pattern of crossing and recombination, choice of population structure, methods of selection, trials decisions, and whether or not to supplement the programme with one or more of the available accessory techniques such as wide hybridization, polyploidy, mutation-induction or biotechnology. These accessory techniques, though sometimes very valuable, are to be thought of as broadly secondary to what we have called mainstream plant breeding, that is, the process of gene-substitution in essentially well-adapted populations. Successful new plant varieties are successful because they are more profitable, first (but only transiently) to farmers and second, to society as a whole, which receives consumer benefits from reduced unit-costs in the form of lower prices. Plant breeding is undoubtedly economically successful but it is only one component of agricultural advance.

2.7 Literature

Plant breeding, textbooks

Allard (1960); Babcock and Clausen (1927); Brewbaker (1964); Briggs and Knowles (1967); Elliott (1958); Hagedoorn (1950); Hayes *et al.* (1955); Hayward *et al.* (1993); Hunter and Leake (1933); Lawrence (1951); Poehlmann (1959); Roemer and Rudorf (1955–62); W. Williams (1964).

Review journal

Plant Breeding Reviews (annual 1983–present).

General articles and books

Åkerberg and Hagberg (1963); Åkerman *et al.* (1948); Bishop (1963); Brookhaven (1956); Frey in Eastin and Munson (1971); Frankel (1950, 1958); Frey (1966); Gowen (1952); Gustafsson (1968); Hanson and Robinson (1963); Janick and Moore (1975); Janossy and Lupton (1976); Riley in Jinks (1970); Leakey (1970); Lupton *et al.* (1972); Moav (1973); Purseglove (1968, 1972); Simmonds (1969a, 1973, 1976a); Sprague (1967); Walker (1969).

History

Bakhteyev (1968); Olby (1966); Reed (1942); Roberts (1929).

Individual crops, monographs and collective treatments

Arnold (1976) (cotton); Atherton and Rudick (1986) (tomato); Bond and Fyfe (1961) (*Vicia*); Briggs (1978) (barley); Bunting and Gunn (1974) (maize in UK); Burton and Powell (1968) (pearl millet); Carnahan and Hill (1961) (forage grasses); Chandraratna (1964) (rice); Chang (1964) (rice); Clifford and Willson (1993) (coffee); Coffman (1961) (oats); Cooke and Scott (1993) (sugar beet); Culbertson (1954) (flax); Davey (1959) (Brassicas); Doggett (1970) (sorghum); Ferwerda and Wit (1969) (tropical perennials); Finlay and Shepherd (1968) (wheat); Gaul (1976) (barley); Gowen (1995) (bananas and plantains); Hanson and Carnahan (1956) (forage grasses); Harris (1991) (potatoes); Hebblethwaite, P.D. (1983) (faba bean); Howard (1962) (potatoes); Hughes *et al.* (1962) (forages); Hutchinson (1959) (cotton); IRRI (1972) (rice); Johnson and Bernard (1962) (soybean); Jones and Lazenby (1988) (grasses); Mangelsdorf (1974) (maize); Nilan (1971) (barley); Quinby and Martin (1954) (sorghum); Rachie and Roberts (1974) (tropical grain legumes); Richmond (1952) (cotton); Rogers (1976) (forage legumes); Simmonds (1962b, 1966) (bananas); Simmonds (1976a) (many crops); Simmonds (1995) (food crops); Smartt (1990) (grain legumes) (1994) (groundnuts); Smartt and Simmonds (1995) (many crops); Smith (1968) (tobacco); Sprague (1955) (maize); Stover and Simmonds (1987) (bananas); Thomson and Wright (1972) (forage grasses); Tysdal *et al.* (1942) (alfalfa, lucerne); Vaughan *et al.* (1976) (crucifers); Wall and Ross (1970) (sorghum); Wallace and Brown (1956) (maize); Weiss (1949) (soybean); Welch (1995) (oats); Whitehouse (1968) (barley); Whyte *et al.* (1959) (forage grasses); Williams (1993) (pulses and vegetables); Williams (1995) (cereals and pseudocereals; Willson and Clifford (1985) (tea); J.W. Wright (1976) (forest trees); Yarnell (1954, 1962, 1965) (vegetable crops).

Chapter 3
Objectives of Plant Breeding

3.1 Introduction

The objectives of a plant breeding programme can usefully be discussed in both biological and economic terms. In this chapter we shall discuss the biological objectives, leaving economic considerations to Chapter 11. Biological objectives are dominated by yield and quality factors, though a few additional features that do not fall under either of these headings can also be recognized. Yield is, in practice, the more important and we shall recognize two main components of it: biomass, the capacity to produce and retain an adequate quantity of vegetable material, and partition, the capacity to divert biomass to the desired product, the crop. Quality, which we shall distinguish carefully from condition, is defined as fitness for purpose and it has many and varied aspects according to crop: chemical, physical, mechanical and aesthetic. The main components of this chapter are summarized in Fig. 3.1 which is, inevitably, rather simplified: we shall see below that yield and quality, biomass and partition are not always so readily separable as might appear. Interactions abound.

3.2 Yield

Biomass

All crops have restricted (though sometimes remarkably wide) ranges of adaptation: bananas and sugarcanes are unlikely ever to be grown in northern Europe and barleys and kales are very improbable crops for the lowland wet tropics. However, extensions of range do occur, as exemplified by the northward movement of wheat and maize in north temperate countries in recent decades. No doubt, plant breeding will bring about many more such extensions but it is well to note that they are exceptional: the great majority of programmes are concerned with improving the adaptation of crops in environments to which they are already basically well adapted. So we shall not be concerned here with crops which do not or hardly grow at a particular site; we shall only be talking about crops which at least grow well enough to have reasonable biomass-potential (Fig. 3.1). (See also Evans, 1993.)

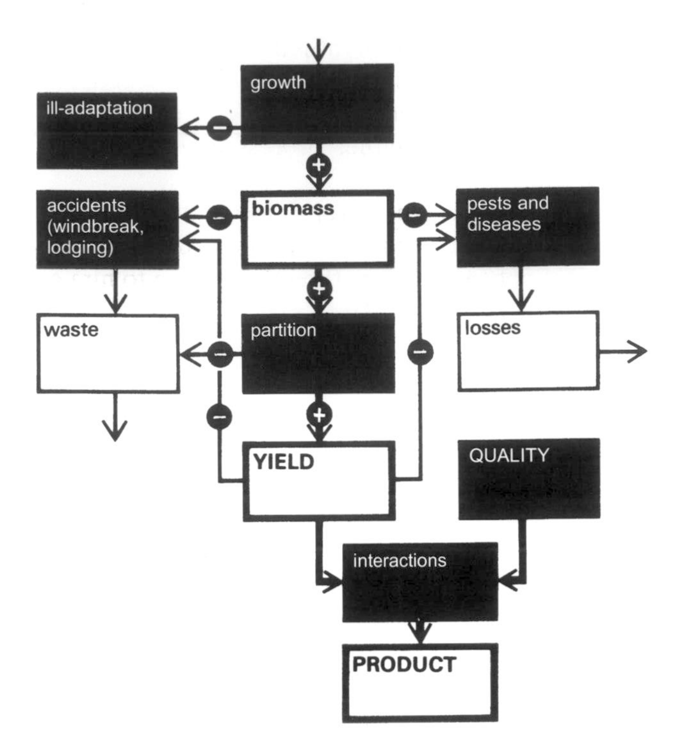

Fig. 3.1 Yield and quality as plant breeding objectives. To advance yield, the breeder seeks to maximize processes marked (+), minimize those marked (–).

The idea of biomass-potential turns out to be a complex one. Briefly we conceive of a climate as providing a certain growing season characterized by time, incident radiation, rainfall and temperature. Superimposed on this is the local environment characterized by topography, exposure, soil, physical and chemical conditions; and superimposed on this in turn is the agricultural environment characterized by choice of site and agronomic practices. So the plant breeder is ultimately constrained by climatic, local and agricultural features. If all these are defined, potential biomass is fairly closely determined. Thus de Wit (in Wellensiek, 1968) found that a range of crops in northern Europe all approached a biomass limit of about 20 t/ha of dry matter in the summer growing season. This represents the limit of what an essentially standard photosynthetic system can produce if well managed in a good season in a favourable site. We must suppose that all crops at a particular place have such a limit which is determined by the environment rather than by the crop itself. We would expect tropical perennials that grow for much or all of the year to exhibit higher gross biomass potentials than temperate crops. Though little investigated, this is no doubt true; thus sugarcane has been recorded as yielding up to 64 t/ha of dry matter, a figure far greater than the European summer production cited above (but not too dissimilar on a per-month basis). During the past four decades, many studies of photosynthetic rates and/or net assimilation rates have been founded on the hope that differences large enough to be significant in the control of biomass (and hence of yield) would be found. They have not and there is no evidence that either process is ever limiting in reasonably well adapted materials.

If the potential biomass of a crop is essentially determined by the total local environment, this might seem to leave little room for manoeuvre by the plant breeder. This is not so: there are three ways open to him of enhancing biomass, as follows (see Figs 3.1 and 3.2).

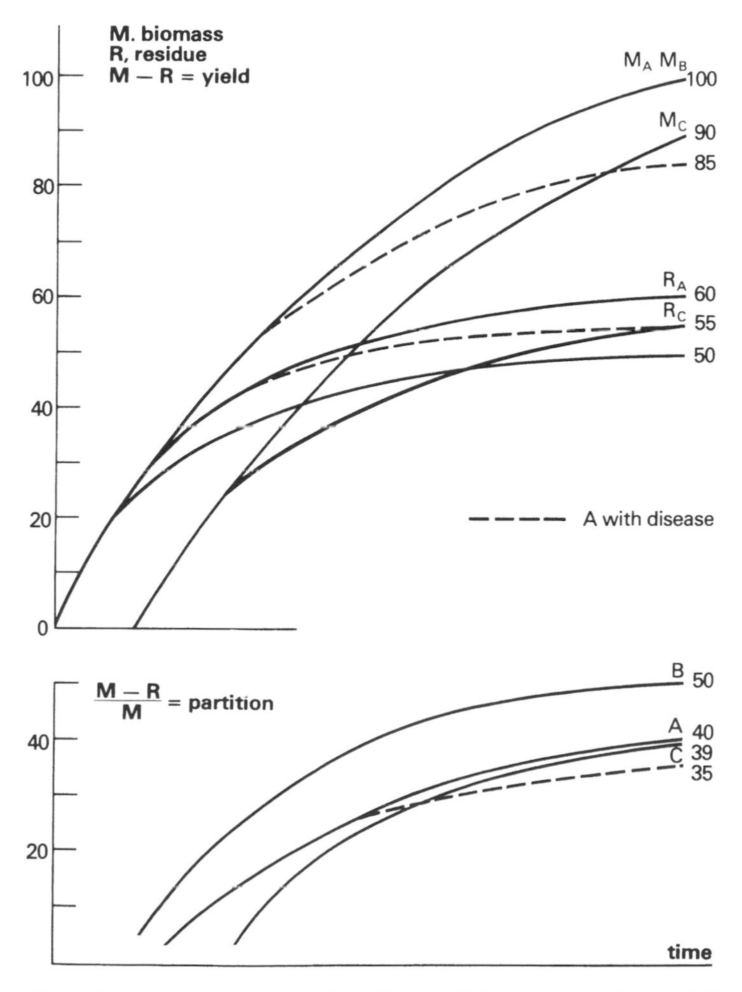

Fig. 3.2 Growth curves to illustrate the effects of biomass, partition and disease on yield.
Notes: Three varieties (A, B and C) of which A and B can be sown earlier and so exploit the season better than C; A and B achieve similar biomass but B shows better partition than A so is higher yielding; disease in A reduces biomass and partition. B is clearly the most efficient of the three in all respects.

First, there is *improved seasonal adaptation.* Anything the breeder can do to extend the growing season, within acceptable agricultural limits, will be worthwhile as providing more time for biomass accumulation. Thus, in temperate annuals, tolerance of cold soil in early spring will permit earlier sowing (surely a factor in the recent adaptation of maize to northern Europe and of adaptation in early potatoes). Similarly, with swedes and sugar beet in Britain, effective selection against bolting to flower by adjustment of the photoperiod–vernalization requirements of the crop, has also permitted earlier sowing. At the other end of the season, the aim must be the latest harvest possible compatible with the agricultural realities of expected weather and cropping sequence. The same principles apply among the tropical annuals such as cereals and cotton but here the adaptation is usually to rainfall rather than to temperature cycles (see, for example, Bunting (1970) on West African sorghums and cowpeas). Among the perennials the opportunities are perhaps more limited but sugarcane provides an example: rapid and vigorous regrowth (ratooning) after cutting is selected for as a determinant of biomass in the next cycle. Whatever the crop, then, the objective is optimal exploitation of the growing season.

Second, we have *tolerance of adverse environmental factors.* The environment of a crop is never perfect, as is evident from the fact that what we take to be maximal yields are very rarely realized. There are always defects of climate, local environment and management. That tolerance of drought and wetness, heat and cold are features of well adapted varieties seems certain but it is hard to give clear cut examples, largely because such adaptation usually emerges from selection for average performance over seasons rather than for specific selection for drought or cold tolerance *per se.* Agricultural experience, however, is convincing that real differences exist in many, probably all, crops. Tolerance of wind is also relevant here, especially in perennials. Thus, no banana withstands a hurricane but dwarf or semi-dwarf clones stand moderately severe winds far better than tall ones and have been selected for this reason (among others). In rubber, spectacular differences in wind resistance are broadly related to crown form, but not (as yet, though this will no doubt come) to plant size.

And, finally, third, we have *pest and disease resistance.* Most pests and diseases do their damage by restricting the biomass of the crop below the level at which compensatory growth is possible. Some (e.g. viruses) cause general stunting, some (e.g. nematodes) limit root performance, some (e.g. vascular wilts) kill individual stems or whole plants and some (e.g. most of the airborne fungi) reduce total photosynthetic activity. Though the basic effect of most diseases is to restrict biomass, there must also be apparent partition effects; thus, as shown in Fig. 3.1, diseases which develop after the essential vegetative skeleton of the plant has been established, must tend to depress the partition towards crop. This is evident for example in relation to the airborne fungal diseases of cereals which conspicuously inhibit grain filling by diminishing the flow of photosynthate. We shall return to this subject in Chapter 7 and it will suffice to note here (what we have already noted in Chapter 2) that disease resistance is often a major, even a dominant, feature of many plant breeding programmes.

Partition

The importance of partition as a factor controlling yield depends upon the crop. In those forages (e.g. grasses and clovers) in which the yield is very nearly the entire overground growth of the crop in season, partition is of, at most, small importance. The forage breeder may be able to adjust flowering time to advantage but must otherwise do what he can with biomass if he is to affect yield. In other crops, indeed in the great majority, partition is a significant feature of yield. In general, it takes the form of enhancement of yield of desired product at the expense of unwanted plant parts (Fig. 3.2). We thus envisage competition for biomass of which three outcomes are apparent, as follows.

First, vegetative-reproductive compensation occurs when vegetative growth is reduced as far as the demands of reproductive performance permit. We noted examples of this process in the earlier history of crops (Chapter 1); many crops have become smaller, shorter-lived and less branched in response to selection for fruit or seed yield. This process continues in the hands of the plant breeder in, for example: several major cereals (wheat, rice, barley, sorghum) which have all tended to reduction of stature, even to dwarfness, in recent decades; in sunflowers, similarly; in several legumes, in which determinate growth has replaced (or is being sought to replace) indeterminate; and in various fruits, including bananas and tree fruits, in which dwarf or semi-dwarf mutants are not only favourable from the partition point of view but also offer other attractive features such as wind-fastness and ease of harvest. The same principle underlies the reduced foliage of okra-leaf cotton (Thomson, 1972) and 'leafless' peas (Snoad, 1974).

Second, reproductive–vegetative compensation, the converse of the preceding, occurs when reproductive performance is suppressed in favour of a vegetative product. This is evident in several tuber crops such as cassava and potatoes, in sugarcane, sugar beet and various vegetable and fodder brassicas. In the first two, flowering and seed fertility are variously reduced; in the last three, flowering is preferably totally suppressed in agricultural crops (though adequate seed fertility must be retained for seed production in seed crops). Flowering is often seasonally controlled (e.g. by temperature and/or day-length) so the breeder's task may become one of adjusting a specific environmental response. Thus, sugarcane has a highly specific day-length demand for flowering, so heavy flowering occurs only over a limited range of latitudes; in these latitudes (but not outside them) non-flowering is an important yield character. In jutes, also, flowering is very exactly day-length mediated, so sowing time must be accurately adjusted (to a few days) if the plants are not to flower young (with disastrous results for fibre production) thus, jute breeders seek to produce day-neutral strains which would yield fibre in any season.

And, third, vegetative–vegetative compensation occurs in potatoes and rubber (and doubtless in other crops too). In both, the breeder's task is to maximize the partition towards yield, leaving the least possible haulm in the one case and the least tree growth compatible with yield in the other. Rubber is a rather unusual crop in that yield is not simply phenotypically determined as what's there at harvest but is rather a response to exploitation; so the rubber breeder determines the potential for partition under exploitation rather than the partition factor *per se.*

Other points about yield

The discussion so far has centred on the fact that plant breeding affects yield through adjustment of biomass or partition or both. This is undoubtedly true but it is surprising how poor our information is on the relative importance of the two major components. The fact is that most plant breeding goes forward – and often very successfully too – on the basis of selection for yield *per se*, with little or no conscious attention to biomass or partition. So, in the absence of specific tests, it is normally impossible to say just what is the biological basis of the marginal yield advances of a few per cent per step which are so characteristic. In retrospect (as in the examples in Table 3.1) the bases of major advances or of several small steps considered jointly may sometimes be fairly clear, but experiments which critically separate biomass and partition effects are, unfortunately, rare. It should be noted that such experiments would be laborious, even if worthwhile; biomass is not easily estimated and the grain/tuber plus straw/haulm data in Table 3.1 are but crude approximations.

Nevertheless, the idea that some understanding of the morphophysiology of a crop should be of help in forming a breeding programme goes back a long way, certainly to the work of Lawrence Balls on cotton in Egypt in the 1920s. His and many subsequent studies on a variety of crops concentrated upon components of yield such as: yield = number of ears per unit area × number of grains per ear × mean grain weight. The hope was that selection for components separately would be more efficient than selection for the outcome, yield itself. In general, this approach failed because of compensatory negative correlations; increase in one component incurs decrease in another. Yield, in short, is a complex terminal outcome of growth to which there are diverse and interrelated developmental tracks. Two cultivars of similar yield may get there by different routes.

More recently, the field has been considerably widened by biometrical and physiological sophistication, the former made possible by computers, the latter by elegant instrumentation. The biometrical approach invokes multivariate statistics, of which index selection is one special aspect. In principle, multivariate methods are designed to cope with several interrelated factors; in practice they have so far failed to have any significant impact on plant breeding, for reasons which we shall touch upon later (Section 5.8). Crop physiology in general has expanded enormously in the past forty years and there is now a copious literature but, again, early hopes that plant breeding would be provided with better defined objectives and improved methods of selection have not been fulfilled. In particular, the notion that photosynthetic or net assimilation rates could be useful predictors of yield by way of enhanced biomass simply has not

Table 3.1 Yield in plant breeding: examples to illustrate biomass, partition and interaction effects.

(1) SPRING BARLEY (Denmark)

		Nürnberg (19C)	Prentice (1904–25)	Binder (1920s)	Carlsberg (1954)	Emir (1966)
Grain (t/ha)	N_1	2.88	3.23	3.32	3.78	3.57
	N_2	3.16	3.63	4.10	4.82	4.62
	N_3	3.09	3.14	3.32	3.88	4.73
Straw (t/ha)	N_1	3.09	3.70	3.34	3.12	3.08
	N_2	4.18	4.95	4.64	4.34	4.32
	N_3	4.66	5.18	5.15	4.94	5.00
Grain per cent	N_1	48	47	50	55	54
	N_2	43	42	47	53	52
	N_3	40	38	40	44	49

(2) MAIZE (USA)

	Yields (t/ha) at three population densities (thousands per ha)				Plant Height	
	29.7	44.5	59.3	Means	(cm)	b
Old OP variety (Reid)	5.15	5.48	4.66	5.10	202	1.00
4 early double crosses, 1929–39	5.78	5.58	4.84	5.40	211	0.98
8 double crosses, 1940–51	6.40	6.67	6.03	6.36	209	1.04
4 later double crosses, 1954–60	6.69	7.31	6.91	6.97	213	1.21
8 recent single crosses, 1965–	7.13	8.22	8.03	7.79	202	0.85

(3) WHEAT (India)

	5 older tall cvs	Dwarf (1) 5 cvs	Dwarf (2) 5 cvs	Dwarf (3) 5 cvs
Height, cm	113	96	82	63
Grain, t/ha	4.78	5.99	6.07	6.22
Straw, t/ha	11.41	11.00	10.49	8.83
Grain per cent	30	35	37	41

(4) POTATOES (UK)

Material	Fresh yields, g/plant Tubers	Tops	Total	Tubers per cent
Adg-unselected	90	903	993	9
Adg-selected (1)	118	914	1032	11
Adg-selected (2)	385	704	1062	34
Tuberosum	387	481	868	45

Contd

Table 3.1 *Contd.*

(5) SUGARCANE (Guyana)

		Plant crop	First ratoon	Second ratoon	Means
Cane yields, t/ha	Noble	88	54	31	58
	Nobilized (1)	93	82	52	76
	Nobilized (2)	127	121	77	108
Sugar per cent	Noble	9.8	11.6	13.5	11.6
	Nobilized (1)	11.1	11.2	12.8	11.7
	Nobilized (2)	12.1	12.0	14.3	12.8
Sugar yields	Noble	8.4	6.4	3.9	6.2
	Nobilized (1)	10.2	9.2	6.7	8.7
	Nobilized (2)	15.3	14.6	11.0	13.6

(6) RUBBER (Malaysia)

	Yield (t/ha, 8 years)
Unselected seedlings	4.1
Three pre-war clones	8.6
Five early post-war clones	11.1
Advanced seedlings	11.0
Leading recent clone (1960 onwards)	13.1
Two high yielders in trials	15.6

		Yield (t/ha-yr)	
	Clone	Control	Ethrel treated
Low responders	A	1.5	1.8
	B	1.0	1.0
High responders	C	1.8	2.4
	D	1.6	2.7

Notes: 1. The spring barley example includes an early nineteenth-century landrace grown from seed preserved in the foundation of a building in Germany for over 100 years. The landrace derivatives, Prentice and Binder, were superior to it in yield but hardly in partition; the modern varieties show improved partition (less straw) and greater N-responsiveness. The reader should plot graphs of these data to appreciate the partition and interaction effects. Source: Sandfaer and Haahr (1975).

2. The maize data comes from a large study made in Iowa with the objective of trying to define the historical yield effects of maize breeding. For explanation of single and double crosses, see Chapter 5. Note the very large yield gains at all plant densities and the evident variety–density interactions. The b in the last column are average regressions that estimate stability of performance over varied environments; the smaller the b the more stable (Section 6.2). The reader should plot these data and try to estimate an optimal density for each variety group by drawing curves by eye. Source: Russell (1974), supplemented by data on plant height kindly provided by Dr. W.A. Russell.

Contd

Table 3.1 *Contd.*

3. The wheat example illustrates improved partition very neatly. The three dwarf groups are major-gene-determined and derive from Vogel's Norin dwarf stocks which were the foundations of the 'green revolution' in wheat (Section 11.4). At the present time, semi-dwarfs are agriculturally favoured but these results might suggest that extreme dwarfs show promise for the future. The reader should plot these data as regressions on plant height. Source: Jain and Kulshrestha (1976).

4. The potato example illustrates a very large partition effect associated with changing day-length adaptation (short-day-demanding to long-day-tolerant). The first three lines represent three generations of selection of Andigena (South American) tetraploids which parallel the early phases of adaptation of the crop to Europe in the eighteenth to nineteenth centuries, by a process of semi-natural selection. The last line represents a random sample of the adapted Tuberosum population which emerged later in the nineteenth century. Note that two generations of selection made good progress but did not effect the entire transition (subsequent generations did). Source: Simmonds (1976b).

5. The sugarcane example illustrates a major biomass effect on cane yield coupled with (probably) a slight improvement in partition towards enhanced sugar content. The three crops represent three vegetative cycles from one planting; note that yields of cane and sugar declined at much the same rate over crops but that the more recent canes, starting from a higher level, maintained an economic return for longer. The three breeding cycles were: noble canes (bred 1900–30) and two cycles of 'nobilized' canes with wild chromosomes, (1) 1920s and 1930s, (2) 1930s and 1940s. Two cycles of breeding roughly doubled yields of cane and sucrose. Source: Sugar Bulletins of the Department of Agriculture, British Guiana, re-analysed by Simmonds (1976c). Another sugarcane example is given in Fig. 11.7.

6. The rubber example illustrates the extraordinary progress that can be made when a highly heritable partition effect is the main feature. Unselected seedlings are the immediate descendants of wild plants. Two cycles of excellent clones more than trebled yields and the next cycle shows further promise. Note that good seedling populations are not much inferior to clones (Section 5.9). The lower part of the table illustrates clonal interaction with yield stimulation by an ethylene-producing chemical. Source: a combination of diverse commercial and experimental data from the Planter's Bulletins of the Rubber Research Institute of Malaysia.

worked. In retrospect this need not surprise us because these rates are only two determinants among many of a very complex process.

Though physiological measurements as such have yet had no noticeable effect on plant breeding, the ideas and arguments have not been without result, though at a relatively unsophisticated level. Thus, one significant outcome of morphophysiological ideas was Donald's (1968) notion of the ideotype, the ideal plant structure which would tend to maximize both biomass and partition. Donald's wheat ideotype was semi-dwarf, non tillering, had few, small erect leaves and a large awned ear; the wheat ideotype of Bingham (in Lupton *et al.*, 1972) is different in several respects, showing that (as would, I think, be generally accepted) there may well be several valid ideotypes for one crop in different places, or even in the same place. The value of this kind of discussion seems to me to be less in the specific guidance offered to plant breeding than in reminding the breeder that yield is a complex terminal product which can be affected by many different processes. And, even if breeding for yield is still, for most crops most of the time, essentially empirical, the simple biomass-partition ideas (now in general currency) may yet be replaced by more refined understanding; the work of Evans (1993) is a substantial step in this direction.

3.3 Quality

Quality and condition

Any agricultural product can be either improved or spoiled by the appropriate luck or management, so the quality of a sample, as perceived by the consumer, is always a compound of two features: inherent quality and condition. The distinction is important and failure to make it leads to much confusion. The plant breeder's task is to adjust the potential quality of the product, defined as fitness for purpose under a reasonably well defined production regime. The plant breeder can do nothing about condition; he can only provide potentially satisfactory quality if the crop and the product are appropriately managed.

An example will help to make this clear. Potatoes in the UK, as elsewhere, are used in a variety of different ways; for example: peeled and boiled, steamed in their skins, baked, fried, crisped and canned. Four or five distinct breeding objectives in relation to quality can be distinguished. Table stock that is to be peeled and boiled should have white flesh (local preference), fairly low dry matter contents (firm texture, non-sloughing) and reasonably free of after-cooking blackening (an inherent chemical feature connected with phenolic constitutents); potatoes that are to be steamed in their skins differ in that a higher dry matter content and hence a floury rather than a firm, coherent texture is desired. Dry matter content and blackening, however, show strong season, site and management (e.g. fertilizer) effects and interactions: the breeder can adjust both characters towards an acceptable or better level of expression in what he conceives to be an average environment but a large element of condition remains quite outwith his grasp: the good quality variety simply produces good samples more frequently than the poor one. Damage done to tubers (bruising, cutting, scuffing) is a feature of quality as perceived by the consumer; this is, in effect, entirely a matter of condition because there is virtually nothing the breeder can do to reduce damage once his prime quality objectives have been determined. Quality objectives for a crisping variety are quite different; so different, in the UK at least, that varieties are not in general interchangeable. A crisper should be yellow in the flesh, high in dry matter (to reduce oil uptake) and should have a low sucrose content after cool-storage followed by warm 'reconditioning'. The last is important if the crisp is to have the desired pale golden colour, rather than the brown produced by too high sugars. As with table stock, condition effects are conspicuous and the breeder can do no more than aim at a variety which produces a technologically acceptable sample (or somewhat better) in average conditions. Perfection is not a realistic objective. Finally, we should note that the preceding refers only to the UK; a similar discussion for continental Europe or the USA or other potato-consuming countries would involve slightly different objectives.

To sum up, quality is potential fitness for purpose under more or less defined circumstances of production and management; condition, good or bad, is adventitious and outwith the breeder's control. Quality demands spring from social circumstances and many crops, as in the potato example above, exhibit multiple quality objectives which often differ from place to place.

Quality characters

The great diversity of quality characters with which plant breeders collectively have to cope reflects the diversity of uses to which crops are put and very various methods of utilization. As Table 3.2 shows, we can recognize four main groups of quality characters according to end-use. They are:

1. Organoleptic, with or without processing; consumer satisfaction as to taste, smell, texture and colour rule;
2. Chemical, at least predominantly; quality is determined by more or less precise industrial criteria, as in oil, sugar and drug plants;
3. Mechanical, as in the fibres;
4. Biological, as in the fodders, in which animal growth is the ultimate criterion of quality.

This list covers the main features but study of Table 3.2 will show that it is perhaps rather too simple because some crops exhibit mixtures of quality criteria.

In discussing the quality–condition problem above in relation to potatoes, we

Table 3.2 Quality factors in plant breeding.

A. CROP PRODUCT ENTERS TECHNOLOGY

Categories	Examples	Quality criteria
1. Food products		
Preservation (canning, freezing and drying)	Many vegetables	Mostly organoleptic assessment after treatment; size and uniformity of canned products; some specific chemical features (e.g. dried potato powder).
Extraction	Oil seeds	Chemical (oil content, specific composition in relation to nutrition and further processing).
Fermentation	(1) Barley	(1) Chemical (extract, N-content in malting barley – see Fig. 3.3);
	(2) Grapes	(2) Organoleptic assay of fermented samples.
Other processing	(1) Wheat	(1) Physicochemical (milling characters, N-content and protein plasticity/elasticity in relation to bread/biscuit use – see Fig. 3.4);

Contd.

Table 3.2 *Contd.*

Categories	Examples	Quality criteris
	(2) Potatoes	(2) Chemical and organoleptic (dry matter and sugar contents; colour, taste, texture of processed samples);
	(3) Tea, coffee, cacao	(3) Organoleptic assay of processed samples
2. Non-food products		
Specific chemical	(1) Sugar plants	(1) Chemical (sucrose content, juice purity);
	(2) Insecticides and drugs	(2) Chemical (active content)
Fibres	Cotton, jute, sisal	Mechanical (length, strength, thickness, in all; also colour and maturity in cotton – Fig. 3.5)
Feed, forage	Grasses, clovers, fodder Brassicas, feed grains; meal residues from oil seeds	Ultimately, growth or productivity of stock, usually forecast by chemical estimates of soluble CHO, *in vitro* digestibility, fibre content etc.; supplemented by estimates of toxic factors (where relevant) and sometimes amino acid composition
Drug	Tobacco	Chemical (optimal nicotine content) and elaborate organoleptic assay

B. CROP PRODUCT CONSUMED (MORE OR LESS) DIRECTLY AS FOOD

Categories	Examples	Quality criteria
Cereals	(1) Maize, sorghum	(1) Locally, for nutritional reasons, enhanced lysine/tryptophan contents
	(2) Rice	(2) Texture, flavour, aroma of cooked product
Pulses	Various peas and beans	Organoleptic characters; high protein, low contents of flatus-inducing and toxic factors
Tubers	Potatoes	Organoleptic assay of cooked samples for texture, taste and colour; texture related to dry matter and colour to phenolic constituents
Fruits	Numerous	Organoleptic assay with or without preliminary cold storage treatments
Vegetables	Numerous	Organoleptic assay

saw that multiple quality objectives within a single crop are common, indeed almost universal. Figures 3.3–3.5 provide further examples, in barley, wheat and cotton respectively. The quality–condition distinction comes over very clearly in Fig. 3.4 and the complexity of technological demands on the plant breeder comes out well in all three.

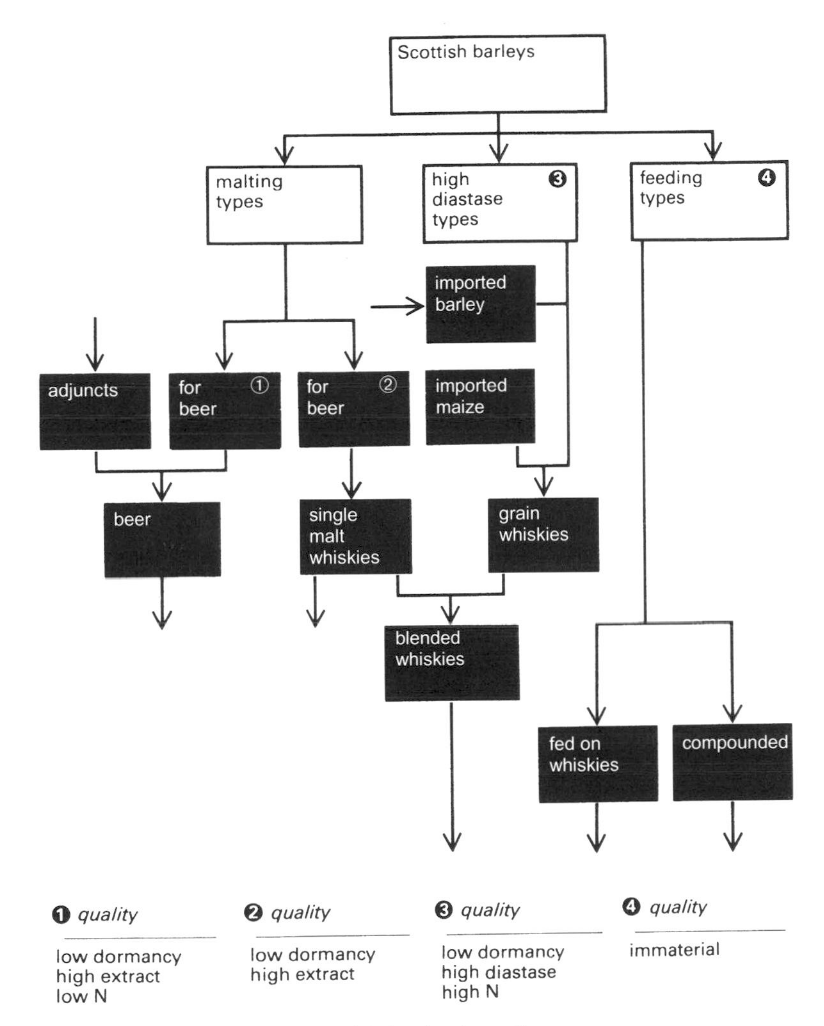

Fig. 3.3 Quality objectives in Scottish barley breeding.
Notes: Four objectives can be distinguished but the requirements for malting for beer and for malt whisky are little different and one variety should serve both purposes. High-diastase barleys, designed to convert the starchy materials used to make grain whisky, are a pretty example of breeding for enhancement of a specific enzyme; they are high-N types, unacceptable for ordinary (beer) malting but potentially valuable for feeding (Hayter and Allison, 1976). Feeding barleys are, in effect, high yielding barleys that are unacceptable for malting; quality is immaterial because they are, in practice, fed on the basis of average composition. Condition effects are most apparent in connection with N-content: heavy N-fertilizing tends to spoil malting quality.

A word of scepticism is appropriate here. Quality is fitness for purpose and standards are defined by what the market wants or says it wants: the normal basis of definition, therefore, is 'what's there already only a bit better'. But tastes for foodstuffs vary widely from country to country and from time to time and technology changes too, often on a time scale far shorter than that of plant breeding. Quality standards for many crops, therefore, are far from immutable; they are partly, even largely, determined by the characters of existing varieties to which people and technology have become accustomed. The plant breeder needs to be wary of breeding for quality characters which no longer matter (Fig. 2.2).

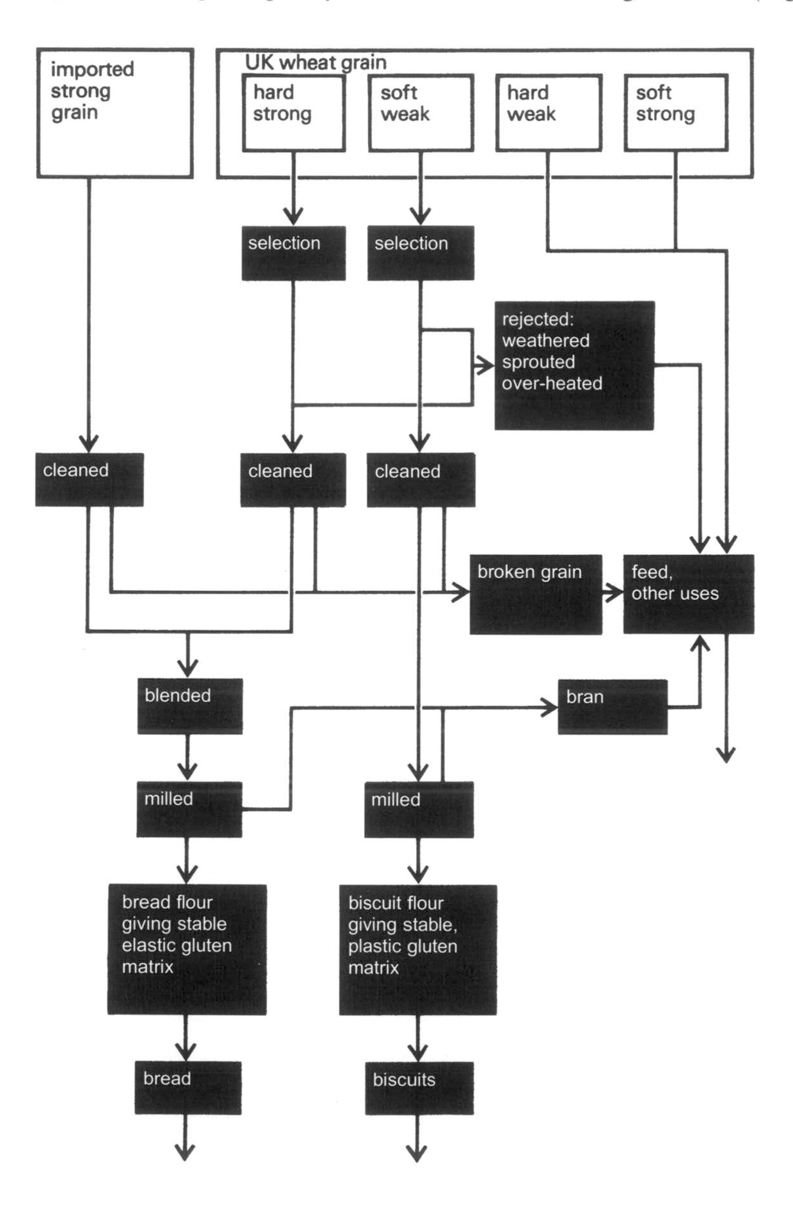

Quality assessment

The basic features of any effective system of quality assessment in a plant breeding programme are that it should be quick, cheap and economical of material. These features are essential because the plant breeder: (*a*) handles thousands of stocks each year; (*b*) is often working against time, having to take keep/discard decisions in preparing planting plans for the succeeding season; and (*c*) will often only have small amounts of material available for testing.

Assay of organoleptic characters is, naturally, by organoleptic methods, on raw or suitably prepared samples. Efforts to characterize flavours chemically have generally (indeed, so far as we know, universally) failed; flavour is yet too subtle for gas chromotography. So the breeder, perhaps aided by a taste panel, looks at, smells and tastes his fruits, vegetables and pulses and takes decisions; usually, he will take care to have control material (two or three established varieties with a range of quality characters) grown alongside his test material and assayed along with it. This is particularly important whenever environmental effects on quality are substantial (as they often are). If the breeder is examining material after processing, he will take care to standardize the preparation procedure as closely as possible. Thus, in the potato breeder's kitchen, tubers will be halved, boiled for a standard time and inspected and scored when cold; or they will be cut into uniform thin slices and fried in a standard oil at a fixed temperature if crisping is the object.

Methods of assessment of chemical quality characters always lay emphasis on small scale and speed. Thus, in sugar plants, sucrose content is what ultimately matters but, for practical purposes, an optical measurement of total dissolved solids in juice (Brix) is sufficiently highly correlated with sucrose to be good enough; so a few drops of a juice and a few seconds work with a hand refractometer usually suffice. In malting barley, low protein content (related to

---

Fig. 3.4 Quality objectives in UK wheat breeding. Based on information kindly provided by Mr B.A. Stewart (personal communication).
Notes: The basic determinants of quality are grain characters: hard *vs.* soft milling (major-gene control, determined in the laboratory by milling samples); and strong *vs.* weak protein (determined in the laboratory by elasticity tests). Of the four combinations two are acceptable for technological uptake (bread and biscuit flour) and two are acceptable only for feed, as shown. For technology, there is a severe selection for condition: sprouted, or incipiently sprouted, grain is high in alpha-amylase, which will hydrolyse gelatinized starch and therefore yields an unacceptably sticky loaf: the breeder can minimize the risks by selection for adequate dormancy and low alpha-amylase levels; overheated grain, purely a condition effect, has denatured protein. Good quality bread flour must be not only hard-milling and have a strong, elastic gluten, but must also have a high gluten content if the CO_2 bubbles generated by the yeast are to be held stably in the matrix, thus giving good bread volume. UK bread wheats are grown at high yield, therefore at low protein content (see Section 5.8); alone, their flour is unacceptable, so it is always blended with imported high-protein material. For most biscuits, a plastic gluten matrix is desired and the protein content of the flour should be about 9 per cent; UK biscuit wheats therefore give satisfactory quality without blending.

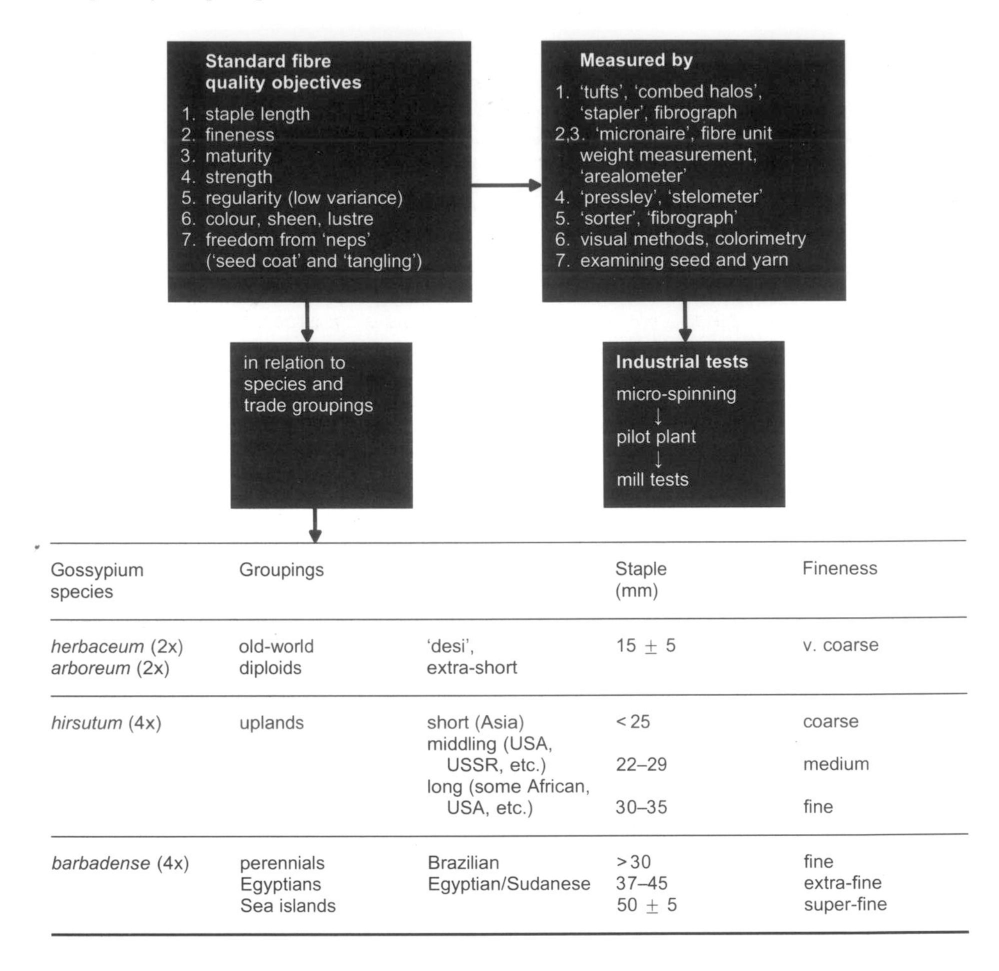

Gossypium species	Groupings		Staple (mm)	Fineness
herbaceum (2x) *arboreum* (2x)	old-world diploids	'desi', extra-short	15 ± 5	v. coarse
hirsutum (4x)	uplands	short (Asia)	< 25	coarse
		middling (USA, USSR, etc.)	22–29	medium
		long (some African, USA, etc.)	30–35	fine
barbadense (4x)	perennials	Brazilian	> 30	fine
	Egyptians	Egyptian/Sudanese	37–45	extra-fine
	Sea islands		50 ± 5	super-fine

minimal haze in the beer) and high extract of soluble (fermentable) oligosaccharides after malting are desired; so the barley breeder estimates, not protein *per se* (which is difficult), but nitrogen content (which is easy and good enough) and soluble carbohydrate ('extract') after micro-malting of a few grams of grain *in vitro*. Similarly, processing potential of wheats is well-enough estimated by quick small-scale physico-chemical tests in the laboratory.

The capacity of the plant breeder to handle large numbers of small samples with adequate precision has improved enormously over the past two or three decades, thanks to the development of improved analytical methods: partition chromotography (paper and thin layer), various form of spectrophotometry (depending on the formation of specific coloured compounds of the substance being assayed), automatic recording devices and automatic analysis have all played a part. So the contemporary plant breeder concerned with chemical characters has access to an array of measurements of a number, complexity and precision which would have been inconceivable to his predecessors. It is well to

Fig. 3.5 Quality in cotton breeding. Above are quality demands made by the market and the methods used by the plant breeder in meeting them; below is a broad classification of the cotton crop in relation to critical lint characters. Based on information kindly provided by Dr J.T. Walker (personal communication).
Notes: 'Lint' is composed of unicellular seed hairs which are spirally thickened at maturity; there is also a layer of short underhairs ('fuzz') which may find uses, but always separately from lint proper. Lint hairs of the cultivated cotton species are spirally twisted, an essential condition for coherence in spinning.

Lint length and fineness are broadly correlated but there are exceptional long-staple, fairly coarse cottons. Fineness/length and lint yield tend to be negatively correlated, so the finest Sea Islands are the lowest yielding and most expensive; the 'middling Uplands' constitute more than half the total trade.

The fibre testing of cotton has progressed during this century from 'hand-pulling' of staple coupled with subjective observation and feel, through various hand sorting and combing techniques to the stage where various forms of 'stapler' and 'fibrograph' utilize automatic sampling and photo-electric devices for measuring length. The 'fibrograph', which exists in complex form with digital output, also measures fibre regularity. These and other such instruments are used by the breeder-technologist under environmental control with internationally agreed standards of humidity and temperature. Standard samples are also used for calibration. The 'pressley' and 'stelometer' instruments are essentially devices for measuring breaking strengths of bundles of fibres standardized by weight. The 'micronaire' estimates fineness indirectly by measuring the permeability of a standard wad of lint to the passage of a standard air-flow. 'Maturity' of lint (cells dried out, walls fully thickened, cells with spiral twist reversals) is an important condition, and any immaturity confounds 'micronaire' estimates of fineness. 'Seed coat neps' are minute tufts of refractory immature lint adhering to the micropylar cap and genetic selection can be imposed against this characteristic. Though such measurements of fibre properties are acceptable for plant breeding purposes, they are rather crudely predictive of industrial performance – regularity, for example, can give a yarn strength in excess of that expected from estimates of average length, fibre strength and fineness alone. Ultimately, 'micro-spinning', pilot scale testing and then large-scale commercial spinning tests on each of several bales are necessary.

---

remember however that these measurements are but a reasonable basis for immediate decision; they are usually no more than forecasts, subject to statistical and other uncertainties, of technological aptitude. We shall return to this point later.

Mechanical assessment of quality, notably in the fibres, rests on measuring characters which will predict the industrial performance of the product. Cotton in the most complex and so provides the best example. Criteria and methods are summarized in Fig. 3.5. The breeder uses visual methods to some extent but relies mostly on various ingenious mechanical devices for measuring length, strength and fineness. Note that maturity is a condition as well as a quality factor. Note also that the breeder in one particular place must aim at a predetermined market standard, rather than at an objective concept of excellence in respect of staple length and fineness; quality is fitness for purpose.

Finally, the quality objectives of forage/fodder breeding programmes are biological in character and would ideally be met by testing animal growth but virtually never are, for good reasons of time and cost. Universally, chemical proxies for nutritional value are sought or the problem is, in effect, ignored. The latter is a perfectly reasonable procedure for crops in which a critical primary objective has to be met and the feeding value of the residual crops and crop residues is secondary (Table 3.2 and Figs 3.3, 3.4); farmers and industrial users can well utilize such materials on the basis of average composition. Among the crops bred as fodders or feeds, *per se*, a variety of methods of predicting nutritive value are current. They are (naturally) all more or less correlated with each other and must, of course, be as quick, cheap and simple as possible. Thus, soluble carbohydrate content, fibre content (negatively related to feeding quality and estimated by chemical or physical methods) and *in vitro* digestibility all have their adherents; the last method uses sheep's rumen liquor (withdrawn from canulated animals) to simulate *in vitro* what happens in the ruminant's gut.

To summarize this section so far, we might say that the plant breeder seeks to predict quality, however complex, by relatively simple measurements or organoleptic assays. Often he will use scaled-down technology as a basis but will sometimes depend on correlations between purely chemical or physical measurement and performance of the product in actual use.

In the end, however, predictions and correlations are not good enough and industrial scale testing is generally essential before a new variety is marketed. So wheats are milled, barleys malted, potatoes crisped, cottons spun, apples stored, strawberries jammed and bananas shipped before final decisions are taken; ideally, fodders would be tested by animal growth but, in practice, this is rarely feasible and most decisions are, in practice, taken on analysis.

3.4 Other objectives

Though the headings yield and quality cover most situations, there are a few examples of objectives which fall neatly into neither category. They are very heterogeneous and are worth brief mention here in the following arbitrarily-ordered list.

First there is the question of uniformity of plant and product. Uniformity is sometimes the secondary consequence of the adoption of breeding plans intended primarily to promote high performance; e.g., pure lines, hybrid varieties, clones (see Chapter 5). Sometimes the attainment of uniformity as a quality feature is a reason for using those same breeding plans independently of any possible yield advantage. So high quality cottons are intensely inbred and uniformity of product is the reason most commonly given for the adoption of hybrid breeding plans in those vegetables in which it is doubtful whether the use of hybrid varieties can be justified by yield advantage. Again, uniformity may have agronomic attractions, as in swedes and sugar beet in which uniform shape, size and posture of roots facilitates mechanical harvesting. Then, there are aesthetic and legal aspects. A well-grown, uniform stand of a crop or a uniform product *looks* attractive independently of any possible economic advantage. The legal requirement (a very recent development – see Section 6.3) varies with crops; it springs from the need to discriminate unambiguously between varieties in the context of plant variety rights. The legal demand may impose considerable burdens on the breeder and may have little or nothing to do with agricultural or technological performance. So the need for uniformity is complex and variable: it has yield, quality, aesthetic and legal aspects.

Second, among clonal crops one occasionally encounters features of propagation that bear on economic value of varieties but are directly referable neither to yield nor quality. Thus, aptitude for rooting or grafting is significant in some tree crops and just the right amount of suckering is an economic feature of banana varieties – enough to permit ready propagation but not so much as to be expensive to prune. The use of clonal rootstocks in tree crops imposes, in effect, secondary breeding programmes or at least the need for specific selection of rootstocks in relation to stock-interactions (as in some pome fruits and citrus). Grapes perhaps provide the best example of deliberate rootstock breeding in the use of hybrids of wild American *Vitis* forms as rootstocks to control the woolly aphis in European vineyards. Rubber clones are normally bud-grafted on seedling rootstocks; clonal rootstocks are infeasible now but may yet become possible; the possibility of budding wind and disease-resistant crowns on to high-yielding trunks ('crown budding') has been known for many years and has recently entered commercial practice in Malaysia, in a few special situations. So, given means of propagating clonal rootstocks on the field scale (which are being sought), rubber breeders a decade hence may be faced with the task of breeding all three components (root, trunk and crown) of tripartite trees; a daunting task, but economically an attractive one.

Third, we have referred above to short stature as a feature that, in many crops, annual and perennial, contributes to improved partition and enhanced yield. There is another aspect to dwarfness, namely economy of harvest in tree crops. Dwarf varieties *per se* and dwarfing rootstocks both contribute. Quicker, cheaper, timelier and a more complete (less wasteful) harvest is the result (so there is a yield aspect of the economy effected).

Fourth, the sugar beet fruit is basically three-seeded and the carpels can not readily be separated by milling. So, if a uniform stand is to be achieved, much singling is necessary. The 'monogerm' character, widely incorporated in modern sugar beets, allows precision drilling to a predetermined stand and hence economy of cultivation.

Fifth, herbicides have only come into general use in technologically advanced agricultures in the past three or four decades. Much experience now shows that crops differ, often dramatically, in susceptibility to different preparations and that there are lesser differences between varieties within crops. So breeding for herbicide tolerance has become a feature of at least some plant breeding programmes; it might be regarded as a contribution to biomass (analogous to much disease-resistance breeding) and so as a component of yield improvement in a changing agricultural environment. The breeder who understood such a programme, however, would probably proceed with caution, reflecting that fashions in herbicides can change rather quickly.

Sixth and finally, elimination of vegetative prickles or irritant hairs occasionally turns up as a breeding objective in diverse crops and has surely been a feature of the evolution of several, e.g. the eggplant and minor *Solanum* fruits. Thornlessness in *Rubus* fruits and avoidance of irritant sheath hairs and sharp leaf edges in sugarcane, which can be very unpleasant for the worker, are current examples.

3.5 Interactions

When a number of genotypes show different responses to a varied set of environments, there is said to be genotype-environment (GE) interaction. This may emerge from the appropriate item in an analysis of variance or from an analogous regression analysis of genotypes on a graded series of environments. Quantitative aspects of GE effects are best discussed in relation to yield trials (Section 6.2). Here, we need only consider what effect the knowledge that GE effects are likely to occur will have on the formulation of objectives.

Yield effects are, in practice, the most important, though GE effects on quality are by no means unknown. Table 3.1. shows that different barley varieties respond differently to fertilizer-N and rubber clones differently to yield-stimulation. The barley example can be generalized. Not only are dwarf cereals in general more efficient in terms of partition than tall ones but they are also more responsive to (or tolerant of) high fertility. The joint adaptation of stature and fertility in cereals has been going on in technologically-advanced agriculture for several decades and it is hard to say whether the varieties came first and permitted the fertilizer or increasing fertility helped to select the varieties. At all events, there was mutual adaptation by exploitation of favourable GE effects which were subsequently quite deliberately adapted to the agronomy-variety packages which later became the basis of the 'green revolution' (Section 11.4); GE effects were thus built into breeding objectives. Even if the need for rather well defined adaptation of this kind cannot be foreseen, GE yield effects of a less specific kind abound. These become apparent in comparing varieties over sites and seasons; some, the more responsive, or less stable, ones vary widely, doing well in good conditions, badly in bad ones; others, the less responsive, or more stable, ones, are less variable, doing neither very well nor very badly. Ideally, perhaps, the breeder would have one or other kind as an objective. In practice, since stability is hard to determine or breed for, and since both kinds of variety have their uses, it is usually more realistic simply to try to exploit to advantage whatever turns up.

Thus we see that, when GE effects on yield or other characters are present, the breeder's objectives may either, in a sense, evade them by defining a narrow range of environments/conditions to which adaptation is sought, or negate them by means of the stable, widely-adapted genotype. A practical compromise, very frequently adopted, is to define sub-environments on the basis of general experience but to be prepared to exploit general adaptation if it emerges.

3.6 Conclusions

The plant breeder is working in a complex system, one in which agricultural, technological and socio-economic factors all play parts and are all in a state of more or less continuous change and interaction. There are indeed very few situations in which plant breeding objectives are clear, simple and stable; rather, they are complex and changeable but nevertheless necessary. Breeding in the hope that something will turn up is unlikely to be profitable. No simple linear system of decision is therefore possible; the plant breeder must define objectives but be prepared to change them as the state of the world and the state of the art change.

Drawing all this together, there seem to be seven points to make by way of conclusion.

1. The breeder needs to take a view of where economic advantage will lie ten to twenty years hence. What weights should be attached to yield and quality? What will the markets (technological and consumer) want? Will they want his product at all or will they want something different?
2. He will also have to take a view of likely environments on the same time scale. Will his crop be managed in the same way as now or will there be moves towards higher populations, more fertilizer, different herbicides and mechanical harvesting? Are new or altered disease patterns likely and, if so, with what implications for his objectives?
3. The breeder will need to recognize the existence of GE effects and decide whether he will aim at widely adapted varieties or will select for adaptation to sub-environments within the relevant area. Experience of past and present varieties will provide at least some guidance.
4. The breeder will need to have some sort of ideotype in view. It will not be too closely defined because knowledge, both physiological and genetic, is never good enough. But he will have to have some sort of 'feel' both for the morphophysiology of the varieties he is trying to produce and for the genetic

potential for achieving them. The perfect ideotype is no good if it can not be constructed. At the very least, the breeder will have a clear idea of whether he is trying to change biomass or partition or both, of the place of disease resistance in his scheme and of the quality factors which he must take into account. And, for reasons which we shall meet later (Section 5.8), he will try to keep the list of desiderata as short as possible.

5. Having defined his objectives, the breeder will be prepared to change them if he must. He will recall that there are other breeders working on the same crop and that they may be cleverer or luckier than he; also that agricultural practices and technology both change so that what seems desirable, even essential, now may be irrelevant ten years hence.
6. The breeder will also reflect that the perfect plant variety never has been bred and probably never will. Successful varieties are merely less imperfect than their predecessors. Husbandry and technology both have considerable capacity for adaptation, so excellent varieties are not only called forth by practical demands but also to some extent determine the practice; in short, there is mutual adaptation.
7. Finally, in determining objectives, the plant breeder will no doubt listen to anyone qualified to comment but must resign himself to receiving nearly as many different bits of advice as he has advisers. Recognition of crucial elements is never simple but the breeder's own judgement on this is probably better than that of most others.

3.7 Summary

In discussing the objectives of plant breeding in this chapter, we saw that the breeder needs clear objectives but that he is working in a complex and changing situation so must be prepared to change them. Objectives can be classified broadly as related to yield or quality (*plus* a few oddments). Yield is promoted by enhanced biomass, of which disease resistance is an important component, and improved partition (more product, less waste). Quality is fitness for purpose and must be distinguished from condition (the product of environmental circumstances outwith the breeder's control). Quality factors are very various and are assayed by four broad categories of test, depending upon crop and purpose: organoleptic (subjective taste, smell, texture, appearance), chemical, mechanical and biological. Much ingenuity has gone into developing quick, simple tests adapted to handling large numbers of samples cheaply. There are often genotype–environment interactions which the breeder needs to take account of. Given the uncertainties, biological, agricultural and economic, decisions about objectives are never simple.

3.8 Literature

Yield, biomass and partition; crop physiology

Austin and Jones (1975); Cannell (1978); Cannell and Last (1976); Wareing and Allen, Monteith in Cook *et al.* (1977); Cooper (1975); Donald (1963, 1968); Donald and Hamblin (1976); Dubois and Fossati (1981); Eastin *et al.* (1969); Frey in Eastin and Munson (1971); Evans (1975, 1993); Gale and Law (1977); Harris *et al.* (1976); Hutchinson (1971b); Jain and Kulshrestha (1976); Lamberts *et al.* (1968); Loomis *et al.* (1971); Lupton and Kirby (1967); Matsuo (1975b); Murfet (1977); Pelton (1964); Russell (1972); Simmonds (1973, 1988, 1994a, 1995a); Snoad (1974); Trenbath (1974); Vogel *et al.* (1956, 1963); Wallace (1985); Wallace *et al.* (1972); Wellensiek (1968); Wittwer (1974); S. Yoshida (1972).

Quality

Baker and Campbell (1971); Baumann *et al.* (1975); Bliss and Brown (1983); Bjarnason and Vasal (1992); Briggs *et al.* (1969); Cook (1962); Fowler and Roche (1975); Hayter and Allison (1976); Hehn and Barmore (1965); Johnson *et al.* (1968); Konzak (1977); McGuire and McNeal (1974); Milner (1975); Munck (1972); Nelson (1969); Peterson and Foster (1973); Thomson and Rogers (1971); Vaughan *et al.* (1976).

Chapter 4
Genetic Aspects: Populations and Selection

4.1 Introduction

The plant breeder alters the genetic composition of crop populations by selection; he is, as we have seen, an applied evolutionist. So this chapter is fundamental to general ideas about plant breeding, divorced from the pressing practicalities of objectives, disease resistance, yield trials, seed multiplication and so forth. In it we shall first distinguish between genotype and phenotype (between potential and achievement as it were) and then go on to consider the basic concepts of population genetics in the plant breeding context. We shall find that the ideas on population genetics must be supplemented, in considering response to selection by the notions of polygenic characters and heritability, the latter itself an extension of the genotype–phenotype distinction. Finally, we shall have to add yet another complication in the form of genotype-environment (GE) interactions.

The fact is that we never know precisely what is happening in a population changing under selection. Fortunately, exact knowledge is inessential (otherwise plant breeding would be impossible) and, at a highly empirical level, as we shall see, there are situations in which it is perfectly reasonable to disregard formal genetics as such and talk in terms of a workable statistical abstraction, namely combining ability.

4.2 Genotype and phenotype

One of the fundamental ideas of biology – and certainly of applied genetics – concerns the relation between phenotype and genotype. The idea and the terminology go back to Johannsen, who, in the early years of the twentieth century, showed that selection for small and large seed in an inbred line of bean (*Phaseolus vulgaris*) was ineffective. There was variation in seed size but it was not heritable. By contrast there were consistent differences *between* different lines which could regularly be recognized despite variability within lines and between samples. So some variability is genetic, some is environmental. The *genotype* (or genetic constitution) determines a certain potential for development, environment determines the developmental track adopted and the *phenotype* is the outcome. Most of the acuter practical problems of plant and animal breeding stem from this distinction. The breeder's task is to construct favourable genotypes which cannot be recognized from mere inspection of phenotype.

We will now explore the subject a little more deeply with the help of a specific example. Major gene substitutions are recognized because they have sharp, consistently identifiable phenotypic effects; conversely, it is because they have such effects that the determinants are called 'major genes'. Their expression is little affected by environment and phenotype is closely, often exactly, related to genotype. They were Mendel's chosen materials in peas (*Pisum sativum*) and their very clarity of expression, free of Johannsen's fundamental but rather confusing ideas, were essential to the Mendelian analysis. Our example concerns a major gene which can be unambiguously classified in segregation but which yet shows an environmental element. In wheat the gene concerned controls chlorophyll-a content of the flag leaf and can be classified visually on chaff colour. Using the symbols *H-h*, the dominant combinations, *HH* and *Hh*, have high content, the recessive *hh* low. Superimposed on genotype, chlorophyll-a content is phenotypically variable in inbred pure lines and genetically homogeneous hybrids, as the top part of Table 4.1 shows.

Table 4.1 A major gene in wheat *Triticum aestivum* (data from Planchon in Janossy and Lupton, 1976). A cross between two varieties differing at the locus *H-h*.

Material	Constitution	Numbers of plants	Mean Cpl-a content	Standard deviation	Joint s.d.
P_1 Capitole	*hh*	19	2.6	0.084	
P_2 Rouge de Bordeaux	*HH*	20	3.4	0.069	0.075
F_1 $P_1 \times P_2$	*Hh*	20	3.2	0.068	
F_2 low	*hh*	13	2.5	0.118	
F_2 high	*H*	37	3.2	0.165	0.150
BC low	*hh*	24	2.5	0.111	
BC high	*H*	26	3.3	0.150	

The F_2 and backcross ratios are satisfactory: *Hh* selfed gives $2(HH + Hh)$: 1 *hh* and $Hh \times hh$ gives 1 *Hh* : 1 *hh* but, within one genotype, content varies, as shown by the estimates of standard deviations in column 5 of Table 4.1. Note that the variability of the segregating generations (F_2 and BC) is markedly greater than that of the non-segregating ones (inbred parental and F_1); this difference implies segregation of a genetic background effect on chlorophyll-a content superimposed on the major-gene (*Hh*) effect. So we can distinguish three components of variation: (1) the major-gene effect, responsible for a large difference between means; (2) an environmental effect apparent as variability within genetically homogeneous families; and (3) another genetic effect apparent only as enhanced variability in segregating families.

Now imagine that the variability due to the environmental-*plus*-genetic background effects was twice as great as observed, with results shown in Fig. 4.1. Discrete differences between parents become a little uncertain in the middle of the range and F_2 and BC distributions become, in effect, continuous; our major gene segregation has all but disappeared and no more than hints of bimodality persist. These results are generalized further in Fig. 4.2 in which the limiting distributions of large samples are plotted, the curves towards which those of Fig. 4.1 would tend if large numbers of plants were measured at very narrow intervals. Despite the continuity of the composite curves it will be seen that ambiguity of classification occurs only in the centres of the distributions. To

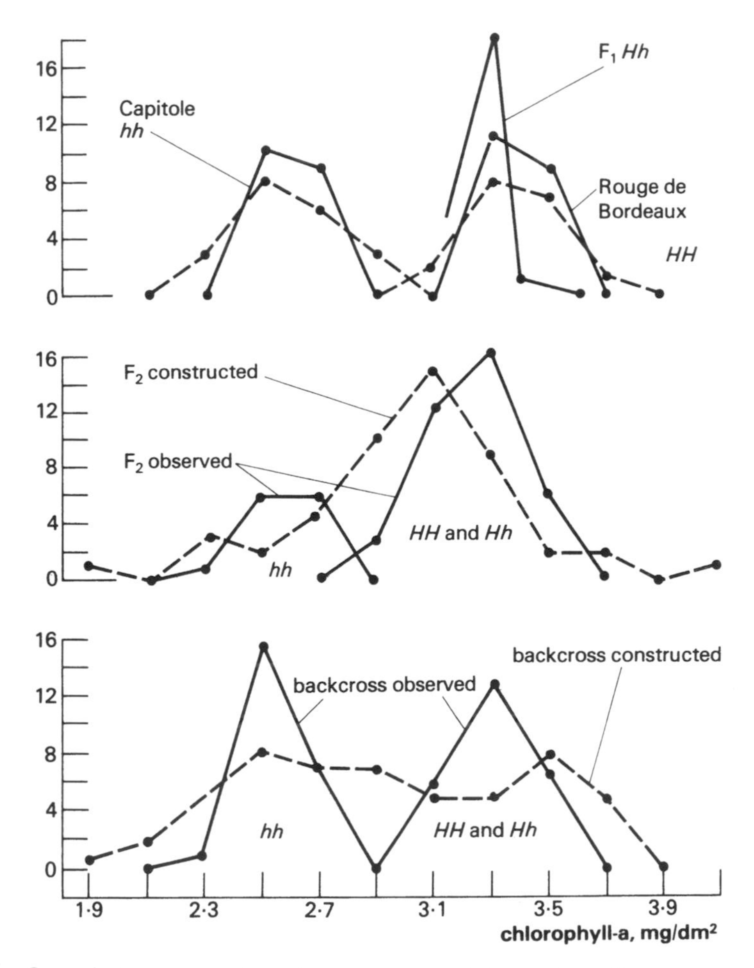

Fig. 4.1 Quantitative expression of a major gene, *Hh*, in wheat (1).
Notes: See Table 4.1 and text. Chlorophyll-a content of flag leaf in parents, F_1, F_2 and BC. Observed distributions are plotted as solid lines; dashed lines are constructed from random deviates with standard deviations twice as large as the joint s.d. given in Table 4.1.

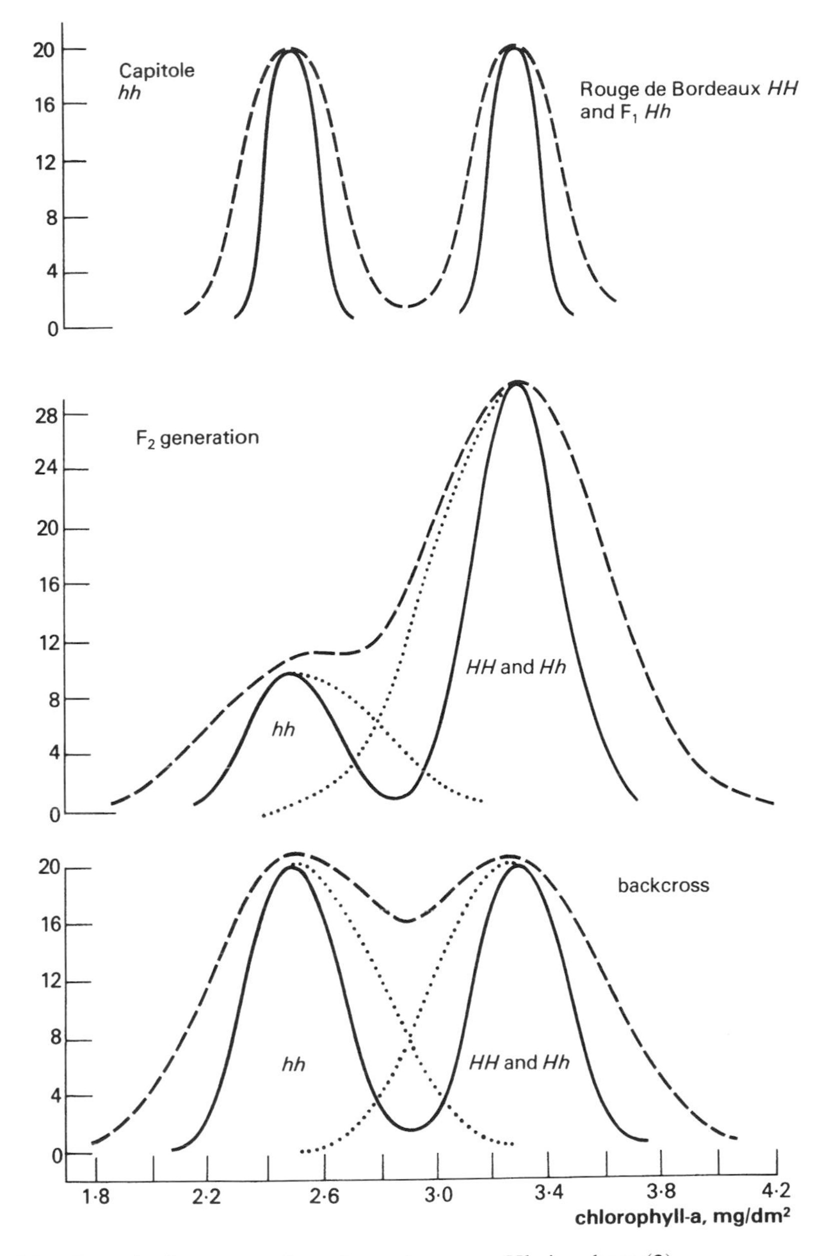

Fig. 4.2 Quantitative expression of a major gene, *Hh*, in wheat (2).
Notes: See Table 4.1 and Fig. 4.1. Distributions expected of large normal samples, with observed and enlarged standard deviations, are shown respectively as solid and dashed lines, as in Fig. 4.1; *H* and *hh* components of the latter are distinguished by dotted lines.

pick plants with low or high expressions would be to pick *hh* or *H* genotypes with high reliability. So the character remains highly 'heritable', a word to which we shall return later.

To summarize, we have, in this example, recognized Johannsen's distinction between genotype and phenotype in respect of a specified major gene; we have also recognized some genetic background control of expression and touched on the idea of genetic and environmental components of 'heritability'.

4.3 Some ideas in population genetics

Introduction

Population genetics is mostly concerned with two main ideas: the behaviour of genes in populations and the evolution of populations by gene-substitution under natural selection. The advanced theory is mathematical and lies beyond the scope of this book, but the basic ideas are simple enough and are fundamental to plant breeding. This is so because, as we saw earlier, plant breeding is a special case of evolution in which natural selection is replaced by a mixture of natural and artificial selection.

We shall start with a brief account of the terminology and assessment of gene expression as a preliminary and then go on to population–genetic ideas *per se*. The starting point is: what happens to alleles in a large diploid random-mating population in the absence of disturbance due to mutation, selection, inbreeding in small populations and linkage? We shall then consider the effects of these disturbing factors and examine a few plant breeding situations in population–genetic terms.

Gene expression

This section is mainly a matter of definitions (Fig. 4.3). First let us remind ourselves that a locus can be known only if at least two alleles have been detected there. The top part of the figure covers the basic nomenclature of all the common situations in terms of two alleles, a_1 and a_2. An allele is said to show dominance, partial or full, if the heterozygote diverges from the mid-homozygote expression; dominance is complete when the heterozygote is identical with the appropriate homozygote. Note, however, that a partial dominant shows an intermediate, though not median, expression and some writers would restrict the word dominant to the extreme case, observing that in all other cases an allele shows at least some heterozygous expression. The ideas of median (or 'additive') expression as against degrees of dominance from partial to full, has taken firm root in biometrical genetics and is generally acceptable in plant breeding contexts.

Two other points must be made here. What we say about gene expression

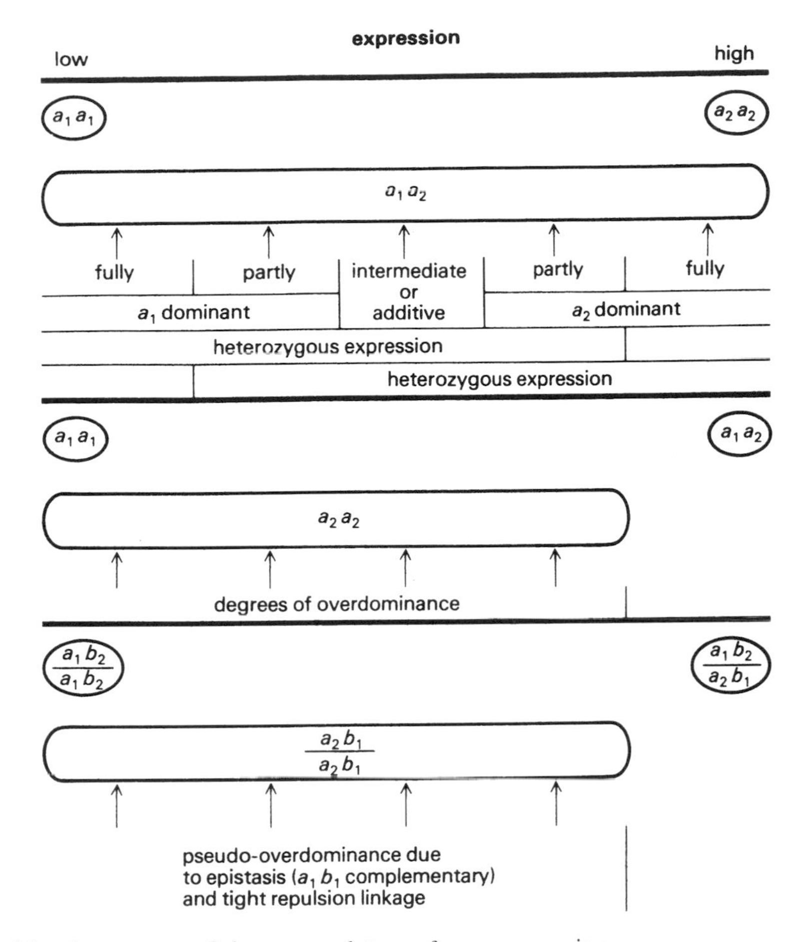

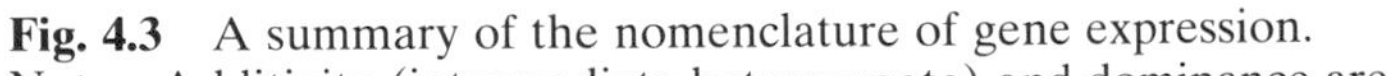

Fig. 4.3 A summary of the nomenclature of gene expression.
Notes: Additivity (intermediate heterozygote) and dominance are shown at the top, overdominance (heterozygote exceeds both homozygotes) in the middle, and epistasis-*plus*-linkage producing pseudo-overdominance at the bottom.

depends both upon what we look for and how we measure it. In genetics we are nearly always concerned with what we regard as the leading effect of the gene: red *vs.* white flowers, tall *vs.* short plants, resistant *vs.* susceptible to disease and so forth. But these features are all end-products of development and dominance or intermediacy at that level, the ultimate phenotypic expression, says nothing about gene expression at the polypeptide level. At this level, in fact, intermediacy is probably the rule (independent allele action) so what we see as dominance is a phenotypic-developmental feature. Nevertheless, for the geneticist and plant breeder, it is an operational reality.

The second point, how we measure gene expression, is a matter of scale. Briefly, our choice of *how* to measure is as arbitrary as our choice of *what* to measure. Obviously, replacement of an arbitrary linear scale of expression (say

x) at the top of Fig. 4.3 by any one of a multitude of possible curved scales (e.g. log x, $\sqrt{x}$) will shift the classification of gene expression. So it is easy to 'convert', so to speak, a partially dominant expression into an intermediate one (or *vice versa*) and biometrical genetics regularly does just this by choice of transformation.

The lower parts of Fig. 4.3 treat overdominance, the situation in which the heterozygote exceeds both homozygotes. Operationally, overdominance presents real problems (as we shall see later) but it is doubtful whether it really exists at the locus level, save in certain very unusual circumstances. Probably, pseudo-overdominance (bottom of Fig. 4.3) is the usual cause and this kind of mechanism is certainly heavily implicated in natural polymorphisms. In effect, the linked loci *a* and *b* are replaced by a single functional locus, *c*, such that

$$c_1 = a_1b_2,\ c_2 = a_2b_1$$

and $c_1c_1 < c_1c_2 > c_2c_2$

Hardy–Weinberg equilibrium

Consider two outbreeding plants each homozygous for a different allele at one locus, *A-a*. Cross them to form a family of heterozygotes, *Aa*, and allow this and later generations to cross at random. What will be the fate of the two alleles *A* and *a* in the absence of selection? How will populations be composed with respect to *A-a*? The reader should solve this problem for himself; it can be done in a few minutes by setting out the possibilities exhaustively. It will emerge that the first and all subsequent randomly crossed populations are composed thus (per cent): 25 *AA*; 50 *Aa*; 25 *aa*. Each allele clearly has a proportional occurrence (referred to as a gene or allele 'frequency') of 50/100 and we can write $p = q = 0.5$ for the gene frequencies. Now repeat the exercise starting with random crossing between 4 *AA* plants and 1 *aa* plant (i.e. $p = 0.8$, $q = 0.2$). You will find that the first and subsequent generations are composed thus:

64 *AA*: 32 *Aa*: 4 *aa*

The rule is that a random mating population reaches equilibrium in one generation at a composition given by the expansion of $(p+q)^2 = (p^2 + 2pq + q^2)$. In the second example above, $(0.8 + 0.2)^2$ gives $0.64 + 0.32 + 0.04 = 1$.

This result can be generalized to several independent loci by the expansion of:

$$(p+q)^2\ (r+s)^2 \ldots$$

If the population is founded on multiple heterozygotes the equilibrium is reached, as before, in one generation; if not, it is somewhat delayed. (The reader should test this for himself by considering two generations of a population founded on 50 *AABB* + 50 *aabb* randomly mated.)

Two results will be obvious. If *A* is dominant to *a*, the latter may be present as an allele at quite a high frequency and yet be rarely visible in the population as a homozygous recessive plant, simply because $q^2 < q$. And heterozygotes attain a maximum at $2pq = 0.5$ when $p = q = 0.5$; all other values of p and q give heterozygosity less than 0.5.

Simple Hardy–Weinberg equilibrium such as we have been discussing is probably never strictly satisfied in either wild or plant-breeding populations. It is disturbed by mutation, by departure from random mating, by differential fitness and by linkage. For our purpose, mutation can be neglected, great as is its importance in the long-term evolutionary theory of large populations. The other three are more important.

Mating system

Population genetics mostly deals with random-mating (i.e. highly outbred) populations. Various kinds of non-random mating are possible, however, of which, in a plant breeding context, inbreeding is the most important. Inbreeding is measured by F, the inbreeding coefficient, which is defined as the probability of identity of alleles by descent. We omit details and note simply that F ranges from 0 for individuals in a large randomly mated population to 1 for the products of prolonged selfing, with all possible intermediate values for less stringent inbreeding.

It can be shown that, with inbreeding, the following genotypic frequencies are approached:

$$[p^2 (1-F) + Fp] : [2pq (1-F)] : [q^2 (1-F) + Fq]$$

This becomes a Hardy–Weinberg equilibrium $(p^2 + 2pq + q^2)$ at $F = 0$ and becomes $p : 0 : q$ when $F = 1$. What does this last result mean? It means, simply, that with $F = 1$, the equilibrium population consists of homozygotes in frequency p*AA*: q*aa*. The reader should now calculate equilibria for various combinations of p, q and F and should also examine the approach to equilibrium with $p = q = 0.5$ when half the population mates at random and half self-pollinates $(F = 0.5)$. How many generations will be needed to attain equilibrium? If the behaviour of the half that self-pollinates is not evident, return to this problem after reading below about the behaviour of inbreeders. We will anticipate here by noting that the modified equilibrium formula shows that any inbreeding leads to increased homozygosis, rigorous inbreeding to the limiting case of no heterozygosis.

Fitness

Darwinian, evolutionary, fitness is defined in relative terms of differential reproduction: the more progeny left, on average, by a genotype in relation to the progeny left by other genotypes, the fitter it is. Conventionally, fitness, symbolized by W, is defined by reference to a chosen genotype (here *AA*):

W_{AA}	W_{Aa}	W_{aa}
1	$W_{Aa}/W_{AA} = 1 - s_{Aa}$	$W_{aa}/W_{AA} = 1 - s_{aa}$

So, if *AA* produces 100 progeny, *Aa* 80 and *aa* 60 the fitness coefficients are $s_{Aa} = 0.2$ and $s_{aa} = 0.4$, therefore the larger the s, the less the fitness, from $s = 0$ (fully fit) to $s = 1$ (lethal).

It can be shown that (in a random-mating population again) the following changes (Δq) in q, the frequency of *aa*, will occur per generation:

(1) *A* dominant, *aa* less fit

$$\Delta q = -\left[\frac{sq^2(1-q)}{1-sq^2}\right] = -\left[\frac{q^2}{1+q}\right] \text{ when } s = 1$$

(2) *A* dominant, *aa* more fit

$$\Delta q = +\left[\frac{sq^2(1-q)}{1-s(1-q^2)}\right] = +[1-q] \text{ when } s = 1$$

(3) *Aa* intermediate in fitness between *AA* and *aa* (least fit)

$$\Delta q = -\left[\frac{\frac{1}{2}sq(1-q)}{1-sq}\right] = -\left[\frac{q}{2}\right] \text{ when } s = 1$$

The reader should play with these equations, starting with $p = q = 0.5$ and taking $s = 0.1$, 0.5, 1.0. He will quickly perceive that a deleterious (i.e. unfit) recessive is fairly quickly reduced in frequency but then declines only slowly, that an unfit dominant is rapidly eliminated and that an intermediate allele is reduced more rapidly than a recessive (because open to selection in the heterozygote). Hence, in outbred populations we should expect unfit dominants or intermediate alleles to be rare (which they are) but unfit recessives to be persistent because protected by their recessiveness (which is also true). By contrast, inbred populations move fairly rapidly towards homozygosis, approaching p*AA* : q*aa*. So unfitness of either allele in relation to the other will expose it to selection and potential elimination. Very unfit alleles, whether dominant or recessive, will have little chance of survival so inbreeders will, in general, have a higher allele-fitness than outbreeders; the latter carry an unfitness load.

Formally, there is one other situation which should be considered here (but is probably better thought of in terms of linkage – see next section). This is overdominance, the case in which the heterozygote is fitter than either homozygote. This leads to an excess of heterozygotes over Hardy–Weinberg expectation and, in classical evolutionary theory, is connected with the development of polymorphisms (which are beyond our scope here). Apart from a few rather special cases – such as sickle-cell haemoglobin in man in relation to malaria – it is doubtful whether true (i.e. allelic) overdominance exists. Pseudo-overdominance due to linked epistatic effects is more general (Fig. 4.1) and linked complexes are certainly the basic feature of natural polymorphisms. We shall return to the overdominance question in plant breeding contexts in later chapters.

Linkage

Genes on different chromosomes, or on different arms of the same chromosome, or even on the same arm but sufficiently distant from each other, segregate independently. In a random-mating population starting from heterozygotes, they come to joint equilibrium in one generation (though from other starting points more slowly). If they are linked, but with some recombination, equilibrium is delayed in a degree related to the closeness of the linkage. In the extreme case of no recombination, of course, the two loci behave as one functional locus and any departure towards joint equilibrium is impossible. It can be shown that even quite loose linkage markedly delays the approach to equilibrium and that close linkage (recombination less than 10 per cent, say) delays it for numbers of generations far greater than would apply in any ordinary plant breeding programme. Note that, at equilibrium, coupling and repulsion gametes will be formed in equal frequency regardless of the recombination rate. So an initial phenotypic association between the two characters, due to linkage, will be replaced by apparent independence at equilibrium. Any inbreeding, by promoting homozygosis, will tend to fix the original linked combinations and close inbreeding *plus* close linkage can reduce recombination to a very low level indeed.

Multiple linkage problems rapidly become extremely complex and all but the simplest usually have to be examined by computer simulation rather than by direct methods. In general, it becomes apparent that, with many loci and many linkages, equilibrium is never, in fact, truly attained. Every allele that enters a population retains a small piece of ancestral neighbouring chromatin even when the allele itself and its more distant neighbours are at equilibrium.

As an exercise the reader could usefully consider a random mating population starting from heterozygotes (AB/ab), all genotypes being of equal fitness. How are the first and second generations composed with recombination frequencies 0.5 and 0.1 and what is their relation to equilibrium?

Drift

In a finite population, any gene frequency will have a sampling variance which is inversely related to population size; so the smaller the population the larger the variance. Therefore, by chance, a gene frequency can drift up or down over a number of generations and the likelihood of it moving a long way is greater in small populations than in large ones. At the limit, an allele may be either fixed ($p = 1$) or lost ($p = 0$) without help from selection for fitness effects. On evolutionary time scales the effect is thought to be quite important, even in relatively large populations. Obviously, alleles already at low frequency are especially at risk of random loss. If populations vary in size over generations, the effective population size from the drift point of view is nearer the smallest than the mean and intermittent very small populations following disasters can be potent sources of random loss or fixation. These ideas have several applications, both in actual plant breeding programmes and in connection with the conservation of genetic variability.

Examples

As examples to amplify the preceding discussion we will consider populations of two contrasted crops: maize (outbreeder, more or less random-mating) and barley (strict inbreeder, $F = 1$). For simplicity we will suppose in the first place that both populations are constructed by crossing inbred lines and that the F_2 is formed by random-mating and selfing of the F_1 respectively. Each F_1 population is heterozygous at n independent loci and, in the absence of linkage, we need not define which parent provides which alleles. As a means of labelling the alleles (but without any implications for the moment as to fitness or breeding value) we will call alleles positive and negative.

The main features of the F_2 populations are summarized in Table 4.2. All gene frequencies are 0.5 and the maize, having started from a multiple heterozygote, attains equilibrium in one step; in the absence of linkage, fitness differences between alleles, mutation or drift, it will have the same average composition (including constant heterozygosity of $2pq = 0.5$ per locus) over subsequent generations. The inbred barley population stands in extreme contrast; it has the same composition in F_2 as the maize but is not in equilibrium and will not be there until reduced to an array of pure lines with zero heterozygosity many generations later.

The numbers implied by Table 4.2 are important and the reader should try putting $n = 1$, 2, 5 and 10 and working through the table systematically to get some feel for them. Even at $n = 10$, numbers become very large and some frequencies very small; the minimum complete F_2 population is $4^{10} = 1.05 \times 10^6$ and the frequency of plants fixed at all positive loci is $(\frac{1}{4})^{10}$ or $= 0.95 \times 10^{-6}$. If a particular homozygote were desired and could be recognized, it can be shown that a population of about 3 million plants would have to be examined to provide a good (90 per cent) chance of detecting at least one occurrence. So very large numbers are involved even when n is quite small: most plant breeders would guess that, in any interesting real situation, n would be larger than 10 (which is less than one locus per chromosome arm in the two crops considered).

Now consider the barley. We saw above that inbreeding leads to homozygosis. The F_1 is heterozygous at all n loci, the F_2 50 per cent heterozygous in respect of any one locus and, evidently, also of all n loci considered jointly. So heterozygosity has been halved in one generation of selfing and it is easily seen that this continues over generations, thus (heterozygosity per cent): 100, 50, 25, 12.5, 6.25, 3.13, 1.56, 0.78.... So it falls to less than 1 per cent of initial at F_8 but

Table 4.2 Genetic structure of an F_2 population. Number of loci n (independent) each segregating for positive (pos.) and negative (neg.) allele.

Genotypes			Frequencies		
A. F_1 gametes					
1. all loci pos.	1	$= 2^n$	$(\frac{1}{2})^n$	$= 1$	
2. pos. and neg.	$(2^n - 2)$		$[1 - (\frac{1}{2})^{n-1}]$		
3. all loci neg.	1		$(\frac{1}{2})^n$		
B. F_2 genotypes					
1. all loci hom.	2^n	$= 3^n$	$(\frac{2}{3})^n$	$= 1$	
2. het. and hom.	$(3^n - 2^n - 1)$		$\left[\frac{3^n - 2^n - 1}{3^n}\right]$		
3. all loci het.	1		$(\frac{1}{3})^n$		
C. F_2 population					
1. all loci hom.	2^n	$= 4^n$	(a) all loci pos. $(\frac{1}{4})^n$	$= (\frac{1}{2})^n$	$= 1$
			(b) pos. and neg. $\left[\frac{2^n - 2}{4^n}\right]$		
			(c) all loci neg. $(\frac{1}{4})^n$		
2. het. and hom.	$(2^{2n} - 2^{n+1})$		(a) no neg. fixation $\left[\frac{3^n - 1}{4^n}\right]$	$= [1 - (\frac{1}{2})^{n-1}]$	
			(b) some neg. fixation $\left[\frac{4^n - 2^{n+1} - 3^n + 1}{4^n}\right]$		
3. all loci het.	2^n			$(\frac{1}{2})^n$	
1. no negative fixation				$(\frac{3}{4})^n$	$=1$
2. some negative fixation				$[1 - (\frac{3}{4})^n]$	

in an infinite theoretical population never quite reaches zero. As heterozygosity falls, so the population approaches the equilibrium state of 2^n pure lines in equal frequency (since $p = q = r = s = \ldots = 0.5$). With $n = 10$, the population at F_8 will therefore contain about 1000 different homozygotes and some near-homozygotes still segregating at one or two loci.

Now let us consider disturbing factors and their likely effects. Fitness differences must occur and must tend to alter gene frequencies away from the initial 0.5. Since both populations were founded on inbred lines, dramatically unfit alleles are excluded but mutation might provide some deleterious recessives; these could be carried along at low frequency in the maize but would, as we saw above, be exposed to severe selection and probable elimination in the inbreeder. Random over-dominance effects, due to interaction of alleles from the different homozygous parents, would emerge as too-frequent heterozygosity in both populations and as prolonged maintenance of it against the pressure of inbreeding in the barley. If the overdominance were due to an epistatic linkage (Fig. 4.3), any recombination would, of course, tend to reduce the apparent overdominance effects on fitness and heterozygosity.

Any linkages between the n loci would delay general equilibrium in both populations and would be likely in the inbreeder, if close, to eliminate some recombinant pure lines altogether from the final potential array of 2^n homozygotes: it is easily calculated that about half the total recombination in an inbreeder takes places in F_1–F_2 and 95 per cent of it by F_4. Drift will surely affect gene frequencies in both populations but, so long as numbers are at least several hundreds in every generation, random loss or fixation is unlikely.

If the simplified maize population we have been considering were replaced by a more complex one derived from mixing two heterogeneous populations (e.g. two open-pollinated varieties), the results would be somewhat different. Equilibrium, even among unlinked loci, would be delayed. Initial gene frequencies would vary widely according to joint frequencies in the parent populations; rare recessives would enter from both sides and would severally have to attain equilibrium in their new genetic environment.

Experimental studies confirm that quasi-natural populations, such as we have been discussing, are every bit as complicated as we would expect. In maize, Sprague and Schuler (1961) examined, after four generations of random-mating, a population that had been purged, by inbreeding and selection, of visible deleterious recessives (e.g. chlorotic plants). They found 20 mutants at an overall frequency of about 8.5 per 100 zygotes, and they were substantially different from the array of mutants found in the original variety at a frequency of about 18.9 per cent. So a mere four generations of random-mating sufficed to accumulate a load of recessives in variety and frequency not far short of the original. Sprague and Schuler could not decide whether exceptionally high mutation rates or overdominant fitness of the heterozygotes was responsible. Certainly, their population did not simply settle down to the simple, expected near-equilibrium among genes already present. Other studies of maize indicate that all open-pollinated varieties carry a substantial load of deleterious recessives, usually in a frequency of 20–30 per cent of zygotes on selfing.

In barley (and Lima beans), Allard and his colleagues have uncovered similar complexity (review by Allard in Baker and Stebbins, 1965). They found that marker genes in complex hybrid populations did not simply go to fixation over generations. Instead, a great deal, of heterozygosity persisted for many generations and, presumptively, indefinitely. Though predominantly selfed, both crops were a little cross-pollinated (barley 1–2 per cent), as revealed by marker gene tests. Indirect estimates of fitnesses (W) from samples of genotypic frequencies suggested a remarkably strong heterozygote advantage which varied somewhat in magnitude with season: so fitnesses were not constant. As in the maize, simple expectation is again far from being realized: the inbreeders remain too heterozygous too long. For this, some cross-pollination (even at quite a low rate) and heterozygote advantage are jointly responsible. The usual interpretation of the heterozygote advantage in fitness would be one of pseudo-overdominance but, as we shall see later, a critical discrimination between this and 'true' overdominance is not operationally possible. This important work shows, in general, that inbreeders are neither so closely inbred nor (in these kinds of populations) so homozygous as had been supposed.

Conclusions – inbreeders and outbreeders

This has been rather a complex section so a summary will not be amiss. Starting from populations heterozygous at many loci, the simplest possible population–genetic assumptions would lead us to expect that: (1) outbreeders, randomly mated, would attain Hardy–Weinberg equilibrium in one generation and would maintain a constant heterozygosity of 2pq per locus; (2) inbreeders, strictly selfed, would approach equilibrium only after many generations when the population would consist of a mixture of homozygous lines. These simple expectations can never be realized because the attainment of equilibrium is disrupted by a combination of: mutation, change of gene frequencies due to differential fitnesses, linkages, drift (random statistical change in gene frequency even to the point of fixation or loss) and variation in mating system (e.g. occasional crossing in an inbreeder). Even with relatively small numbers of genes, numbers of different genotypes in populations become very large; add to this the interacting factors of mutation, fitness, linkage and drift just mentioned and it becomes apparent that even seemingly simple plant breeding populations defy detailed interpretation.

Nevertheless, an important distinction does emerge from this discussion, namely the distinction between inbreeders and outbreeders (Table 4.3). The former are adapted to a predominant homozygosity, the latter to heterozygosity. Anticipating here for a moment Section 4.4, on polygenes and the effects of inbreeding, the outbreeders are intolerant of enforced inbreeding because of the load of more or less deleterious recessive genes that they carry. Some are overt Mendelian major genes (as in the maize example), the majority presumably polygenes of individually small effect. The inbreeders, by contrast, though they may show some inbreeding depression are tolerant of prolonged selfing and must be assumed to have been largely freed, by past selection, of deleterious recessives. The inbreeder–outbreeder distinction assumes, as we shall see in Chapter 5, great practical importance in plant breeding.

Table 4.3 Outbreeders and inbreeders.

Outbreeder	Inbreeder
Has crossing mechanism, approaches random-mating	Closed flowering, approaches regular selfing
Individuals heterozygous at many loci	Individuals approach homozygosity
Variability distributed over population	Variability mostly between component lines
Carries deleterious recessives	Deleterious recessives tend to be eliminated
Intolerant of inbreeding	Tolerant of inbreeding
Much heterozygote advantage (epistasis? overdominance?)	Less heterozygote advantage?
Adaptable (evolutionarily flexible) at some cost to immediate adaptation	Closer adaptation, less flexible

4.4 Oligogenes and polygenes

Introduction

Mendel's choice of a number of major genes, each readily classifiable by a reliably expressed phenotypic effect, was essential to his success. Only thus could the fundamentally particulate nature of heredity have been demonstrated with the necessary clarity. But he and his contemporaries knew perfectly well that a great deal of biological variation is not discontinuous and not amenable to any simple particulate interpretation. Johannsen's demonstration that some, at least, of the continuous variation was environmentally caused went some of the way to resolving the problem but not all, because some, too, was equally clearly genetic; Galton recognized this in his human genetic studies and Darwin thought of very small, continuous genetic changes as the basic steps in evolution. The resulting unseemly (and, in retrospect, unnecessary) squabble between the Geneticists and the Biometricians was not resolved until the early 1920s by which time East had shown experimentally that continuous variation was perfectly compatible with Mendelian segregation and Fisher and Wright had established the essential mathematical bases of a new subject, biometrical or quantitative genetics, the genetics of continuous variation.

Most of the variation with which the plant breeder has to cope and the characters he has to select are continuous in expression: Mendelian genes and discontinuities are the exception. Biometrical genetics is the genetics of continuous variation and the purpose of this and the next two sections is to sketch the main features of it. We shall see that the principles and ideas are crucial for the understanding of plant breeding, in practice rather more at the qualitative than at the quantitative level. The reader should be warned that the treatment is highly condensed and selective; for fuller exposition he should turn to the standard texts (e.g. Falconer, 1989; Kempthorne, 1957; Mather and Jinks, 1971; Wricke and Weber, 1986).

Polygenic variation

We will start by reconsidering the wheat example introduced in Section 4.2 above. *H-h* is a good Mendelian major gene readily classifiable on colour of chaff. *H* determines red chaff and also high chlorophyll – a content of the flag leaf; whether the double effect is due to pleiotropy or tight linkage does not matter for the present purpose. Suppose the chaff colour element were lacking, so that simple visual classification were impossible. What would the investigator conclude? If the phenotypic variation in the character were as observed experimentally (Figs 4.1, 4.2), he would conclude that he was dealing with a dominant major gene which could be virtually unambiguously classified in segregation. But if the environmental variation were at the hypothetical higher level, he might suspect the correct interpretation from considering means and shapes of distributions but could not prove it without recourse to further breeding tests (e.g. by sampling selfed F_3 families). Hence we see that even a single major gene, the expression of which is subject to environmental variation, can generate continuous variation which could be quite hard to interpret; the Mendelian clarity has been lost.

Now consider what would happen if the quantitative character we were interested in were controlled, not by one locus showing dominance (as in the wheat example), but by three loci with individually small additive effects (Fig. 4.4). The word 'additive' was mentioned above (Fig. 4.3) and we shall return to it later. Here it means that loci and alleles within loci are supposed to act independently so that any *A*, *B*, *C* scores 1 regardless of the rest of the genotype. Thus the heterozygotes are intermediate: *AA* (2), *Aa* (1), *aa* (0). Note that: the F_1 is intermediate between the parents and backcross means intermediate between parents and F_1; the F_2 distribution is markedly bell-shaped and transgressive of both parents; eight pure lines could be isolated by subsequent inbreeding ($2^3 = 8$, see Table 4.2), spanning the F_2 range. If the character studied were subject also to environmental variation (indicated by the continuous curves in Fig. 4.4) formal genetic interpretation would be difficult and any genetical complications such as dominance, unequal locus effects, linkage, etc., would make it impossible.

Now imagine that the situation in Fig. 4.4 is replaced by one in which genetic control is exerted by six loci, each with only half the additive effect. Thus *A*, *B*, *C*, *D*, *E*, *F* would each score $\frac{1}{2}$ and we have: *AA* (1), *Aa* ($\frac{1}{2}$), *aa* (o). It is easily seen that this would leave means and ranges unaffected but would subdivide the scale of measurement, multiply the numbers of classes to be distinguished in each generation and increase the possible emergent pure lines to $2^6 = 64$. This

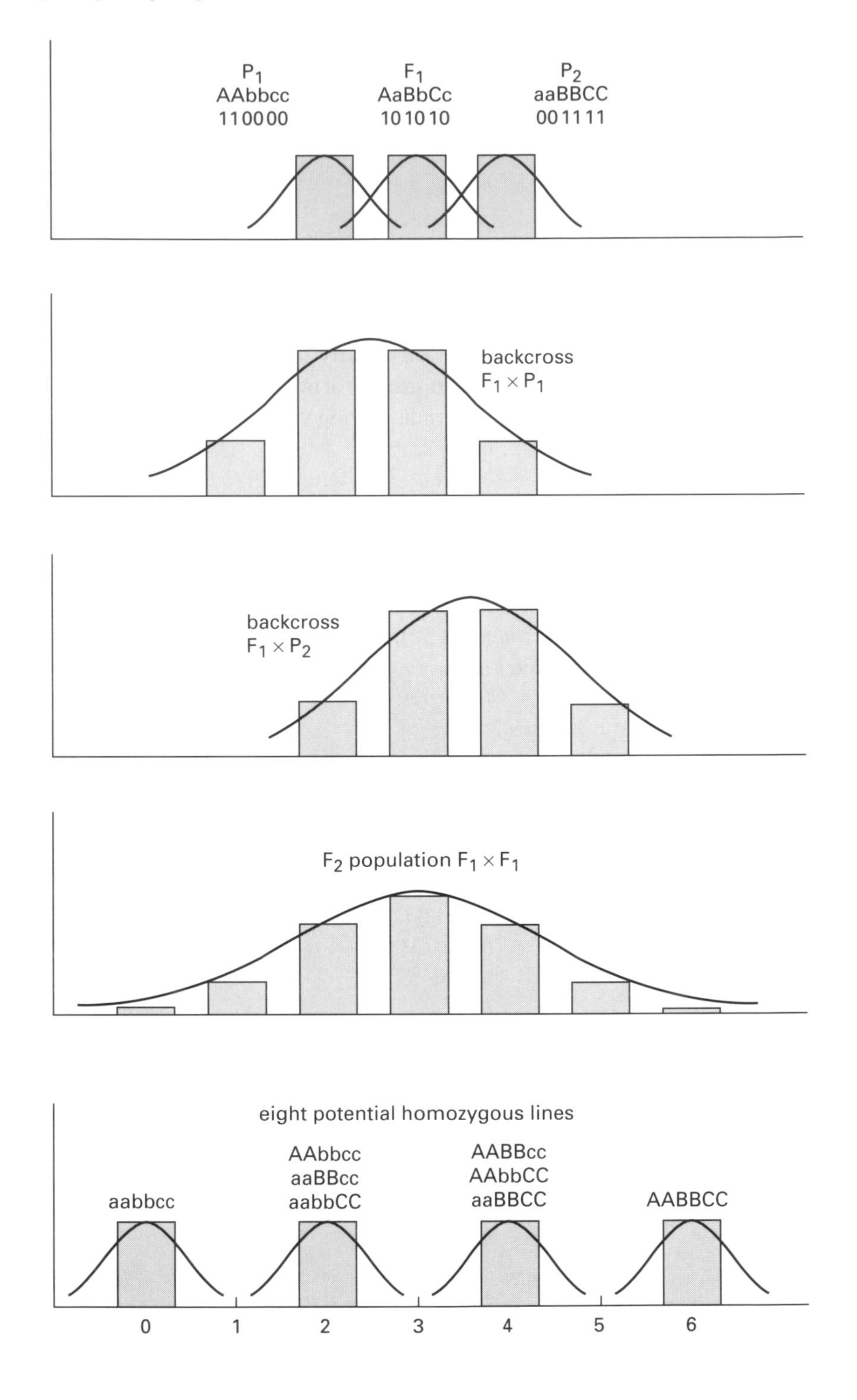

P1
AAbbcc
110000
F1
AaBbCc
101010
P2
aaBBCC
001111
backcross
F1 × P1
backcross
F1 × P2
F2 population F1 × F1
eight potential homozygous lines
aabbcc
AAbbcc
aaBBcc
aabbCC
AABBcc
AAbbCC
aaBBCC
AABBCC
0
1
2
3
4
5
6

Fig. 4.4 Continuous variation generated by three additive loci.
Notes: See Table 4.4 and text. Note that, since the parents were not extreme genotypes, segregation is transgressive. Additivity is assumed, so distributions are symmetrical. The curves indicate the sorts of distributions that would be obtained if an environmental effect were added.

---

point is generalized in Table 4.4. Frequencies in it are derived from successive expansions of the monohybrid ration (1:2:1) and can readily be written down as alternate lines in a Pascal triangle. (The reader should try this for himself and verify the genetic distributions in detail up to $n = 3$.) It will be seen that, as n rises, so this 'binomial' distribution rapidly takes on a smooth bellshape which approximates to a statistically 'normal' form (see bottom of Table 4.4).

The example we have just been discussing is very like the classic case of corolla length in *Nicotiana longiflora* on which East founded his multiple factor hypothesis. His studies are well summarized by Allard (1960). East crossed two extreme pure lines and examined means and distributions in P, F_1, F_2 and in samples of F_3–F_5. He found he had to postulate about five additive loci to account for the observed segregations. He did not detect transgressive segregation because his parental lines were already extreme selections; our example (Fig. 4.4) includes it as a feature because of its importance in a plant breeding context.

We see, therefore, that there are two factors contributing to the continuous phenotypic variation that is ubiquitous in both wild and cultivated populations of plants and animals. These are: (1) environmental variation and (2), genetic variation brought about by the segregation of several–many genes of individually small effect. Distinguishing between these two sources is the basic operational problem of biometrical genetics.

A character controlled by several or many loci having effects too small to be individually distinguished is said to be polygenic. By transfer, the word polygene refers to the genes themselves and it is to be contrasted with oligogene, which is the same as a major gene (i.e. a Mendelian-type determinant). Polygenes are sometimes also called minor genes. Probably the commonest nomenclature would be polygene-major gene.

Complications

The example we have been considering (Fig. 4.4) is extremely simplified; it assumes only three loci, independent in action and with alleles strictly equal and additive in effect. One can readily imagine a great many complications. Loci might have different effects ($A\text{-}a \neq B\text{-}b \neq C\text{-}c$); there might be any degree of dominance, varying between loci ($Aa \neq (AA + aa)/_2$ to the extreme ($Aa = AA$); there might be interactive effects between loci (epistasis) such that, for example $AB \neq AC$ or BC (an effect itself complicated by dominance); and there might even be over-dominance such that $Aa > AA$ and aa (Fig. 4.1).

In short, we could go on making up examples like Fig. 4.4 indefinitely. Results

Table 4.4 Expected F_2 frequency distributions for 1–6 loci having additive effects. At bottom, normal curve fitted to n = 6 distribution.

n	2^n	Frequencies in various classes of expression																				
		Low										Mean										High
1	2	1										2										1
2	4	1					4					6					4					1
3	8	1				6		15				20				15		6				1
4	16	1			8		28		56			70			56		28		8			1
5	32	1		10		45		120		210		252		210		120		45		10		1
6	64	1	12		66		220		495		792	924	792		495		220		66		12	1
n = 6, per cent		0	0.2		1.7		5.4		12.2		19.3	22.4	19.3		12.2		5.4		1.7		0.2	0
normal fit		0	0.3		1.6		5.2		11.8		19.5	22.9	19.5		11.8		5.2		1.6		0.3	0

would depend on assumptions as to the constitutions of the parents and as to gene action. Means would turn out (generally) not to be exactly intermediate between the parents and distributions would be skewed rather than symmetrical. (The reader could here usefully consider the full dominance case for all three loci ($AA(2) = Aa(2) : aa(0)$) when the F_1 (score 6) would exceed both parents and segregating generations would be quite strongly skewed but the emergent homozygotes would be as shown in Fig. 4.4).

However different the results in detail, all sets of assumptions would show a certain family resemblance to each other in which, at least, segregating generations varied more widely than non-segregating ones. But, given some environmental variation superimposed on the genetic, the problems of distinguishing between small differences in generation means or between a little more or a little less variability or between normality and a little skewness become, in practice, insoluble. Add to this the complications of scale whereby the use of a chosen transformation can shift means and variances and alter skewness and it will be apparent that even simple cases of quantitative variation lie, in practice, beyond formal analysis. The best that biometrical genetics can do is to achieve a formal statistical partition of variation such that, for any one experiment, we can say that this or that portion behaves *as though* it could be attributed to the activity of additive polygenes, this or that to dominance, epistasis, interactions and so forth.

More about polygenes

Since polygenes are not amenable to the formal certainties of Mendelian analysis, the evidence that they are indeed responsible for polygenic characters is mostly rather indirect, though, in total, overwhelmingly strong. First, we know that major genes and polygenes intergrade in expression; no sharp distinction can be drawn between them. Second, reciprocal crosses generally agree closely, excluding, in effect, cytoplasmic determination. Third, the facts of segregation and transgression can only be matched by well known Mendelian processes. And, fourth, polygenes show linkage, both among themselves and, more important, with major gene markers. In several organisms (e.g. *Drosophila*, mouse) marker studies show decisively that polygenic activity is well spread over the genome and the same conclusion emerges in wheat in which, thanks to elegant cytogenetic control, individual chromosomes can be assayed for polygenic activity (see Section 8.3).

Three other points are worth making about polygenes. First, how many can be identified? Linkage studies in *Drosophila* provide the best answers, in the range 2 to 20 and other studies suggest more, even very many more. In the vast majority of cases, though, we simply have no secure knowledge of numbers. The best we can say is that they are probably often numerous but sometimes fewer in number and larger in individual effect than we usually suppose. Second, do they differ in any essential feature from major genes? The answer seems to be:

probably not. We must suppose (it can hardly be proved) that genes show a distribution of phenotypic effect from small to large and that those of large effect are the minority, the major genes. And, third, what is their evolutionary significance? The answer to this is evident from the second point. They are by far the most numerous class and are of fundamental evolutionary (and practical) importance. It is safe to assert that any character which varies genetically in populations, natural or artificial, will show a substantial element of polygenic control and will respond to selection. Collectively, many studies of the genetical architecture of variable characters (especially in *Drosophila*) all point towards the idea of maintenance, in outbred populations, of stores of potential genetic variability in the form of linked polygene complexes. Such complexes are thought of as bearing mixtures of plus and minus alleles, so would be conservative in stabilizing the phenotypic expression around an adaptive mean and yet maintain the genetic potential for response to selection in new directions. In these terms, linkage is not merely an obstruction to the attainment of Hardy–Weinberg equilibrium; it is a vital component of adaptation and, from the breeder's viewpoint, an agency which both reduces and prolongs response to selection.

Inbreeding and heterosis

In a large randomly mating population the probability that two alleles in an individual are identical by descent approaches zero and the population is fully outbred ($F=0$). The inbreeding coefficient will depart from 0 in several circumstances, namely: (1) if the population is small so that there is a likelihood of mating between relatives; (2) if there is assortative mating between like phenotypes; (3) if there is any tendency to assortative mating between relatives (e.g. sib-crossing or, at the extreme, self-pollination). All three are relevant to plant breeding but it is the third which concerns us here. Any inbreeding, however caused, has the same result: it increases homozygosity and decreases heterozygosity. So the phenotypic effect of inbreeding must somehow be interpreted in terms of heterozygosity.

In practice it is found that to inbreed any normally outbred organism has fairly dramatic effects, especially upon fitness characters. There is always a decline in general vigour, size and fertility and the decline is often such that, with close inbreeding, some or many inbred lines become non-viable; they simply die out (Fig. 4.5). This 'inbreeding depression' is accompanied, of course, by greater phenotypic uniformity as homozygosity progresses. Species differ somewhat in response to it: in maize, some lines are lost and the survivors yield, at best, hardly 50 per cent of the source population; in lucerne (alfalfa) all lines are lost in two or three generations; some cucurbits, by contrast, are rather tolerant of inbreeding.

The genetical interpretation of inbreeding depression is, broadly, fairly clear. If the character studied (e.g. seed yield in maize) were truly additively deter-

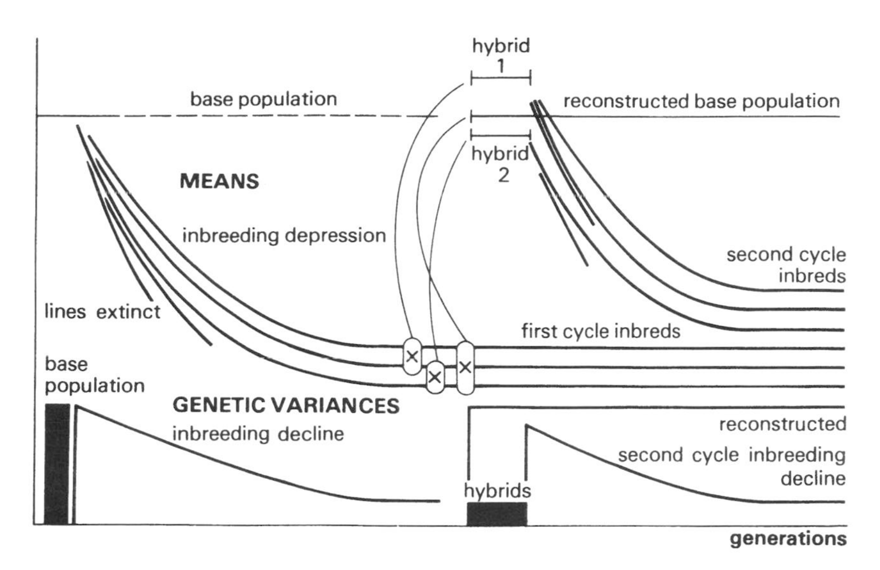

Fig. 4.5 Inbreeding depression and heterosis.
Notes: The diagram was constructed with maize in mind but is quite general. Note concurrent inbreeding depression (due to exposure of recessives) and decline of genetic variance (due to increased homozygosis). Crossing restores vigour (heterosis) to or near the base population level and random-mating among lines also restores genetic variance. But F_1 crosses between inbreds, though highly heterozygous/heterotic, have low genetic variance and are nearly uniform. 'Second-cycle inbreds' (see Section 5.5), having been purged of some deleterious recessives in the first cycle, commonly show slightly reduced inbreeding depression.

mined (i.e. exactly intermediate heterozygotes), there could be no depression, as consideration of the mean value of a single locus will quickly show. Dominance, including overdominance, must be invoked and this amounts to assuming that inbreeding depression is caused by the inevitable fixation of inferior recessive alleles (Table 4.2, bottom). If this were correct it would follow that hybrids between pairs of inbred lines should at least show marked improvement and that recombination among an array of inbreds should restore the original population (Fig. 4.5). This, in fact, happens and it seems clear that the dominance interpretation of inbreeding depression explains the main features of the situation. We shall see later however that it does not explain them all.

The restoration of vigour and fertility to (or near to) the original level by crossing inbreds is called hybrid vigour or 'heterosis', the latter word invented by Shull in the context of his maize studies. Heterosis is thus simply the converse of (or complement to) inbreeding depression. The usual practical definition of heterosis is of vigour, yield, etc., that exceeds the higher parent and often greatly exceeds it; in maize, the yield of hybrids is of the order of the *sum* of the yields of the inbreds.

Heterosis is not confined to crosses between inbred lines of outbred species.

It turns up also in crosses between pure lines of inbreeders (see Fig. 4.11) and between outbred populations of diverse origins (often, e.g., among land-races of maize). Such cases can normally be plausibly interpreted in terms of different dominance contributions from the two parents. Heterosis in crosses between pure lines of inbred cereals is indeed quite common and we shall see later that a considerable controversy centres around its significance for plant breeding (Section 5.5).

Single-locus heterosis ($Aa > AA, aa$), is not so readily interpreted.It has turned up on many occasions in diverse crops, sometimes in the form of measurements of homozygotes and heterozygotes at known major gene loci (e.g. chlorophyll mutants in barley), sometimes in the form of estimates of fitness of heterozygotes in complex populations (examples in Section 4.3, maize and barley). On the face of it, overdominance is the obvious explanation but it is not a necessary one and there is a certain reluctance to accept it on general grounds of physiological implausibility. Apparent overdominance could be due to linked dominants or epistatic effects of neighbouring loci which, in the limit, could be genetically irresolvable. The point is important because, if one believed in true overdominance at the major-gene locus level, one would have to accept it also at the polygenic level; and, contrariwise, one would have to deny it at both levels.

The main facts of inbreeding depression and heterosis could as well be interpreted on an overdominance as on a dominance basis but they provide no foundation for choice. One additional point, however, is relevant: the dominance hypothesis leads to the inference that inbreeding depression could be circumvented by recombination among inbred lines, leading to the construction of a line carrying all the relevant dominants and in the homozygous state. Such a line would be the equal of the best possible hybrid combination among inbreds. Experience in maize is that cyclical improvement of inbreds has gone far, but not that far: the best inbreds are still well short of the best hybrids. This suggests, but by no means proves, the persistence of some residual overdominance. Overdominance is not the only genetical complication. Epistasis between genes from parental lines interacting favourably in the hybrids could be involved. This, like dominance, should be fixable in inbreds unless impeded by linkage.

Thus there are, broadly, two schools of thought on the subject. One (e.g. Mather and Jinks, 1971) would appeal to dominance *plus* some genetical complications, but not to overdominance, as the basis; the other (e.g. Lerner, 1950, 1954, 1958) would regard overdominance as important and widespread and the essential basis of heterosis. Shull, the inventor of the word, would have been with the latter school because he thought of overdominance as a consequence of hybridity *per se.* We are nowhere near a resolution of the problem because no critical test is evident. Probably, a widely acceptable current view would be that dominance is the essential basis but with genetical complications that could sometimes include true overdominance. The matter is of practical importance because, as we shall see later, it bears on whether or not we should breed 'hybrid varieties'.

4.5 Heritability

Partition of variance

Consider a complex trial in which several varieties (V) are compared in randomized blocks over several years (Y) and places (P). In the full analysis, a hierarchy of interactions emerge after the main effects have been taken out: first order (VY, VP, YP), second order (VYP) and error (based on residual comparisons between replicates within sub-experiments). If the experiment were made more complex by the addition of, say, spray–no spray or irrigation–no irrigation treatments, a whole new layer of interactions would be introduced as well as an extra main effect. Notice that the interactions are of two kinds, variety × environment and environment × environment; the first relates to the constancy (or otherwise) of varietal differences over environments and is an example of the genotype × environment effects (hereafter GE) mentioned earlier (Section 3.5). Now it is a statistical, though not an obvious, fact that a variance of a main effect, as it appears in an analysis of variance table, can be regarded as a compound of variance due to that main effect itself *plus* components due to all its interactions *plus* residual 'error' or 'noise'. So, in the experiment we are discussing, the varieties line could be broken down into bits attributable to: differences between varieties freed of interaction (V_G say), to GE interactions (V_{GY}, V_{GP}, V_{GPY}) and to error. The data could also be made to yield an estimate of total variance among varieties (V_V, say), whence a variance ratio, V_G/V_V, is calculable. This ratio is a particular kind of 'heritability' which estimates the fraction of total variance among lines attributable to genetic differences between them. Most heritabilities are of this nature; they are variance ratios.

Biometrical genetics takes this process of variance-splitting a stage further by enquiring into components of total genetic variance (V_G). The usual starting point is to consider a single polygenic locus with homozygotes having values $-a$ and $+a$ about a mean of zero and heterozygote with value d; so if $d=o$, the effect is strictly additive and, if $d=\pm a$, fully dominant. Then, with a number of assumptions, expressions for expectations of mean squares that can plausibly be attributed to additive effects (conventionally symbolized V_A) dominance (V_D), and various genetic interactions can be developed. These expressions all relate to specific generations or samples of them (e.g. F_2, BC, F_3, F_4, F_5 lines and so on). A grcat dcal of ingenuity has gone into developing mating plans which will permit comprehensive breakdown of genetic variance in this way. Environmental (i.e. non-genetic) variance (V_E) must also be considered. To estimate this some non-segregating material (such as inbred parental lines or clones) is necessary; sometimes it is desirable to recognize more than one kind of error (e.g. between and within families). Always there is the assumption that environmental variances so estimated can legitimately be applied to different materials in the same experiment.

So, in a suitably designed experiment, it is possible to extract estimates of genetic variance and its components, of environmental variance and of GE interactions (Fig. 4.6). This is analogous to the variety trial discussed above and indeed the two could, in principle, be combined into one gigantic genetic experiment over places and years; it would yield interactions of horrible complexity. In short, there is almost no limit to the number of interactions and components that can be recognized if experiments are made big enough. Contrariwise, simple experiments inevitably leave interactions hidden.

An experiment that distinguished an additive component (V_A) from total

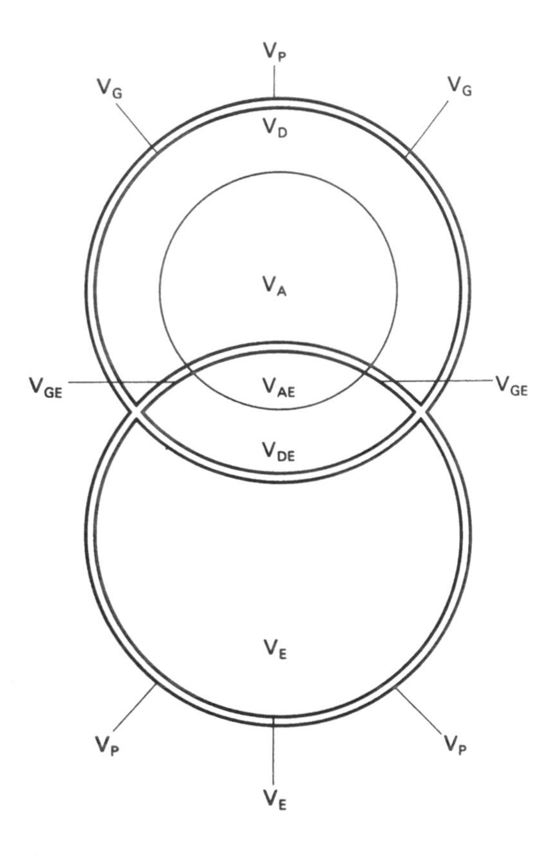

Fig. 4.6 Diagram to illustrate the idea of components of variance in biometrical genetics.
Notes: V, variance, with subscripts as follows:
G, genetic, with A, additive, and D, dominance, components; E, environmental; P, phenotypic, with G, E and GE components; GE, genetic × environmental, with AE, additive × environmental and DE, dominance × environmental components. Appropriate experiments could seek to distinguish additional E elements (e.g. sites and seasons) and G elements (e.g. epistatic variance), with all the appropriate EE and GE interactions.

genetic variance (V_G) would yield two heritabilities, as follows. First, there is the ratio:

$$\frac{V_G}{V_G + V_{GE} + V_E} = \frac{V_G}{V_P}$$

and second the ratio:

$$\frac{V_A}{V_G + V_{GE} + V_E} = \frac{V_A}{V_P}$$

The latter ratio, since V_A is at most equal to V_G, is smaller than the former: it may be called the 'narrow sense' heritability in contrast to the 'broad sense' type, V_G/V_P. So we have distinguished two kinds of heritability based on components of variance. We shall return later to their possible utility and consider now a simple example of the kind of calculations involved.

In Table 4.5 it will be seen that hybrid generations lie between the parents (but are not exactly intermediate) and that segregating generations are more variable than parents and F_1. So genetic segregation is apparent. In the lower part of the table note that F_2 data alone would permit identification only of $V_G(=V_A+V_D)$; the addition of backcross data allows the solution of two simultaneous equations which separate V_A from V_D. Heritabilities are rather high and the narrow differs little from the broad, reflecting a small V_D (i.e. $V_G - V_A$) component. Note that this is a simple experiment which did not permit the recognition of interactions. Note also that the three estimates of V_E are rather variable and that there is suspicion that the F_1 was less variable than the parents; the assumption of a common value for V_E *is* an assumption and one which affects subsequent calculations. The reader should try recalculating heritabilities assuming $V_E=5$ and 11. And how to interpret the apparently rather different variances of the two backcrosses?

Estimation of genetic variance and its components suffer from a statistical weakness which has been treated by Gilbert (1973). Briefly, many statistical estimates and procedures (e.g. means and t-tests) are 'robust' in the sense of being little affected by departure from normality; variances, however, are unrobust and therefore unreliably estimated. Furthermore, in comparison with means, they are also inaccurately estimated because the variance of a variance contains a term in V^2 so, the bigger they are, the less accurately are they known. The partition of variance makes things worse because small differences between inaccurate numbers are liable to be very inaccurate indeed; as a result, components of variance sometimes turn out negative. Similarly, variance-component ratios (and therefore heritabilities) are poorly estimated and can also turn out on occasion to be negative.

Parent–offspring regression

We now turn to an alternative approach to heritability. Consider a reasonably diverse array of parents measured for whatever character the experimenter is

Table 4.5 Experiment on heading time in wheat. Ear emergence measured in days from an arbitrary origin. Based on Allard (1960, Table 10.3) by permission of the author and publishers, Messrs. Wiley.

Item	P_1	P_2	F_1	F_2	B_1	B_2
Means	12.99	27.61	18.45	21.20	15.63	23.38
Numbers	159	148	171	552	326	314
Variances	11.036	10.320	5.237	40.350	17.352	34.288

Expectations:

$V_{F2} = V_A + V_D + V_E$

$V_{B1} + V_{B2} = V_A + 2V_D + 2V_E$

Estimates:

$V_E = (V_{P1} + V_{P2} + V_{F1}) \div 3 = 8.864$

$V_{F2} = 40.350$

$V_{B1} + V_{B2} = 51.640$

Hence: $V_A = 29.060$, $V_D = 2.426$

Heritabilities:

(1) broad sense

$$\frac{V_A + V_D}{V_A + V_D + V_E} = \frac{V_G}{V_P}$$

$$\frac{29.060 + 2.426}{40.350} = 0.78$$

(2) narrow sense

$$\frac{V_A}{V_A + V_D + V_E} = \frac{V_A}{V_P}$$

$$\frac{29.060}{40.350} = 0.72$$

interested in. If samples of progeny from these parents are measured they will commonly bear at least some resemblance to their parents: so the progeny of high parents will themselves tend to be high but will diverge from their parents as a result of genetic segregation and of environmental effects on both generations. At the low extreme the same will apply; the progeny sample will again tend to diverge towards the general mean. This was the basis of Galton's use of the word 'regression': progeny of extreme parents tend to regress towards the mean. If parents and progeny agreed perfectly, the regression would have a slope of unity but there would be no regression in Galton's original sense (Fig. 4.7).

At the other extreme, if there were no similarity between parents and offspring, the result would be a random scatter of points on the diagram, with no

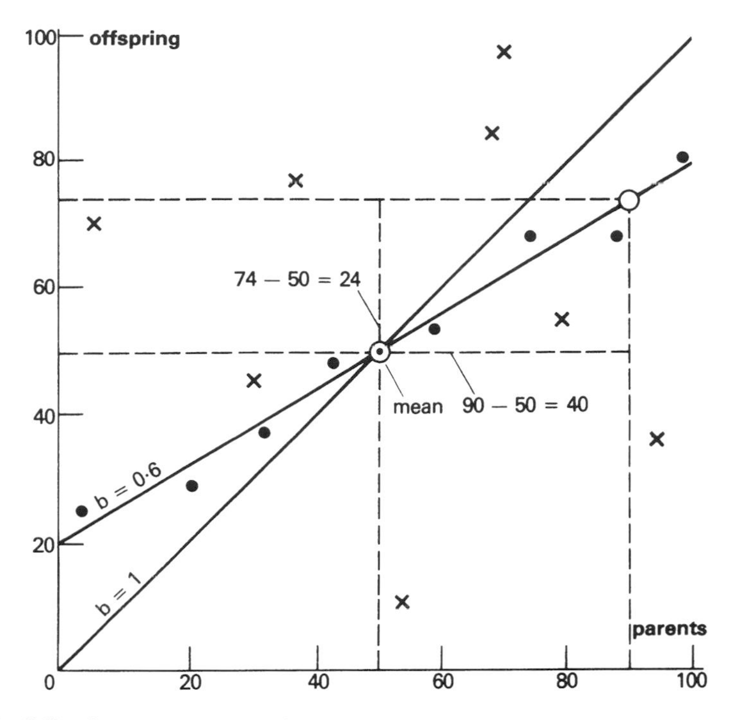

Fig. 4.7 Offspring–parent regression as a measure of heritability.
Notes: Three cases are distinguished: b = 1 (limiting case, perfect heritability, parents predict offspring very accurately); b = 0.6 (dots •) (high heritability, Galtonian 'regression' apparent); no regression (crosses ×) (zero heritability). The relation to 'realized heritability' may be appreciated by considering the dashed lines: a sample of parents with mean 90 (S = 40) will have progeny with mean 74, an advance of R = 24; so R/S = 0.6 = b.

evident regression. This would only happen if there were environmental effects, but no heritable genetic ones.

So, close parent–offspring similarity (high regression) suggests a large genetic effect and a small environmental one. In fact it can be shown that b, the regression coefficient, itself estimates the narrow sense heritability, V_A/V_P. This is true if parental values are means of both parents (i.e. are mid-parents); if only one parent is known or relevant (as in some animal experimentation or poly-crosses) then $b = \frac{1}{2}(V_A/V_P)$.

Regressions of this kind are essentially empirical: they describe parent–offspring relationships directly. Thus they do not depend upon genetic assumptions, though interpretation of b as a narrow-sense heritability does involve the idea of separating V_A from the rest. Statistically, such regressions are not free of problems but they do at least avoid the troubles associated with partition of variance. In general, then, as Gilbert (1973) pointed out, they offer a more secure approach to heritability than the partition-of-variance.

Offspring–parent regressions were developed by animal geneticists and are increasingly used by plant breeders. An example in maize is given by Robinson *et al.*, (1949) and of disease resistance by Simmonds (1991a).

Heritability in general

The conventional symbol for heritability is h^2 even though it is a ratio rather than the square of a quantity. Conventionally also, h is written for the square root of h^2 when necessary. Narrow sense heritability (i.e. V_A/V_P) is usually understood to be implied unless the context states otherwise. Sometimes, subscripts can conveniently be used to differentiate between the two, e.g. h^2_N and h^2_B.

The very nature of a heritability makes it clear that any estimate is specific both to the material under study and to the structure of the experiment. Parents that differ little in respect of the character examined cannot generate large V_A or V_G in their progeny; parents that differ widely may do so. So the numerator of a heritability is partly a function of parental difference but partly also a function of experimental design in separating V_A and V_G from each other and from interactions. The size of the denominator is determined both by genetic differences and by V_E; obviously, it can be reduced by an experimental design that reduces the estimate of V_E, e.g. the use of bigger plots, more replication and so forth. Offspring–parent regression, similarly, depends upon the magnitude of genetic differences between parents and upon control of error; bigger plots, larger families, more replication and repeated measurements on individuals (as practised by the animal breeders) will tend to increase the numerator (covariance) of the regression coefficient and reduce the denominator (parental variance).

So any estimate of heritability is unique. It is a property of a specific population in a specific experiment. Therefore the numerical use of estimates out of context is illegitimate though it is sometimes reasonable to consider general orders of magnitude of heritabilities in a wider framework. Thus, a great many studies of reasonably diverse crosses in plant breeding material show collectively that the heritability of yielding ability is often low or very low, even approaching zero. Because, as we shall see below, heritability is one component of response to selection, this is an important conclusion; it means that selection for high yield, the dominant feature of most programmes, will be difficult and laborious. But this fact is empirically well known to plant breeders already, so the heritabilities merely confirm the familiar.

Finally, the word 'repeatability' deserves a brief comment because it has been used in two different ways. For animal breeders and geneticists, it relates to repeated measurements of the same character on each individual, yielding a variance ratio of the type: between individuals/total phenotypic variance. So it is, in effect, a special kind of heritability and it can be used to improve an estimate of narrow sense heritability by reducing the denominator, V_P (Falconer, 1989). In plant breeding contexts, the word nearly always relates to broad sense heritability of characters among a group of fixed entities such as clones or pure lines and the interest in it lies in comparing characters rather than improving estimates of h^2. Thus the sugarcane breeder may remark that sugar content is more repeatable than cane yield (i.e. generally shows higher V_G/V_P and design his selection schedules accordingly.

4.6 Response to selection

General

Selection is differential reproduction of genotypes in a population so that gene frequencies change and, with them, genotypic and phenotypic values of the character being selected. In any genetically heterogeneous population, wild or cultivated, natural selection is directed automatically towards the enhancement of net reproductive fitness (Section 4.3), the fundamental process of Darwinian adaptation. Natural and semi-natural populations approach complex equilibria in which overall fitness is as high as the varied demands of differing sites and seasons, complex genetic control and the long-term demands of adaptability allow.

The reproductive performance of a population as a whole thus tends to become stabilized at a level which is adequate for maintenance but still allows the retention of a great deal of variability and of wide differences in fitness between individuals. Clearly such a population can express no character at its highest possible level; the expression is always subordinate to the long-term demands of reproductive performance. So characters become stabilized around means with distinct genetic and environmental tails on each side and this is as true of reproductive performance as of other characters. This is the result of what may be called 'stabilizing selection'; note that 'directional selection' for a character, up or down, will shift the mean and the variability to new levels while 'disruptive selection', up *and* down, will tend to produce two new populations with distinct stabilizing norms.

Now let us apply these ideas to plant breeding. The plant breeder starts with a genetically heterogeneous population; clearly, there is no sense in starting with a homogeneous one. By selection he attempts to impose new fitness values on the genes governing the characters in which he is interested. Thus the breeder selecting in F_2 for the phenotype governed by the dominant gene *H* in our wheat example (Section 4.2) would, in effect, write:

W_{HH}	W_{Hh}	W_{hh}
1 $(s_{HH} = 0)$	1 $(s_{Hh} = 0)$	0 $(s_{hh} = 1)$

and, if he did progeny tests in the second generation, could substitute $S_{Hh} = 1$ and $W_{Hh} = 0$. So he could fix *H* and eliminate *h* in two steps, changing gene frequencies from 0.5 to 1.0 and 0 in the process. The complementary process, of fixing *hh*, would take only one step. This is the essence of artificial selection: change of gene frequency under new values of W. With a well-expressed major gene the process is simple; with polygenes it is not because they show small individual effects and relatively large environmental ones. So, for an individual polygenic locus, W will never be far from 1 nor s from 0. We can never know actual values but the very fact of response to selection shows that they diverge from neutrality. This uncertainty, of course, merely brings back the idea of heritability in a new guise. We cannot usefully talk about polygene frequencies but we can about polygenic characters at the phenotypic level.

Before turning to the idea of a 'character' there is one further point to make about the direction of selection. In natural populations, as we saw above, stabilizing, disruptive and directional selection can usefully be distinguished, all subordinated to the overall demands of reproductive fitness. All three kinds of selection have been extensively examined in genetical experiments but, for the plant or animal breeder, only the last kind, directional selection, is of practical concern. The breeder seeks a one-way response whether for milk, egg, grain or tuber yield, for disease resistance, sugar content, flavour or whatever. Nevertheless, however accurately controlled the environment in which the breeder's stocks are raised, some natural selection, at least of the stabilizing and directional kinds, must persist in respect of reproductive fitness. We will return to this question later in connection with correlations and merely recall here that the idea of 'pure' artificial selection is an abstraction.

Characters

In natural populations selection acts ultimately upon but one complex 'character': reproductive fitness of individuals over varied environments, seasons and generations. The breeder's aims may be related to natural reproductive fitness but are never identical and are sometimes very different indeed.

The best definition of a 'character' for the breeder is what he says it is. From the point of view of selection, unit characters controlled by major genes present no problems (as we saw above) and may be ignored. Polygenic characters are, by definition, genetically and biologically complex and, in practice, defy efforts to disassociate them into simpler components. It is no practical simplification to regard grain yield of a cereal as the product, ears per hectare × grains per ear × mean grain weight because the components are nearly as complex as yield itself. Transformations are sometimes desirable on purely statistical grounds (e.g. to equalize variances) and are sometimes used to enhance the V_A fraction in dealing with characters the scales of which are suspected of being multiplicative. This is a legitimate procedure in biometrical studies but the breeder is interested in overall worth, yield, disease resistance, stature or whatever, not in their logarithms or square roots.

So the practical approach is to proceed empirically; to recognize that the breeder's character will usually be biologically complex; but that, if it is, the fact will be apparent in low heritability or (see below) low GCA.

Response in one generation

We saw above that response to artificial selection was basically a matter of changing gene frequencies by imposition of new values for fitness, W; but that, for polygenic characters, this was not, in practice, a very useful idea. Instead, we must proceed at the phenotypic level at which the following equation applies:

$$\overline{X}_o - \overline{X}_p = R = ih^2\sigma$$

Where $\overline{X}_o$ is the mean phenotype of the offspring of selected parents, $\overline{X}_p$ the mean phenotype of the whole parental generation, R the advance in one generation of selection, h^2 the appropriate heritability, σ the phenotypic standard deviation of the parental population and i is the 'intensity of selection', a statistical factor that depends upon the proportion of the population selected to be parents. If there were such a thing as a fundamental equation in plant breeding this would be it and the reader must understand it, at least in a qualitative sense, even if, as a practising plant breeder, he never has occasion to use it numerically.

Figure 4.8 summarizes the meaning diagrammatically. Note how, with a middling heritability, the selected tail of the parent curve will contain some elements which are genetically excellent and some which are phenotypically but not genetically outstanding. So the offspring will 'regress' towards the parental mean. If the heritability were unity ($V_A = V_P$, no environmental variation) progress would be perfect and the offspring mean would equal that of the selected parents; if it were zero there would be no progress at all and $R - 0$. Progress also depends upon $i\sigma$, sometimes symbolized as S, the selection differential; $i\sigma$ is measured in the same units as X so i is clearly a dimensionless number; it is, in fact, a purely statistical quantity that arises from the properties of the normal curve and depends upon the fraction, k, of the total population, N, of parents selected. The factor i has been extensively tabulated (e.g. by Becker, 1975) but, so long as N is about 20 or more and k is in the range 0.1 to 0.001 (reasonable plant breeding values) the following empirical equation is good enough and helpful in the present context:

$$i = 1.13 + \log_{10} 1/k$$

Clearly, there are decreasing returns to an increasing selection rate since to multiply (1/k) tenfold less than doubles i.

The response equation we have been discussing must not be applied uncritically though its general form is quite standard and the point we have just made about selection rate in relation to i is also general. The difficulty, as usual, centres on heritability. Narrow sense heritability would be appropriate in the case of pair-crosses of selected parents in a segregating maize population or an

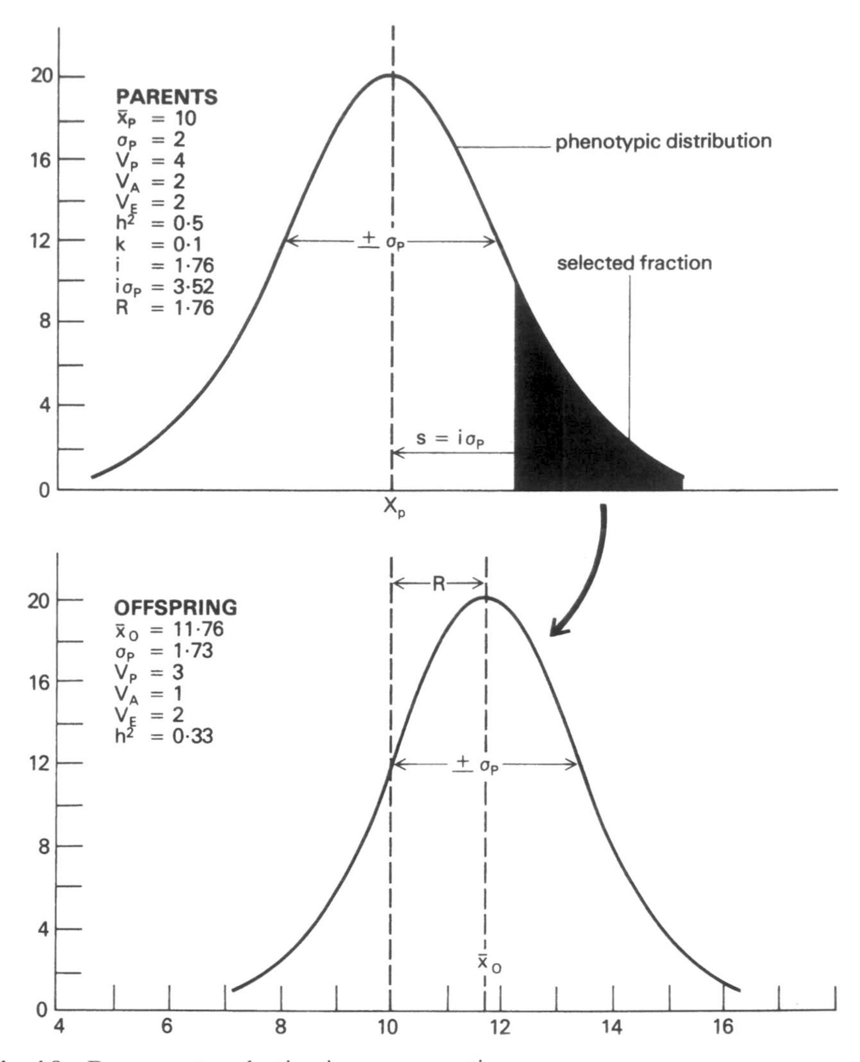

Fig. 4.8 Response to selection in one generation.
Notes: The reader should work through this example numerically and calculate the results of a second round of selection. The assumption of $V_A = 1$ in the offspring is arbitrary and is made merely to illustrate the expectation of declining genetic variance and heritability.

F_2 of an inbred cereal and would also be appropriate if open pollinated cobs of outstanding maize plants were selected (though with lower h^2). If the inbred cereal had gone to fixation so that it consisted of numerous pure lines, or if we were dealing with a population of clones, the broad sense heritability would be the appropriate one to use.

Returning then to the response equation and recalling the formulae for broad and narrow sense h^2, namely V_G/V_P and V_A/V_P we can write:

$$\overline{X}_o \propto \overline{X}_p + \log 1/k \, . \sqrt{h^2} \, . \, \sigma$$

where σ is either σ_A or σ_G, depending upon which heritability is appropriate. This says that excellent selected offspring depend upon high parental mean, upon selection rate (but with a diminishing returns basis), upon heritability and upon the appropriate genetic variance. So the breeder would do well: (*a*) to seek excellent families (high $\overline{X}_p$) derived from diverse parents (high V); (*b*) to maximize $\sqrt{h^2}$ by good control of V_E; and (*c*) to select intensely but, recognizing the diminishing returns aspect, be prepared to sacrifice intensity of selection in favour of enhancing $\overline{X}_p$ or V by exploring more diverse parents and progeny. Contrariwise, if progress in a breeding programme is poor, we can invert the argument and use the equation to help to enquire why. To do so at least states the essential questions: does the fault lie primarily in the genetic base (affecting $\overline{X}_p$ and V), should selection be intensified (increase log l/k) or could heritability be improved?

The response equation has been used many times in genetic studies to predict response in one generation. It works well – as it must – so long as h^2 is reasonably well estimated; if it does not predict well, then there is usually reason to suspect h^2. Alternatively, if selection were intense, the area under the tail of the curve becomes critical, just where variances are weakest, so estimates of h^2 and V adequate for less intense selection could break down here.

In practical plant breeding, despite the theoretical importance of the equation, it is rarely applied numerically because appropriate data for estimating the genetic parameters are rarely available, could be got only at considerable expense and would need to be estimated anew in each generation.

Finally a brief comment on 'realized heritability' is appropriate. It will be evident that the response equation can be rewritten:

$$h^2 = R/i\sigma = R/S$$

so that response can be used to estimate heritability *a posteriori* and without recourse to any components-of-variance calculations. Obviously, this has something in common with offspring–parent regression (Fig. 4.7) though the statistics are different.

Continued response

The simple picture of response that we have been considering involves change of gene frequencies to the point of fixation of favourable alleles. So response will involve a decline in both the heritability and the phenotypic standard deviation. Ultimately, at fixation, no further response will be possible and the limit will have been reached; the emergent population will be different from (better than) the source population but less variable.

A great deal of genetical research on this subject has been done, with results too complex for more than brief summary here. Shortly, response does indeed decline, as predicted, and the population generally reaches a stable plateau. However: (*a*) response is sometimes inexplicably long-continued; (*b*) apparently usable genetic variance can sometimes be detected at the plateau stage; (*c*) a plateau is sometimes succeeded by a jump to a higher level; and (*d*) recombination among isolated plateau sub-populations sometimes procures a new response and a move to a new, higher plateau. Delayed release of polygenic variability by the breaking of linkages is usually invoked to explain (*a*) and (*c*); persistent heterozygosity to explain (*b*); and chance unfavourable fixation (i.e. drift) to explain (*d*) (Fig. 4.9).

These generalizations have all been drawn from experiments with outbreeders (notably *Drosophila*, mice and maize). For close inbreeders under artificial breeder's selection for performance (and uniformity) items (*a*), (*b*) and

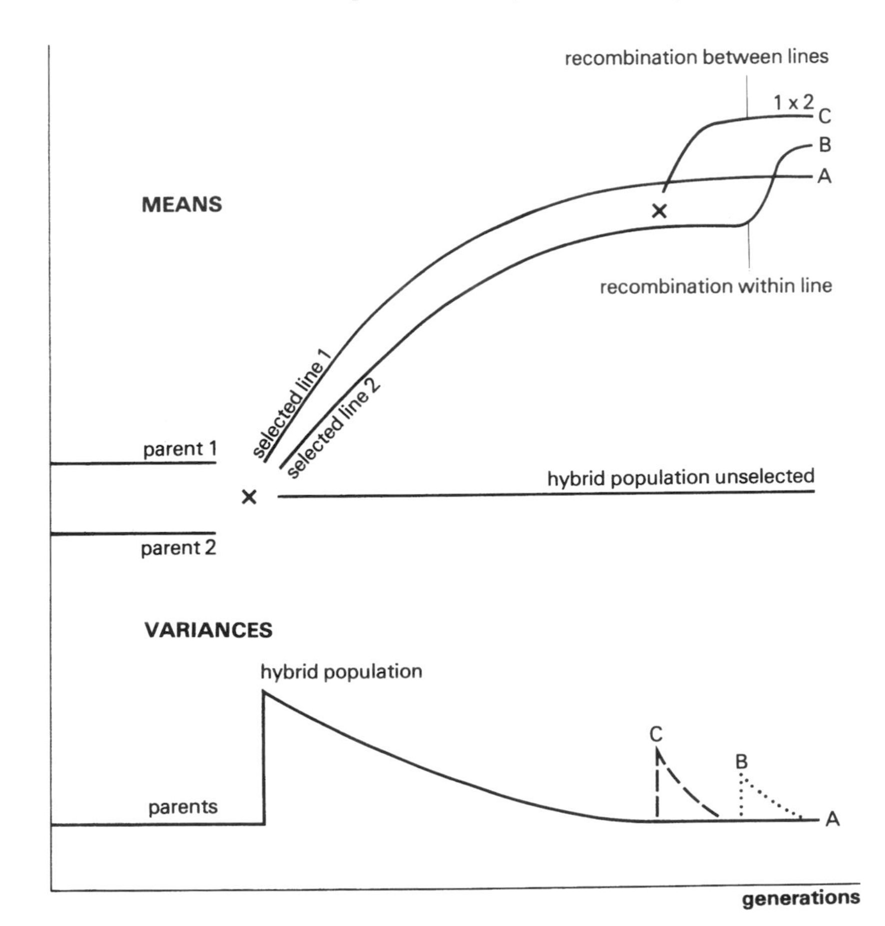

Fig. 4.9 Generalized diagram to illustrate the main features of selection continued over generations.
Notes: Case A is the simple, general one, often more or less realized in practice. Cases B and C represent renewed responses consequent upon release of new variability (see text).

(*c*) are irrelevant; selfing ensures a high degree of fixation and the breeder does the rest. (Selection in a slightly outcrossed inbreeder, however, might well present a rather different picture – see Section 4.3.) Effect (*d*), however, could well occur in an inbreeder and very likely does. Indeed it seems improbable, *a priori*, that the selected pure lines that emerge about F_7 from a diverse F_2 ever contain the best possible line (with a frequency of only $(\frac{1}{2})^n$, and see bottom of Table 4.2 for random fixation of unfavourable alleles). We shall return to this subject later (next section and Section 5.3).

In summary, long-term response in outbreeders (the idea of selection limits – Robertson, 1960) presents considerable genetical problems and the basic response equation, which is conceptually fundamental at the one-generation level, is little help in thinking about them. There is no means of predicting long-term response even if general experience and genetical reasoning suggests a curve of declining slope. We have, however, extracted one point of practical concern, namely that drift effects may procure the fixation of negative alleles, so that, in inbreeders and outbreeders alike, there may well be merit in exploring the possibilities of recurrent recombination.

Examples

An example which brings together a number of features we have been discussing in this chapter is summarized in Table 4.6 and Fig. 4.10. It is based on laboratory simulation of response to selection in an inbreeder, starting from an F_2, heterozygous at 10 differential polygenic loci. Starting assumptions were that all 10 loci were equal in effect, independent and individually and collectively additive (no dominance, no interactions). Scores assigned were 60 for genetic background and *AA*(4), *Aa*(2), *aa*(0). Parents were unequal with

Table 4.6 Simulation of response to selection in an inbreeder, assuming additive gene action at 10 loci. Based on experiments by students in the Edinburgh School of Agriculture.

	Heterozygosity		Genotype			Phenotype		
	H	H_S	G	G_S	V_G	P	P_S	V_P
F_1	100.0	–	80.0	–	–	–	–	–
F_2	51.0	51.0	80.9	83.4	16.8	81.1	93.2	56.0
F_3	30.0	22.5	83.7	87.4	14.9	84.4	98.2	52.0
F_4	11.6	15.5	87.1	89.5	15.0	88.4	100.5	52.7
F_5	8.2	10.5	89.3	91.3	7.2	89.3	101.8	52.2
F_6	5.2	4.0	91.2	93.2	7.1	91.1	105.4	70.1
F_7	2.4	1.0	93.1	94.8	5.3	93.3	103.4	40.1
F_8	0.5	1.0	94.8	95.2	3.2	96.2	106.8	42.9

Note: See text and Fig. 4.10. The subscript (s) in H_S, G_S, P_S, denotes the selected fraction of the appropriate population.

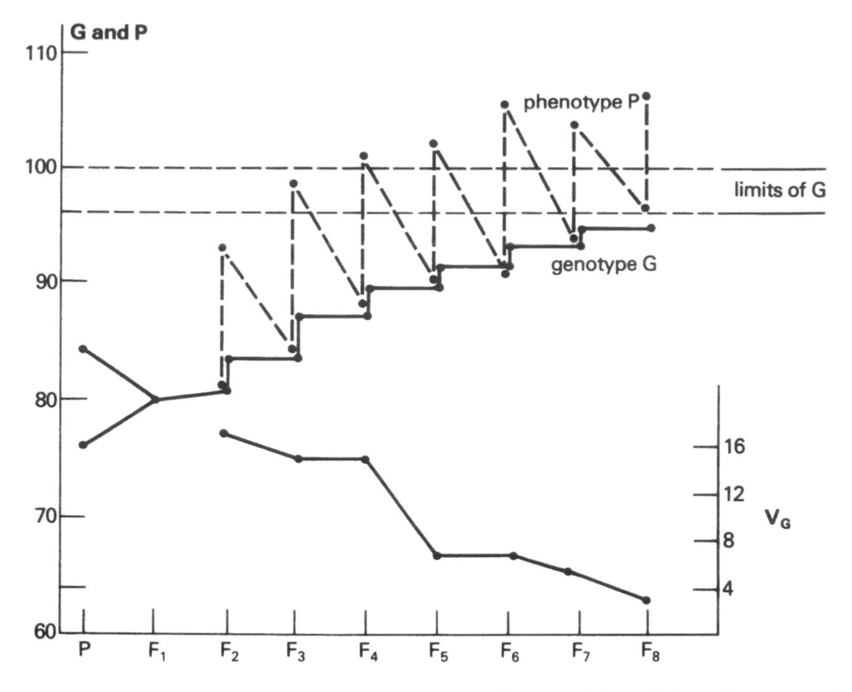

Fig. 4.10 Mechanically simulated selection experiment with an inbreeder, assuming n = 10 and additive gene action.
Notes: See Table 4.6 and text. Note the intermediate F_1, the positive response of G and P to selection and the declining V_G; also the zig-zag pattern generated by phenotypic selection in the presence of V_E.

$6 \times 4 + 60 = 84$ and $4 \times 4 + 60 = 76$; the F_1 was, obviously, intermediate, with score $10 \times 2 + 60 = 80$. The F_2 and subsequent generations (200 plants each) were generated by drawing pairs of coloured counters from a relatively large pool (with replacement). A random error was applied to each genotype with the aid of a mechanical device (tables could, of course, have been used); V_E having been arbitrarily predetermined. In each generation the top 10 per cent of phenotypes were identified and the next generation was constructed on their genotypes.

Now consider the results. Heterozygosity fell, with selfing, to near zero in eight generations. The reader should plot the data against expectation (see Section 4.3) and examine the fit; what does the apparent discrepancy in F_3 signify? Is it real? Genotypic and phenotypic means in each generation agree quite well (as they should) but what are the differences due to? Note that, in each generation, $P_s > P$ and $G_s > G$ but that the former difference is much greater. Why? This leads to the striking zig-zag effect in Fig. 4.10 (Galtonian regression again). Genetic variance fell to near zero as G and P rose but V_P declined relatively little. Why? (Consider the meaning and the actual values (tabulate them) of $(V_P – V_G)$; note that the difference fluctuates considerably and speculate why this might be so.) Since the model is strictly additive, V_G/V_P ($=V_A/V_P$) is the narrow sense heritability. For $k = 0.1$, $i = 1.76$ and $R = 1.76 \times \left[\frac{16.8}{56.0}\right] \times \sqrt{56.0} = 3.95$ in the F_2, so F_3 expectation is $81.1 + 4.0 = 85.1$, in comparison with the observed 84.4. Alternatively, the realized heritability is $R/S = \left[\frac{84.4 - 81.1}{1.76 \times \sqrt{56.0}}\right] = 0.25$. The reader should repeat these calculations for each generation, finally plotting observed against expected responses.

Now consider selection limits. The possible limit at the start was $100 = (60 + 10 \times 4)$. The perfect homozygote might have emerged in F_2 but, with a probability of $(\frac{1}{4})^{10}$, not surprisingly it did not; indeed there were no total homozygotes in F_2 and this is not surprising either (see Table 4.2). By F_6 all material was fixed in at least one unfavourable allele so, from that point on, attainment of the limit of 100 became impossible. The experiment stopped at F_8 with a possible (in effect, certain) limit of 96 ($9 \times 4 + 60$) in several selections, each fixed in different unfavourable alleles: so crossing, recombination and selection would have circumvented the drift which had occurred and permitted a step to the limit beyond the plateau evidently approached in F_8 and to be attained in F_9–F_{10} (see Fig. 4.9).

Finally we must consider briefly what would have happened in the absence of selection and with downward selection. The latter is simple because the pattern is symmetrical: the result would have been a downward curve, a mirror image of the upward one, approaching a value of 60 (but perhaps not quite reaching it). In the absence of selection, gene frequencies would have stayed at $p = q = 0.5$, the mean would have stabilized at 80 (why?) and the population at F_8 would have consisted of an array of near-homozygotes ranging genotypically from 60 to 100 but probably attaining neither limit.

To summarize the main features, we have: (*a*) additive effects leading to F_1 and F_2 means intermediate between unequal parents; (*b*) a diminishing slope of response towards a plateau; (*c*) a characteristic phenotypic zig-zag pattern; (*d*) declining V_G and h^2, towards zero at fixation; (*e*) attainment of a limit less than that theoretically possible; and (*f*) the possibility of attaining the theoretical limit by renewed recombination.

Another version of the same experiment is summarized in Fig. 4.11. Here, complete dominance at all 10 loci was assumed with *AA*(6), *Aa*(6), *aa*(0) and base 40. Parents were $(6 \times 6 + 40) = 76$ and $(4 \times 6 + 40) = 64$. The F_1 was pronouncedly heterotic ($10 \times 6 + 40 = 100$). The same kind of response to selection as in the preceding example was apparent and the selection limit (100) was attained; heterosis, in fact, was fixable. Note the very similar phenotypic zig-zag, mirrored this time by a parallel genotypic behaviour (Why? Consider the phenotypic similarity of *AA* and *Aa* but their different breeding behaviours.) Note also that, even though complete dominance at all loci was assumed (*AA* = *Aa*), there was good response to selection, so a part of the genetic variance must have been additive in the formal sense of being responsive to selection; once again, additive

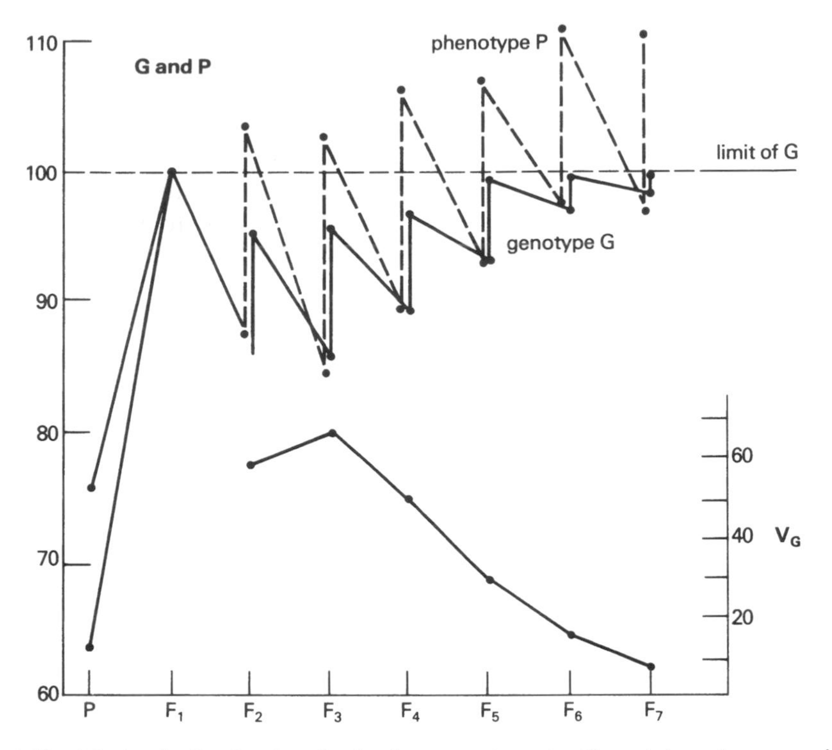

Fig. 4.11 Mechanically simulated selection experiment with an inbreeder, assuming n = 10 and dominant gene action.
Notes: See text and compare Figure 4.10. Note the F_1 heterosis and the evident effect of segregation of recessives in each generation. As in the additive case, there is good response to selection and V_G declines.

variance does *not* imply additive gene action. Finally, what would have been the results of downward selection and of no selection? The reader should be able to answer these questions for himself, at least in general terms, in the light of the preceding discussion of the additive case; first think about end-points and then consider why the F_2 is less than the F_1 and scores, as it does, about 87.

These two examples we have just been examining are extremely simplified. No real cross of any breeding interest is ever likely to segregate for as few as 10 non-interacting loci of relatively large, uniform and equal effect. Biological systems are not so simple. Any real experiment would be fraught with all the interactions, variable errors, undetected linkages, uncertainties of estimation and interpretational doubts inevitable in dealing with any real biological system. Nevertheless, if the reader really understands these 'experiments' in the light of the preceding text (and has done the examples) he will have a basic feel for what artificial selection is all about; he will not understand it fully – no one does – but he will have some sense of the biological nature of a process which, after all, works quite well in practice.

Correlated characters

Our discussion so far has centred on selection for a single arbitrarily defined 'character'. In plant or animal breeding, however, several or even many characters often have to be handled together. No difficulties of principle arise if the characters are genetically independent though we shall see later that there may be very considerable practical difficulties related to the demands of sheer numbers. Non-independence (i.e. correlation) of characters, and especially of fitness characters, is not uncommon and most breeders meet it fairly frequently in one form or another.

Correlations may be purely environmental (e.g. within genetically uniform stocks) or phenotypic (i.e. a compound of environmental and genetic). Genetic correlations can be separated from phenotypic ones by analysis of covariance by methods analogous to those used for the partition of variance. Details lie beyond our scope. Environmental and genetic correlations may be quite different in size and even have different signs; phenotypic correlations are more or less intermediate.

Genetic correlations arise from linkages that have not reached equilibrium and from pleiotropy, the latter being far the more important. Formally, pleiotropy implies that polygenes do more than one thing, at the phenotypic (but not the molecular) level. Conceptually, it is perhaps easier to recall that development of a living organism is a complex process in which every stage depends in some measure upon what has gone before, so genes that have some early activity are very likely to affect terminal processes and products. At all events, correlations (presumed, if not always proved, to have a genetic basis) are common and have to be taken account of in plant breeding.

We shall examine some examples later (Section 5.8) and here make only three rather general points about the implications of correlations for selection. First, a positive correlation between desired characters is favourable to the breeder because it helps selection rather than hinders it. By contrast, a negative correlation will hinder the recovery of recombinants high in both characters and, if close, may forbid altogether the simultaneous expression of both characters at a high level; in this case some sort of an economic compromise (perhaps by index selection – Section 5.8) will be inevitable.

The second point is this. Any strong selection applied to a character must tend to change genetically correlated characters too; this is 'correlated response' which is perhaps most characteristically manifested in a decline of fertility in stocks subjected to prolonged selection for certain other characters (but not all). We saw examples of this in Chapter 1 (Table 1.3): clonal crops selected for vegetative characters (e.g. tuber crops, sisal, sugarcane) nearly always show some, and often much, derangement of flowering and seed fertility mechanisms. This almost certainly reflects correlated responses during the evolution of the crops concerned. Incidentally, the breeder, since he must select for at least enough seed fertility to raise the progeny he needs, must thereby evoke the same mechanism as a hindrance to his own selection.

Third and finally, it may sometimes be practical to substitute for the 'real' character in which the breeder is interested a proxy character which is highly enough correlated to make selection effective but is easier to measure. A great many quality characters come in this category (examples in Section 3.3).

4.7 Combining ability

Introduction

We saw above that heritabilities calculated, as they nearly always are, as variance ratios, suffer from drawbacks arising from assumptions and the non-robustness and inaccuracy of variances and their components. These defects introduce uncertainties into any quantitative treatment of response to selection and the uncertainties increase when more than one generation is considered. So biometrical genetics, regrettably, has little to offer the breeder in the way of quantitative guidance (as distinct from qualitative understanding) as we believe this chapter will have shown. It is natural therefore to enquire whether there might be an alternative approach, preferably free of uncertain genetic assumptions and stronger statistically.

Ideas on combining ability go back to maize breeders in the USA in the 1930s and Sprague and Tatum (1942) provide an early systematic exposition in the context of evaluation of inbred lines of maize for crossing. The essential idea is to consider a systematic set of crosses between a number of parents and enquire to what extent variation among crosses can be interpreted as due to statistically additive features of the parents and what must be attributed to residual interactions. In effect this means writing, for each cross (parents A, B), an equation:

$$X_{AB} = \overline{X} + G_A + G_B + S_{AB}$$

where $\overline{X}$ is the general mean, G_A and G_B are the general combining abilities (hereafter GCA) of the parents (the additive bits) and S_{AB} is the statistically unaccounted for residual, or specific combining ability (SCA). Evidently, the G here are measured as deviations from the general mean; some authors omit the $\overline{X}$, thus in effect estimating the G each increased by $\frac{1}{2}\overline{X}$.

If this trial were suitably replicated, another term E (error) could be added to the equation and more elaborate experiments still would permit the identification of various interactions. Statistical details are beyond our scope however, and our treatment will be confined to the combining abilities themselves.

Examples

Consider Fig. 4.12 and Table 4.7. Diallel crosses (all combinations of N parents but with the various possible systematic restrictions indicated) and M × N

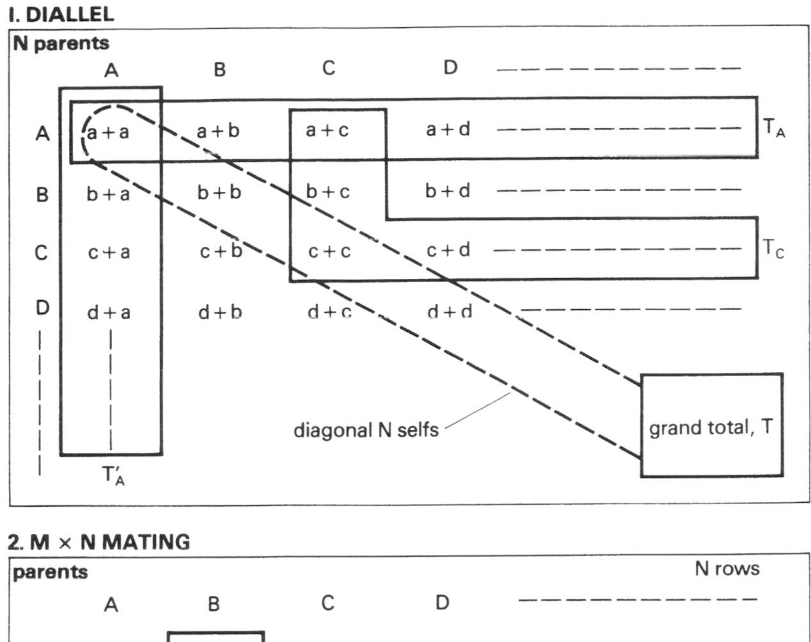

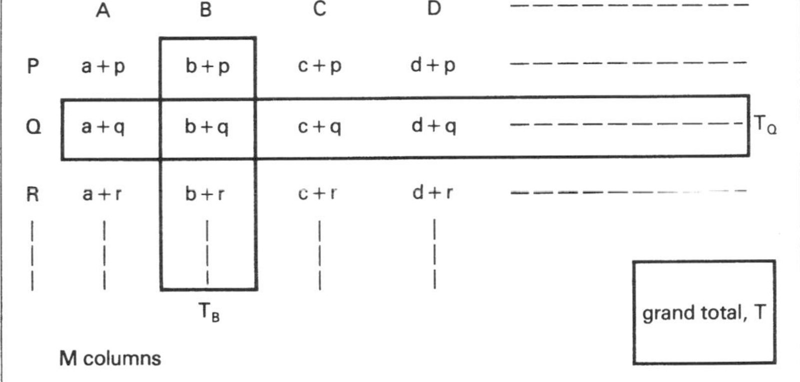

Fig. 4.12 Mating patterns for estimation of combining abilities.
Notes: See Table 4.7 and text. Parents indicated by capital letters (A, B, C etc), gametes by lower case letters (a, b, c etc), progeny by appropriate combinations (a + a, a + b etc). T, with appropriate parental subscripts, are marginal totals.

matings (two separate groups of different parents) are the two basic patterns. The formulae given in Table 4.7 for estimating GCA are not obvious but an example will help to make the principle clear. Consider the half-diallel without parents or reciprocals). A marginal total, say T_A, will be made up thus:

$$T_A = (N-1)\overline{X} + (N-1)G_A + (G_B + G_C \text{---})$$
$$= (N-1)\overline{X} + (N-2)G_A + (G_A + G_B + G_C \text{---})$$

but the last term is zero and the general mean is the grant total (T) divided by the number of cells in the table, so that:

Table 4.7 Combining abilities; estimation of GCA parameters (see Fig. 4.12).

Type of cross	Structure and GCA
DIALLEL	N parents in all combinations. Parents, A, B, C, etc., with gametes a, b, c, and combining abilities G_A, G_B, etc., measured from general mean so that $(G_A + G_B + \cdots) = 0$. Marginal totals T_A, T_B, T_C, etc.; grand total T; row and column T_A distinguished as T_A distinguished as T_A and T'_A where necessary.
(a) complete N × N with reciprocals and selfs	$G_A = \left(\frac{T_A + T'_A}{2N}\right) - \frac{T}{N^2}$
(b) half diallel, with parents, no reciprocals	$G_A = \left(\frac{T_A + aa}{(N+2)}\right) - \frac{2T}{N(N+2)}$
(c) reciprocals but no parents (diagonal missing)	$G_A = \left(\frac{T_A + T'_A}{2(N-2)}\right) - \frac{T}{N(N-2)}$
(d) half diallel, no parents no reciprocals	$G_A = \left(\frac{T_A}{(N-2)}\right) - \frac{2T}{N(N-2)}$
(e) various incomplete patterns	No simple solutions. Incomplete designs are available and random incompleteness can also be handled (see text)
M × N MATING	Two groups of parents A, B, C, etc. (M in number), and P, Q, R, etc. (N in number); combining abilities G_A, G_B, etc., and G_P, G_Q, etc., measured from general mean. Assume that $(G_A + G_B + \cdots) = 0 = (G_P + G_Q + \cdots)$. There are M × N crosses between groups, none within. Marginal totals T_A, T_B, T_C, etc., and T_P, T_Q, T_R, etc.; grand total T.
(a) Complete M × N	A, B, C parents: $G_A = \frac{T_A}{N} - \frac{T}{MN}$
	P, Q, R parents: $G_P = \frac{T_P}{M} - \frac{T}{MN}$

$$\overline{X} = \frac{2T}{N(N-1)}$$

substituting and rearranging, it follows that:

$$G_A = \frac{TA}{(N-2)} - \frac{2T}{N(N-2)}$$

Clearly the first term estimates the combining ability from zero and the second adjusts for the mean. GCA values having been calculated thus, the expectation for any cell in the table is given by the general mean plus the two relevant G values; and the residual difference from observation is the SCA relevant to that combination.

As a numerical example we shall now consider an M × N mating in rubber (Table 4.8 and Fig. 4.13). The reader should first verify several GCA values for himself, using the formulae in Table 4.7; note that both sets sum to zero (as they must). A specific example of fitting the constants to an item in the table with resultant calculation of SCA by difference is given in the Figure. The fact that the regression of observed values on expected has a slope of 1 has no biological meaning: it is a consequence of the way the constants are calculated. But that the correlation coefficient is high does mean something, namely that GCA fitting accounts for a large share of the difference between crosses; alternatively, that the SCA contribution is small. If the SCA contribution had been larger, there would have been more scatter about the line and smaller r. So, in these data, GCA accounts for much, SCA for little. The same conclusion emerges from the analysis of variance in Table 4.8 where it will be seen that the proportion of sums of squares between crosses accounted for by GCA is identical with r^2. No satisfactory estimate of error is available; if it had been, the Remainder (SCA) item could have been tested. Significant SCA or not, GCA, in this experiment, is clearly far more important.

For an example of combining ability treatment of a diallel cross the reader should examine Table 4.9 and should do the calculations suggested in the notes to that Table. Sensible conclusions can be drawn without statistical analysis but

Table 4.8 Combining ability for yield in rubber, *Hevea brasiliensis* (Ross and Brookson, 1966).

Parents	**A**	**B**	**C**	**D**	**E**	**Row totals**	**GCA values**
P	10.1	19.7	20.5	10.7	10.7	71.7	+0.15
Q	14.1	18.4	22.5	15.1	12.9	83.0	+2.41
R	8.3	13.8	14.4	10.3	11.4	58.2	−2.55
Column totals	32.5	51.9	57.4	36.1	35.0	212.9	GM =
GCA values	−3.36	+3.11	+4.94	−2.16	−2.52	0.01	14.19

Item	d.f.	SS	MS
Parents A–E	4	169.19	42.30
Parents P–R	2	61.68	30.84
Remainder	8	27.49	3.44
Total	14	258.36	18.45
Fraction of Total SS attributable to parents =			0.894
Variance ratio, parents/remainder =			11.19

Note: Mean yields (lb dry rubber) per tree per year over 15 years for 15 families in a 5 × 3 crossing pattern; with analysis of variance. See Fig. 4.13 and Table 4.7.

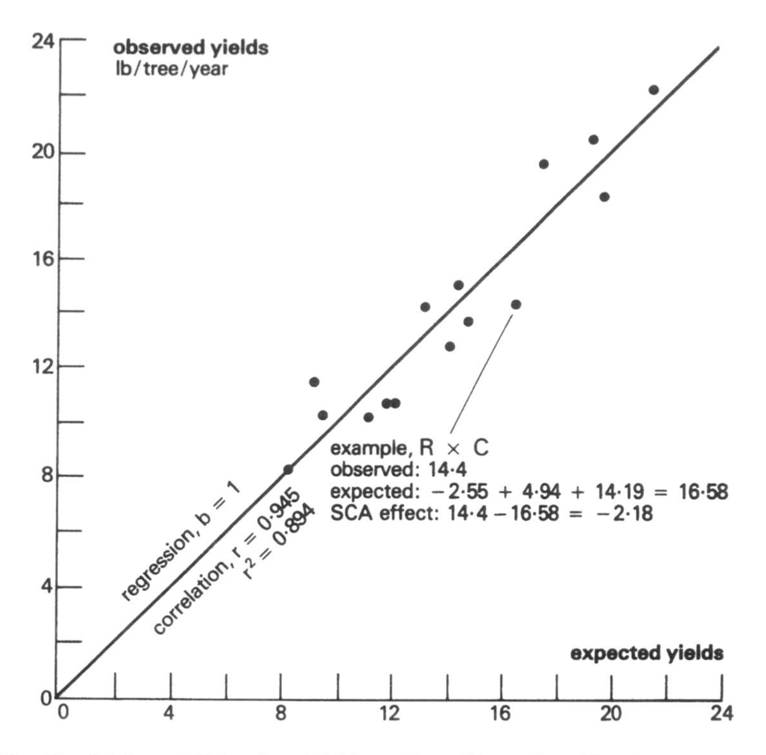

Fig. 4.13 Combining abilities for yield in rubber, *Hevea brasiliensis*.
Notes: See Table 4.8 and text. Observed yields of progenies plotted against yields predicted from parental GCA. Deviations of points from the line, vertically or horizontally measure SCA, which are negative to the right of the line (o < e), positive to the left (o > e).

correlations between observed and expected (as in Fig. 4.13) would be helpful; the reader with greater statistical expertise who wished to do fuller analyses should consult Griffing (1956a, b).

The examples we have been considering are complete in the sense that all the cells in the tables are filled. Sometimes there are reasons for using designedly incomplete diallel designs (Fyfe and Gilbert, 1963) but these are beyond our scope. More often, random incompleteness is encountered in real plant breeding data which would thus look like Table 4.9 but with gaps representing missing crosses. With certain restrictions (Gilbert, 1967) such tables can be analysed as though they were complete and combining abilities calculated; the GCA equations quickly become complicated and solution by computer is necessary. The results are less reliable, of course, than they would be with complete data but the method at least permits some sensible quantitative treatment of figures which would otherwise have been unanalysable. An example by Gilbert, Dodds and Subramanian (1973) on rubber shows that the method can yield guidance on choice of parents and of crosses not yet made which is simply not apparent from inspection; it is virtually impossible to reach

Table 4.9 A diallel cross of beans, *Phaseolus vulgaris*. F_3 yields (dg/5m row) of a complete 6 × 6 cross. Diagonal (selfed) entries in bold face. Based on Evans (in Milner, 1975, Table 7) by permission of the author and publishers, Messrs. Wiley.

Parent	**A**	**B**	**C**	**D**	**E**	**F**	**Row totals** (T_A, etc.)
A	**31**	33	40	32	29	21	186
B	39	**38**	40	35	33	24	209
C	40	39	**40**	36	34	26	215
D	32	39	39	**33**	28	17	188
E	30	32	29	25	**18**	16	150
F	21	21	25	22	15	**11**	115
Column totals (T'_A, etc.)	193	202	213	183	157	115	T = 1063
GCA values	2.05	4.72	6.14	1.39	–3.95	–10.36	GM = 29.53

Notes: Refer to Table 4.7 and verify the GCA values tabulated above. Discard the diagonal selfs, recalculate marginal totals and GCA values. Take means of reciprocals and reanalyse twice again, with and without diagonal. For the half diallel with selfs (case (b) in Table 4.7), construct GCA expectations for each of the twenty-one cells and plot observed against expected. What conclusions can you draw about the relative contributions of GCA and SCA, about reciprocal differences and about inbreeding depression? What parents and crosses are most promising?

sensible conclusions merely by looking at a large incomplete diallel table – which is a not infrequent outcome of plant breeding measurements.

General features

The attraction of combining abilities is that they provide an empirical summary of complex observations and a reasonable basis for forecasting the performance of yet untested crosses but yet make no genetical assumptions. Being based on first degree statistics (totals, means), they are statistically robust and, being genetically, so to speak, neutral, they are equally applicable to inbreeders (as in the *Phaseolus* example) and outbreeders, whether seed propagated or clonal (as in the rubber example).

There is something in common between combining abilities and offspring–parent regression: both are essentially empirical predictive procedures free of genetical assumptions. The similarity is more than superficial because the proportion of total variance taken up by GCA is near to a narrow sense heritability; it is not identical however and should not be used as such. Thus, in the rubber example (Table 4.8), if parental clonal yield data were available, one might expect the offspring on mid-parent regression to be not too far away from (probably rather less than) 0.89. The interested reader might usefully consider Table 4.9 in this connection: it yields an offspring–parent regression which can be compared with the r^2 calculated from combining abilities.

Like heritabilities, it will be obvious that combining abilities can be used numerically only in the context in which they were calculated; GCA values are relative and depend upon the mean of the chosen materials. This point draws attention to a weakness of the M × N mating; we have to assume that sums of row and column GCAs are equal to each other and to zero. This is a shaky assumption if M or N is small, so a diallel pattern is generally more reliable.

In actual plant breeding, combining abilities have found their principal use in predicting the performance of hybrid populations of outbreeders, often in the form of test-crosses or polycrosses. In principle, they are equally applicable to inbreeders or clones and one may expect to see ever-increasing use of randomly-incomplete diallel data now that computers are generally available to do the sums.

4.8 Heterogeneity, heterozygosity and stability

We turn in this section to a brief survey of a complex area of research which lies somewhere between population genetics and ecology. It primarily concerns phenotypic interactions in heterogeneous populations, and goes on to general ideas of stability of performance. The simplest possible assumption about the overall performance of a heterogeneous population is that it will be predicted by the sum of the independent performances of its components. This assumption is often perforce made in population and biometrical genetics studies but it is rarely justified.

Physical mixtures

Most experiments have been done with simple physical mixtures of pure lines of inbreeders such as wheat, barley, rice and lima beans. Consider first the rather numerous experiments carried for one or two years only and based on equal representation of two or a few different varieties. Collectively they indicate that the mixtures, on average, yield a little more than the means of the constituent lines when grown alone and are, as judged by variances, more stable in performance too. There are exceptions but the trend is fairly clear. An example from wheat is given in Table 4.10: every mixture slightly exceeded the mid-constituent and mixtures showed lower levels of all three GE interactions than lines. So, to generalize, heterogeneity with respect to inbred lines often generates both favourable (but quite small) phenotypic interactions in the population and a certain tolerance of environmental variability; the population shows enhanced buffering or homeostasis.

Another aspect of physical mixtures which has had some attention relates to the question of what happens over several or many generations of bulk propagation of pure line mixtures (without hybridization). Generalizing from a number of experiments (mostly on barley and wheat in the USA), it turns out

Table 4.10 Nuding's experiment on wheat mixtures. There were four varieties (M, T, C, R) and six mixtures, grown in six places (P) for three years (Y). Based on Tables 1 and 3 in Simmonds (1962) by permission of the Cambridge University Press.

Entry	Mean yields (t/ha) of lines and mixtures					
	MT	MC	MR	TC	TR	CR
Line 1	3.51	3.51	3.51	3.63	3.63	3.76
Line 2	3.63	3.76	3.08	3.76	3.08	3.08
Mixture	3.60	3.74	3.48	3.79	3.51	3.56

	Analysis of variance (d.f. in brackets)			
	V	VP	VY	VPY
Lines	158(3)	25(15)	25(6)	16(30)
Mixtures	30(5)	9(25)	7(10)	7(50)
Between	62(1)	7(5)	9(2)	3(10)

that, though individual weak lines take a long time to die out, mixtures rather quickly come to be dominated by one (occasionally by two) components. On examination, it usually appears that the dominant survivors are, agronomically, the best locally adapted components of the original mixture. So natural selection in mixture tends (but not always) to favour the same genotypes as the plant breeder. This is encouraging for the idea (see Section 5.3) of bulk handling of inbred material. Presumably (though this has not been investigated) such many-generation mixtures would show the population buffering effects described above.

Heterozygosity and stability

We saw above that there is abundant evidence, in many organisms, of heterozygote superiority leading for example, in Allard's barleys and lima beans, to prolonged persistence of heterozygosity in populations in which it would have been expected to decline, even disappear. Whatever the interpretation (overdominance *vs.* linkage effects) the fact that heterozygotes are often (not always) fitter than homozygotes is well established. They are also often more stable in the face of environmental variation, that is, they are homeostatic or well-buffered.

In crop plants, this buffering effect has been noted fairly frequently. In maize, inbred lines show larger errors and GE effects than hybrids (Shank and Adams, 1960); in barleys, Finlay (in Broekhuizen *et al.* 1964) found that F_2 hybrid populations were more stable than their parents, doing relatively better in bad (i.e. dry) years than in good ones.

Heterozygosity therefore often (but not always) promotes stability of performance. Stability however is a feature, not only of heterozygosity, but also of general genetic constitution. This is evident from comparisons of the same inbred lines that were used in the maize and barley experiments just mentioned. Inbred lines themselves differ in responsiveness to environment and a vast amount of general agricultural experience all over the world confirms this for many inbred crops. Similarly clones, too, vary in adaptability though here we can never know whether we are seeing the effects of heterozygosity or of specific genotype or both.

We shall return to this question of varietal adaptability several times below (Sections 5.9, 6.2) and merely conclude here that genotypes differ in their capacities for homeostasis and that heterozygosity is one feature that commonly promotes it.

Complex populations

The work we have been discussing permitted the separate recognition of heterogeneity, heterozygosity and specific genotype as components both of performance and of stability. What of more complicated populations; for example, of natural outbreeders such as maize or herbage grasses, of later generation hybrid bulks of inbreeders (Section 5.3), or even of double-cross (as against single-cross – Section 5.5) maize hybrids? Two examples will have to suffice to illustrate the problems we must expect to encounter in the analysis of such populations.

Allard and Adams (1969) extracted eight random lines from a bulk hybrid population ('composite cross' – see Chapter 5) of barley. They then tested various mixtures and found an array of remarkably consistent favourable interactions: the best mixtures significantly exceeded the better component, and the average increase was about 6 per cent, ranging up to about 13 per cent over the mid-component and this with lines that were inherently well-adapted good yielders. These are much more dramatic effects than have been found in the mixtures of random pairs of pure lines described above (e.g. Table 4.10). The difference lies in the genetic history of the material. Allard's and Adams's barley lines had evolved together in a recombining population and had evidently evolved mutual adaptation in the process. The authors write of 'ecological combining ability'. Their source population was in effect a reconstructed 'landrace' (Section 5.3). So we can infer that landraces in general (now much rarer than they were) are much more than the mere mixtures of pure lines that they were once thought to be. They show continued crossing and recombination, maintain much heterozygosity (Section 4.3) and must be assumed to benefit both in performance and stability of performance from their heterogeneity and heterozygosity. We shall return later to the question of whether by rejecting the use of landraces the plant breeder is not neglecting a possibly useful kind of population.

Another example comes from maize, in which it has been shown that double-cross hybrids (Section 5.5) are more stable than single crosses. The difference was apparent in a large series of trials summarized by Jones (1958) in which the means of the two groups were equal but coefficients of variation were 12.3 and 21.4 per cent respectively. Later studies summarized by Weatherspoon (1970) point to this same general conclusion. Double crosses are more heterogeneous but slightly less heterozygous than single crosses, so heterogeneity seems to be the cause.

Conclusions

Populations are not simply mixtures of genotypes independent of each other in terms of performance and stability of performance. There are complex interactions such that, in general, both characters are promoted by heterogeneity and level of heterozygosity as well as by specific genotypic constitution. This is a complex area of research in which we yet have only the beginnings of understanding. Allard's idea of 'ecological combining ability' is an important one with implications for plant breeding procedures and strategies in relation to selection methods, population structure and disease control. This is an area of evolutionary thinking that has yet hardly touched plant breeding practice but probably should do so.

One other point must be made. The work we have been reviewing also has implications for biometrical genetics. If individual genotypes and mixtures show individual GE effects (including errors which are simply the highest order GE effects available) how is one to estimate the appropriate V_E essential for partition-of-variance and heritability calculations?

4.9 Summary

The expression of major genes is little or not affected by environment and classification of segregates is unambiguous; hence the rigorous clarity of Mendelian genetics. Members of the (much larger) class of minor genes (polygenes) are not individually classifiable in segregation; they affect characters quantitatively rather than discontinuously and their expression is more affected by environmental factors. Hence arises the distinction between genotype (genetic potential for development) and phenotype (the actual outcome of environment acting upon genotype). Most continuous variation in living organisms (including the plant breeder's populations) is brought about by polygenic action modified by environment. Polygenes cannot, in general, be individually recognized but the evidence that they exist and are numerous (based on statistical considerations and linkage) is persuasive.

Plant breeding is concerned with genetic change in populations so the leading ideas of population genetics are relevant. They start from the notion of a large, diploid, random-mating population and the way in which individual loci reach equilibrium in it. Attainment of equilibrium may be disturbed by: any degree of inbreeding (tending to reduce heterozygosity); differential fitness of alleles (based on relative numbers of offspring produced); linkage (retarding recombination and the attainment of equilibrium); and drift (random change of gene frequency even to fixation/loss of alleles, especially in small populations). All these ideas are relevant to plant breeding populations in various contexts.

Polygenic variation is the basic feature of the material with which the plant breeder works. Accordingly, his task is to construct and isolate good genotypes from knowledge only of phenotypes. Biometrical genetics seeks to aid this process by identifying the genetic mechanisms that underly phenotypic variation. It does this by partition of phenotypic variance (V_P) into genetic (V_G), environmental (V_E) and interactive (V_{GE}) components; then by further partition of V_G into additive (V_A) and dominance (V_D) portions; and then by yet more elaborate subdivision. This leads to the idea of 'heritability' which is a variance ratio that estimates the fraction of observed variation ascribable to a genetic source, thus: V_A/V_P (narrow sense heritability) and V_G/V_P (broad sense heritability). The idea of heritability is important in relation to likely response to selection but all partitions of variance suffer from statistical weaknesses and parameters based on them are rarely well estimated. Heritabilities can also be estimated from offspring–parent regressions and the idea of statistically (not genetically) additive combining abilities is often helpful in providing means of functional analysis of breeding data free of genetical assumptions.

It should be noted that the additive genetic variance, V_A, is the portion that behaves 'as though' it were due to additive polygenes and would be responsive to selection: it does not imply that the genes concerned are actually acting additively (i.e. have exactly intermediate heterozygotes). In general, partition of variance methods can give little secure information on gene action.

Most outbred organisms, and some inbreeders, show inbreeding depression, the former to the point of extinction of inbred lines. Crossing restores vigour and the recovery is called 'heterosis', shortly defined as superiority of a hybrid over the best parent. The main facts are explicable on the basis of dominant polygenes, the depression being caused by the fixation of inferior recessive alleles. Overdominance, however, might also be involved though this (and, indeed, the very existence of overdominance) would be disputed by some workers.

Response in one generation of selection depends upon the variability of the source population, the appropriate heritability of the character being selected and the intensity of selection (a statistical factor determined by the fraction chosen to be parents). Continued response over generations is not predictable at the start and poses many problems; selected populations usually reach plateaux, genetic variance declining in the process. Selection for two characters simultaneously is easy if the characters are independent or positively correlated but difficult, even impossible, if negatively correlated. 'Correlated response', the obverse of this problem, occurs when selection for one character produces a change in another, apparently unrelated, one.

There is evidence that different genotypes in a population often interact, in ways yet far from understood, to produce slightly higher means and lower variances: that is, heterogeneous populations tend to be more productive and more stable. Stability of performance is also often a feature of heterozygosity and of specific genetic constitution.

4.10 Literature

Population genetics

Baker and Stebbins (1965); Crow and Kimura (1970); Ewens (1969); Kojima (1970); Lewontin (1974); Li (1955); Mettler and Gregg (1969); Srb (1973); S. Wright (1968, 1969, 1977).

Biometrical genetics

Bailey and Comstock (1976); Becker (1975); Comstock and Robinson (1948); Cornelius and Dudley (1974, 1976); Darrah and Hallauer (1972); Demarly (1977); Dudley and Moll (1969); Ehdaie and Cress (1973); Falconer (1989); Freeman (1973); Gardner *et al.* (1953); Gardner and Lonnquist (1959); Gilbert (1961, 1973); Gowen (1952); Griffing(1956a, 1975); Hammond and Gardner (1974a, b); Hanson and Robinson (1963); Hardwick and Wood (1972); Hayman (1954); Hill (1975); Hogarth (1968); Horner and Weber (1956); Jinks (1955); Johnson *et al.* (1955a, b); Kearsey (1993); Kempthorne (1957); Knight (1970, 1973); Lerner (1950, 1954, 1958); Lyrene (1977); Marani (1975); Mather (1971); Mather and Jinks (1971, 1977); Moll and Stuber (1974); Pederson (1969, 1972, 1974b); Robertson (1960); Robinson *et al.* (1949, 1951, 1955); Sinha and Khanna (1975); Walker (1969); Wricke and Weber (1986); S. Wright (1968, 1969, 1977).

Combining ability

Anderson *et al.* (1974); Busbice (1970); Comstock *et al.* (1949); England (1974); Fyfe and Gilbert (1963); Gilbert (1958, 1967, 1973); Gilbert *et al.* (1973); Griffing (1956b); Kronstad and Foote (1964); Plaisted and Peterson (1959), Plaisted *et al.* (1962); Rinke and Hayes (1964); Russell (1969); Sharma *et al.* (1967); Simon *et al.* (1975); Sprague and Miller (1952); Sprague and Tatum (1942); Tai (1976); Tysdal and Crandall (1948); Wellensiek (1947, 1952).

Crop population problems, heterogeneity and heterozygosity

Adair and Jones (1946); Allard (1961); Allard and Adams (1969); Allard and Bradshaw (1964); Allard and Hansche (1964); Allard and Jain (1962); Allard *et al.* (1968); Allard and Kannenberg (1968); Allard and Workman (1963); Bal *et al.* (1959); Borlaug (1959); Brim and Schutz (1968); Browning and Frey (1969); Busch *et al.* (1976); Christian and Grey (1941); Clay and Allard (1971); Donald (1963); Early and Qualset (1971); Erskine (1977); Fehr and Rodriguez (1974); Frey and Maldonado (1967); Funk and Anderson (1964); Gedge *et al.* (1977); Giles *et al.* (1974); Jain and Qualset in Gaul (1976); Grummer and Roy (1966); Gustafsson (1953, 1954); Hamblin and Rowell (1975); Hanson (1970); Harding and Allard (1968); Harding *et al.* (1966); Harlan and Martini (1929, 1938); Harlan *et al.* (1940); Helgason and Chebib in Hanson and Robinson (1963); Hinson and Hanson (1962); Imam and Allard (1965); Innes and Jones (1977); Jain (1961); Jain and Allard (1960); Jain and Jain (1962); Jain and Marshall (1967); Jain and Suneson (1964); Jennings and de Jesus (1968); Jennings and Herrera (1968); Jensen (1952, 1965, 1966); Jensen and Federer (1964, 1965); Jones (1958); Jurado-Trovar and Compton (1974); Kannenberg and Allard (1967); Kannenberg and Hunter (1972); Khalifa and Qualset (1974, 1975); Laubscher *et al.* (1967); Laude and Swanson (1943); Lin and Torrie (1971); Marshall and Allar (1974); Marshall and Brown (1973); Mumaw and Weber (1957); Norrington-Davies (1972); Nuding (1936); Pahlen (1968); Pfahler (1964,

1965); Phung and Rathjen (1976, 1977); Probst (1957); Qualset and Granger (1970); Rasmusson and Byrne (1972); Reich and Atkins (1970); Richmond and Lewis (1951); Riggs (1970); Ross (1965); Rowe and Andrew (1964); Roy (1960); Sakai *et al.* (1958); Sandfaer (1954, 1970); Schutz and Brim (1967); Schutz *et al.* (1968); Schutz and Usanis (1969); Shank and Adams (1960); Simmonds (1962a); Sprague and Schuler (1961); Stevens (1948); Suneson (1949, 1951, 1956, 1960, 1969); Suneson and Ramage (1962); Suneson and Stevens (1953); Suneson and Wiebe (1942); Thompson (1977); Trenbath (1974); Tucker and Harding (1974); Walker (1963); Weltzien and Fischbeck (1990); Wiebe *et al.* in Hanson and Robinson (1963); Workman and Allard (1962, 1964).

Chapter 5
Breeding Plans

5.1 Introduction

This chapter is concerned with breeding plans; that is, with the choice of parents and of crossing and selection patterns. Logically, the choice of parents might seem the best starting point but, in practice, it is better to expound crossing and selection schemes first and then to consider the source and choice of parents in the wider context of breeding strategy. The essential question is: starting from reasonably well adapted parents and having defined his objectives, how does the breeder move from there to the production of a new variety potentially worthy of release to agriculture? The answer must vary with the crop and its mating system as well as with social circumstances but, nevertheless, even very diverse crops have much in common, as this chapter will show. So we are concerned here with the very broad pattern of what we have called 'main-stream' plant breeding, not at all with the accessory techniques, such as mutation induction, polyploidy and so forth, that form the subject matter of Chapters 8 and 9. The reader might find it helpful here to refer again to Fig. 2.1, p. 28; we shall be elaborating on the lower central part of that diagram.

5.2 The four basic schemes

General

The essential features of the four basic breeding schemes are summarized in Table 5.1 and Fig. 5.1. Inbreeding annuals are bred as inbred pure lines (IBL), outbreeding annuals as open-pollinated populations (OPP) or as hybrid varieties (HYB) and perennials (which are effectively all outbreeders) as OPP or as clones (CLO) – depending upon aptitude for vegetative propagation. The IBL scheme for inbreeders is founded directly upon their tolerance of selfing leading to homozygosity (Table 4.3); all three schemes for outbreeders either maintain a high degree of heterozygosity throughout (OPP and CLO) or restore it in the final phase (HYB). Apart from a little ambiguity as to the borderline between OPP and HYB, the four schemes are quite distinct.

Note that three out of the four crop populations (OPP is the exception) that emerge are mono-genotypic or effectively so: single pure line, single clone,

Table 5.1 The four fundamental populations in plant breeding.

Mating system	Life cycle, propagation	Population type	Characteristics
Inbreeding	Annual seed-propagated	Inbred pure lines (IBL)	Homogeneous, homozygous; isolated by selection of transgressive segregates in F_2–F_7 generations of crosses between parental IBL
Outbreeding	Annual, biennial or perennial seed-propagated	Open-pollinated populations (OPP)	Heterogeneous, heterozygous; constructed by changing gene-frequencies by selection (population improvement) or by making synthetics via parental lines or clones; verges on HYB
Outbreeding	Annual or biennial, seed-propagated	Hybrids (HYB)	Homogeneous, highly heterozygous; constructed by crossing inbred lines selected for combining ability; verges on OPP
Outbreeding	Perennial or quasi-annual, vegetative	Clones (CLO)	Homogeneous, heterozygous; isolated by selection of transgressive clones in subsequent vegetative generations of F_1 between heterozygous parental CLO

single hybrid. These are the only three biologically conceivable ways of producing homogeneous populations. Note also, in the light of discussion in Chapter 4, that CLO and HYB can completely exploit all genetic variability available, including overdominance (real and pseudo-); IBL are a little less efficient in the sense that they can exploit all, *except* overdominance. By contrast, OPP can exploit all but at less than maximum potential; their very heterogeneity implies that some constituent genotypes must be less than ideal so they must tend to be somewhat less efficient than CLO or HYB, unless the heterogeneity itself can compensate (Section 4.8).

Historical

The broad scheme just outlined is of quite recent origin. A century and a half ago, in north temperate countries, the inbreeders (e.g. wheat, barley, oats) were

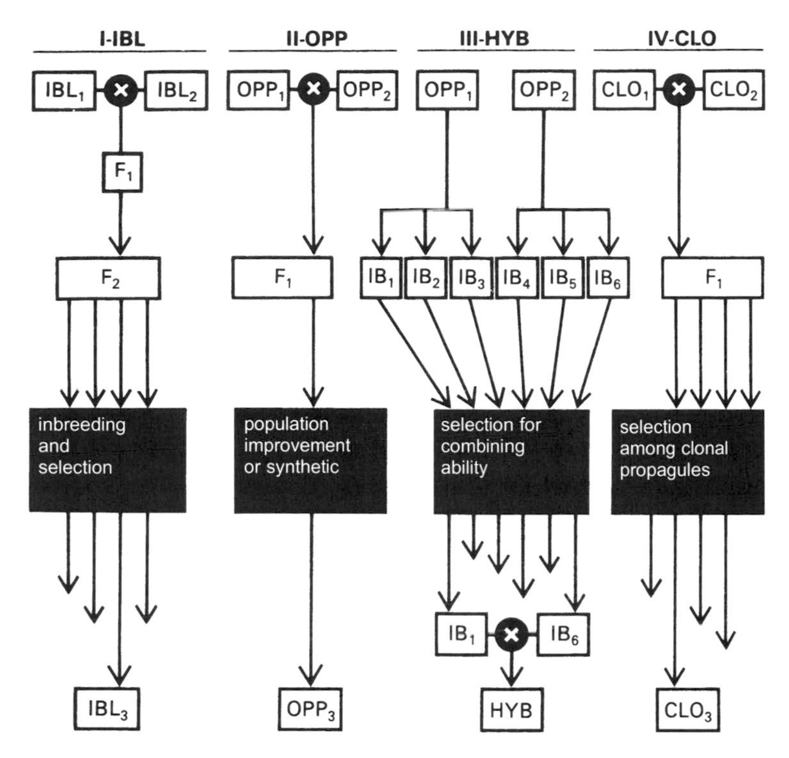

Fig. 5.1 The four fundamental populations in relation to breeding methods.
Notes: See Table 5.1. The widths of the boxes roughly reflect variability; thus the F_1 of an IBL and HYB are uniform but the F_1s of OPP and CLO (parents respectively variable-heterozygous and uniform-heterozygous) are variable.

'landraces', locally adapted as populations but quite variable and (as a matter of inference rather than direct knowledge – Sections 4.3, 4.8) far from being a simple mixture of pure lines. Pure line breeding started when enterprising farmers isolated components of landraces as distinct, narrowly-based, inbred varieties. Lacking direct evidence, it is reasonable to assume that the new pure lines were indeed superior to their predecessors; at all events, they had largely displaced the landraces by the end of the nineteenth century, in the technologically more advanced sectors of agriculture. Elsewhere, the landraces persisted, though in declining amount, in the remoter corners of north temperate countries and, more widely, in the tropics. Many local tropical rices, sorghums and pulses are still of this character, though fast disappearing.

The open pollinated populations of such crops as maize, rye, brassicas, sugar beet and many forages changed little in general character with the advent of plant breeding. They have become more intensively bred but their populations have not changed fundamentally in structure and the essential idea of progeny testing as an adjunct to improvement was current in the middle of the last century, thanks to the work of de Vilmorin on sugar beet.

'Hybrid varieties' (HYB) are essentially an invention of American science-based plant breeding in the early years of this century. Though HYB can be thought of as a highly specialized adaptation of OPP breeding methods (which is both historically and biologically sensible) it seems better to regard it as being a fundamentally new pattern of plant breeding with, as we shall see, applications to inbreeders as well as outbreeders. As a matter of perspective, it is worth noting that, though the concept goes back for many decades, the actual use of hybrid varieties on any scale is quite recent and still geographically restricted. Hybrid maize was introduced in the USA in the late 1930s and has, in effect, spread outside that country only since the late war. HYB in other crops are almost entirely post-war developments. So, on the world-scene, HYB is yet relatively unimportant, being confined to few crops and relatively small, though expanding, areas.

The practice of clonal propagation is ancient, indeed in some crops probably as old as agriculture. A clone is potentially immortal and some existing ones (e.g. of bananas, dates, apples, bread-fruit and sugarcane) may be many hundreds, possibly even a thousand or more years old; we can rarely know with certainty but a few fig clones known to classical Rome certainly survive to this day. What plant breeding has done to clonal crops is to systematize the production of segregating generations rather than rely upon selection among chance seedlings; it has also encouraged the exploitation of clones (when biologically feasible) in crops that previously existed as OPP (for example, rubber generally and some tea, coffee and cacao).

Plant breeding always operates in a specific social-agricultural context and the changes in crop populations that we have been discussing were not brought about by plant breeding alone but rather by changing agricultural systems of which change due to plant breeding is one component. In total, the result has been: (1) to displace variable landraces by uniform IBL (nearly universally); (2) to intensify breeding of OPP, having recourse to fewer but more thoroughly selected parents; (3) to displace some OPP by CLO when biologically feasible to do so; (4) to displace other OPP by HYB (a fundamentally new kind of crop population); and (5) to replace traditional mixed clonal plantings by the uniform cultivation of few clones over large areas.

All these changes were good in the sense that the new populations were more productive than the old and better adapted to the changing environment provided by the farmer (Table 3.1). But they were all accompanied by enhanced uniformity and a reduced genetic base in the producing populations. This trend is an inevitable consequence of intensive breeding for high performance and is not bad in itself. Furthermore, uniformity is often intrinsically desirable from the point of view of the husbandry and utilization of the crop (Section 3.4). But uniformity can incur hazards, especially from disease (Section 7.3). Similarly, the narrowing of the genetic base consequent on successful plant breeding means, at the very least, that deliberate action to conserve the genetic variability lost from actual agricultural populations is necessary in support of future breeding (Section 9.3). Thus the very success of plant breeding in the past 100 or so years has brought its own problems. Forty years ago these problems were hardly even recognized; now, they have entered general awareness and no well founded plant breeding system ignores them.

5.3 Inbred lines

General

The breeder's task in exploiting a cross between two pure lines is to isolate segregant pure lines which are positively transgressive of the parents in respect of the complex character 'general worth'. We shall discuss selection procedures later (Section 5.8) and here merely assume that the breeder will follow what is, in effect, the universal practice of selecting in early generations for relatively heritable visual characters such as general habit or conformation and some disease resistances and defer to later generations any serious selection for characters that are poorly heritable and/or difficult to assess (such as yield and many quality features).

A cross between two markedly different pure lines produces much variability in the F_2 (Table 4.2). Under selfing, this variability is rather quickly resolved into differences between a lot of nearly pure lines among which, the breeder hopes, are the transgressive ones he wants. (The reader may here find it helpful to refer again to the examples in Section 4.6.) The basic question before the breeder is: will he apply strong or indeed any selection in the early segregating generations when heterozygosity is high and lines have not attained individuality or will he defer selection long enough that his task becomes, in effect, solely a matter of selection between fixed lines? Table 5.2 summarizes the possibilities. On the left is what might be called the 'classical' pedigree system; single plant selection is practised up to near-homozygosis; detailed records permit the elimination of near-relatives (on the grounds that, as between two, one is likely to be about as good as the other); final selection is, inevitably, between families. The other extreme, on the right is the so-called 'single-seed descent'. On this scheme, no single-plant selection is practised at all; the object is to get through the early generations with all speed, using small crowded plants in the glasshouse, and taking two or three generations per year if possible. Only when a reasonably high degree of homozygosis is assured is the population treated more generously and a large number of lines, one per plant in the preceding generation, established. In between the extremes, but in outlook nearer to single-seed descent than to pedigree, are diverse bulk breeding schemes which all rely on zero or minimal early selection until homozygosity is well advanced; if any selection is practised it will be confined to characters which are highly heritable and easily handled such as stature, seed size or (some) disease resistance. All three schemes finish up at the same point, at F_5–

Table 5.2 Breeding of inbred crops.

	Pedigree	Bulk	Single-seed descent
F_2	**Maximal variability, high heterozygosity, heterosis persists:**		
	spaced planting; single plant selection (SPS)	F_2 bulk	single seed taken from each plant of large sample
F_3–F_6	**Declining heterozygosity and heterosis; lines assume individuality:**		
	family-rows with SPS: progressively replaced by plots and family selection; near-relatives eliminated;	bulks, with or without some artificial selection	single seed descent; two glasshouse generations per year
	quality tests and yield trials started	large sample of near-homozygous lines isolated and selected	
F_5–F_8	**Surviving lines effectively homozygous; variability greatly reduced:**		
	final family selection; detailed yield and quality evaluation; purification started	selection and yield and quality evaluation completed; purification started	
F_7–F_{12}	**Only one-few genotypes survive, as pure lines:**		
	further trials over diverse places and seasons; pure stocks multiplied; naming and release		

F_8, with intensive evaluation of emergent lines and the beginning of purification.

How to choose? If early selection were effective, the pedigree system would be attractive because it would greatly reduce the numbers of lines for later evaluation. In practice, early selection for characters of low heritability is predictably inefficient, as a great many experiments go to show. So the pedigree breeder is liable to end up with a few lines which do not include the all-important high yielders which might have been there potentially in the tail of the distribution. This argument is generally accepted so, in practice, most IBL populations are handled by one or other of the various bulk methods, with increasing attention being paid to the possibilities of single-seed-descent. The diverse bulk methods proposed all have this in common: they allow the population to run to at least F_4–F_6 and then emphasise between-line selection (so improving heritability) and the earliest possible use of replicated trials. Extra time and effort spent here are compensated by savings in the early generations of bulk propagation.

Natural selection will undoubtedly be working in the bulks but there is no reason to think that high yielding genotypes will tend to be eliminated – rather the reverse, indeed (Section 4.8); there may well be some persistent heterozygosity (Section 4.3) but this will readily be coped with by later purification. So the breeder will be looking at something approximating to a random sample of the pure lines potentially generated by that cross. The sample generated by single-seed-descent will be even nearer to random but the difference from a bulk procedure is unlikely to be large. The attraction of single-seed-descent is the speed with which early generations can be passed; but it is costly in glass-house space and, if done in the field, would lose the advantage of speed.

In adopting the general principle of late selection, the breeder is, in effect, posing himself the same problem that faces the breeder of clonal crops, that of isolating from a large but effectively fixed population a few of the best genotypes. Selection in CLO and IBL is therefore a question of isolating what is already there; this contrasts sharply with the situation in OPP and HYB, where the breeder is concerned with continual change in gene frequencies.

Genetic features

Table 5.2 suggests that breeding practice is founded on the assumption of effective fixation of lines about F_5. What does theory have to say about this? Consider (see Table 4.2) an F_1 heterozygous at n loci; heterozygosity (h) after g generations (g = 0 at F_1) will be:

$$h = (\tfrac{1}{2})^g$$

The probability of homozygosity at n loci will be:

$$p = (1 - h)^n$$

Hence, after g generations:

$$p = [1 - (\tfrac{1}{2})^g]^n$$

This relation generates a set of curves which the reader should compute for himself with g = 2, 4, 6, 8, 10 and n = 5, 10, 20, 50. For example:

$$p = [1 - (\tfrac{1}{2})^4]^{20} = 0.28$$

So 28 per cent of lines will be wholly homozygous at F_5 with n = 20 and many more will be nearly so. Even with n = 50, half the lines will be homozygous at F_7 and, by F_{10}, the size of n hardly matters. So the breeders' practice of assuming essential homozygosity by F_5 is obviously well founded (though we should perhaps add the qualifications that linkage and heterozygote advantage would delay the approach).

Now consider the related questions of numbers, drift and selection limits. At F_2, the frequency of zygotes that have already suffered some negative fixation (Table 4.2) is given by $[1-(\tfrac{3}{4})^n]$; with n = 20 this is about 99.7 per cent. So, even at this early stage, plants having the potential of generating the best possible line are rare. In this situation, the breeder would need to grow an F_2 of about 1000 plants in order to have a 95 per cent chance of retaining at least one of the favourable plants. If n were larger or if a better chance were desired, a larger number still would be needed. (As a rule-of-thumb, if the probability of a rare (<0.1) desired event is p, the number of events that should be sampled to give a 95 per cent chance of at least one occurrence is $3(\frac{1}{p})$, here 3×333.)

For want of exact genetic knowledge, no precise guide is possible but it is at least clear that an F_2 should be large, certainly in the hundreds, better a few thousand. This corresponds with normal breeders' practice. In later generations, numbers are hardly less daunting. In a random array of pure lines the probability of the 'perfect' line fixed in all positive loci is $(\frac{1}{2})^n$ which is roughly one per thousand for $n = 10$ and less than one per million for $n = 20$. The great attraction of efficient single plant selection now becomes apparent; if the truly best few per cent of the F_2 could only be captured, the chance of approaching the limit later would be immeasurably enhanced. In practice, they cannot and it is plain that IBL breeding in general can never be expected actually to attain the limit; there must always be some random fixation of negative alleles. (See Fig. 4.10 and text discussion of it.) Any such drift effect would be enhanced by linkage.

Just how big is the gap between limit and achievement we do not know and have yet no means of knowing. An interesting and perhaps useful indication might perhaps be obtained from experiments suggested by Fig. 4.9. If selected lines do indeed suffer from negative fixation, then recombination among them should generate a new plateau. Obviously such recurrent recombination could be advocated in principle as a regular adjunct to the handling of IBL material. But before this were done it would be necessary to show that in fact established methods were significantly defective and that recurrent recombination was favourable. In reality however, even if this were shown to be true, the practice would be unlikely to be widely adopted. Breeders are mostly competing with other breeders handling the same or similar crosses and success goes to him who gets the variety out first. The next cycle of crosses, using derivatives, will probably repair the defects of the first. So speedy production of good varieties which are short of the best possible is probably in the interests both of the breeders themselves and of farmers. But it would still be valuable to know more about limits of a cross and deficiencies of selection.

One final point about selection limits. We shall see below (Section 5.5) that the frequent occurrence of yield heterosis in inbreeders such as small-grain cereals, tomatoes and cottons has prompted a widespread interest in the development of hybrid varieties (HYB) in such crops. Our discussion in Section 4.4 showed that only overdominance is unfixable by efficient inbreeding; heterosis due to dominants should be fixable. So, on the common view that true overdominance is rare or non-existent, the frequent observation of heterosis would indicate that F_1 performance should provide an immediate *minimum* estimate of the potential limit of a cross. The very fact that HYB schemes are so frequently proposed for inbreeders suggests, either, that limits by IBL procedures are rarely approached or that real or pseudo-overdominance is rather prevalent. There seems to be no good means of deciding which is true but it is clear that we need a much better understanding than we yet have of IBL breeding plans in relation to limits.

Selection between crosses

Our discussion so far has centred on the methods by which a breeder may handle a single cross. He may well be generating scores or even hundreds of crosses a year, knowing well that he does not have the resources to carry them all in quantity to late generations; the worst, at least, must be eliminated and the earlier the better. It may be possible to discard some on unexpected weaknesses of habit or disease reaction or whatever; but many will look acceptable and the breeder may be confident enough of field characters and yet have no idea where to seek the all-important yielding ability.

In principle, an empirical test is simple: test all crosses as (say) F_5 lines and estimate the cross means and between-line variances. This would immediately identify the crosses with the greatest potential and predict the advance to be expected under selection (without making critical genetical assumptions). Unfortunately, this is simply impracticable. Large samples of lines for each cross would have to be used if variances were to be adequately estimated and replication over sites and seasons would be necessary to minimize GE effects. The experimentation required for any sizeable programme would be terrifyingly large. Simpler alternatives, such as yield trials of F_2 or F_3 bulks, would be attractive but entail unacceptable genetical assumptions; the highest yielders could well be heterotic combinations of no greater potential than lower-yielding intermediate (additive) combinations. Any overdominance (real or pseudo-) would be especially deceptive. Experiments designed to test methods of predicting the potential of crosses have, inevitably, given conflicting results; there can be no *general* method that does not include both means and variances between lines.

This is a rather discouraging conclusion but it seems inescapable. In practice, breeders discard as many families as they dare on general field characters and are guided by general experience, instinct and 'eye' in effecting some kind of reasonable balance between numbers of surviving families and intensity of exploitation of each. Most breeders would agree, I think, that selection is often little better than random and none would care to bet that he had never thrown away an excellent family or line.

Long-term bulks

We saw above that some degree of bulk management is a very general feature of IBL schemes as they are practised nowadays. As originally proposed, bulk breeding went rather further than is implied in Table 5.2. It was thought of as a means of generating a sample of pure lines at F_8–F_{10} onwards, rather than as a labour-saving preliminary to a near-pedigree treatment. As such it was simply too slow and has never, we think, been seriously practised.

The underlying assumption, that a bulk would quickly settle down to being a mixture of pure lines, has however been challenged by the work on complex inbred populations mentioned above (Section 4.3). Hybridity and recombination persist on a very considerable scale and, as we shall see below (Section 5.9), natural selection for fertility produces long-continued yield improvements in bulk populations of inbreeders. There is, therefore, a place for long-term bulks (or 'composites') but its place is at the strategic level of plant breeding not at the operational level, where time always presses.

Purity and mass selection

New IBL varieties are, nowadays, habitually pushed to a state of extreme uniformity by a rigorous process of purification, so that the variety traces to a single plant, tested in later generations for homozygosity (Sections 6.3, 6.4). This practice is the joint outcome of seed certification schemes, wherever more critical standards apply, and plant varieties rights schemes, in which uniformity is a condition for acceptability. There is, however, no compelling reason to think that uniformity is biologically necessary, or even desirable. We saw (Section 4.8) that heterogeneity can, at least sometimes, enhance performance and stability. Certainly, many breeders have thought this to be so and, some decades ago, IBL varieties were often mixtures of related lines which were intensely selected for uniformity in agronomic features but no more. This end-point may be reached either by stopping purification while there is still some residual heterozygosity left, or by mixing late-generation lines selected for agronomic similarity, or by prolonged mass-selection for trueness-to-type. The last was the process by which the first near-IBL populations were extracted from the landrace cereals; Harland (1944) provides an example in the Peruvian Tangüis cotton; and swedes in Scotland are traditionally mass-selected.

The practice of deliberately leaving a little variability in IBL varieties developed by pedigree or bulk-pedigree methods may, as we saw above, have some biological advantages; it also has the great attraction of speed because purification takes time. It emerges, therefore, that legislation intended in the interests of farmers and breeders may well have disbeneficial effects for both.

There is one crop in which uniformity seems to present especial problems, namely cotton. Here, varieties destined for highly specialized quality markets are rigorously inbred but still retain some residual heterozygosity more or less indefinitely. So it is common experience that some variability persists, whatever the intentions of the breeder (Justus, 1960; Thomson, 1973). Its origin is not known; homoeologous pairing (the commercial cottons are allotetraploids) has been suggested but seems unlikely; new mutations might contribute – compare the maize example of Sprague and Schuler (1961) discussed in Section 4.3; perhaps the most plausible explanation is the one that involves the preservation of substantial linkage blocks that normally persist unchanged but recombine under enhanced chiasma frequencies promoted by homozygosis. This might suggest that, though predominantly selfed and certainly highly tolerant of inbreeding, the tetraploid cottons have something of the outbreeder about them. Sorghum might present a similar picture. Perhaps significantly, HYB breeding has been successful in sorghum and promises well in cotton, if seed production problems could be overcome (Section 5.5). Though some cottons are closely inbred (with results noted above) many others are not; indeed the practice of quite deliberately retaining or compounding variability in commercial varieties has persisted here more widely and for a longer time than in any other comparable crop (Arnold in Leakey, 1970). Given a fairly high rate of cross-pollination the result seems to be not only some heterogeneity but a good deal of heterozygosity as well and there is evidence for substantial heterozygote advantage and adaptive change under natural selection (Walker, 1963). Such populations begin to resemble OPP; once again, cotton seems to have outbreeding features despite its pollination habits.

The example of cotton naturally provokes the question as to whether maize inbreds also tend to retain potential heterozygosity. They do (e.g. Fleming, 1971) and to the extent of posing problems for maintenance. Again, we have the contrast with rigorous inbreeders, where purity of IBL offers no real difficulty.

5.4 Open-pollinated populations

General

The improvement of OPP of such crops as rye, many maizes and sugar beets, herbage grasses and legumes and tropical tree crops such as cacao, coconuts, oil palm and some rubber, depends essentially upon changing gene-frequencies towards fixation of favourable alleles while maintaining a high (but far from maximal) degree of heterozygosity. Uniformity in such populations is impossible and trueness-to-type in an OP variety is a statistical feature of the population as a whole, not a characteristic of individual plants; all three other breeding populations, IBL, CLO and HYB, it will be recalled, are homogeneous (or virtually so).

It is convenient to recognize two main types of OPP (Fig. 5.2). First, there is the situation in which a population is changed *en masse* by the chosen selection procedure. The outcome is an improved population which is indefinitely propagable by random-mating within itself in isolation. It is, in effect, a version of the source population with changed gene frequencies and, presumably, always on a narrower genetic base. Following Sprague (in Frey, 1966) we may describe this process as population improvement (PIM for short). Second, the synthetic variety (SYN) attains the same end result as population improvement but is not itself propagable as such; it has to be reconstructed from parental lines or

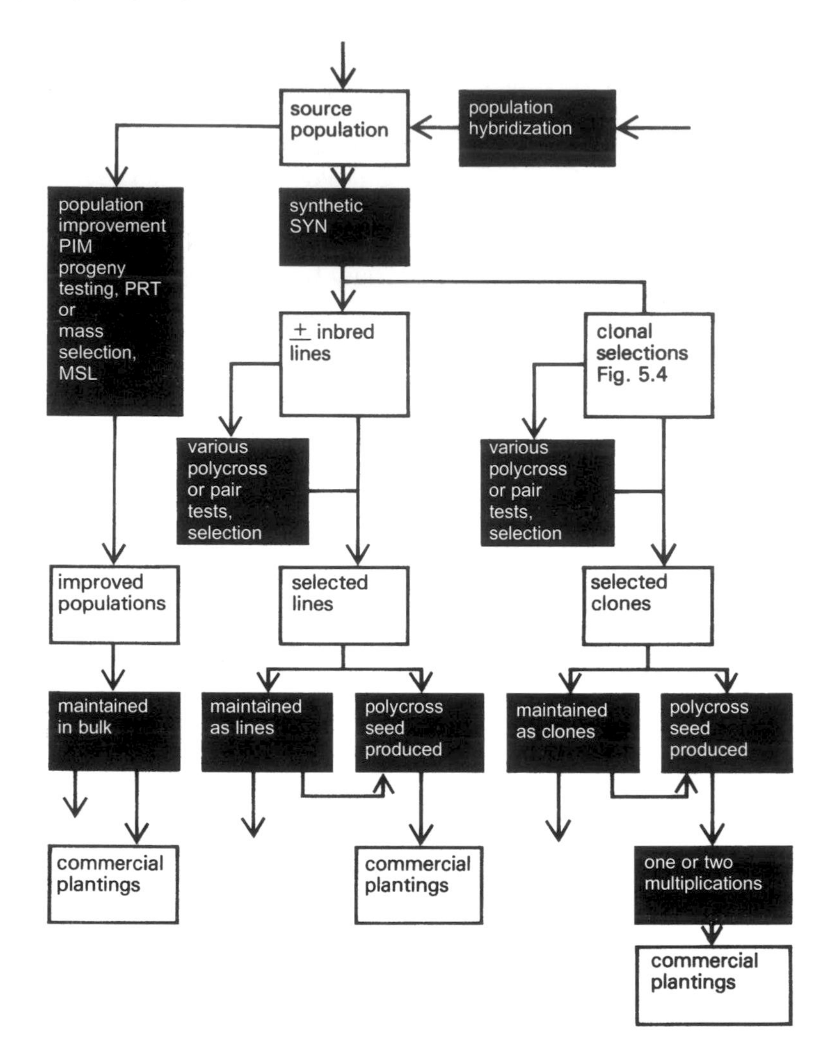

Fig. 5.2 Open-pollinated populations: population improvement and synthetics.
Notes: This diagram is quite generalized covering population improvement (PIM) at the left (treated in detail in Fig. 5.3) and synthetics (SYN), at the centre and right (Fig. 5.4).

clones. The word 'synthetic' has been used in quite various ways in the plant breeding literature. Some authors would use it of any OP variety constructed from combining inbreds, lines or clones as parents, even if the variety were indefinitely propagated without recourse to parental stocks; this overlaps our definition of population improvement. Other authors would restrict the meaning as we propose to do. In the American maize literature, the word has a somewhat specialized significance; there, it normally applies to experimental (rather than commercial) populations compounded from inbred lines randomly mated; they may be transient (when the nomenclature SYN_1 SYN_2 SYN_3, etc., is usual) or long-continued, when they become, in effect, specialized OPP (e.g. the widely used 'stiff-stalk-synthetic' called SSS).

Population improvement methods fall naturally into two groups, those based on purely phenotypic selection (mass selection MSL) and on selection with progeny testing (PRT). These are explored in the next section. Thus, summarizing the slightly confusing nomenclature of OPP breeding, we have: PIM (varieties propagable as such) contrasted with SYN (varieties regularly reconstructed from parents), with the former based either on MSL or PRT.

Population improvement

Population improvement methods are summarized in Fig. 5.3 which is based in part on those listed by Sprague (in Frey, 1966), supplemented by others from the literature. All the methods given in Fig. 5.3 have been tried at one time or another in maize in which breeding methods have been systematically explored in a way that no other crop can begin to match. Maize is peculiarly well adapted to elegant progeny-testing techniques because of the ease with which it can be reliably both selfed and crossed. Several of the plans listed would not be feasible in crops such as some Brassicas, beets and rye in which self-incompatibility forbids selfing or in which pollination is laborious.

Mass selection is the simplest and historically by far the most widely practised form of population improvement. It depends for success upon relatively high heritability, i.e. upon regression of offspring on parents. As usually practised, OP seed is taken from selected plants which means that any heritability is halved, the maternal parents alone being known. The alternative, of crossing selected parents, would improve the heritability and further improvement could be had by reducing the error component by selecting on a grid against immediate neighbours. So there are several variants of mass selection (and others could no doubt be invented). They all have in common a weakness when heritability is low but the attraction of a one-generation cycle.

The progeny-testing procedures that make up the rest of the figure represent compromises between rigour in the control of parentage and improvement of heritability and the inconvenience and delay of longer cycles. Half-sib (1) and (2) are ear-to-row methods as practised in maize but widely applicable in principle to other species. Half-sib (3) and (4) are variants of the so-called 'top-cross progeny test'. They are well-suited to maize but workable only when crosses can be reliably made in quantity. Half-sib (5) is the 'ear-to-row-to-ear' method of Lonnquist (1964), an elegant variation of half-sib (1) that, alone among all these methods, tries to take GE effects into account by diversifying the test-plot environments. The full-sib and S_1 performance tests improve further on the control of parentage. Finally, the reciprocal recurrent selection procedure summarized at the end of Fig. 5.3 is an altogether more elaborate operation designed to improve two populations simultaneously with respect both to performance and to their mutual combining ability, each with the other.

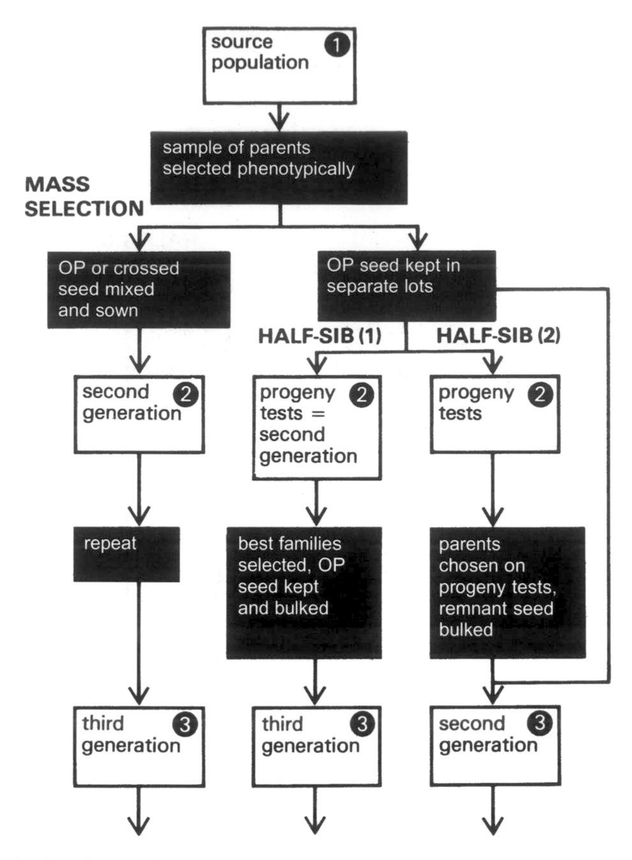

Fig. 5.3 Methods of population improvement in OPP: mass selection and various progeny-testing techniques.
Notes: The diagram is in five parts, the first relating to mass-selection (MSL) (no progeny testing), the rest to various PRT methods, in order of increasing complexity and rigour. The numbers set into the boxes are years, on the assumption of annual life cycle. Most methods are based on maize practice but are applicable in principle to other crops if pollination systems permit. 'Recurrent selection' methods, ranging from mass-selection (MSL) to 'reciprocal recurrent selection' (RRS), are included (see text).

We shall see below that it was invented in relation to HYB breeding in maize but it is nevertheless one of the several methods of population improvement that appeals to PRT data and so logically belongs here.

Around the end of the nineteenth century and in the first decade of the twentieth it became evident to US maize breeders that their populations were not responding to selection for yield by the then conventional mass selection and ear-to-row procedures. Indeed, this failure was one stimulus to the development of HYB varieties in the crop. These were very successful, and only in the last fifty odd years has there has been renewed interest in population improvement. Paradoxically, it turns out that OP maize varieties respond well to selection for improved yield; it was not so much the basic methods of ninety

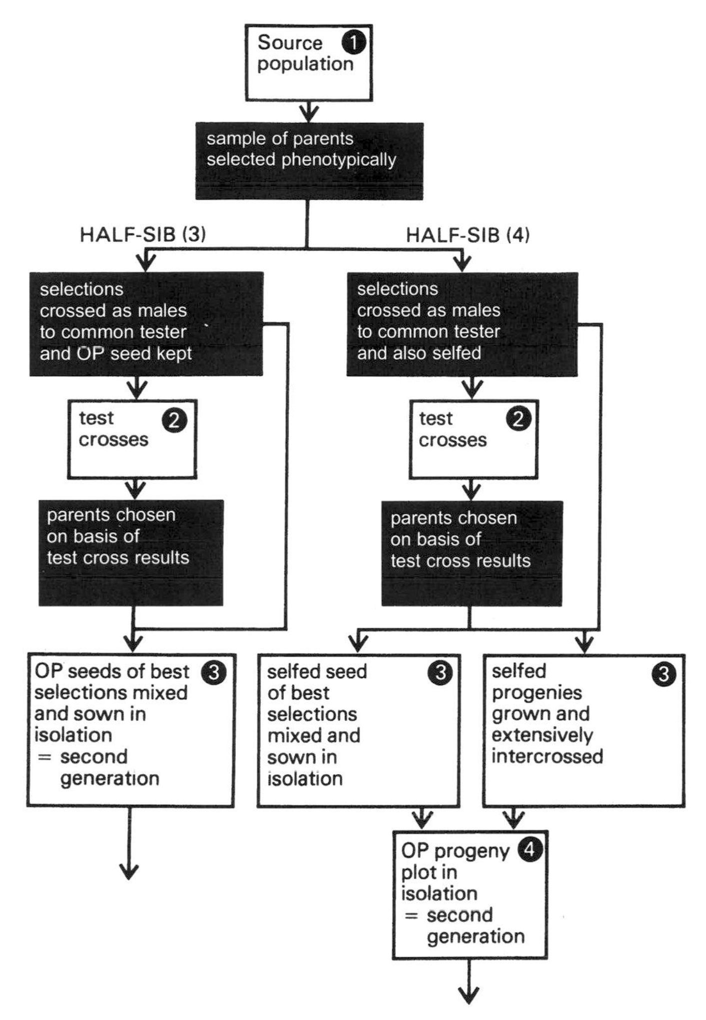

Fig. 5.3 *Contd.*

years ago that were inadequate but rather the way in which they were used. All the methods in Fig. 5.3 have had at least some success in advancing yield (examples in Table 5.4). Most dramatic, perhaps, was the demonstration by Gardner (1961) that mass selection, provided it was carried out on a grid system ('stratified mass selection'), was an effective means of improvement. Evidently, additive genetic variance (i.e. the part that responds to selection, with no necessary implication as to additive gene action) was not wanting in OP maize, in which it had long been supposed to have been exhausted.

The implications of these findings for maize breeding programmes will emerge below. Meanwhile we may remind ourselves that all the methods of Fig. 5.3 have been shown to be effective in maize and that there is no reason to think they would be any less effective in other crops, provided only that the character

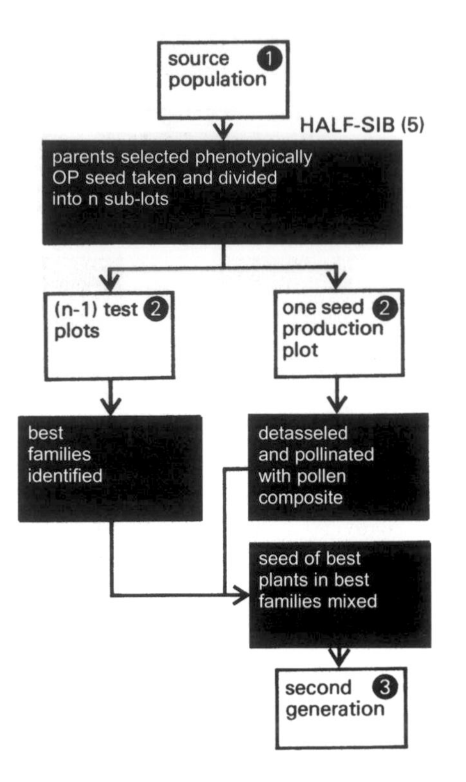

Fig. 5.3 *Contd.*

selected shows at least some heritability. No objective means of choosing one method out of the several suggests itself. Any choice must be a compromise between rigour and expenditure of time and effort. Nor, unfortunately, do the comparisons of methods which have been made in maize help very much; even when it can confidently be asserted that this method is better than that one, the conclusion will strictly apply only to one population in one experiment, not to other populations or generations; and, contrariwise, standard errors are usually such that seeming similarity of response could cover a large real difference. So practical empiricism must rule and every breeder will have to decide for himself what sort of position he will take up in deciding upon the necessary compromise between rigour and speed.

Recurrent selection

The various forms of recurrent selection were developed by American maize breeders during the 1930s and 1940s, though the original ideas had emerged earlier. The objective of all schemes was, ultimately, the improvement of inbreds for HYB production but the methods, as we saw above, are essentially those of population improvement and do not have to be linked to HYB programmes.

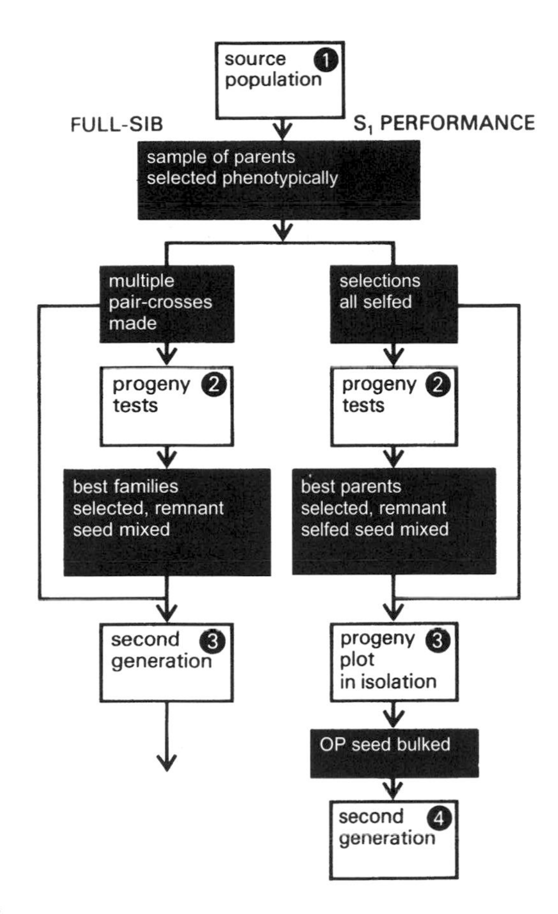

Fig. 5.3 *Contd.*

The four schemes are as follows (Fig. 5.3). Simple recurrent selection is essentially mass-selection with one or two years per cycle. Recurrent selection for general combining ability emerged from work on the early testing of inbreds; in effect, it is a half-sib progeny test procedure, using a widely based variety as tester. Recurrent selection for specific combining ability is similar but uses instead an inbred line as tester. Reciprocal recurrent selection aims at the mutual genetic adaptation of two populations, thus exploiting both GCA and SCA. Another technique developed by the hybrid maize workers with the improvement of inbreds in view is gamete selection. This will be mentioned later (Section 5.5) but we will note here that it, too, like the recurrent selection procedures, is applicable to population improvement.

Synthetics

The construction of synthetic varieties (using the word in the restricted sense defined above) is summarized in Fig. 5.2. Whether parents are (more or less

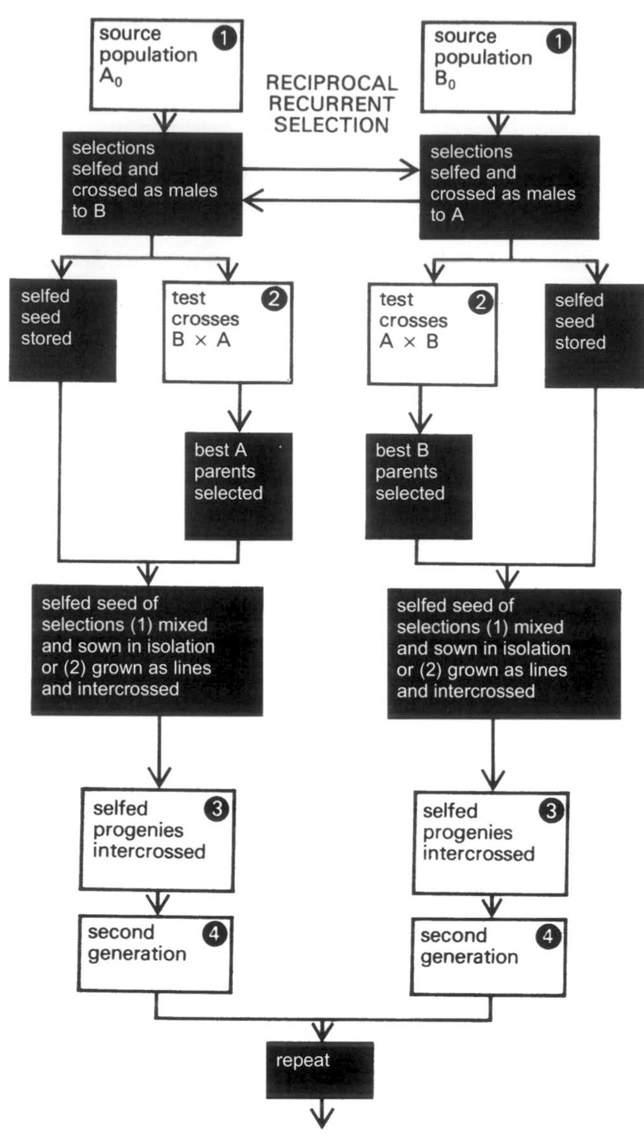

Fig. 5.3 *Contd.*

inbred) seed-propagated lines, as in some sugar beet and beans (*Vicia*) or clones, as in herbage grasses, clovers and alfalfa, makes no difference in principle. Parents are selected on general combining ability, sometimes by test crosses or topcrosses, more generally by polycrosses. Parental seed lines may be deliberately inbred (e.g. by selfing or sib crossing) but, even if they are not, selection within lines during maintenance will ensure that some inbreeding occurs. Clonal parents will, of course, remain unchanged and highly heterozygous. Both systems offer the possibility of making slight improvements in varieties by substituting better lines or clones as they become available. This was certainly a common practice in the past but will presumably be inhibited in future by plant varieties rights rules (Section 6.3).

Whether a synthetic can go straight from the parental seed production plot to the farmer (i.e. as SYN_1 seed) or must first undergo one or two cycles of multiplication depends on seed production and the scale of demand for seed. In practice, grasses and clovers are generally multiplied once or twice and are thus considerably removed from the original synthetic. In synthetic sugar beets, SYN_1 is normal as it is also in cacao and rubber, where quite small clonal seed gardens can supply large planting areas.

The development of SYN varieties in grasses and clovers is explored more fully in Fig. 5.4 in which the essential analogy with population improvement methods (Fig. 5.3) is brought out. Mass selection (on the left) and various levels of refinement of testing are possible. In actual practice, progeny testing now seems to be all but universal, with a general preference for polycrosses, because of their operational simplicity and obvious relevance to the objective, namely exploitation of GCA in a synthetic. In principle, an indefinite number of cycles of improvement could be undertaken but, with a two-year life cycle, the demands on time become formidable; allowing for preliminary screening of sources and official trials and formalities, a new grass or clover synthetic is not likely to emerge in much less than twenty years from the start, even with only one cycle of progeny testing and parent selection.

Polycrosses are undoubtedly convenient but they have their drawbacks. Any substantial degree of self-pollination, especially if it varied between entries, or non-randomness of cross pollination could cause bias. In practice probably no polycross is perfect and many may be very imperfect, however, Simmonds (1986) considers designs for the field layout of rubber or any other polycrosses which are suitable for general use and unconstrained by unnecessary mathematical conventions.

The numbers of parental lines or clones that enter a synthetic vary but not very widely. If few, there could be some deleterious inbreeding at the seed production stage; if many, it is unlikely that enough excellent high-GCA parents would be available, so performance would suffer. In practice, numbers in the range 3–15 are commonly mentioned and our impression is that 5–10 would be about average.

Combining ability

All progeny-testing procedures depend upon the estimation of combining abilities, with emphasis nearly always on GCA. It may not be apparent to the reader how our discussion of combining abilities in relation to diallel or $m \times n$ matings (Section 4.7) relates to the test-cross or polycross matings which are usual in PRT operations. In fact, a simplification is necessary.

Consider n lines or clones being tested; they will yield n families, where the

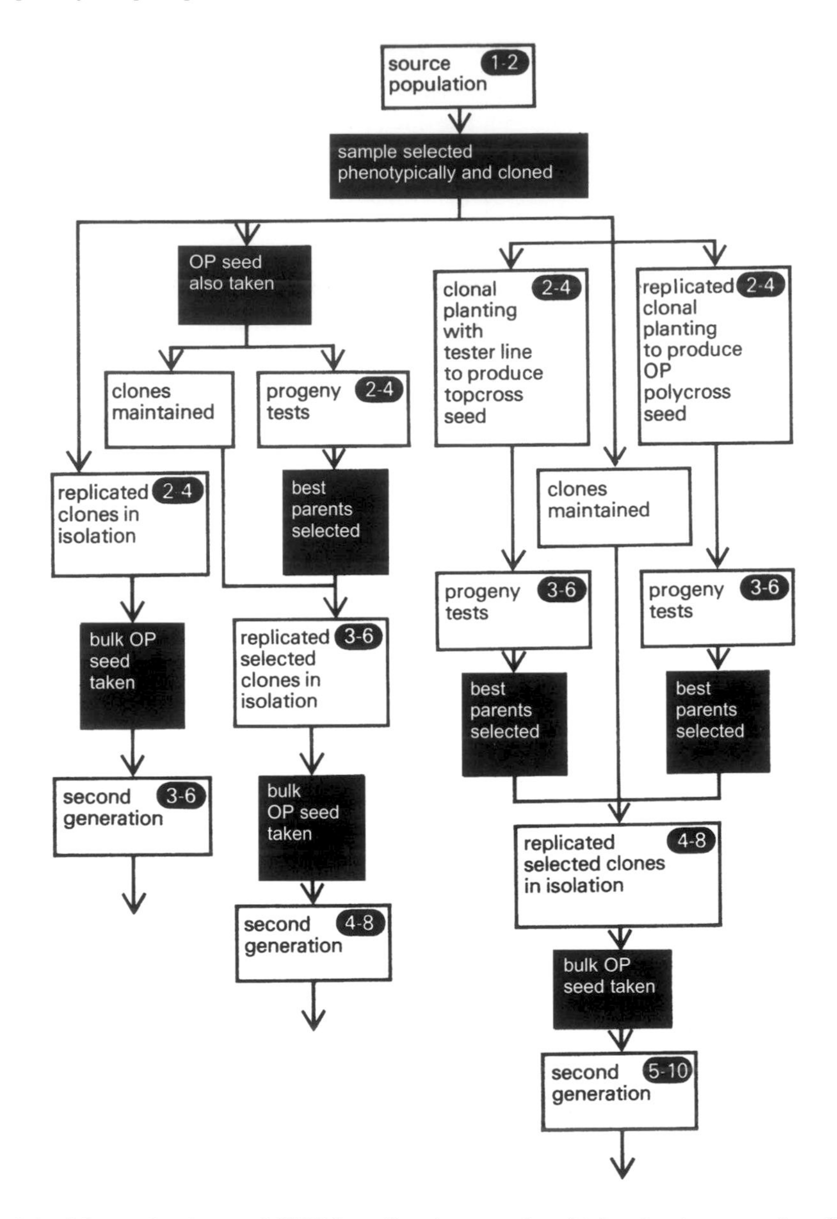

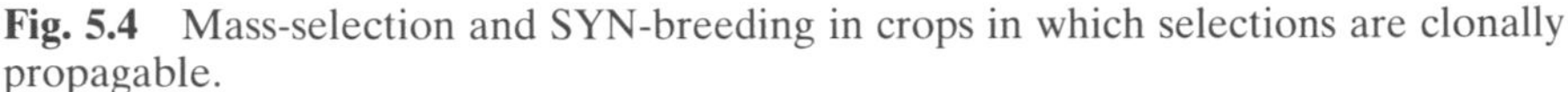

Fig. 5.4 Mass-selection and SYN-breeding in crops in which selections are clonally propagable.
Notes: There are, essentially, three patterns: mass selection without PRT (left), PRT based on OP seedlings from original population (centre) and PRT based on top-cross or polycross seedlings (right). The numbers set against boxes are years, on the assumption of one- to two-year life cycle. Most forage grasses and legumes will have a two-year cycle; tree crops, such as rubber and cacao, a much longer one.

other parent is either a tester or else all other (n–1) parents themselves acting randomly as polycross males. In either case we have, in effect, an n × 1 table in which it is easily seen that the GCA of the tester (G_T) must be zero whence, writing $C_A = (G_A + S_{AT})$:

$$Y_{AT} = (G_A + S_{AT}) + G_T + \overline{Y}$$
$$C_A = \overline{Y} - Y_{AT}$$

Thus, with only a single tester SCA cannot be separated from GCA and we have to use instead C, the average combining ability, which is simply the difference of a cross from the experiment mean. This, of course, is the commonsense estimate of the breeding value of a parent and it is the commonest one to emerge from progeny testing work.

If diallel or m × n data are available, then, naturally, somewhat more subtle choices may be open to the breeder. Consider Table 4.8 on rubber. The GCA values readily provide estimates of the yields of SYN populations based on any combination of parents. The reader should calculate these figures for the 2, 3, 4, 5 best parents and note how yields decline as numbers grow and the less good parents are included. This procedure ignores SCA. But six out of the ten best combinations were tested as families so the SYN expectations can be recalculated to take account of SCA by using family data rather than simple GCA prediction. As the reader will find, the adjustment makes little difference in this case because SCA effects are small; had they been large, the choice of parents could become quite tricky. Several of the clones included in Table 4.8, incidentally, are among the parents of the high yielding SYN seedling populations mentioned in Table 3.1 (6).

5.5 Hybrid varieties

General

In the last part of Section 4.4 (and see especially Fig. 4.5) we surveyed the main features of inbreeding depression and heterosis as revealed by a great many experiments with outbreeding organisms, mostly animals such as *Drosophila*, mice and chickens but equally characteristic of outbred plants such as maize. We saw that inbreeding depression was generally attributed to the fixation of unfavourable recessives and that heterosis was simply the converse, unfavourable recessives fixed in one line being covered, so to speak, by dominants from the other. If this were all there were to it, heterosis should be fixable in true-breeding, homozygous lines. In general, it is not and the question arises whether this is purely a statistical difficulty of fixing numerous dominants each of small effect in relation to V_E or whether some appeal must be made to overdominance or pseudo-overdominance reinforced by linkage (Table 5.3). Since, in any practical situation, the number of loci prob-

Table 5.3 Gene action in relation to hybrid varieties. Genotypes and arbitrary genetic values. To illustrate the several possible sources of apparent overdominance and the phenotypic potential of homozygotes. Under epistasis, three cases are distinguished: (i) simple epistasis, (ii) with additive effect, (iii) with dominant effect.

Genetic situation	Inbred parents	Hybrid	Derived homozygotes
Unlinked dominants	AAbb × aaBB →	AaBb →	aabb
	2+0 0+2	2+2	0+0
		↘	AABB
			2+2
Linked dominants	Ab/Ab × aB/aB →	Ab/aB	none –
	2+0 0+2	2+2	pseudo-overdominance
Overdominance	AA × aa →	Aa	none
	2 0	3	true overdominance
Linked epistatic	Ab/Ab × aB/aB →	Ab/aB	none –
	(i) 0+0 0+0	0+0+2	pseudo-overdominance
	(ii) 2+0 0+2	1+1+3	
	(iii) 2+0 0+2	2+2+2	
Unlinked epistatic	AAbb × aaBB →	AaBb →	aabb
	(i) 0+0 0+0	0+0+2	0+0+0
	(ii) 2+0 0+2	1+1+2 ↘	AABB
	(iii) 2+0 0+2	2+2+2	(i) 0+0+2
			(ii) 2+2+2
			(iii) 2+2+2

ably exceeds the number of chromosome arms, linkage effects seem likely *a priori.* Table 5.3 shows that either dominance or inter-locus interaction (epistasis) *plus* linkage can readily simulate overdominance. True overdominance is certainly rare at the major-gene level; it may or (more probably) may not play a part at the polygenic level.

The general principle underlying HYB varieties will now be apparent. Starting from a hypothetical OP population in equilibrium with a locus *Aa* at $p=q=0.5$ (Section 4.3) and genetic values shown in Table 5.3 ($AA = Aa = 2$, $aa = 0$), the population average will be $6/4=1.5$ but a pure hybrid population (*Aa*) will have a value 2.0, a gain therefore of 0.5 per locus. Under selfing and with many loci, no line will escape substantial fixation of recessives; hence inbreeding depression. However, random pairs of lines will only rarely have just the right combinations of dominants and recessives to complement each other, so the average HYB will be near to the source population and outstandingly good or bad combinations will be rare. If there were much overdominance, true or pseudo-, excellent combinations would be rarer still. This follows from consideration of one locus and of the four hybrids that could be produced from crossing the two possible homozygotes (*AA* and *aa*), namely: 1*AA*, 2*Aa*, 1*aa*. With dominance, 3/4 are favourable but, with any kind of overdominance, without recombination, only 2/4. With n loci the best possible inbred × inbred pairs would have frequencies $(3/4)^n$ and $(\frac{1}{2})^n$, which, with n = 20, say, are equal to about 3×10^{-3} and 10^{-7} respectively. (The reader should check this for himself and try also putting n = 10, 50.) Pseudo-overdominance with some recombination would obviously produce intermediate figures. Perfection is not to be expected but, on any reckoning of this kind, excellent HYB will be rare. Conversely, the undoubted difficulty of producing vigorous and productive inbreds from OPP is intelligible on the basis of numerous dominants *plus* some pseudo-overdominance, with or without true overdominance. And this seems to be about as far as we can go; exact diagnosis lies beyond the current reach of genetical analysis.

In practice, therefore, the problem before the HYB breeder resolves itself into: (1) isolating numerous inbred lines; (2) testing combinations of inbreds to identify excellent ones; (3) devising short-cut methods of achieving (2) without resort to the fearsome labour of crossing everything with everything; and (4) developing methods of making HYB seed on the commercial scale. Discussion of item (4) is deferred to Section 6.4. Items (1)–(3) are treated in this section with, naturally, especial reference to maize where nearly all the experience lies. At the end of this section we will consider applications to other crops and express some doubts and questions.

To conclude this section it is convenient to place here two Tables (5.4 and 5.5) that illustrate a number of important points about maize breeding in general and HYB maize in particular. One (Table 5.4) has already been referred to earlier in this chapter and both will attract recurrent reference below. They merit close study for their bearing on: interpopulation heterosis (5.4), the relative places of OPP and HYB methods in maize improvement schemes (5.5), the use of backcrossing methods in maize improvement schemes (5.5), the use of backcrossing for inbred improvement (5.5 and see Section 5.6) and the nature of yield heterosis (5.5).

Short history of hybrid (HYB) maize

Around the turn of the last century in the USA there was a general awareness that decades of corn shows, mass selection and ear-to-row selection were not producing results in the form of improved yields. In retrospect, it became apparent that the corn shows were positively dysgenic in effect because they selected for single large ears rather than for two or three smaller ones at higher yield ('prolificacy'); and the selection procedures, though sound in principle, were ineffective in practice (as we saw in Section 5.4). That the impasse was more apparent than real does not affect the fact that it was an impasse. The first indications of a way out came from the knowledge, well established by the end of the century, that variety × variety crosses often showed heterosis and there were practical proposals by Beal and others to exploit such combinations commercially, hybrid seed to be produced by

Table 5.4 Maize breeding in Kenya: population improvement and inter-population heterosis. Grain yields in t/ha. Based on several tables in Eberhart, Harrison and Ogada (1967) and Darrah, Eberhart and Penny (1972) by permission of the publishers, Messrs. Springer (*der Züchter*) and the Crop Science Society of America (*Crop Science*).

Comparisons		Results			
A.	Population improvement in Kitale maize	KI 3.87	KII 4.01	KIII 4.40	
B.	Population crosses, Kitale × Ecuador 573, and early HYB	KII 4.32	KIII 4.62	KII × Ec 573 5.18	KIII × Ec 573 5.68
		Three K-derived HYB: 5.43–5.36 Two K-lines × Ec 573: 5.68, 6.18			
C.	Composites (= OPP) derived from KII × Ec 573	KII 5.09	Ec 573 3.87	KII × Ec 573 6.27	
		Four composites: 5.12–5.75 Comp. B × Comp. C: 5.98			
D1.	Improvement of source populations by two methods	KII 4.69	(1) 5.04	(2) 4.94	
		Ec 573 3.36	(1) 4.66	(2) 4.56	
D2.	Improvement of derived populations KII × Ec 573	Composite 5.06		(1) 6.22	
		Variety cross 6.52	(1) 7.02	(2) 7.68	
		Topcross 6.90	(1) 6.64	(2) 7.68	
E.	Hybrids	KII control – 4.01			
		K-inbred single cross × Ec 573 5.70		improved version 6.67	
		Population cross KII × EC 573 5.27		improved (1) 6.13	improved (2) 6.44

Notes: This programme started in 1955 and was based on the well-adapted local Kitale population; a large sample of tropical American maizes yielded the outstanding Ecuador 573, strongly heterotic with the Kitale population. The data clearly illustrate the following points: (1) excellent response to within-population improvement both of source populations and of crossed derivatives; (2) interpopulation heterosis at all levels of improvement; (3) effect of parental improvement on hybrid performance. The first commercial hybrid was KII × Ec 573 (the population cross) which was quickly outclassed.

detasselling the female parents in mixed isolation plots. These proposals came to naught but the underlying idea, of exploiting heterosis in controlled crosses, was soon to be taken up.

The first proposals were made by G.H. Shull in 1909 on the basis of his genetic studies which were, in effect, following up Darwinian ideas on inbreeding and heterosis. Shull perceived that exceptional inbred × inbred combinations could outyield the source population, even though the average was no better. Furthermore, such hybrids would have the attraction of uniformity within the stand and of exact (or nearly exact) reproducibility from year to year. The concept, as it stood then, failed because few inbreds even approached 50 per cent of the OPP yield and hybrid seed would have been impossibly expensive (besides being often poorly formed).

Shull's proposals took nearly fifty years to come to practical fruition. Meanwhile, they were very successfully displaced by the idea of double-crosses, suggested by D.F. Jones in 1918. Since commercial seed production was the limiting factor, small supplies of expensive F_1 hybrid seed could be 'stretched' by making double-cross HYB varieties of the type (A × B) × (P × Q). In practice this worked very well. Double-crosses were economical in seed production, reproducible from year to year and uniform enough; they were an enormous success and, starting in the 1930s, dominated the US acreage by the 1940s with result for yields indicated in Table 3.1.

For many years all seed was produced by detasselling the female lines and Mangelsdorf (1974) says that, at the peak, as many as 125 000 people on any one day were engaged in the detasselling operation. By the late 1960s, detasselling had largely ceased, replaced by the use of cytoplasmic male sterility (CMS – Section 6.4). CMS ideas go back to the 1930s and the all-important Texas cytoplasm was identified in 1938. The great practical success of this very elegant piece of applied genetics was, unhappily, qualified by an unforeseen phytopathological side effect (southern leaf blight – Chapter 7).

Right from the start, it was recognized that a necessarily limited (even though large) sample of inbreds could not be expected to capture the full genetic potential of the source populations and that provision would have to be made both for continued isolation of new inbreds and for cyclical recombination and improvement of the old. This idea led straight to 'recurrent selection' (Section 5.4), with two important consequences. First was the realization (since the 1950s) that OPP maizes were far more susceptible of improvement by MSL and PRT methods than had earlier been supposed (examples in Table 5.4); and, second, it led to sufficient improvement in the performance of the inbreds themselves to permit a return to Shull's original concept of the single-cross hybrid. By the 1960s, single had largely replaced double-crosses, with results indicated in Table 3.1. Hence there emerges a general maize breeding strategy comprising: (1) maintenance and improvement by OPP methods of source populations; (2) isolation of new inbreds and improvement of old ones (by back-crossing methods – Section 5.7); (3) suc-

Table 5.5 Earliness conversion in maize. Data from D.L. Shaver (personal communication).

Identity	A Anthesis, days	B Leaf number	C Plant height (cm)	D Ear height (cm)	E Ear length (cm)	F Kernel row number	G Ear number	H Kernel weight (g)	K Yield (t/ha)
9	95	20	147	62	16	16.3	0.97	0.24	4.02
9E	83	16	102	33	15	13.9	1.00	0.24	2.67
9 × 9E	87	18	133	54	17	15.6	1.00	0.26	4.79
Hy	94	19	121	62	13	14.9	0.98	0.22	2.53
HyE	87	16	94	40	13	16.4	1.02	0.16	2.25
Hy × HyE	88	18	110	52	16	16.3	0.96	0.17	2.60
41	101	19	141	68	18	12.3	1.02	0.19	2.82
41E	99	19	132	63	16	12.1	0.97	0.21	2.99
41 × 41E	103	20	142	69	17	12.3	0.98	0.22	3.36
38	103	21	167	89	19	15.3	1.00	0.21	3.38
38E	97	19	153	68	20	13.6	1.02	0.24	3.35
38 × 38E	98	20	163	83	21	14.4	1.02	0.24	4.31
9 × Hy	86	20	190	103	18	18.4	0.99	0.26	8.18
9E × Hy	83	18	170	85	17	17.5	1.00	0.28	7.79
9 × HyE	84	19	172	78	17	18.7	1.00	0.25	7.51
9E × HyE	81	17	149	61	16	18.2	1.01	0.27	7.16
41 × 38	95	21	200	120	21	13.7	1.03	0.29	7.63
41E × 38	93	20	191	109	21	13.9	0.99	0.27	7.30
41 × 38E	91	20	195	110	21	13.5	1.03	0.32	8.09
41E × 38E	90	19	188	103	21	13.5	1.02	0.25	7.80

Lines, L	98	20	144	70	17	14.7	0.99	0.22	3.19
Lines, E	92	18	120	51	16	14.0	1.00	0.21	2.82
L × E	94	19	137	65	18	14.7	0.99	0.22	3.77
Hybrids LL	91	21	195	112	20	16.1	1.01	0.28	7.97
Hybrids LE	88	19	182	96	19	15.9	1.01	0.28	7.67
Hybrids EE	86	18	169	82	19	15.9	1.02	0.26	7.48

Notes: These data were developed from a large experiment in which the four outstanding inbreds Wf9, Hy2, Oh41 and 38-11 were converted to earliness (E) by ten generations of backcrossing after a cross to the very early Gaspé Flint, with selection only for earliness (see Section 5.7). Note the correlated response of characters A–D (all expressions of earliness), and their varied relations to heterosis.

The other characters recorded, E–K, are all connected with yield: E, F, G, H, can be regarded as yield components and the multiple of all four is a rough estimate of yield. The reader should calculate this quantity and plot scatter diagrams against yield itself, K. Note that, within lines, the components show no or little heterosis but that yield itself is clearly heterotic (related presumably to the small chromosome segments introduced by backcrossing). The hybrids show marked heterosis for earliness, vigour, yield components and yield. Note that the inbreds are old and outstandingly good ones (the four parents of US 13) but that, even so, their yields are less than half those of their hybrids. All four were 'public lines'.

cessive improvement of single-cross HYB by parental improvement; and (4) CMS-based seed production.

No account of hybrid maize would be complete without reference to two significant socio-economic features. The first is that, until recently, the USA lacked any plant varieties rights legislation. So IBL, OPP and CLO had no attractions for commercial breeders; once out, they were free for all and seed could be multiplied and sold by anyone, without profit to the breeder. By contrast, maize (the most important crop in the USA), bred as HYB rather than as OPP, offered an inbuilt economic protection; the breeder could retain complete control of the inbreds and sell only the HYB seed, confident that farmers would have to return each year for fresh F_1 seed. So HYB maize, once it had been well started by the state-supported experiment stations (whose inbred lines were widely used for many years), became a predominantly commercial operation, and a very large, powerful and successful one indeed. The total resources put into HYB maize must have been enormous by any ordinary plant breeding standards, but economically justified by the results; excellent HYB seed sold cheaply.

The other socio-economic feature was this. Hybrid maize was a component of the transition from traditional agriculture to intensive, technology-based agriculture. So the hybrids were not only intrinsically high yielding but were specifically adapted to the increased plant populations and rising fertility levels of the times (see Table 3.1). Since the late war, HYB maize has been widely adopted outside the USA but only in those countries (for example throughout Europe) with agricultures and agricultural infrastructures sufficiently advanced to take advantage of it. Much of the world's maize is still of an OPP character and likely to remain so for a good many years to come. Thus, population improvement is, surely rightly, the foundation of the CIMMYT programme in Latin America (Section 10.4); elsewhere, Kenya provides a good example of a country that is making the transition and in which, therefore, HYB and improved OPP maizes co-exist in their own particular agricultural niches (Table 5.4). It is an interesting reflection that it was research that was essentially peripheral to hybrid maize that generated the means of population improvement now being widely adopted *per se.*

HYB maize breeding practice

The main features of HYB maize breeding are summarized in Fig. 5.5. This is somewhat idealized in the sense that few programmes will observe indefinitely a rigorous separation between two (or more) distinct source populations. Nevertheless, unrelatedness of inbreds is a significant component of heterosis in their hybrids, as was recognized long ago in considering the order-of-pairing question in double crosses (see below); the Kenya programme (Table 5.4) provides a nice example of the orderly exploitation of complementary populations at both OPP and HYB level. So the general principle of seeking

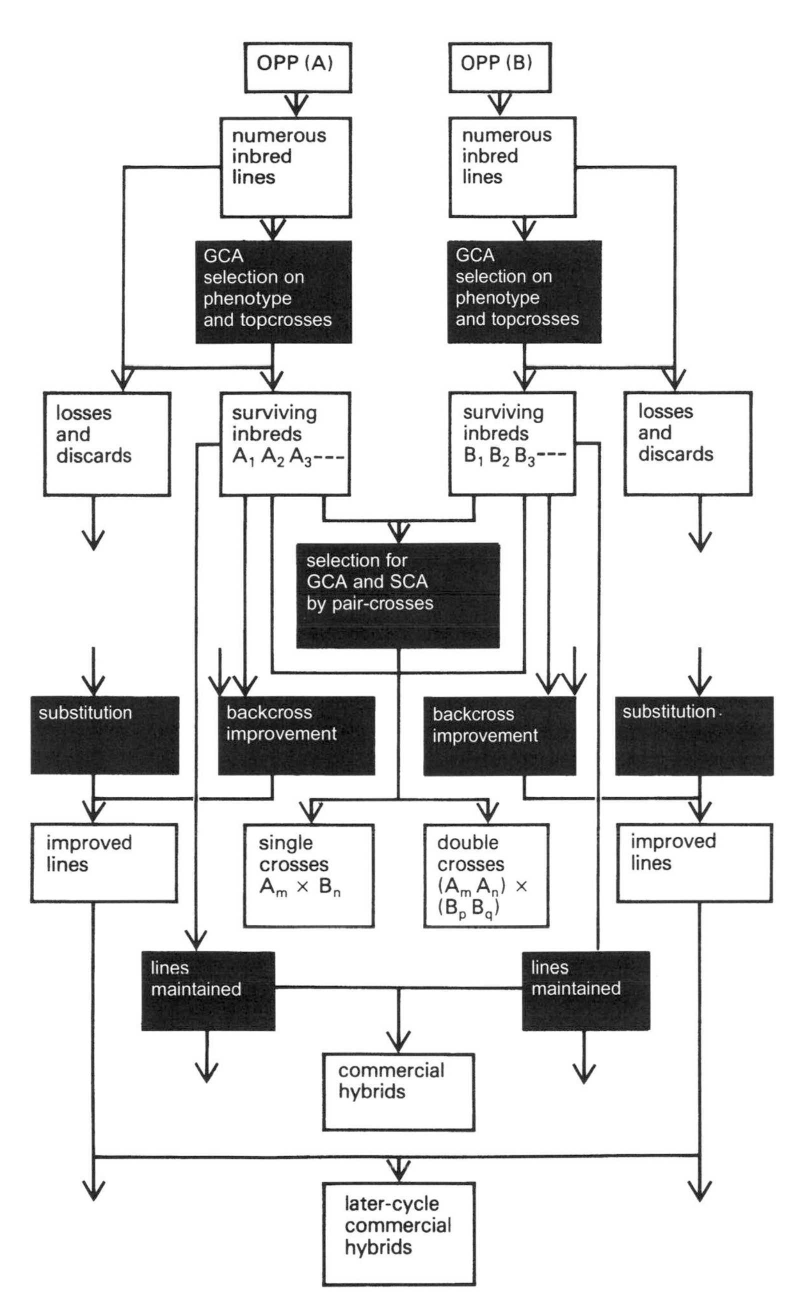

Fig. 5.5 Basic HYB maize breeding practice.
Notes: The separation into two complementary streams of distinct origin is fundamental in principle but maybe not very closely observed in practice in long-established programmes; it is very clear however in the Kenya programme (Table 5.4). The principles of GCA-first-SCA-later and of parental improvement by backcrossing and substitution emerge clearly. See also Fig. 5.6 for a strategic view of a HYB programme and Fig. 5.13 for the joint exploitation of HYB and PIM methods.

complementary inbreds from unrelated sources is sound in practice though often qualified, especially in later generations of inbred improvement when pedigrees tend to become rather entangled.

We saw above that good inbreds are rare, generally yielding less than half as much as the source populations. The wastage is accordingly great, partly because many lines simply die out and partly because many others must be discarded untested for intrinsic weaknesses. Kiesselbach estimated in the late 1940s that, up to then, some 100 000 inbreds had been isolated of which about sixty were in actual use; recently, some 70 per cent of the US maize planting was based on no more than half a dozen inbreds. Even the truly outstanding survivors yield, on average, a little less than 50 per cent of their single-crosses (Table 5.5). The practical question therefore is: how are large, even huge, numbers of inbreds to be reduced efficiently to the few that are to survive as parents?

A very large body of work on this question, spread over four or five decades, may be summarized as follows. First, phenotypic selection among inbreds is effective for characters of relatively high heritability (high GCA) and even moderately effective for yield; thus the yield correlation over inbreds between performance in selfs and crosses is usually found to be small (about 0.2) but positive and large enough to be useful. The phenotypically worst can fairly safely be discarded and regularly are. Second, primary selection among the survivors can be securely based on average combining ability as estimated in top-crosses. A 'top-cross' is a cross to a generalized tester stock appropriate to the objective. Such testers will always be at least fairly widely based genetically; they may be OPP, synthetics or single-or double-cross hybrids according to circumstances. For example, a breeder developing inbreds under A in Fig. 5.5 would probably use OPP (B) as top-cross tester. Similarly, a breeder seeking a specific replacement in a double-cross would probably use the relevant complementary single-cross as the tester. Both steps described so far, phenotypic selection and top-cross testing, depend essentially upon GCA and they work only because there is, in fact, a substantial GCA component of cross performance. The third step is identification of the outstanding combinations, those depending upon SCA on top of GCA; this, the difference between good and excellent, is what makes HYB varieties economically feasible. Here, there is no alternative to pair-wise testing, either in diallel form or, if the breeding population is differentiated into two main streams as in Fig. 5.5, in an m × n type pattern. If double-crosses are the object, it is well established that their yields can be satisfactorily predicted from the means of the four single-crosses that will not actually be used in making the double, thus: if inbreds A and B are related and also P and Q, the double-cross (A × B) × (P × Q) can be predicted from A × P, A × Q, B × P, B × Q. This also illustrates the order-of-pairing rule: the last cross must be the widest (for obvious reasons of maximizing heterozygosity and heterosis).

So the development of HYB maize goes in three phases: phenotypic selection among inbreds, top-cross testing for GCA, pair-wise testing for SCA. It is generally agreed that all yield tests should be replicated over sites and seasons, as well as within trials, in order to minimize the ever-present GE effects. The effort required, though much reduced, remains formidable.

To get some idea of the numbers involved consider the following. Among n lines there are n(n–1)/2 possible crosses, direction disregarded; there are n!/4!(n–4)! possible double-crosses but three times as many if order-of-pairing is not predetermined. If the inbred population is divided into two sub-populations of separate origin, each of size n/2, there are $n^2/4$ crosses within origins and n(n–2)/4 between (sum = n(n–1)/2); between-origin double-crosses made from within-origin singles will number$[n(n-2)/4]^2$. The reader should try putting n = 100 into these expressions when he will quickly appreciate the practical importance of the short-cut selection methods founded on the GCA-first-SCA-later principle. He should then try assuming that n is successively reduced to fifty by phenotypic selection and to ten by top-cross testing. How many crosses and how many yield tests would be required, assuming five sites and two seasons for each, in order to arrive at a calculation of the probable best double-cross? Note that, at fixed n, some economy will accrue from differentiation into two streams of distinct origin; why and how much?

A continuing HYB programme advances by substitution of new inbred parents that are better than their predecessors. These may be developed by: (1) backcross procedures which leave the inbred but little changed (example in Table 5.5 and see also Section 5.7); (2) by fresh isolation of new inbreds from the same genetic source (OPP, hybrids, synthetics, etc.); this is the second-cycle improvement towards which recurrent selection procedures were essentially directed; (3) isolation of new inbreds from wholly new sources (which will generally be heterotic in relation to the original sources. These methods are illustrated in Fig. 5.6 which brings out the cyclical nature of long-term HYB programme.

A method of inbred improvement which is, in effect, a special version of (1) above is called ‘convergent improvement’. A single-cross hybrid is backcrossed as donor parent to its two inbreds, with selection, generation by generation, for field characters in the progeny. The result should be improvement of both inbreds and hybrid. The method has not been much used.

‘Gamete selection’, which has special bearing on item (3) above, is intrinsically attractive but has yet, seemingly, to realize its full potential, perhaps because HYB maize breeders have not yet felt the need greatly to expand the genetic base. The technique depends upon the simple fact that if, in an OPP, superior gametes have a frequency p, superior zygotes will have a frequency p^2 ($\ll p$). An inbred is crossed to the OPP that is to be sampled, F_1 plants are selfed and outcrossed to an appropriate tester and the testcross results are used to indicate which selfed lines should be continued. The procedure resembles the half-sib (4) method of PIM in Table 5.3. If p were, say, about 0.1 or less the method, though laborious, could have considerable attractions over the large-

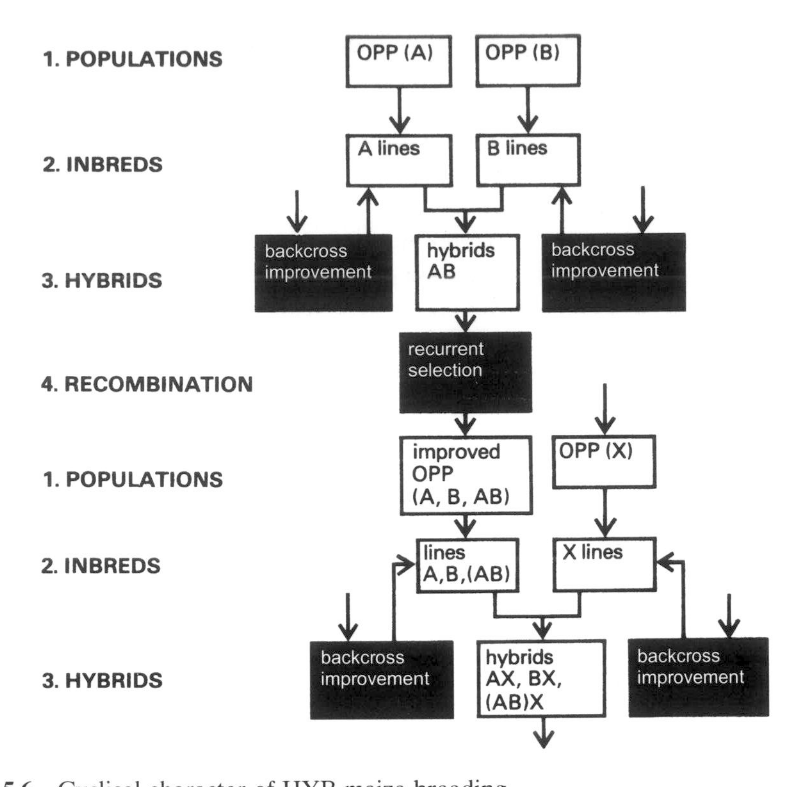

Fig. 5.6 Cyclical character of HYB maize breeding.
Notes: The diagram shows that, as one cycle of improvement approaches limits, so renewed recombination and supplementation of the genetic base (OPP(X)) by exotic materials becomes necessary. The new inbreds (X) will be genetically complementary to (i.e. heterotic with) the old and will be developed on the same GCA-SCA principles.

scale isolation of primary inbreds. Mangelsdorf (1974) argues cogently for its use as a basic approach to systematic broadening of the genetic base of the crop.

A more direct use of the same principle has been developed by Chase (in Gowen, 1952). By using a pollinator carrying a suitable dominant seedling marker crossed on to the stock to be sampled, rare maternal progeny of parthenogenetic origin can readily be recognized. Their frequency varies widely around an average of about 10^{-3}. These progeny are usually sterile maternal haploids (= monoploids) (Section 8.4) which, however, yield fertile, doubled sectors in fairly low frequency (variable but normally about one in ten). Lines so isolated start as perfect homozygotes so this is a means both of direct gamete-sampling and of producing 'instant inbreds'. Test-crossing shows that such lines are equivalent in combining ability to a random sample of conventional inbreds. The method is elegant and fast but not universally applicable. Several hundred lines have been isolated thus from several thousand primary haploids and a few have become parents of commercial hybrids.

Paternal haploids are also known from suitably marked stocks but are, at best, rare (on the order of 10^{-5}). They provide a neat means of converting a fertile inbred into a male-sterile one by transferring its nucleus into a CMS cytoplasm.

HYB breeding in other crops

In no other crop is HYB breeding so well developed as it is in maize. The reasons for this are complex, and include the following: the historical accident that ideas on HYB developed at just the time at which progress by other methods seemed poor; the great economic weight of the crop in the USA and the research and development drive, both public and private, that is generated; and, perhaps most important of all, the happy biological accident that a maize plant can be emasculated by a single tug of the hand. If maize had not been diclinous and easily detasselled, HYB would have been delayed at least twenty years and perhaps much longer; as we shall see later (Sections 6.4, 7.3), HYB maize in the late 1960s depended upon a single cytoplasm for seed production.

Hybrid sorghum, with seed production based also upon a CMS system, displaced IBL in the USA in the period 1957–61 and has since been used fairly widely elsewhere, notably in South Africa. The crop is essentially inbreeding but (like cotton) has a fairly high rate of random outcrossing. One result of this is that single-cross hybrids were adopted from the start because inbreds are good seed producers and there is therefore no need for double-crosses.

Other crops in which HYB varieties have had some local impact include: onions in the USA, sugar beet in the USA, some *Brassica* crops (notably Brussels sprouts and a kale) in Europe and Japan, glasshouse tomatoes in Europe, castor in the USA and cottons in India. We will anticipate Section 6.4 by noting here that these crops illustrate almost the complete range of possible HYB seed production methods: onions and beets use CMS systems, *Brassica* crops an incompatibility system, castor a nuclear-gene system and HYB tomatoes and cottons depend upon hand-emasculation of female lines. All methods are expensive and several are, at least intermittently, technically troublesome and unreliable. The reasons for the, at least partial, success of HYB varieties in these crops are various and include, probably: (1) the economic interests of breeders in the absence of varieties' rights legislation; and (2) field uniformity (especially in the vegetables). Dramatic yield advantage for HYB varieties in these crops is not evident. Thus European beet breeders seem to be essentially satisfied with the yield of SYN sugar beets and the great California tomato trade stays with IBL.

Among other annual crops there is hardly one for which HYB varieties have not been at least mentioned as a possibility and, on several, a great deal of work has been done. This statement is true of inbreeders and outbreeders alike. Though we have tended to think of HYB procedures as being specific to outbreeders, one notes the very successful exception of sorghum. The traditional genetical argument is that outbreeders are adapted to high levels of heterozygosity, with much pseudo-overdominance, and that both performance and uniformity are promoted by the use of HYB breeding schemes. Inbreeders, by contrast, display an essentially homozygous balance, do not suffer inbreeding depression and are capable of excellent performance, as well as, of course, perfect uniformity, in the homozygous state. There are, however, many situations in which it is simply not true to say that inbreeders suffer no inbreeding depression. The by-now classic data of Immer (1941) illustrate the point. The average yields of six barley crosses (all promising ones) over five generations were (per cent): parents 100, F_1 127, F_2 124, F_3 113, F_4 105. There must be, by now, in the literature, hundreds, perhaps thousands, of observations to the effect that the F_1 in an inbreeder often outyields its parents and that later unselected generations decline. There has probably been a tendency to overestimate the magnitude of the effect (see below) but it nevertheless seems certain that responses that correspond with inbreeding depression and heterosis in outbreeders occur also and quite commonly at that, in inbreeders. But it does not follow from this that HYB varieties of inbreeders are necessary to achieve the highest levels of performance. In the absence of true overdominance or of pseudo-overdominance *plus* tight linkage, IBL should be just as effective as HYB in promoting yield. Uniformity, of course, is not at issue, so any *biological* argument for the use of HYB in inbreeders must rest upon yield considerations. Thus we return to the unresolved question (Section 5.3) of the efficiency of selection among IBL. Does random fixation of unfavourable alleles dominate the outcome of a cross? If so, how could IBL breeding be made more efficient? Or is it perhaps efficient when viewed over cycles of crosses even though ineffective within a single cross?

If HYB varieties of inbred crops were to be biologically justified, therefore, it would have to be on grounds of yield heterosis unfixable in homozygotes, not on grounds of uniformity. Opinions, naturally, differ on the subject. Committed 'hybridists' of which there are many, would argue that yield heterosis is there for the exploiting, that HYB varieties are economically attractive to breeders, that the technical problems, especially of seed production, are challenging and that the biological question is irrelevant. Sceptics, of whom there are perhaps fewer, would argue that IBL breeding techniques ought to be susceptible of improvement, that there is no good evidence of unfixable heterosis, that the economic attractions of HYB for breeders are counterbalanced by high cost of seed to farmers, that technical skills could find more worthy objectives and that the biological question is all-important. And they would probably add that even the yield heterosis so often recorded may be substantially over-estimated as a result of bias due to space-planting of scarce materials and the unavoidable use of large F_1 hybrid grains borne on poorly set ears. This argument certainly has some force in respect of wheat and barley but presumably none in other crops (e.g. cotton) in which hand-crossing is much easier and more productive.

Perhaps we should say that our own view is one of qualified scepticism. It may be that HYB wheats, barleys or cottons – the crops which have had most attention – are not biologically necessary in the sense of being intrinsically superior to the best IBL but would nevertheless be agriculturally desirable from the point of view of flexibility and speed of variety production; after all, n IBL represent n(n–1)/2 potential HYB. So the biological question remains a good one which certainly deserves an answer but the practical outcome, regardless of that answer, is likely to be HYB if good hybrid seed can be produced cheaply enough and non-HYB if it can not. We shall see later (Section 6.4) that seed production is the key limiting factor: in wheat and cotton the reliability of the CMS system and of pollination are in doubt while, in barley, lacking a workable CMS system, recourse is had to cytogenetic techniques of great elegance but doubtful efficacy.

Doubts and questions

We saw above that a HYB variety can be regarded as a rather specialized kind of SYN, with few inbred parents rather than several outbred ones and controlled rather than random crossing at the seed production stage. The similarity becomes even more apparent when we recall that double-crosses are not uniform and that HYB maize, so called, may even sometimes take the form of population-crosses (e.g. Hybrid 611 in Kenya – Table 5.4); also that proposals have often been made for 'quick' hybrids based on minimal inbreeding. By contrast, the narrowly based two-clone cacao SYN mentioned above (Section 5.4) could be classified as a kind of HYB.

The realization that SYN and HYB overlap and that refined modern methods of population improvement are very powerful indeed (Section 5.4, Table 5.4) prompt the question: are HYB varieties really necessary, even in outbreeders? In trying to answer this question, the undoubted practical success of HYB maize is irrelevant; a huge effort has gone into it over a period of nearly eighty years; population improvement is recent in origin and small in scale by comparison, yet it is evidently capable of rates of advance which are at least comparable and may well be achieved more cheaply. Darrah, Eberhart and Penny (1972) found no evidence of decline of genetic variance in their experiments in Kenya (Table 5.4) so continued improvement is in prospect.

A priori, it seems likely that excellent HYB would usually be the best performer because of the efficiency with which it exploits SCA but that good OPP might be but little or not inferior, especially if their inevitable genetic heterogeneity were to generate favourable phenotypic interactions in the population: this last point seems to be potentially important (Section 4.8) but we are essentially ignorant about its relevance in outbreeders. At all events, it is clear that, whatever the situation at the limit, HYB methods should be regarded as complementary to OPP/PIM rather than competitive; they offer the means of using SCA in population in which GCA has already been effectively exploited by PIM methods. Sprague (in Frey, 1966) has convincingly argued thus and most new maize programmes adopt the principle.

This conclusion leaves open the question of whether or not we ever really need HYB varieties for the highest performance. If great uniformity is required, the answer must clearly be yes; if it is not, the answer is probably yes, but that empirical experience alone can tell. In practice, we may never get a good answer because the economic stimulus to adopt HYB breeding is great; if hybrid seed can be made cheaply enough, HYB varieties will be bred, whatever the best overall strategy might be. Our discussions of inbreeders and outbreeders therefore arrive at the same point; not the least of the many merits of HYB varieties is that their success prompts important questions about the efficiency of population improvement and the breeding of inbred pure lines.

5.6 Clones

General

The clonal crops are all, of course, perennials, though several (notably the tubers such as potato, cassava and sweet potato) are – indeed have to be – treated agriculturally as annuals and replanted at each crop cycle. At the other extreme are the long-lived tree crops such as rubber, mango and the rosaceous top fruits which come to bearing late and last for several decades. In between are the herbaceous or shrubby crops such as sugarcane, bananas, pineapples, strawberries and *Rubus* spp. which last for a few years, exceptionally for a decade, before they must be replanted. Whatever the life cycle chosen (which is clearly determined by a mixture of biological and agricultural factors) all clones are potentially immortal. Some, indeed, are known to have survived for at least two millenia (e.g. certain figs) and other crops (e.g. dates, grapes, citrus, sugarcane, sisal and bananas) could perhaps offer similar examples of great age if we knew enough about them. At the other extreme, clones of crops troubled by infectious, systemic viruses, notably the potatoes, have until recently been unlikely to survive for more than a decade or two, whatever their potential for longevity. All this implies that there is no such thing as 'clonal degeneration', a subject to which we shall return briefly below.

Methods of propagation are very various. Details do not concern us here but a few examples may be useful: buddings and graftings on various rootstocks (clonal or seedling) in the rosaceous top fruits, citrus, mango, avocado, grape and rubber; leafy cuttings in cacao, pineapple, sweet potato and strawberry; leafless stem cuttings in cassava and sugarcane; bulbils in sisal; lateral shoots in bananas; tubers in potatoes; and modern micropropagation methods.

The general approach to breeding of clonal crops will be apparent. The breeder generates segregating families of seedlings between which, both as seedlings and as vegetative descendants, he practises selection. Vegetative propagation ensures that the genetic span of a family is fixed and displayed at the outset; this is never true of the seed-propagated crops we have been considering above.

Breeding methods

In effect, all clonal crops are, by nature, perennial outbreeders (see Table 1.4). They are intolerant of inbreeding, individuals are highly heterozygous and outstandingly good ones may be presumed to represent favourable heterotic combinations. If maize were clonally-propagable, there would be no need for HYB varieties because the top fraction of an OPP, propagated as CLO, would be genetically equivalent to them. In a sense, therefore, CLO breeding is easy and quick because all the genetic variability is instantly fixed; both GCA and SCA can be as fully exploited as the size of the family and the efficiency of selection permit. The situation in a late generation of an IBL cross which has been bulk-propagated without selection or subjected to a single-seed-descent procedure would not be dissimilar to that found in the first year of a cross between heterozygous clones. In the IBL, as we saw above (Section 5.3), the breeder is faced with the task of isolating a few of the best (nearly) pure lines from among the many that confront him; similarly, the CLO breeder's task is also one of isolation from what is already there and fixed. The difference lies in the fact that the potential of the CLO family is determined at the outset whereas that of the IBL changes as inbreeding progresses, as unfixable heterotic effects are lost and as random negative fixation occurs.

Breeding of CLO therefore reduces, quite simply to the crossing of heterozygous clonal parents and selecting in the F_1 seedlings and in subsequent vegetative generations with the object of isolating one or a few best segregates. As selection proceeds, the number of surviving zygotes falls and, concomitantly, the numbers of clonal plants per surviving zygote rises. At the end, the breeder will have on hand a considerable clonal bulk of each of a few survivors of a family that may initially have been numbered in thousands.

The problems of selection are very much the same as those faced by an IBL breeder and the approach to them is the same: select as weakly as possible at the beginning and only for characters known to be highly heritable; intensify selection only when substantial quantities of individual clones are available to reduce the effects of V_E; rely for ultimate decisions upon trials replicated over sites and seasons. These are sound principles which, as most CLO breeders would, I think, agree, are not always closely observed in practice. The temptation to reduce numbers severely in the early stages is great, even though selection then must often be but little better than random. There is probably an optimum population size that will maximize genetic progress for a fixed effort but this has not, I think, been explored.

The problem of choosing parents or of choosing which families to exploit intensively is, in principle, easier than in IBL because, once again, genetic variability is fixed from the start. It would be easy to estimate family means and between-plant variances, even (though less reliably) V_G of family samples in an early clonal generation for whatever characters the breeder was interested in. The result would identify the families having the greatest potential – something which can not be done, with any confidence, as we saw, in an IBL. However, if such tests were done thoroughly enough to provide good estimates of variance, they would be laborious. In practice, such a procedure is rarely employed but the principle makes an appearance, usually in the form of growing first a small sample of a family and using a visual impression of it to decide whether or not to go on to a much larger sample for thorough exploitation. This procedure, in effect, bases decision upon family mean general worth but with some reference to the idea of variance, based on occurrence of seemingly outstanding individuals. This is admittedly subjective but it is quick, cheap and genetically sensible: the experienced potato or sugarcane breeder gets a good idea of family potential thus. However a more formal treatment is emerging (Simmonds, 1996).

Much the same arguments apply to choice of parents as to choice of families and, indeed, one is but an extension of the other. To the extent that there are GCA components of the various facets of performance (and there nearly always are), parents that are themselves at least reasonably good performers, if not actually excellent varieties, are the natural basis for all except special-purpose crosses. Empirical experience of family assessment will then indicate the relative importance of GCA and SCA. If the former predominates, the better the parents, the better the families and the breeder will know which few parents to exploit intensively. If SCA predominates, large numbers of small, exploratory test-families spread over many possible parents would be indicated.

A word of warning is perhaps necessary here. All breeders make some (often inexplicit) assumptions along these lines by the very act of choosing to exploit few or many parents. But, given a substantial body of family means, even if they are only subjective scores, it is remarkably hard simply to look at such data (usually in randomly incomplete diallel form) and make sense of them. This is a situation in which formal combining ability analysis, even of quite crude breeding data, can be very helpful. We saw above (Section 4.7) that such analysis in rubber revealed an unexpectedly high GCA component of yield and showed, further, that some of the most promising crosses had not in fact been made.

The essential features of CLO breeding are summarized in Fig. 5.7. Whether explicitly or inexplicitly, the breeder seeks to make high GCA combinations by selecting parents on phenotype and progeny performance but hopes for the unpredictable positive SCA which makes the difference between a good family and an excellent one; experience, with or without formal analysis, alone can tell whether it is better to exploit relatively few parents intensively or large numbers on an exploratory basis. As we shall see later (Section 5.9) there are hazards in restricting the numbers of parents too severely and Fig. 5.7 shows that even moderate SCA has to be reckoned with.

Finally, we must note the existence of two unorthodox approaches to the breeding of clones which have had little attention but probably deserve more. First, there is the idea of clonal mass selection in sugarcane developed by Brown and his colleagues (Brown, Daniels and Stevenson, 1971; Brown and Daniels,

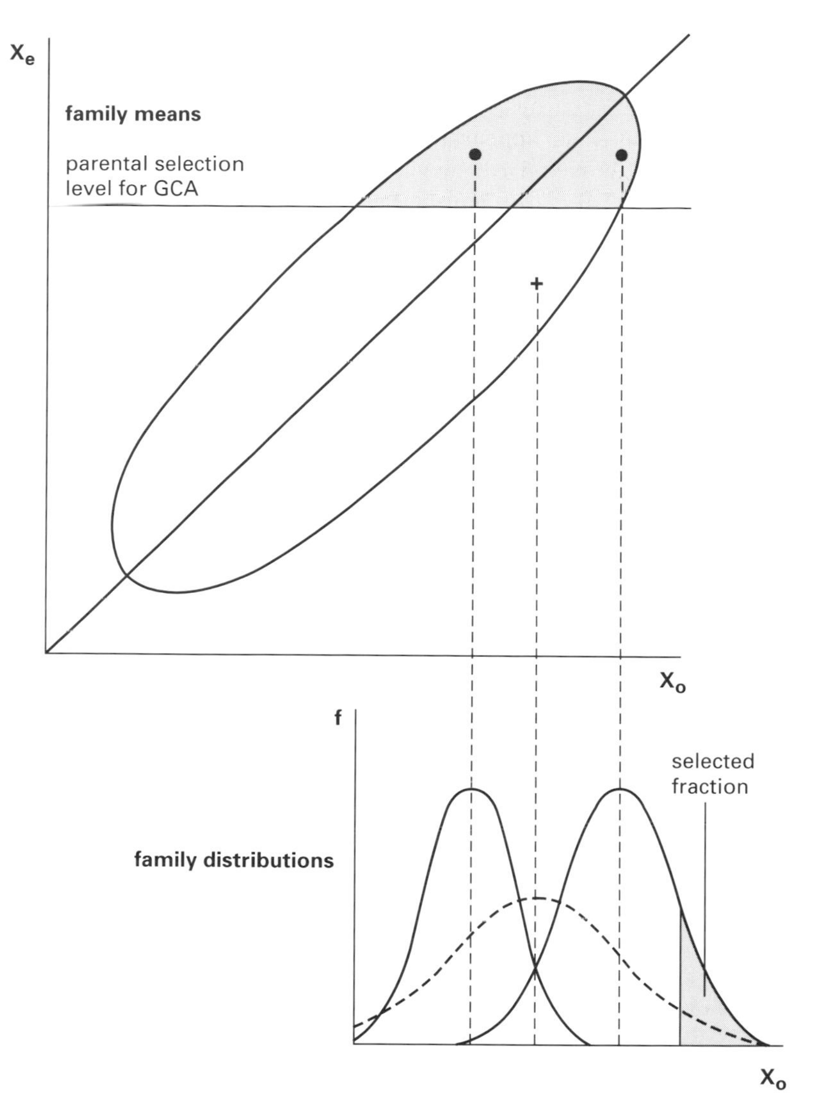

Fig. 5.7 Combining ability in CLO breeding.
Notes: X_e are family means expected on basis of parental GCA, X_o observed family means (compare the rubber example, Fig. 4.13). The slope, b, is unity as it must be from fitting GCA. Deviations from regression are a combination of SCA and error, the latter here assumed to be zero, for illustrative purposes. Note that – bottom part of diagram – the value of selected progeny will depend not only on parental GCA but also on SCA and variance within families (which is generally unpredictable).

1973). They set up a population of 534 clones, newly raised from seedlings of good parentage, and mass selected, over vegetative generations, for yield and sugar content. They use the phrase 'mass selection reservoir' (MSR) to describe the population. After three generations, sub-populations contained 30–50 clones each and exceeded the standard variety in sugar yield (as they had done all along). Though the experiment was laborious, the method looks as though it would be reasonably economical in practice and the population certainly responded. It is not yet established whether the good yields obtained are to be attributed to the presence of some outstanding clones, yet to be isolated as such, or to favourable phenotypic interactions (Section 4.8). At all events, there are interesting implications for clonal breeding techniques.

Second, there is the idea of practising some inbreeding in the development of clonal parents. No clonal crop could stand intense inbreeding, of course, but a moderate degree of inbreeding ($F = 0.5$, maybe) should sometimes be possible without incurring too severe losses or too much sterility. The most favourable result would be outstandingly heterotic F_1 families that exploited SCA effectively and had high means but (presumably) reduced genetic variances (Fig. 5.7). Proposals along these lines were made by Krantz and his colleagues years ago for potatoes but failed because of severe inbreeding depression. Spangelo *et al.* (1971) have recently made similar proposals for strawberries. There is no evident means of predicting the utility of the approach. Success would depend partly upon tolerance of partial inbreeding, the nature of exploitable CA and the effort that had to be put into preparatory breeding and testing. One suspects that the method might turn out to be elegant but uneconomic of time and effort.

Reproductive derangement

In annual crops produced for seed (for example, the cereals and pulses) there is intense selection for reproductive normality; the breeder rarely encounters any seed or pollen sterility and then only in out-of-the-way stocks. In annuals or biennials produced for a vegetative product (for example, sugar beet, forage brassicas, leafy vegetables) there is generally intense selection against flowering in the production crop – an aspect of partition (Section 3.2) – but good seed fertility must be maintained by the breeder to meet the demands of economical seed production. Here, too, any sterility is a rare feature strongly disfavoured by selection. And the same is true of perennials propagated clonally for seed, such as some cacao and coffees, or fruits highly dependent on good seed set.

In virtually all other clonal crops, a degree of reproductive derangement is present and sometimes so highly developed as to forbid normal sexual reproduction altogether. This feature has already been referred to above in relation to crop evolution (Section 1.6, Table 1.3) and we must now outline implications for the plant breeder. The following paragraphs can be no more than a sketch because a full treatment would be too long for the present purpose. Nearly every crop has its own peculiar complexities which can only be properly interpreted against the background of the history of that crop itself; furthermore, our knowledge, especially of some of the tropical clonal crops such as yams, aroids, sisal and bread-fruit is yet poor. Summaries, crop by crop, and references will be found in Simmonds (1976a) and Smartt and Simmonds (1995).

We may start by distinguishing two main groups of clonal crops: (1) those producing a vegetative product and (2) those producing a fruit. The former nearly always show reduced flowering developed, presumably, by a correlated response to selection for vegetative yield. The extremes are the yams, aroids and sweet potatoes in which many cultivars never flower and often cannot be induced to do so – so far as imperfect knowledge goes. Cassavas and potatoes show much reduced flowering in relation to their less-selected relatives; some of the latter never flower in the field but can be induced to do so in the glasshouse by horticultural tricks such as detubering or grafting on tomato rootstocks. In sugarcane, flowering depends on day-length, and therefore on latitude; it is unimportant at very low (0°) or high (20°–30°) latitudes but can be a major nuisance at middling latitudes (5°–15°); so breeders can ignore it in some places but must select against it in others and, wherever they are, have to find breeding plots in sites that favour flowering so that crosses can be made. Superimposed on this tendency to suppress flowering, there is also a good deal of sterility in these crops. This is probably sometimes also an aspect of correlated response to past selection but polyploidy (sometimes high), as in the sugarcanes, potatoes, sweet potatoes, yams and agave fibre crops, and complex interspecific hybridity, conspicuously in the sugarcanes, are also involved. Cytoplasmic male sterility certainly occurs in the potatoes, may be suspected in the sugarcanes and could be present in other crops; once present in a few good parents, it inevitably tends to increase in frequency in the breeding population. In both these crops, crosses are almost as often made on the basis of whether the parents are crossable as whether the combination is especially desired. Finally, rubber is an interesting case. Its flowering and fruiting are naturally weak and hand-pollination is a troublesome and expensive operation; the crop is only about three generations away from the wild state but already the first hints of clones even less seed-fertile than usual are beginning to emerge.

In summary, none of these clonal crops grown for a vegetative product is reproductively normal; flowering is suppressed and fertility reduced; polyploidy and hybridity contribute to sterility and generate uninterpretably erratic segregations; the breeder's task is never simple.

Turning now to the fruit-producing group of clonal crops, there is, of course, no question of suppression of flowering but reproductive peculiarities and sterility problems are hardly any less. Polyploidy occurs in some apples, pears *Rubus* fruits, bananas, strawberries and bread-fruit; hybridity, fairly well-defined in the *Rubus* fruits, strawberries and bananas, less well-defined elsewhere abounds. Some degree, at least, of pollen and seed sterility is usual. In general, it looks as though selection favours the vegetative aspect of fruit growth at the expense, as far as physiologically possible, of seed production. At the extreme the bananas are vegetatively parthenocarpic and many clones are so seed-and pollen-sterile that they simply cannot be bred from at all; the pineapple is also parthenocarpic but self-incompatible, so that a clonal planting is seedless even though fruit would be seedy if pollinated. Parthenocarpy or related behaviours (not yet thoroughly investigated) occur also in some grapes, citrus and bread-fruit; the fig shows a persistence and growth of unpollinated syconia which is not strictly parthenocarpy but is closely analogous to it. Many mangoes and citrus produce polyembryonic seedlings (in effect, clonal pseudo-seedlings identical with the mother plant) which can be a great nuisance to the breeder, who needs to select among sexual progeny. One aspect of reproductive irregularity of unknown importance concerns the breeding system: among the *Rubus* fruits, the raspberries have moved from self-incompatibility to compatibility at the diploid level while the blackberries are extraordinarily complicated by a mixture of polyploidy, hybridity and subsexuality; in the strawberries and grapes, a basic dioecism has been modified towards hermaphroditism, presumably as a response to selection for adequate fruit setting in isolated plants.

In short, therefore, all the clonal crops grown for a vegetative product or for a fruit (as distinct from seed) present flowering/fertility problems, often very acute ones. Often, potential parents cannot be used at all or, if they can, cannot be crossed in all desired combinations. Regular diploid inheritance can rarely be assumed and formal genetic interpretation, Mendelian or biometrical, is commonly impossible or, at least, in appropriate. The essential simplicity of clonal breeding schemes that we noted above is, therefore, overlain by formidable practical difficulties. Clonal breeding has had its share of outstanding successes (Table 3.1) but it is rarely simple in practice.

Clonal degeneration

'Clonal degeneration', with the implication of a mysterious biological decline inherent in vegetative propagation, has often been referred to but never substantiated. When properly investigated, it always turns out to be due to somatic mutations or disease, especially virus disease. Somatic mutation is, of course, of universal occurrence and some recurrent mutations can occur in frequencies high enough to be quite troublesome. Thus, in potatoes, 'bolter' mutants occur in some cultivars at frequencies around 10^{-3}. Since the plants are late in maturity and tend to bear small tubers, they can all too easily multiply in seed stocks. Though viruses are the principal disease problem in the maintenance of clonal stocks and were responsible for the 'running out' of potato varieties in the years before the causal agents were identified, other more or less cryptic diseases can be involved. Thus ratoon stunting disease (RSD) of sugarcane gives every appearance of 'clonal degeneration', was for years attributed to virus causation but has recently been shown to be caused by a small (and refractory) bacterium. It is transmitted in planting material but can be controlled by heat treatment.

So, in general, the risk of multiplying mutants or cryptically diseased material is ever-present and may occasionally be quite serious. This is essentially a problem for certification or plant-health authorities or for seed growers or for the farmers themselves rather than for the plant breeder. If a disease were the cause, the breeder might be able to develop a usable resistance (if this were a realistic objective); selection for low mutability is conceivable but unlikely.

There is also, of course, a more positive aspect to mutation, natural and induced, in clones, which we shall touch upon later (Section 8.5). Some mutants are desirable and the diversity of all old clonal crops has been enhanced by them. In the short term clones are fixed; in the long term, they are not. This, however, is peripheral to the point we are considering here. 'Clonal degeneration' can always be resolved into pathological and/or genetical components.

Progress in sugarcane breeding is presented in Fig. 5.8, perhaps a fitting conclusion to this section.

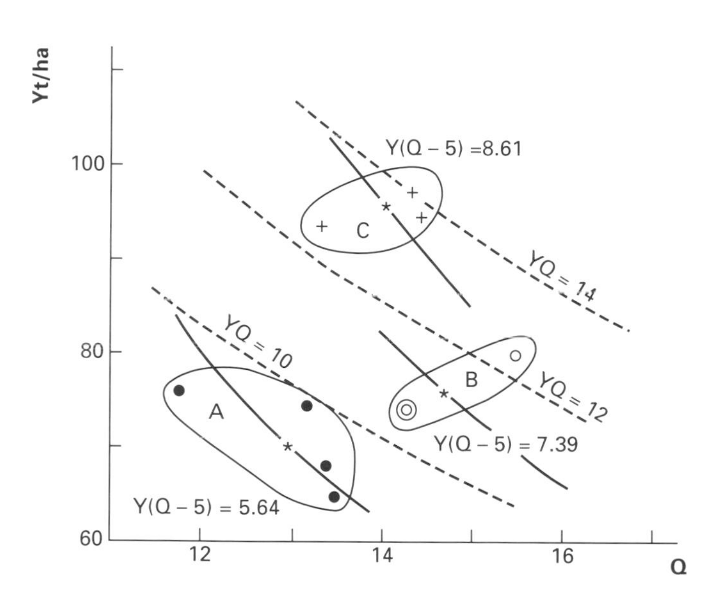

Fig. 5.8 Progress in sugar cane breeding.
Notes: The figure, data from Copersucar (1987, p. 231) summarizes three cycles of selection in sugarcane in Sao Paulo State, Brazil. Four old varieties (A) are compared with three more recent ones (B) and with the three latest (C), dominant at the time of writing. Data are from commercial productions from huge areas of 307 kha from 28 sugar mills in the year 1986–87. Y is cane yield in t/ha and Q is sucrose per cent cane, so that YQ is t sucrose/ha. The equations in Y (Q-5) are economic selection indices (cf. Fig. 10.6). Enormous gains, in both sugar yield and index, are apparent. Note that the two possible criteria of selection are different and that the economic index favours Q. The breeder is always selecting towards the top right hand of the diagram. Note that both Y and Q are high and that this is one of the most productive sugar industries in the world. This is one of the best available examples of the gains due to plant breeding because all data are measured, the samples are from agriculture (not trials) and the samples are huge.

5.7 Backcrossing

General

Backcrossing is sometimes treated as though it were a 'method' of plant breeding parallel with mass-selection or pedigree handling of IBL. It is better regarded as a useful adjunct to standard breeding plans, a special technique to be applied to special problems.

The principle is simple enough. If, having made a cross, the F_1 hybrid and subsequent generations are crossed recurrently to one parent, the other (the 'donor') parent's genetic contribution to progeny will be halved in each generation and will ultimately become vanishingly small. If, however, the donor parent contributes a selectable character, this may be maintained by selection in the face of backcrossing. The outcome is a stock that approximates the 'recurrent' parent in genetic constitution but carries one or a few (selectable) genes from the donor *plus* (unavoidably) one or more neighbouring blocks of chromosome.

The procedure is summarized in Fig. 5.9. The reader will note that the rate of reversion to the recurrent genotype follows the same rule as the increase of homozygosis under selfing: alternatively, that the donor parent's contribution declines as does heterozygosity, halving in each generation.

Applications

Backcrossing programmes may be inter-or intra-specific; they may transfer dominant or recessive genes; the recurrent parent may be constant or it may be varied; and the technique is equally applicable, though with variation in detail, to IBL, OPP or CLO. In practice, by far the commonest use has been the intra-specific transfer of dominant disease resistances, mostly to rusts and mildews, in the inbred cereals. Hundreds of such transfers must have been made, with varying ultimate success, depending upon the adaptability of the pathogen (Sections 7.2, 7.3). Inter-specific transfer of dominant resistances is sometimes equally simple and effective, as in the transfer of cotton blackarm resistance from *Gossypium barbadense* to *G. hirsutum*, where sterility barriers are slight. More distant combinations (as in the tobaccos and wheats) may present special problems arising from the persistence of segments of foreign chromosome or from pairing difficulties; they often demand special cytogenetic techniques to which we shall return later (Section 8.3).

Though simple dominant disease resistances in IBL have been by far the commonest object of backcross programmes, any character that shows some heterozygous penetrance can be handled, though maybe somewhat weakened in expression. Table 5.5 provides a nice example of what the author called 'earliness conversion' of maize inbreds. After crossing four excellent lines with the very early Gaspé flint, he backcrossed for ten generations with selection

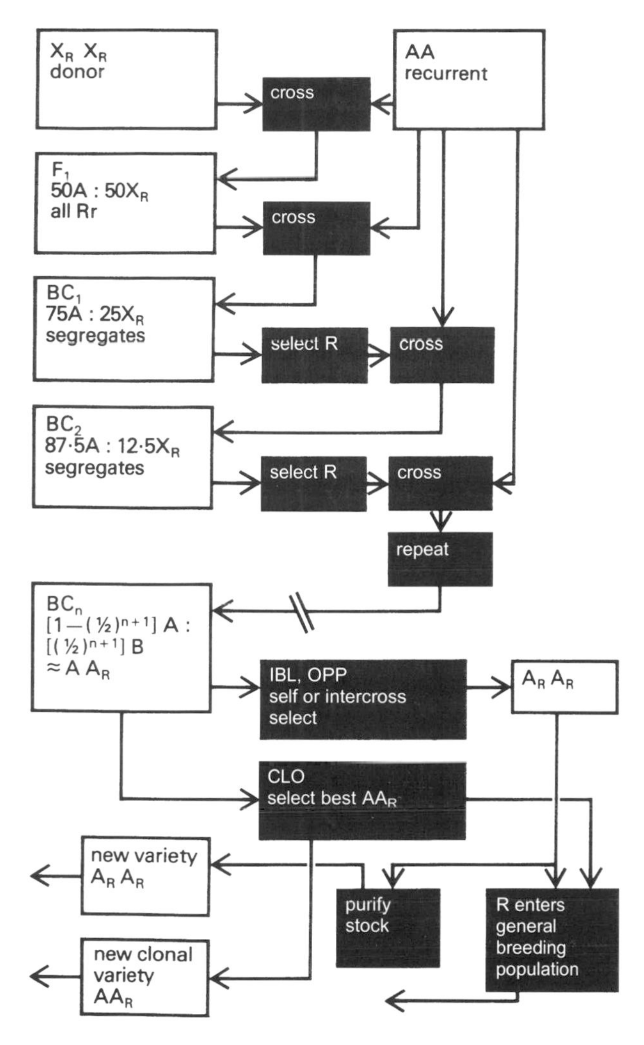

Fig. 5.9 The backcross technique as applied to the transfer of a dominant gene. Notes: The donor parent is XX, carrying the dominant R, the recurrent parent is AA. Constitution with respect to parent is indicated in the boxes on the left. If the parents are IBL, homozygosity is approached at the same rate as under selfing; if the parents are OPP or CLO, heterozygosity will remain high throughout but different CLO will have to be used over generations to minimize inbreeding. At the end, progeny tests will be necessary to isolate RR in seed-propagated crops but CLO may be left heterozygous.

only for earliness. The earliness genes were dominant enough that selection was effective and the resultant single crosses were a few days earlier and somewhat shorter in stature, but with a (variable) yield penalty. In general, maize inbreds can be handled just as though they were IBL and backcross improvement has been a frequent feature of HYB maize programmes.

In its simplest form, a backcross programme in an IBL goes to effective homozygosity at about the sixth or seventh generation, using a constant IBL as recurrent parent. Many cereals cultivars, especially in North America, have been through a whole series of improved versions by this means. Sometimes, however, backcrossing may be stopped early and stocks selfed and finished off by pedigree methods, according to the judgement of the breeder. Sometimes, again, an early selfed generation is interjected, with intense selection for the plant type of the recurrent parent. Allard (1960) gives an interesting description of the breeding of Baart 38 wheat by a complex backcrossing programme that introduced two resistances (to rust and bunt) more or less concurrently.

In clonal crops when a dominant gene is being transferred, there will be, to avoid inbreeding, not one but several different recurrent parents and the product can be left heterozygous for the gene transferred. R-genes, giving specific resistances to blight in potatoes, provide a good example. This was actually an inter-specific transfer (though an easy one) from *Solanum demissum* to *S. tuberosum*. It worked genetically but was a disastrous failure.

Whatever the nature of the population (IBL, OPP, HYB or CLO) and whatever the ultimate source, the transferred gene or genes, enter general breeding currency and may or may not find regular use in normal breeding programmes. Since many disease resistances fail (Chapter 7) our inbred cereals and potatoes often carry unsuspected resistance genes, relics of past backcrossings.

Recessive genes can, in principle, be transferred by interjecting a cycle of inbreeding and selection between each backcross, or by test-crossing or by making each backcross on a scale sufficient to ensure (statistically) that the gene is retained. The first is slow and all are laborious. In practice, backcrossing is used for dominants or at least for characters with adequate heterozygous expression.

5.8 Selection

Characters

The plant breeder selects for individual characters, sometimes singly, sometimes two or more at a time. But his objective is the improvement of the aggregate of individual characters that constitute 'general worth'. The questions that arise, and which we shall explore in this section, clearly are: (1) what sorts of and how many characters does the plant breeder need to cope with; and (2) how can he, in practice, reconcile the sometimes conflicting demands of selection for the several or many characters that contribute to general worth?

Characters are always numerous, they vary in importance and they vary also in ease of classification. The examples in Table 5.6 show that a breeder is likely to have to deal with between 10 and 30 characters. Some are easily and reliably scored visually (potato tuber shape and colour, stature in barley and kale,

Table 5.6 Characters for selection in six crops, arbitrarily chosen to include examples of: temperate and tropical, seed-propagated and clonal, and food, feed and industrial products.

Crop	Breeding characters
POTATO	**Field:** strong sprouting, early growth and ground cover; correct maturity; resistances to late blight, to viruses (Y, X, leaf roll, others), to wart disease, to eelworms, to insects **Tubers:** good yield; shape, smoothness, regularity of size, colour; resistance to mechanical damage; lack of cracking and second growth; appropriate dormancy; resistances to tuber diseases (blight, scab, gangrene, skin-spot, others); good storage characters (delayed sprouting, disease resistance) **Quality:** flesh colour; flesh texture; lack of enzymic browning; specific gravity (dry matter content); lack of after-cooking blackening; texture on cooking; flavour; chip/crisp colour after frying; reducing sugar content; 'reconditioning' capacity after cool storage; low glycoalkaloids
BARLEY	**Field:** strong seedling growth; restricted tillering; erect leaves (perhaps); short stature; correct maturity; strong straw; strong neck and ears (non-shedding); non-sprouting; resistance to fungi (mildew, yellow rust, *Rhynchosporium*) **Grain:** high yield; non-splitting; appropriate dormancy (low for malting); appropriate N-content (low for malting, high for feeding); malting features (high 'extract'; diastase activity – see Fig. 3.4)
KALE	**Field:** strong seedling growth, quick cover, appropriate stature; resistance to clubroot and mildew; strict bienniality; winter-hardiness; high yield of leaf/stem (according to type); high dry matter content; low fibre; adequate seed production **Quality:** high digestibility and palatability; low toxic (S-containing) constituents
COTTON	**Field:** vigour; correct maturity in relation to seasonal cycle; resistances to insects (boll weevil, various bollworms; jassids, others); resistance to bacteria (black arm/angular leaf spot); resistance to fungi (*Verticillium* wilt); high yield of seed cotton (balance between boll number and boll size); aptitude for hand picking/mechanical harvest; high ginning out-turn (lint per cent of seed cotton) **Quality:** length, strength, fineness, uniformity, colour and lustre of lint (Fig. 3.5); maturity; lack of seed coat neps; gossypol content of seed (glandlessness)
RUBBER	**Field:** vigour (sustained under tapping); wind resistance (related to branching habit); aptitude for grafting (including crown budding); high latex yield (partition effect, related to large latex vessel system); latex vessel 'plugging' in relation to chemical stimulation **Quality:** latex colour; molecular weight of polymer (possible character for the future)
SUGARCANE	**Field:** quick, strong germination of planting pieces; early ground cover; drought/cold tolerance; resistance to diseases (smut, mosaic, leafscald, pink disease, red rot; sereh disease, others); erectness (non-lodging); high cane yield (balance between cane numbers, length, thickness); non-flowering habit; appropriate fibre content; lack of prickly hairs (for hand-harvest); aptitude for mechanical harvesting; good 'ripening'; tolerance of fire; quick and prolonged ratooning **Quality:** high sucrose content in juice; low impurities

flowering in sugarcane, latex colour in rubber); others require laboratory measurements such as organoleptic assay of cooking quality in potatoes, mechanical measurement of fibre quality in cotton and chemical estimation of malting quality in barley and of toxic constituents of kale; disease resistance will sometimes be visually estimable in the breeding plots but will often demand special glasshouse or field tests, with or without deliberate infection; yield can only be effectively estimated by trials repeated over sites and seasons.

Characters vary in importance. Of those listed in Table 5.6, some would be simply irrelevant to particular breeders at particular times and places. Thus the potato or barley breeder's objectives would determine whether or not he had to pay any attention to processing characters; winter hardiness in kale might be irrelevant if the local winters were always mild; the pests and diseases of a crop are always more or less localized and no breeder ever has to tackle all of them; the sugarcane breeder working at some latitudes need not bother about flowering but, elsewhere, would have to take it very seriously. Any working list of characters is therefore local and closely related to specific objectives.

When the list is determined, characters still vary in importance. Nearly always, yield comes first; at least, if a new variety is not a good yielder by the standards of the time, it must have remarkable merits in other respects if it is to succeed. The only good way of judging the importance of a character is by formal economic analysis (Section 11.3), which is very rarely done; however we shall see below that subjective judgements as to importance sometimes have a place.

In general, the values attached by the breeder to various levels of expression of a character are not linearly related to expression (Fig 5.10). Thus there must always be a level of yield below which a stock would be regarded as worthless, so curve A in the figure drops to zero well short of the origin, even though linear above. Disease resistance would commonly follow the pattern of curve B; extreme susceptibility would be as unacceptable as low yield but extreme resistance would offer little advantage over fairly high resistance and the exact shape of the curve would depend upon the local intensity of attack; this would apply to most of the diseases and pests mentioned in Table 5.6. Again (curves C and D), a middling rather than a high level of expression of a character is sometimes needed; thus maturity is often very narrowly defined by the local environment (as in cereals and cotton), sometimes more loosely defined (as in potatoes) but is sometimes irrelevant (as in kale and rubber). In cotton, selection for lint length is always towards a narrowly defined optimum for the character except in the Sea Islands in which an extreme expression is sought (Fig. 3.5 and curve E in Fig. 5.10). In sugarcane there is a local optimum fibre content. Finally, (curve F) threshold characters are of discontinuous, all-or-nothing nature; contents of toxic constituents (e.g. glycoalkaloids in potato tubers) often have to be treated thus, the threshold between worthless and acceptable being determined by decree based on nutritional considerations. The content of erucic acid in rape seed oil for human consumption provides

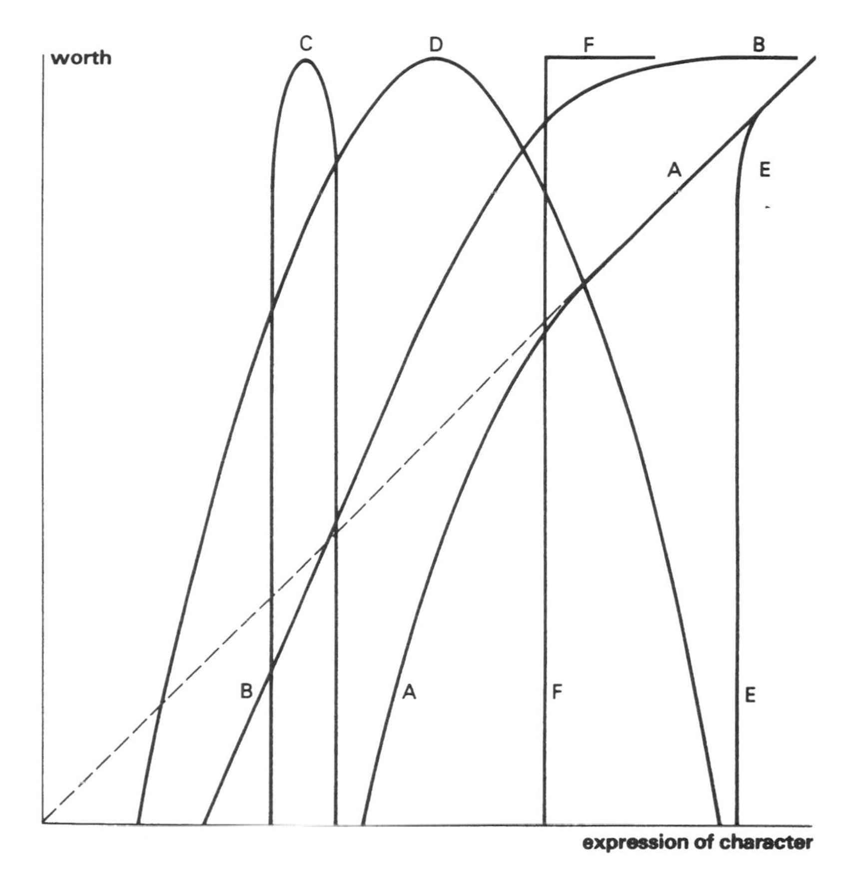

Fig. 5.10 Expression and worth of characters for selection.
Notes: The relation is rarely linear but may be nearly so over short ranges (A and B); more often there are intermediate optima (C and D) or threshold or near-threshold effects (E and F). For examples see text. See also Fig. 7.6 for further treatment in a disease-resistance context.

another example; the threshold has been, or is being, set by law in many countries at a level based on medical judgement.

In conclusion: the plant breeder's characters that contribute to 'general worth' are numerous, they vary in importance and their individual worths are non-linearly related to expression (though sometimes linear over short ranges).

Modes of selection

The animal breeders (e.g. Lush, 1945; Lerner, 1950) have long distinguished three main modes of selection: tandem, independent culling levels (equivalent to 'truncation' selection for characters independently) and index. In discussing these three basic modes of selection we will assume that expression and worth of a character are linearly related. So, in the light of the preceding discussion, some characters cannot be treated at all in these terms and others can be treated only over the range in which they are nearly linear or over a wider range but only after suitable transformation.

Tandem selection occurs when a population is improved serially over generations, first for one character, then for another and so on. It is agreed to be inefficient and has never been a deliberate feature of plant breeding programmes. Something rather like it, however, must take place whenever a character is so improved by several generations of breeding that it can be neglected or at least relegated to a low level of importance. An example is the adaptation of potatoes to long summer days in temperate countries (Table 3.1); ill-adaptation to long days must have been a dominant feature of the early years of potato breeding; the contemporary breeder now merely has to throw away a few per cent of extremely late segregates, or none at all. Similarly, whole populations of dwarf cereals may need no selection for this character.

Selection by independent culling levels is the selection of two or more characters simultaneously, at levels independently determined for each character (Fig. 5.11). This can also be called 'truncation' selection in respect of each character separately. If two characters are normally distributed and uncorrelated, the result is a more or less triangular sector cut out of the distribution; if three or more characters were considered, then the diagram would, of course, become three-or-more-dimensional.

Index selection selects simultaneously but not independently. In effect, the breeder writes an equation in the two characters, X_1 and X_2, draws the curve on the diagram and selects above the line. The nature of the equation embodies his assumptions. Simple addition of the two scores (straight line, slope 45°) implies that the characters are independent in economic effect and of equal importance; to multiply the score for one character by a constant (>1) implies that it is more important than the other and the result is a straight line with slope not equal to 45°. A multiplicative assumption produces an equation with the product X_1, X_2 in it and a curve rather than a straight line; examples might be provided by boll number × boll size as an approximation to yield in cotton or fresh yield × sucrose content for sugar yield in sugar beet or sugarcane. The result of index selection, whatever the equation chosen, will always be rather different from the result of using independent culling levels because segregates poorish in both characters are excluded and marked superiority in one is allowed to compensate for moderate inferiority in the other (Fig. 5.11).

Index equations may be constructed on a purely economic basis or on a genetical one or on both. If two or more characters were uncorrelated and of similar heritability, the economic equation would be appropriate. It would take the general form (Fig. 5.11):

$$b_1X_1 + b_2X_2 = C$$

which can clearly be rewritten

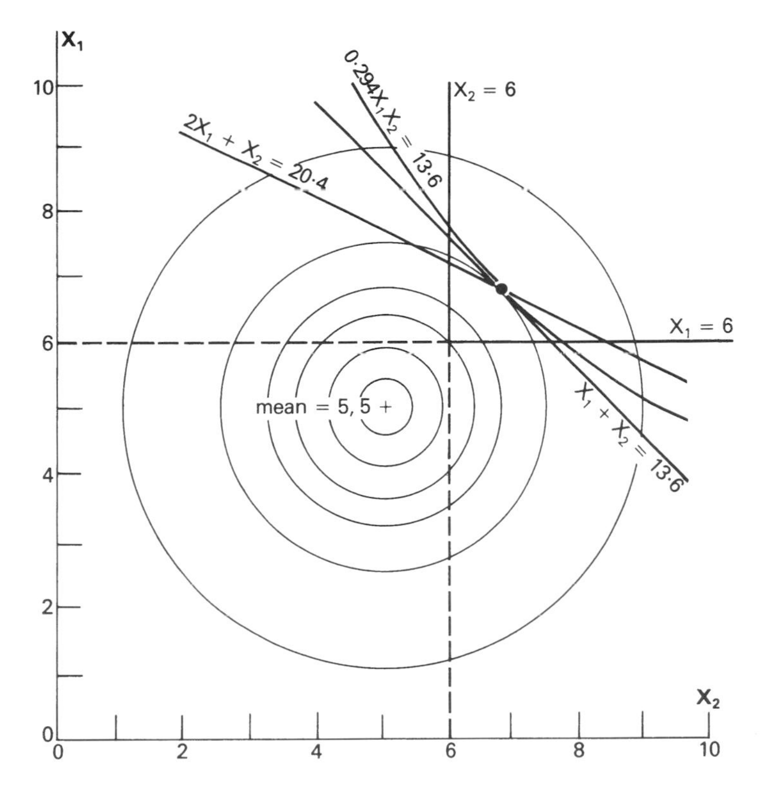

Fig. 5.11 Truncation selection ('independent culling levels') and index selection for two independent characters.
Notes: There are two independent characters, bivariate-normally distributed. Truncation selection (at $X_1 = X_2 = 6$) removes a more or less triangular sector different from the sector removed by any one of the three index selections illustrated. See text.

$$bX_1 + X_2 = I$$

or, if the scale of X_1 were adjusted,

$$X'_1 + X_2 = 1$$

the last operation being equivalent to multiplying the scale by b so that the index selection line takes on a slope of 45°. The original weights, b_1 and b_2, would come from economic studies and would obviously require periodic revision.

A genetic selection index is more complex. It starts from the assumption, either that the characters are equally important and need no economic weighting, or that the weights have been applied and the scales adjusted accordingly. Then the problem becomes one of calculating an equation that maximizes the correlation between the index value, I, and genetic merit. It turns out that the appropriate equation is a kind of partial regression that takes into account the heritabilities or and correlations between the characters. Details lie beyond our scope here; for a general treatment, the reader may refer to Lerner (1950) or Falconer (1989). Superficially, the equation takes the same form as the economic equations mentioned above, generally:

$$b_1X_1 + b_2X_2 + b_3X_3 \ldots = I$$

Since I is in arbitrary units, any equation in two terms can always be put in the simplified form:

$$bX_1 + X_2 = I$$

If $b > 1$, this would indicate that more genetic weight attaches to character 1 than to character 2, e.g. by reason of higher heritability. Diagrammatically, selection by a genetic index looks the same as by an economic index (Fig. 5.11); the breeder selects beyond a line drawn obliquely across a scatter-diagram.

We may now enquire what selection procedures are actually employed in plant breeding programmes. The answer, nearly always is: a blend of independent culling levels (i.e. truncation selection) and (inexplicit) index selection. This solution is perhaps as much imposed by circumstances as deliberately chosen. Recall that the plant breeder normally has to cope with many (10–20) characters, with large (even very large) numbers of plants and with acute restraints on time and resources. He is always pressed by circumstances to discard as much as he reasonably can as quickly as possible. So characters of reasonably high heritability that can be scored visually (or otherwise easily) are first selected early in the programme; if only one character is treated at a time, the procedure is essentially one of truncation; if several, the breeder might use independent culling levels but, much more probably, he would use an intuitive index, balancing good qualities against moderate defects. Later in the programme, when numbers have been greatly reduced, information about characters of low heritability, or ones that require tests or measurements rather than observation, will start to come in. The same blend of selection methods will obtain. Some characters can be cut off at a predetermined level; others will enter another intuitive index. Others again may be truncated at first but treated by index later: for example, lodging in cereals, blight resistance in potatoes, sugar content in sugarcane. In the last stages of a programme, when there is a good deal of information about a few survivors, the intuitive index takes over completely. Barring accidents (which do happen, however) there will be no characters so bad as to justify immediate discard on that character alone. Final judgements have to be taken on a balanced view of all the characters.

It might be asked why, if plant breeders in fact make extensive use of inexplicit, intuitive, indices, explicit indices are not developed and used instead. The answer is simply that they would usually be too laborious. In the early stages of a programme hundreds, perhaps thousands, of plants or lines have often to be dealt with in a few days; many of them will be manifestly useless; for the breeder to commit himself to thousands of measurements, a great deal of bookkeeping and heavy calculations for the sake of a probably modest gain in efficiency, but at the cost of other work forgone, would be unreasonable.

In practice, though genetic indices have been tried occasionally (e.g. in cotton by Manning, 1956; in soybean by Brim, Johnson and Cockerham, 1959) they have had no significant practical impact and are unlikely to do so. The situation in the later stages of a programme may be rather different. Here, the information about survivors is rather good, heritabilities are hardly relevant and a strictly economic index could replace a genetic one. This sounds attractive but really satisfactory economic indices are by no means easily arrived at; in practice, they have very rarely even been calculated, let alone used though we shall meet an example of one later (Section 11.3). To look to the future, it seems that economic indices ought to have a place at the variety selection level but that genetic indices are unlikely, for good operational reasons, to displace the truncation-plus-intuitive-index selection traditionally practised by the plant breeder.

Correlations

Correlations between characters are a frequent feature of plant breeding. They may arise from linkage or from developmental genetic interaction, with or without a purely phenotypic component. The developmental-genetic kind is by far the commonest. We shall consider first the immediate consequences for selection and then discuss some examples and probable causes, with reference to longer-term effects.

The essential points are summarized in Fig. 5.12, in which it should be noted that the directions of the scale correspond with the directions of selection, so

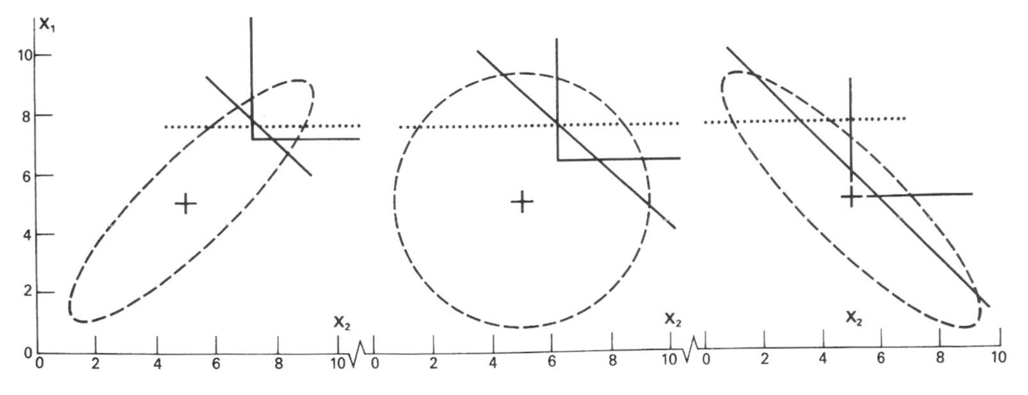

Fig. 5.12 Selection for correlated and uncorrelated characters.
Notes: Three cases are shown: uncorrelated (see Fig. 5.11) in the middle; and positively and negatively correlated, LHS and RHS respectively. The dotted line shows that selection for a single character (X_1) will always be more intense than selection for two or more characters. Positive correlation is favourable for intensity of selection, whether by truncation or index, negative correlation unfavourable.

that a positive correlation is favourable, a negative one unfavourable. The effect, at a predetermined overall intensity of selection, of a positive correlation, is to raise the levels of selection for both characters; a negative correlation does the reverse. Indeed, a close negative relation could confine the breeder to selecting very near the mean of both characters. Note that, if character 2 could be disregarded, character 1 could be more intensely selected, regardless of any correlation; the difference would be small for a positive correlation but large for a negative one. Multiple selection always reduces the intensity of selection for individual characters and negative correlations reduce it still further.

Well established examples of correlations due to linkages are few. They will tend to disappear as recombination goes on over generations and would not be expected, if negative, to present a prolonged obstacle to progress. Developmental genetic correlations are certainly far commoner and are sometimes simply interpretable in *a priori* terms.

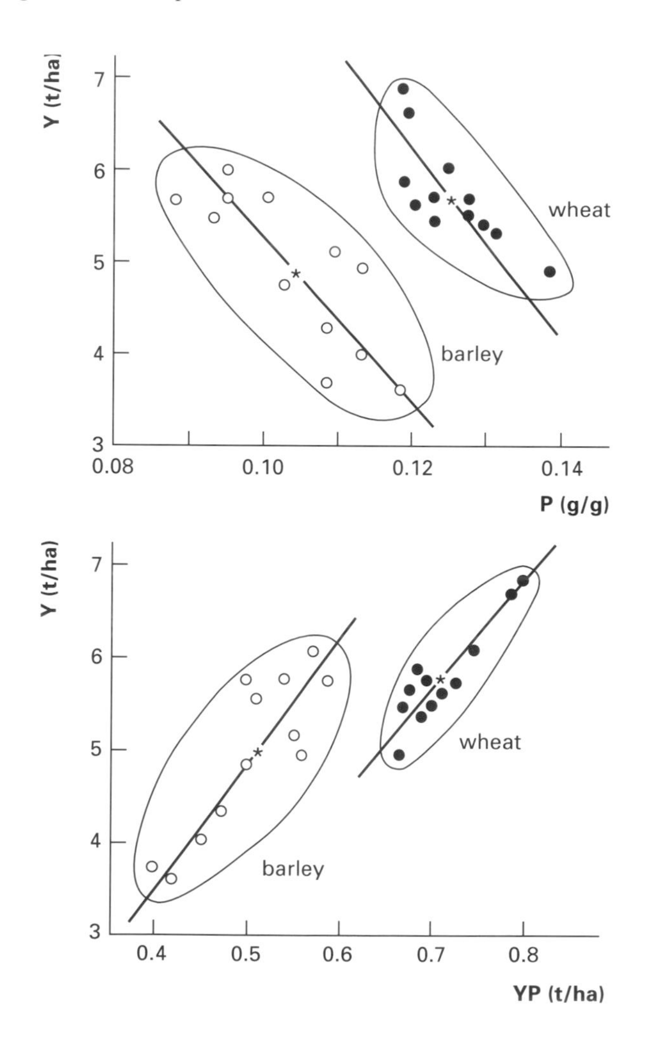

Thus early-maturing genotypes grow for a shorter time than late-maturing ones, often have fewer vegetative nodes and so are shorter in stature and lower yielding. Table 5.5 provides an example in maize; a five-to six-day gain in maturity reduced leaf number, stature, several components of yield and yield itself in lines and hybrids alike (though hybrids differed in the yield penalty incurred). It is, in effect, universal experience that early maturity is negatively correlated with yield, though not necessarily with stature. Similarly, fresh yield is negatively related to dry matter content in fodder crops such as kales and swedes; any array of well-adapted genotypes has much the same dry-matter producing potential (on a fixed time scale, of course).

These negative correlations are, in a sense, obvious and we should be very surprised if they did not occur; they merely imply that there are limits to yield per unit time. Some correlations are less obvious but hardly less regular. In the inbred cereals, it has often been observed that plant stature and yield are positively (though not strongly) correlated yet, as we saw in Section 3.2, reduction of stature and improved partition of assimilate towards grain yield has been a general feature of the recent breeding history of wheat, barley, rice and sorghum (examples in Table 3.1). This is an example of a genetic-developmental correlation that has been effectively broken by prolonged selection; nevertheless, *within* crosses the negative correlation persists and Gale and Law (1977) make the interesting suggestion that in selecting semi-dwarf wheats for yield the best procedure would be to select for tall stature within the appropriate semi-dwarf mutant class.

Fig. 5.13 Negative relations between grain yield and protein concentration in grain in two cereals. Data from Dubois and Fossati (1981) for wheat in Switzerland and from Weltzien and Fischbeck (1990) for barley in Syria. Also given, from the same data, are illustrations of the strong positive relations between gross grain and protein yields in cereals.

Notes: Regressions of the kinds shown are universal in cereals of all kinds, all over the world (Simmonds, 1995). No rational physiological interpretation is available, but the empirical facts are overwhelming (Simmonds, 1995). The regressions given here are in 'functional' form and do not have to be strictly linear; in fact, several more or less sensible curves (quadratic or hyperbolic) can also be fitted (but not interpreted). Note that the fits of Y to YP are not wholly satisfactory because both sides contain Y and a tendency to positive correlation is therefore constructed. In the literature, data are rarely given in a form in which Y on YP regressions can be calculated and the published information is very defective in both quantity and quality. The four lines fitted here are given by the following equations:

barley $Y = 14.3 - 90.3\,P$
wheat $Y = 17.0 - 89.6\,P$
barley $Y = -2.1 + 13.8\,YP$
wheat $Y = -2.7 + 11.8\,YP$

For the plant breeder, the implications are clear (and apply to all cereals): (1) high Y and high P are impossible; (2) breeders (see text) can aim at one or the other or compromise; (3) high Y could be encouraged by correlated response to selection for low P; (4) high YP, needed in some circumstances for good social reasons, can be encouraged by ignoring protein and simply aiming for high yield.

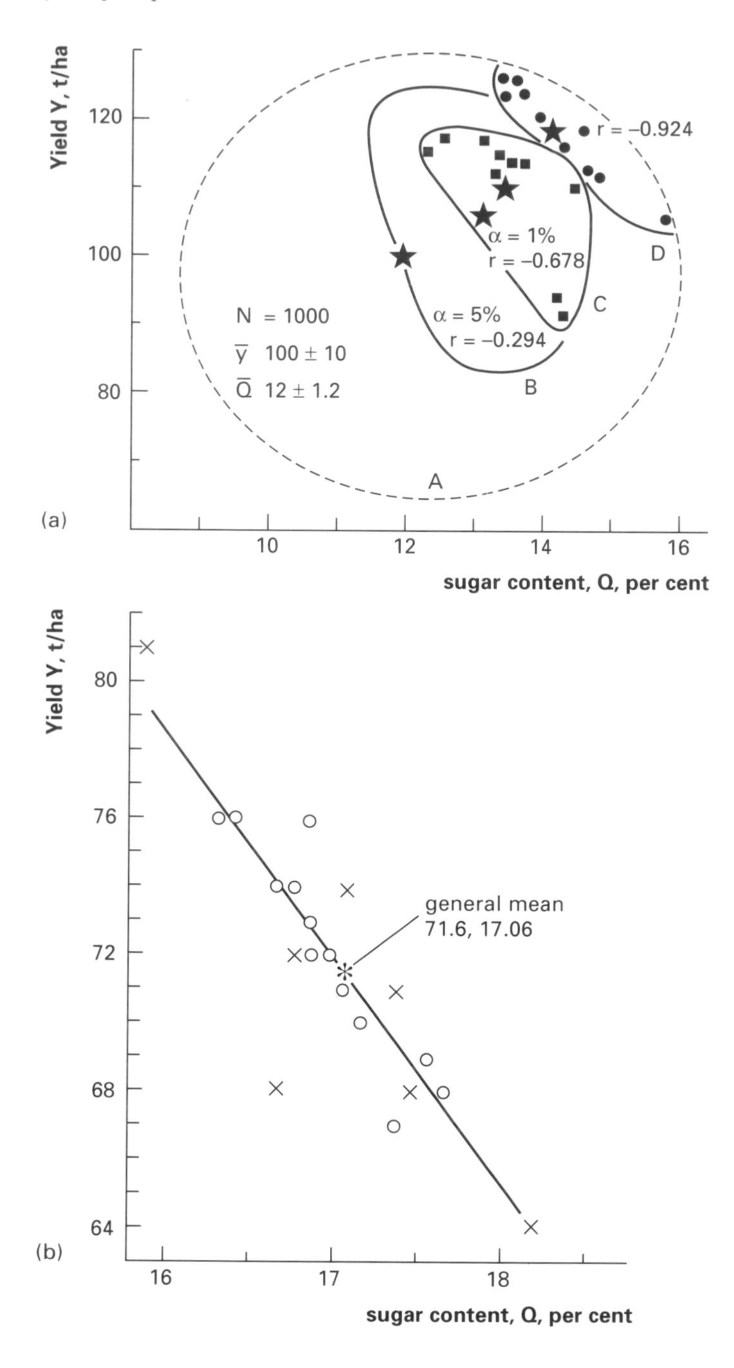

Yield Y, t/ha
120
100
80
r = –0.924
α = 1%
r = –0.678
D
C
α = 5%
r = –0.294
B
N = 1000
ȳ 100 ± 10
Q̄ 12 ± 1.2
A
(a)
10
12
14
16
sugar content, Q, per cent
Yield Y, t/ha
80
76
72
68
64
general mean
71.6, 17.06
(b)
16
17
18
sugar content, Q, per cent

In cereals and legumes there is well-night universal experience of a negative correlation between seed yield and protein content (Fig. 5.13). The genetic and environmental components have not often been well-distinguished but a general failure to make progress in both characters suggests that the former is dominant. The relation is best established in wheat and barley, with interesting consequences for plant breeding. In bread wheat, high protein content is essential for quality (Fig. 3.4), so high-quality wheats are low yielding and there is no sign that the plant breeder can make any more than a small change in one character at a fixed level of the other. European wheat breeders have generally bred for high yield at the expense of protein; breeders in the high-quality wheat areas of North America and Australia have accepted protein content, and other quality features, as dominant. The market uses blends of flour from various sources, according to its needs. The situation in barley is biologically similar but economically quite different. Here, low protein content is desired for malting (Fig. 3.3) so the breeder is under no restraint in selecting for yield; from his point of view, a negative correlation is turned into a positive one because one scale is reversed. The breeder of feeding barley, however, aiming at both yield and protein content would meet the same problem as the wheat breeder. In both crops, at all but exceptionally high levels of yields, the highest protein yield is always given by the highest yielding line; so, paradoxically, protein yield would be advanced by selecting for low protein content.

We have no understanding of the biological basis of this correlation beyond stating the obvious, that some kind of cost attaches to protein accumulation. The cost appears to be too high to be 'energetic', in any crude sense (Simmonds, 1981). To break the correlation would be a good achievement, socially and economically; meanwhile, it is simply a fact of life for the plant breeder, favourable or unfavourable according to his circumstances. The breeder must beware of false or spurious correlations – 'pseudo-correlation' as epitomised in the data presented in Fig. 5.14. They do exist but are statistical artefacts.

Fig. 5.14 'Pseudo-correlation' of yield and sugar content in sugar cane and sugar beet. From Simmonds (1994), by kind permission of the *International Sugar Journal*.
(a) shows 'pseudo-correlation' as a statistical artefact of bivariate selection for the product of uncorrelated components, Y (yield) and Q, sucrose content. (b) shows a strong negative regression of Y on Q in a German sugar beet trial of advanced varieties.
Notes: (a) is a computer simulation based on means and heritabilities plausible for cane: Y = 100 t/ha, Q = 12 per cent, CV 10 per cent for each, overall heritability about 30 per cent. Sample of 1000 reduced by successive selection to 50 and 10 survivors. 'Correlation' falls from zero to –0.294 (envelope B), to –0.678 (envelope and points C); D shows the result of 'best possible' response to selection (i.e. perfect heritability). Joint means are shown by stars. Response is in the expected direction but is far less than maximal. (b) shows a strong negative regression in sugar beet at the trials stage, reflecting very efficient selection in preceding populations having initial correlations between Y and Q erratically around zero. Not all sugar beet trials are so dramatic as this but good data are hard to find.

Numbers

If a plant breeder selects among N plants for a character at rate s he will keep Ns selections. If now he selects among the survivors for a second character at the same rate he will keep Ns^2 selections and so on, to Ns^m for m characters. In order to finish up with just one final selection, clearly $N = (1/s)^m$. Let us take m = 10 and 20, which are superficially reasonable figures (Table 5.6), and $s = 10^{-1}$ and 10^{-2} (weak and middling selection respectively). Then N in the four cases is 10^{10}, 10^{20}, 100^{10} and 100^{20}. The example is quite artificial in the sense that a strict succession of independent cullings at a fixed selection rate would never in fact be practised. However, plausible modifications that take into account the possibility that some index selection is included, that major genes may be selected at quite high rates or that there may be correlations between characters alter the arithmetic without altering the conclusion. Simultaneous selection for many characters, at even moderate rates, necessitates very large numbers of plants, numbers far larger indeed than most programmes could ever hope to handle.

This kind of calculation would seem to lead towards the rather discouraging conclusion that plant breeding is impossible. Fortunately, this is not so, for two reasons. First, in any particular population, some characters will be broadly satisfactory and will need either no selection or but little; this will tend to reduce m or s or both in the equation above and so also to reduce N. Second, selection levels are never, in practice, set arbitrarily at the start and maintained unaltered. The breeder will always be prepared (with some intuitive index in mind) to slacken selection on less important characters if, by doing so, he will still retain a selected population that lies in an agriculturally reasonable range; his aim is useful and usable varieties, not perfect ones.

The effect of the above rather discouraging arithmetic, therefore, is not to inhibit plant breeding; rather it is to enforce realization of the fact that perfection is not a realistic objective. There is an obvious and important extension of this conclusion, namely, that with fixed N, the fewer the characters the better the progress. Intense selection for a single character is likely to be very successful but will not produce the balanced phenotype characteristic of a good variety. Contrariwise, any addition to the list of characters which must be handled reduces the effective intensity of selection for the rest. Thus a new disease which must be selected at, say, 10 per cent will either reduce the number of emergent selections by a factor of 10 or demand an initial population 10 times as large (i.e. 10 N) to reach the same end-point. Thus realism in setting plant breeding objectives will be served by putting m and s as low as possible.

Efficiency

Consideration of the response-to-selection equation (Section 4.6) shows that a broad-sense heritability is a direct measure of efficiency of selection in that generation. For the all-important yield character, heritability, especially at the single plant level, is nearly always very low and selection correspondingly inefficient. This is true whatever kind of plant or breeding system we are considering, annual or perennial, inbred or outbred. Heritability will be somewhat, but perhaps not much, higher when unreplicated lines or clonal rows are being compared. So it is natural to enquire whether there are economical means of improving selection efficiency. In principle, this can be done by minimizing the effects of V_E in the denominator of the heritability. Three methods have been suggested. The first and, in practice, the commonest, is to intersperse numerous standard varieties in the plot and select, whether visually or on actual measurements, against them. The nearer the standard the more efficient the selection, so numerous standards are implied. The method is effective but expensive in space and labour (Briggs and Shebeski, 1969, 1970, 1971; Briggs *et al.*, 1969). The other two methods involve selecting against neighbours rather than standards, either on a grid or against a moving average of neighbours. Grid selection implies a random distribution of superior genotypes in the plot and has been quite widely used (e.g. in maize and cotton: Gardner, 1961; Verhalen *et al.*, 1975). The idea of a moving average as a standard is akin to that of a grid but is more recent and relatively little explored (Townley-Smith and Hurd, 1973); it implies extensive measurement and computation rather than visual selection.

There is no doubt that efficiency of selection can be improved by one or other of these methods but every programme will have to decide empirically whether the gain is worth the cost in space and labour. One should note that any comparison of unguarded plots or rows leaves genotypic differences confounded with competitive effects, so any improved selection could be as much for competitive ability as for the desired performance in pure stand; furthermore, diverse GE effects of site, season or disease could also affect the outcome. In the final outcome, good comparisons of genotypic potential are usually made in proper replicated trials with guarded plots (Section 6.2). However, Spoor and Simmonds (1993) have clearly demonstrated the potential value of carefully conducted pot trials in making such comparisons (Table 5.7 and Fig. 5.15).

5.9 Breeding strategy

Choice of parents

The parents used in a plant breeding programme generally fall into two categories: locally adapted varieties which are selected because they are locally adapted and expected therefore to contribute immediately to overall performance of progeny; and varieties chosen for a particular attribute without regard to local adaptation. The former are nearly always more numerous and more important in terms of overall progress. The latter, because they very generally lack local adaptation, typically enter a breeding programme by way of a backcross procedure or some variant of it. Disease resistances, earliness and

Table 5.7 Mean grain yields of barleys in pot trials; g dry grain per pot. From Spoor and Simmonds (1993). See also Fig. 5.15.

N(L)	O	6.2	4.6	8.1	6.30
	M	8.3	4.7	9.5	7.50
	sem	1.00	0.84	0.68	
N(H)	O	12.3	10.4	17.1	13.27
	M	16.1	11.7	19.4	15.73
	sem	2.03	0.68	1.07	–
	b(O)	0.88	0.91	0.95	–
	b(M)	1.12	1.09	1.05	–
General		0	9.78, M	11.62	
means:		N(L)	6.90, N(H)	14.50	

Notes: Three of a larger number of trials in which Old (O, land-race) cultivars were compared with Modern (M) dwarf cultivars at two levels of nitrogen fertilizing, low (NL) and high (NH). For experiments 1 and 2, interactions (O, M) * (NL, NH) were significant at the 5 per cent level, expt 3 marginal. In all three, the 'regressions' b(O) < 1 < b(M), as expected; modern cultivars were more 'responsive'. The 'regressions' are scaled mean differences between nitrogen treatments. Generally, N(H) > N(L), M > O and differences within M and O groups all significant, most of them highly so. But in every experiment, there was a substantial overlap between cultivars such that many O > M, that is, O had genetic potential for high yield if imaginatively used. The fourth column are means.

quality characters are often managed thus but can only enter the programme effectively after having been transferred into a locally well adapted stock.

Locally adapted stocks, therefore, whether introduced as such or deliberately developed to exploit specific characters, are the basic parental material. The underlying assumption, of course, is that phenotype and breeding value are correlated, which is another way of saying that we expect a substantial GCA component or heritability of general worth. The assumption is supported by a great deal of empirical experience; the best parents are nearly always either good varieties or near-varieties and excellent varieties nearly always leave successful progeny. However, it is universal experience that SCA effects are frequent; outstanding combinations can rarely be predicted purely on parental phenotype or even GCA and, contrariwise, good parents can produce poor progeny.

So a working strategy begins to emerge. The breeder will exploit as wide a range of phenotypically acceptable parents in as great a range of crosses as he can manage. He will be looking both for generally good parents (high GCA) and excellent specific combinations (GCA *plus* SCA); Table 5.4 provides a good example (local × Ecuador 573 maize in Kenya). As information accumulates, some parents will be promoted, others discarded or retired to a collection. New locally adapted varieties produced by other breeders (local and foreign) will be taken in as parents as early as possible. Promising stocks in the breeder's own programme (often materials which never quite achieve commercial status) will also be incorporated. Thus there will be a continual turnover, new parents entering from diverse sources as older parents go out.

This flux of parental material (summarized in Fig. 5.16) has very different time scales in different crops, depending both on life cycle and intensity of breeding activity and so on varietal turnover. Thus, there is nowadays intense cereal breeding activity throughout northern Europe, new varieties are very numerous and a parent is unlikely to last for as long as a decade. At the other extreme, parents in tree-crop breeding programmes are generally several to many decades old. (This, incidentally, is a subject which has never been properly explored; we have no quantitative knowledge of rates and patterns of parental turnover in various crops in relation to breeding histories.)

The sources of well adapted foreign stocks entering a breeding programme with the immediate prospect of being used as parents may be few or many. Generally, the best sources will, (predictably) be places elsewhere with homologous environments. Thus northern European wheats and barleys have much genetic history in common, are often rather widely adapted over the area and breeding programmes have many common parents. Similarly, the environments of the UK and New Zealand are similar enough that exchange of varieties and parents of cereals, grasses and brassicas is frequent and effective. By contrast, the more continental environment of North America is such that American wheats and barleys are little use in Europe, and *vice versa*. On the other hand, western North American barleys do well in South Australia and so do North African ones, in reflection of the Mediterranean-type climate of the three areas; but northern European barleys are not adapted there (Finlay and Wilkinson, 1963). We may generalize by saying that whenever a crop is widely spread over varied environments and long-enough established to have differentiated genetically, then homologous environments elsewhere will provide sources of parents of immediate potential utility. If, however, a crop is widely spread but relatively little differentiated, useful parents may come from almost anywhere; the tropical sugarcanes provide an example of a crop in which varieties have gone round the world in a truly remarkable fashion and Asian, American and Pacific canes occur in nearly every list of parents.

The physiological bases of adaptation are not well understood but this is no impediment to its exploitation because the breeder recognizes it well enough for practical purposes when he sees it. Maturity period, day length response, vernalization requirements, drought resistance and disease resistance are certainly all components. We shall see below that a deficiency in one aspect of adaptation can effectively disguise the breeding potential of a stock, so the traditional concentration upon locally adapted parents, reasonable and practical though it is, by no means exploits the full potential.

The genetic base

We have seen that all programmes concentrated upon locally well-adapted parents, that are themselves either at or near local varietal standard in performance. This is genetically reasonable and indeed, in practice, unavoidable, given that there always exist more (often far more) potential parents than a programme can exploit effectively. The question naturally arises, however: is this a sufficient strategy from the point of view of long-term progress? There is reason to think not.

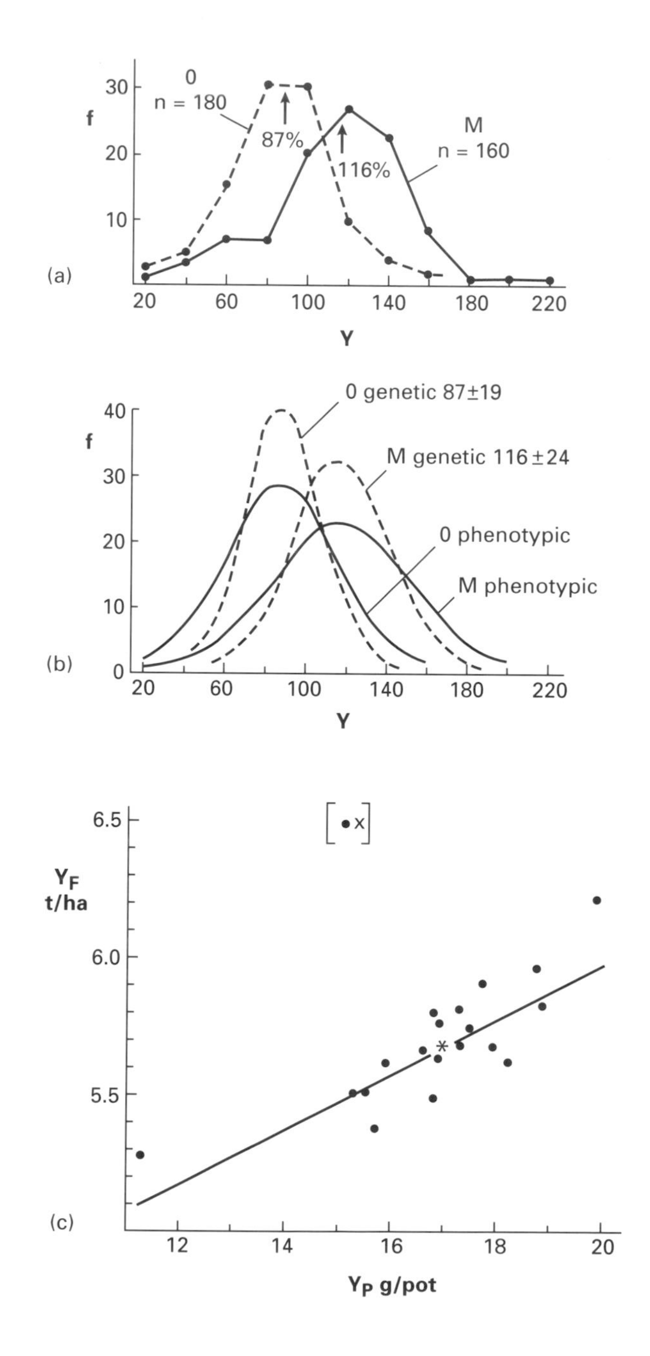

The universal practice of using the best products of the preceding generation as current parents, coupled with much interchange of parents between breeders working in the same area must lead to some narrowing of the genetic base, to a degree of inbreeding (evidenced by relationships between parents) and, ultimately, to restriction of progress. Obviously, there are two kinds of evidence that bear upon the diagnosis of a narrow genetic base. First, there is historical or pedigree evidence that demonstrates or implies that parents tend to be related to each other. Thus many of the millions of *arabica* coffee trees in South America descend from one mother tree grown in the Amsterdam botanical garden in 1706; had the species not been tolerant of inbreeding, the history of the crop would have been very different. Rubber was established as a crop in Malaysia from (probably) twenty-two seedlings from Brazil and African oil palm in Southeast Asia from a mere four trees grown originally in Java. After two or three cycles of very successful clonal improvement, rubber breeders in Malaysia find it necessary to avoid crosses between known relatives, while recognizing that apparently unrelated clones may yet go back to a common source. In sugarcane, the genetic base of the crop was effectively reconstructed by breeding in the 1920s and 1930s, with the result that all modern canes go back to a dozen or so ancestors and a few outstanding parents figure in virtually every pedigree. Similarly, modern soybeans in the USA emerged from highly successful breeding in the 1920s and all trace back to the few (about eleven) lines that emerged from a large collection from eastern Asia. European barleys seem diverse (certainly there are hundreds of names) but virtually all are two-row types that trace back to a limited series of isolates from the nineteenth-century landraces, principally Isaria, Hanna and Gull (Lein in Broekhuizen *et al.*, 1964). North American corn-belt maize started from an interpopulation cross of northern flint by southern dent and has passed through the extreme bottleneck of HYB breeding; a dozen or two inbreds of which only about six are much used, produce the crop and recurrent improvement procedures must ensure that there

---

Fig. 5.15 Use of pot trials in assessing seedling lines of cereals, for plant breeding or genetic resource conservation purposes. (a) and (b) yield comparisons of old (O, mostly land-race) varieties with modern ones (M, mostly dwarfs). Mostly barleys, but some wheats and oats were studied, too. All experiments were done in an outdoor cage in Scotland. See also Table 5.7.
Notes: Frequencies (per cent) of O and M lines plotted against mean pot yields. There was a large overlap between O and M, indicating that many O had similar yield potential to modern, M, varieties. The diagram includes smoothed, normal distributions showing that 'shrinking' of phenotypic values to estimate genetic values preserved the overlap very clearly. Pot trials were also compared with field trials in Scotland in the same year, as shown in (c). Pot trials yields (abscissa) predicted field performance (ordinate) remarkably well, with the sole exception of one variety, which was quite atypical as to field results; in other years, that variety was only middling, as predicted by the pot results. Clearly, pot trials give far better information than would usually be expected. They have not yet been systematically exploited.

is much relationship between new IBL being developed (Section 5.5). North temperate potatoes emerged in the eighteenth–nineteenth centuries from a few (perhaps a dozen) tetraploid Andigena clones introduced to Europe from the Andes very much earlier, plus a few later (nineteenth century) additions

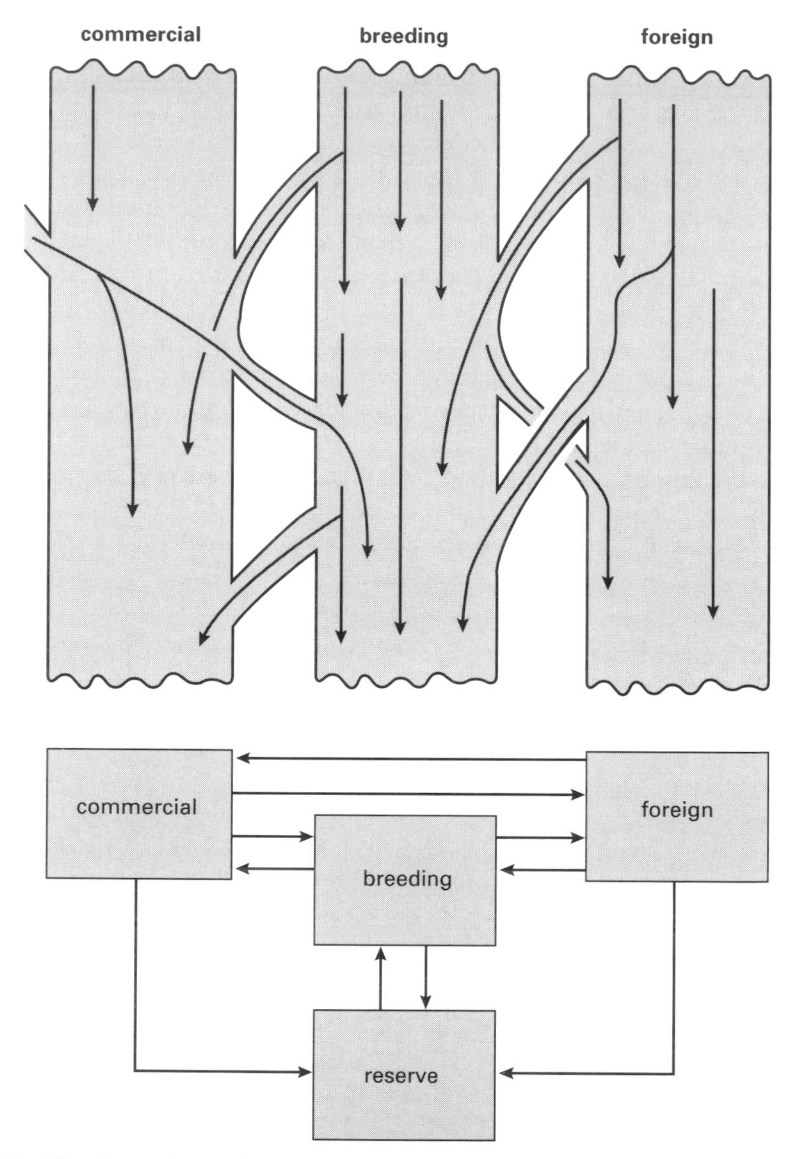

Fig. 5.16 The flow of genetic material in a plant breeding programme.
Notes: The top part illustrates the three main sources of parents in relation to time (vertical axis); the bottom part shows that, if a breeding reserve designed to supplement the genetic base of the crop were included, the diagram would have to be conceived three-dimensionally.

(Simmonds, 1976b). The swedes and rapes (*Brassica napus*) are allotetraploids that probably originated in cultivation in Europe about 200 years ago from crossing between *B. oleracea* and *B. campestris*. The cross is very difficult and is suspected to have occurred only very rarely; if so, the genetic base of the whole species is narrow (see Section 8.3).

The list could go on; the examples above are merely a sample. The second kind of evidence on the genetic base comes from a consideration of progress. Whenever well-founded breeding programmes with realistic objectives fail to make progress there is, I think, *a priori* reason to enquire into the genetic base. Thus, poor breeding progress in north temperate potatoes generally, in North American soybeans and to some degree, in sugarcane, all crops in which excellent breeding progress was succeeded by a 'plateau', has been attributed to defects of the base which, as we have seen, are perfectly intelligible in historical terms. In other crops in which the base seems narrow but progress satisfactory, we have to enquire as to what is 'satisfactory'. It might be that good progress could have been better or that a decline is incipient. North American HYB maize still seems to be making good progress (Table 3.1) but fears for the future have been repeatedly expressed (Committee on Genetic Vulnerability, 1972).

We now enquire what can be done about it, how some long-term provision for the genetic base can be built into breeding strategy. First, the breeder can make a deliberate effort to augment the flow of unrelated foreign parents into the programme (Fig. 5.16, RHS). This makes sense but is not likely to be, of itself, sufficient. In addition, there is need to find some economical means of exploiting the genetic variability that is there but not available in locally well adapted genetic backgrounds; in most variable and widely spread crops this is probably by far the largest fraction of the variability. The underlying assumption is that at least some stocks do indeed have genetic potential as parents in places to which they themselves are ill-adapted. This is obviously true from consideration of the fact that excellent contemporary varieties sprang from ancestors that would not now be even considered as parents; the very successes of plant breeding make the point (Table 5.7, Fig. 5.15).

More generally, performance of a foreign stock may be but a poor guide to breeding potential if adaptation is dominated by a single critical physiological feature. Day-length response is the obvious example; a great many crops, including nearly all the more important ones (Section 1.6), vary in this respect. Introductions to different latitudes (given that responses are adjusted to minutes rather than hours) very generally fail because they flower too early or too late or not at all. In effect, this is an epistatic interaction inhibiting performance; the true breeding potential of a foreign stock having the wrong response can never be determined by inspection and rarely (depending upon genetic control) even by progeny test. In practice, we are still remarkably ignorant of the genetics of day-length response and its evolutionary history. It is certain however that it imposes, in some crops, a severe limitation on the exploitation of foreign stocks and has sometimes (as in maize and soybeans in North America) been responsible for current restriction of the genetic base. One might suspect that vernalization requirements would sometimes present a similar picture in crops in which they may be critical (e.g. winter cereals and brassicas).

Whatever the limiting feature of adaptation, there must always exist some genetically sensible steps that the breeder can take to work up ill-adapted foreign materials into a usable state; to develop, in short, the populations referred to as 'reserve' in Fig. 5.16. The methods available must depend upon the breeding system and the biology of the crop, so will vary widely. Mass selection or semi-natural selection of composites of foreign materials which are known or thought to have the capacity for adaptation may be appropriate. Examples are provided by the Californian composite crosses of barley (see Sections 4.3, 5.3) which have provided a remarkable flow of parents to breeding programmes over the years; by alfalfa composites mass-selected for general performance and disease resistance (Hanson *et al.*, 1972); by Andigena potato composites mass-selected for day-length adaptation, general performance and disease resistance in the UK and the USA (Simmonds, 1976b); and by rubber seed gardens recently set up in Malaysia as the first step in a mass selection programme designed to broaden the genetic base.

In other situations, however, mass selection might be inappropriate. Thus in sugarcane, with a vegetative product and highly erratic sexual reproduction, base-broadening has taken the general form of crossing and backcrossing wild and primitive forms into commercial stocks, while maintaining essential parental control in the process.

If the foreign stocks that are to be explored are homogeneously ill-adapted, they may altogether lack the capacity to adapt to a critical environmental feature such as day-length. In this case, extensive backcrossing (of the appropriate day-length response genes into foreign stocks) would be a necessary preliminary to exploitation by either pedigree or mass selection methods; some knowledge of the genetics of day-length response would clearly be essential. This is the situation faced by North American maize and soybean breeders (see Mangelsdorf, 1974, for discussion in the maize context).

General consciousness of the strategic importance of the genetic base in plant breeding is historically quite recent. It has grown out of the realization that successful plant breeding tends to generate both pathological crises (Chapter 7) and the need to conserve the genetic variability that is threatened by its very success (Section 10.3). Deliberate measures to expand the genetic base – i.e. to develop breeding reserves which shall generate an orderly long-term parental flow – are not yet numerous. In some crops, they are no doubt yet unnecessary and in others the need may be there but unapparent. Plant breeding, however, is probably progressing faster now than ever before and the need for strategic breeding reserves can only grow. We may, I think, expect to see steady development along these lines (Fig. 5.17).

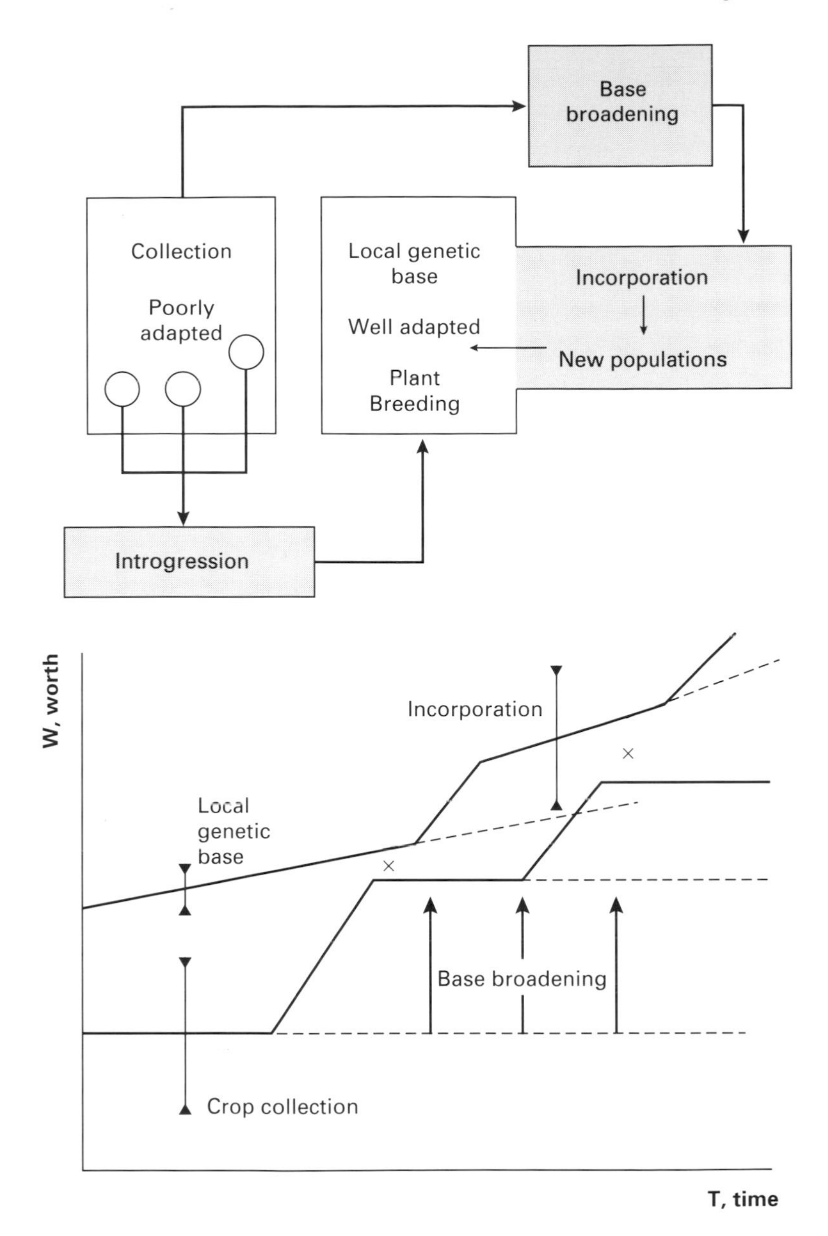

Fig. 5.17 A general view of 'Introgression' and 'Incorporation' as distinct but complementary ways of using crop genetic resources to enhance plant breeding. From Simmonds (1993, Figs 1 and 2) by kind permission of the Editor of *Biological Reviews of the Cambridge Philosophical Society*.

Changing methods

Clonal crops, at least those with erratic reproductive systems, are very unlikely indeed ever to become anything else; but breeding methods adopted in other crops are likely to vary with time and place. Methods, therefore, are not immutable and they change according to economic and social demands and biological circumstances.

Perhaps the most conspicuous trend is the tendency for HYB to displace OPP or SYN in outbreeders and even IBL in inbreeders. The trend is not uniform and HYB maize, sorghum, onion, sugar beet and a few vegetables coexist with OPP, IBL and SYN populations in other countries; HYB wheats and cottons attracted much research but have yet to reach agriculture in any significant way; HYB varieties in many other crops are at least conceivable. We saw above (Section 5.5) that there are biological grounds for scepticism about this trend but that further extension of HYB breeding is very likely in agricultural systems that can stand expensive seed, provided always that the formidable practical problems of reliable seed production can be overcome. Sometimes, HYB coexist with OPP or SYN or IBL in one country, a state which might turn out to be transient or could, conceivably, be long continued. The need for comprehensive breeding plans to meet this situation has been recognized by several workers in recent years, especially in the context of agricultural development in technologically less advanced countries. Figure 5.18 is based on the Kenya maize programme referred to in Table 5.4 (see also Eberhart, Harrison and Ogada, 1967). The scheme recognizes that there will be peasant farmers who want good OPP and others who can afford expensive hybrid seed but there will also be a place for relatively quick and cheap population HYB. The scheme is flexible and pays due attention to the genetic base; with a continual augmentation of sources it should meet both local agricultural needs and any crises that might arise. Its strength lies in a reasonable diversity of methods and materials. In principle, a very similar diagram would apply in any situation in which HYB methods penetrated a crop previously bred as OPP, SYN or IBL. Thus, if HYB wheats developed in Europe or the USA or HYB cottons in Africa, there would be a transitional period, which might be indefinitely prolonged, in which the methods co-existed and were, indeed, complementary; excellent IBL would sometimes be both varieties and parents.

Many perennial crops could not, as we have seen, be conceivably bred as anything other than CLO. But others, those which are reproductively more or less normal, may present a choice between SYN or OPP and CLO (Fig 5.19). Again, the populations are complementary, with good CLO acting both as varieties and parents. Rubber provides a good example with clones dominant but SYN not far behind in performance (Table 3.1). In cacao, clones never attained such importance as in rubber and anyway are tending to be displaced by the outstanding SYN (genetically not far from HYB) that usually seem to

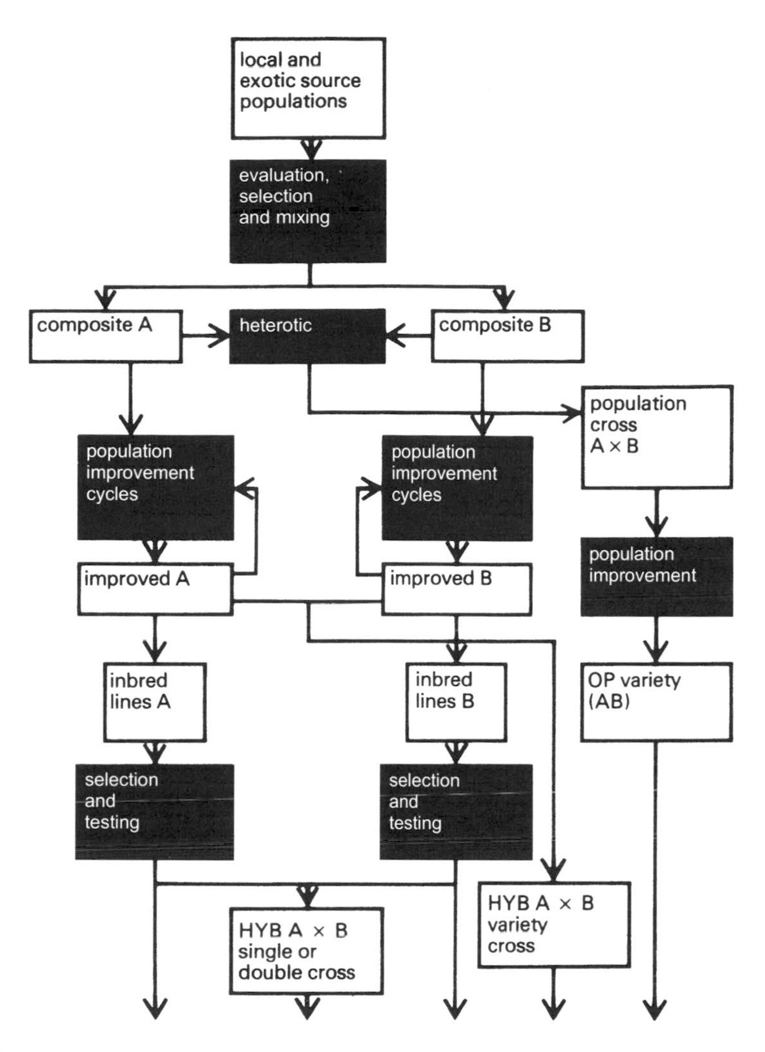

Fig. 5.18 OPP, PIM and HYB as complementary breeding methods in an outbred crop. Notes: Based on maize in Kenya (Table 5.4) but widely applicable in principle wherever complementary methods can co-exist. Two genetic pools (populations A and B) produce successively improved OPP and several grades of HYB.

exploit interpopulation heterosis. In tea and coffees, likewise, CLO and SYN coexist and it seems uncertain whether CLO offer any marked general advantages; there was certainly disagreement about this point in tea. Finally, there are a few crops in which clones would be attractive *a priori* but are not yet possible. When and if clonal propagation becomes possible, as in the oil palm recently, such crops might, with advantage, move towards the CLO-SYN situation. But it would be for investigation first whether CLO offered any major advantage over the excellent interpopulation crosses now being widely used in these crops.

In short, while trends might suggest that HYB and CLO are the inevitable

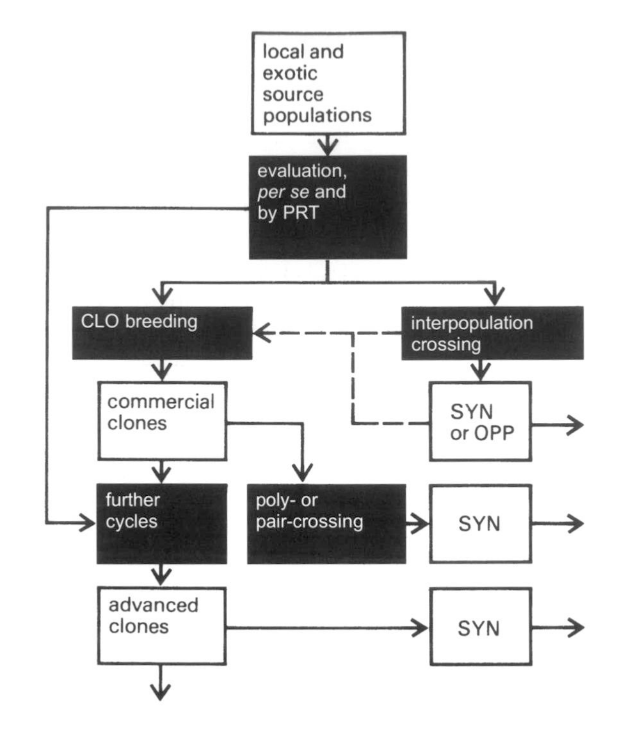

Fig. 5.19 CLO and SYN as complementary breeding methods in perennial outbreeders. Notes: Rubber, cacao, coffees are examples in which excellent CLO can be both cultivars and parents of SYN. The scheme does not apply to strictly clonal crops (those not propagable by seed such as potatoes, sugarcane, bananas, many other tubers and fruits); nor to those perennials which are not propagable clonally. The latter are treated as SYN (or interpopulation HYB) but clonal propagation (see text) would bring them within reach of the scheme.

breeding end-points of annuals and perennials respectively, there is room for scepticism. The long-term needs of agriculture and flexibility of breeding response would probably be better met by the preservation of a judicious balance of methods.

GE effects and stability

Genotype-environment (GE) interactions occur when two or more genotypes are compared in different environments and found to differ in their responses. The existence of such an interaction is shown by a significant item in an analysis of variance, for example sites × varieties or years × varieties or years × sites × varieties. Formal statistical trials are thus normally necessary for detection. However, an analysis of variance can only detect GE effects; it tells us nothing else about them. If the trials are extensive, covering a wide range of environments and therefore a wide range of yields, it is possible to regard the individual trial means as, in a sense, summarizing the environments. This procedure provides a scale of environmental quality which is not otherwise available (though conceivable in principle). If now, results for individual varieties are plotted, trial by trial, against trials means, regression lines will emerge. The strictly average variety will tend to agree with the means, so its line will pass through zero and have a slope of unity. Interacting varieties, those responsible for the GE items in the analysis of variance, will have slopes not equal to one; a 'stable' variety, one unresponsive to environments, will have a low slope ($b < 1$), an unstable or responsive variety a steep slope ($b > 1$). Obviously, the whole set of regressions must be distributed around a mean of $b = 1$ and any measure of stability or responsiveness of individual varieties is strictly relative to the other varieties tested; no independent measure emerges and, indeed, none is yet possible. A numerical example is given in Table 6.2.

The approach just outlined was first suggested by Yates and Cochran (1938) and later exploited for plant breeding purposes by Finlay and Wilkinson (1963) working on barley in Australia. The latter authors were concerned to analyse adaptation (measured by varietal mean) and stability (measured by the regression slope, b) of varieties in the context of screening foreign barleys for use as parents in Australian breeding programmes. Eberhart and Russell (1966, 1969) suggested that, from the point of view of varietal stability, the residual deviations from regression should also be taken into account, as well as the regression itself. The general idea of studying regressions on trials means as a measure of GE effects has been extensively used by biometrical geneticists and Hill (1975) reviews the matter from the plant breeding viewpoint. Endless statistical elaboration is possible.

An extension of the idea which promises to be useful is to consider regressions of variety on variety rather than of variety on means. This leads, as we shall see later when discussing trials systems (Section 6.2), to methods of comparing varieties which, unlike means, are undisturbed by GE effects. But, in the present context, I think we have to say that, while the method has certainly proved valuable in identifying interactive varieties and has sharpened our ideas on stability and adaptation, it has not had any significant impact on the practice of plant breeding. The reason for this is that extensive trials data are required and these, in the nature of things, become available only when a variety is nearly finished or even already being marketed. What, then, can the plant breeder do about GE effects in the earlier phases of a programme when he is conscious that they are likely to occur but cannot estimate them well?

The early generations of a breeding programme are virtually always, and for good practical reasons, grown at one site, the breeder's own home plots. Thereafter, there is always some attempt to diversify the sites, both for selection and for preliminary trials. Sites are chosen on the basis of offering contrasted climates, soil types or diseases. The last, incidentally are not usually considered in a GE context but it is reasonable to do so; given environments with a variable incidence of a disease, a resistant and susceptible variety will show a GE effect. Selection decisions will usually be taken on an *ad hoc* basis, normally with preference for the stock which seems – often, admittedly, on limited evidence – to do well over sites and seasons and is therefore presumptively widely adapted; but more erratic stocks which sometimes do very well will not be lightly discarded. When the biological basis of the GE effect is effectively understood (as with disease resistances), the breeder will simply take account of the character in his intuitive selection index (Section 5.8). On balance, then, there must be some selection pressure towards general adaptation or stability of performance; most breeders would agree that a high mean and low regression slope – if the two could be combined – would be a desirable combination. But, since the breeder may be unwise or unlucky in his choice of sites and can do nothing about seasons, selection for stability is unlikely to be very efficient. In practice, finished varieties differ in their patterns of adaptation and the trials system (Section 6.2) and end-users sort them out accordingly. The study of GE interactions *per se* has so far been fruitless.

Progress and efficiency

The requirements for prolonged progress in a plant breeding programme have all been identified above and merit a summary here. All programmes can be thought of as having two main elements: (1) the making of arrays of potentially useful populations and the identification among them of those actually worthy of intensive exploitation; and (2) the exploitation itself, the principal subject matter of this chapter. Element (1) demands a genetic base adequate to generate the excellent populations resulting from high GCA *plus* SCA that will virtually always be the sources of good new varieties. Progress under element (2) is connected with the response-to-selection formula discussed in Section 4.6, that is to $ih^2\sigma$; it is therefore promoted by large numbers (hence intense selection, high i), by high heritability (hence experimental efficiency, good control of error) and by phenotypic variability – which will reflect parental diversity and thus relates back to the need for a wide genetic base. Thus one conceives of a continuous flow of good parents and overlapping, multi-stage selection processes within populations. The basic requirements of prolonged progress are common to all crops and methods and may, therefore, be listed as: (*a*) good parents, (*b*) good derived populations, (*c*) adequate numbers, (*d*) adequate genetic variability, (*e*) efficient selection.

The idea of efficiency is one that has been but little explored and only rather vague general comments can be made on it. Let us start by supposing that the resources available to a programme are fixed and that this is equivalent to fixing costs and the numbers of plants raised. Efficiency might then be defined as the genetic progress made, evading the difficult question of how this is to be measured, per unit of expenditure or per million plants. But what time scale are we to choose? A short one would imply concentration on quick varieties at the expense of provision for longer term progress; a long one would imply the reverse. If the choices were for a 'reasonable' (though undefinable) balance between varieties soon and progress later, we should still face the problem of what is the best allocation of effort between and within populations. Somewhere, there must indeed be an optimum for any given combination of crop, situation and objective but it could only be calculated on the basis of foreknowledge, or acceptable predictions, of genetic parameters of hybrid populations yet neither made nor even conceived. In practice, of course, all programmes, by doing what they do, embody decisions but we really have no idea whether they are anywhere near optimal or not. The best that can be done is to ensure that a good flow of diverse segregating populations comes forward and to look to sources if there is not at least a 'reasonable' frequency of excellent ones among their number.

Progress under selection within the chosen populations is somewhat better understood. The animal breeders have long paid attention to the relative efficiencies of different methods of selection and England (1977) has extended their calculations to several plant breeding situations. The underlying problem is, given a number (fixed or not) of plants, how best are resources to be allocated as between selection (so enhancing i) and progeny tests (so improving the relevant h^2)? Reasonable answers can generally be given, provided basic information about heritability is available. (We touched on this question above, Section 5.4, in relation to choice between mass-selection and progeny-testing schemes.) Though some such questions can be answered along the lines indicated by England, many others remain; thus the questions of optimal balance between bulk and pedigree procedures and limits of progress in IBL breeding (Section 5.3) are largely unresolved.

Finally, the question of numbers deserves a rather more general comment. All plant breeding is founded on the generation and isolation of rare recombinants, 'rare' being a number of order 10^{-4} to 10^{-6}. So large numbers of plants are required and these are further increased by the need to grow many replicates of individual genotypes for progeny testing, trials, collections and so forth. Any need to select more stringently or to add new characters for selection (Section 5.8) can only increase the need for numbers still further. However, since i is proportional to log n (Section 4.6), a diminishing returns rule applies. Merely to increase numbers of plants selected will by no means ensure a proportional increase in response and could actually be deleterious if other resources were stretched, so that selection efficiency fell. In short, there must be an optimal balance between numbers and other resources put into the programme; given this balance, an increase in numbers must always favour progress though not proportionally. Large numbers, *if well used*, will always be favourable.

Is there, then, a 'correct' or 'optimal' overall size for a breeding programme? One can conceive of answers in economic (financial appraisal or cost-benefit) terms but one cannot conceive of good answers being arrived at thus. In practice, the size of any programme is determined by a mixture of historical accidents and informed judgement; if economics enters the process, it does so at the intuitive level.

5.10 Summary

There are four basic breeding schemes: inbred pure lines (IBL), open-pollinated populations (OPP), hybrid varieties (HYB) and clones (CLO). The first is appropriate to inbred, seed-propagated crops, the second and third to outbred seed-propagated crops and the fourth to outbred perennials. Historically, OPP and CLO are ancient; IBL emerged during the last century or so from heterogeneous 'landraces' and HYB is a very recent invention, developed in maize but now being ever more widely used.

In breeding IBL, crosses are made between parental IBL and selection is practised over subsequent generations of self-pollination, with the objective of isolating one or more transgressive segregants. Effective homozygosity is attained at about F_5–F_7. Selection can be applied from the F_2 onwards ('pedigree' methods) but the modern tendency is to defer selection until lines have attained some individuality ('bulk' methods). The question of whether or not IBL breeding ever closely approaches the 'best possible' recombinant line poses yet unsolved problems. In breeding OPP, a distinction is drawn between population improvement (PIM) in which the emergent varieties are indefinitely propagated as closed populations and synthetics (SYN) which are regularly reconstructed from selected source materials (which may be either seed-propagated lines or, in perennial crops, clones). In either case, the emergent variety is at least somewhat heterogeneous, it is widely enough based to avoid inbreeding depression and individuals are highly heterozygous. Population improvement (PIM) proceeds by mass selection (MSL) of phenotypically good parents or by progeny testing procedures (PRT), of which there are many variants; the former is quick but inefficient if heritability is low, the latter slower but more efficient. So, in choosing methods a balance must be struck. 'Recurrent selection' methods are variants of MSL and PRT. SYN methods depend upon the isolation of numerous potential parental lines or clones and their testing in various combinations for combining ability (GCA and SCA). Usually, about 5–10 parents enter the commercial variety, but there are exceptional cases in which reliable crossing at the seed-production stage permits the use of only two parents without hazard from inbreeding. Such SYN verge on HYB, inasmuch as they are narrowly based and exploit interpopulation heterosis.

Hybrid varieties (HYB) were developed in maize in the USA during this century and have been agriculturally important for about sixty years. They depend upon the isolation of very numerous inbred lines and the identification of the rare combinations between them that exploit exceptionally high GCA *plus* SCA. The parental lines being highly inbred, their F_1 hybrid populations are homogeneous or nearly so, uniform and highly heterozygous. Earlier maize hybrids were 'double-crosses', a compromise enforced by the weakness of even the best inbreds; recent ones are 'single-crosses'; a reflection of general inbred improvement over the years. Long-term progress is secured by the substitution into hybrids of wholly new inbreds or, perhaps more commonly, of improved lines developed by 'recurrent selection' or backcrossing. 'Gamete selection' and the use of maternal and paternal haploids are specialized techniques that depend upon the superb genetic control possible in maize. Maize is the outstanding example of successful HYB breeding but the method is being ever more widely used in a variety of other crops, and with marked success in a few. Though typically applicable to outbreeders, there has been interest in the possibilities for inbreeders, especially wheat. Evidence for F_1 yield heterosis is not wanting but interpretation is not unambiguous. It is simply not known whether HYB are intrinsically superior to IBL in inbreeders in significant agricultural characters. The outcome is yet in doubt but the economic pressures are such that, if seed production problems (the practical limiting factor) can be overcome, HYB breeding in inbreeders is likely to be widely adopted.

Clonal crops are, in effect, all outbreeders and clonal varieties are highly heterozygous though, of course, homogeneous. CLO breeding is basically a matter of selecting among the vegetative descendants of the variable F_1 families produced by crossing heterozygous parents; it is efficient, in the sense that it exploits all the genetic variability. The best families for intensive selection are those that combine high GCA from both parents plus a measure of the unpredictable SCA. Clonal crops grown for a seed product are (indeed, obviously, must be) reproductively more or less normal; others, those grown for vegetative products or for fruits that are not critically dependent upon normal seed fertility, all show varying degrees of reproductive derangement which is sometimes severe enough to be a serious impediment to breeding. Though a clone is, in one sense, biologically fixed, somatic mutation is an inevitable source of some variability, at least in the longer term. There is no evidence for 'clonal degeneration', for which a pathological cause is usually to be sought.

'Backcrossing' is a technique whereby one or a few genes from a 'donor' parent is/are transferred by repeated crossing and selection to a 'recurrent' parent. It is an adjunct to the four basic breeding plans, especially useful for the transfer of dominant genes from unadapted, foreign stocks to adapted parents already in a breeding programme or to established commercial varieties.

A breeder working with a specific crop is likely to have to pay attention to about 10–30 characters of varying heritability and economic importance; they vary widely in ease of determination and the level of expression of any one is rarely linearly related to worth. In a population undergoing selection, characters may be selected independently or jointly; joint selection of two or more characters simultaneously is called 'index-selection' and it may be based on economic or economic-plus-genetic considerations. In practice, a blend of independent ('truncation') selection plus intuitive (rather than explicit) index-selection predominates. Correlations between characters are common and usually of a physiological-developmental kind; they are occasionally favourable to the breeder but (more often, if genetically unbreakable) unfavourable, imposing compromise selection standards on the characters involved. Theoretically, simultaneous selection for many characters imposes dauntingly large demands for total numbers of plants/line/clones selected; in practice, numbers of characters and intensities of selection are kept as low as possible, in recognition of the fact that the realistic objective is useful varieties rather than perfect ones.

Breeding strategies for various crops have much in common regardless of breeding methods currently employed. The following elements are important: the means adopted to ensure a flow of new and diverse parents into the programme; the relative weights accorded to identifying good parental combinations and to exploiting them; the development of longer-term breeding reserves to supplement the genetic base; the recognition that breeding methods themselves are not fixed but may well change over time, the commoner trends being for HYB to displace OPP or IBL and CLO to displace OPP (but opposite changes may occur and contrasted methods may sometimes be complementary rather than competitive). Genotype-environment (GE) interactions, for yield and other characters, appear as differing responsiveness to varied environments. Some genotypes are thus more or less stable than others. By selecting for average performance over environments there is some preference for the stable genotype. In practice, finished cultivars vary in this respect but find appropriate agricultural niches. Overall progress (loosely defined) is favoured by large numbers of plants for selection (but with a diminishing returns aspect) and efficient allocation of resources within the programme.

5.11 Literature

IBL, Inbreeders, cereals

Adair and Jones (1946); Allard and Jain (1962); Allard *et al.* (1968); Atkins (1953); Atkins and Murphy (1949); Baker (1971); Baker and Leisle (1970); Bhatt (1973, 1977); Boyce *et al.* (1947); Briggs and Shebeski (1969, 1970, 1971); Briggs *et al.* (1969); Broekhuizen *et al.* (1964); Busch *et al.* (1974, 1976); Chang and Tagumpay (1970); Chebib *et al.* (1973); Christian and Grey (1941); De Pauw and Shebeski (1973); Dixon (1960); Donald (1968); Donald and Hamblin (1976); Eagles and Frey (1974); Eagles *et al.* (1977); Finzat and Atkins (1953); Fischer and Kertesz (1976); Frankel (1939, 1947); Frey (1954, 1962, 1967, 1968, 1972); Frey and Horner (1955); Frey and Huang (1969); Frey and Maldonado (1967); Gale and Law (1977); Various authors in Gaul (1976); Geadelmann and Frey (1975); Grafius (1972); Grafius *et al.* (1952, 1976); Guitard *et al.* (1961); Gustafsson *et al.* (1971); Hamblin and Donald (1973); Hamblin and Rowell (1975); Harlan and Martini (1929, 1938); Harlan *et al.* (1940); Harrington (1932, 1937, 1940); Hayes and Paroda (1974); Horner and Frey (1957); Immer (1941); IRRI (1972, 1977); Jain (1961); Jain and Allard (1960); Jain and Jain (1962); Jain and Kulshrestha (1976); Jensen (1952, 1965, 1966, 1970); Johnson and Aksel (1959); Johnson *et al.* (1968); Kaufmann and McFadden (1960, 1963); Khalifa and Qualset (1974, 1975); Khush and Coffman (1977); Knott (1972); Knott and Kumar (1975); Kronstad and Foote (1964); Laing and Fischer (1977); Lupton (1961, 1965); Lupton and Whitehouse (1978); Marshall (1976); McGuire and McNeal (1974); Palmer (1952); Park *et al.* (1976); Peterson (1958); Pfahler (1971); Purdy *et al.* (1968); Rasmusson and Cannell (1970); Redden and Jensen (1974); Reitz and Salmon (1968); Rosielle *et al.* (1976); Rosielle and Frey (1975); Roy and Murty (1970); Sakai *et al.* (1958); Salmon *et al.* (1953); Sampson (1972); Sandfaer and Haahr (1975); Sidwell *et al.* (1976); Smith and Lambert (1968); Struthman and Steidl (1976); Suneson (1951, 1956, 1960, 1969); Suneson and Stevens (1953); Syme (1972); Thompson and Whitehouse (1962); Townley-Smith and Hurd (1973); Vogel *et al.* (1956, 1963); Walsh *et al.* (1976); Whitehouse *et al.* (1958).

IBL, inbreeders, legumes

Baihiki *et al.* (1976); Beg *et al.* (1975); Boerma and Cooper (1975a, b, c); Brim (1966); Brim and Stuber (1973); Byth *et al.* (1969, 1969); Caldwell and Weber (1965); Cooper (1985); Empig and Fehr (1971); Erskine (1977); Fehr and Rodriguez (1974); Gedge *et al.* (1977); Hamblin (1977); Hamblin and Morton (1977); Hammons (1976); Hartwig (1972); Hinson and Hanson (1962); Johnson and Bernard (1962); Johnson *et al.* (1955a, b); Lin and Torrie (1971); Luedders *et al.* (1973); Paschal and Wilcox (1975); Thorne and Fehr (1970).

IBL, inbreeders, cotton

Bilbro and Ray (1976); Bridge *et al.* (1971); Culp and Harrell (1973); Feaster and Turcott (1973); Harland (1944); Innes and Jones (1977); Justus (1960); Knight (1945); Manning (1956); Meredith and Bridge (1973); Quisenberry and Kohel (1971); Siddig (1967); Thomson (1972, 1973); Verhalen *et al.* (1975); Walker (1960).

IBL, inbreeders, other and general

Bailey and Comstock (1976); Duncan *et al.* (1963); Horner and Weber (1956); Marani (1975); Matzinger and Wernsman (1968); Pederson (1969, 1974); Riggs and Snape (1977); Snape (1976); Snape and Riggs (1975); Stam (1977); Williams (1960).

PIM, population improvement, maize

Arboleda-Rivera and Compton (1974); Brown and Allard (1971); Darrah *et al.* (1972); Dudley (1974); Eberhart *et al.* (1967, 1973); Gardner (1961); Gardner *et al.* (1953); Gardner and Lonnquist (1959); Genter (1976a, b); Genter and Eberhart (1974); Goulas and Lonnquist (1976); Hallauer (1970, 1972); Hallauer and Sears (1969); Hammond and Gardner (1974a, b); Harris *et al.* (1972, 1976); Horner (1968); Horner *et al.* (1973); Lonnquist (1964, 1968); Lonnquist and Gardner (1961); Lonnquist and Lindsay (1964); Lonnquist and McGill (1956); Mangelsdorf (1974); McGill and Lonnquist (1955); Moll *et al.* (1977); Moll and Stuber (1971); Obilane and Hallauer (1974); Paterniani and Lonnquist (1963); Penny and Eberhart (1971); Robinson *et al.* (1949, 1951, 1955); Russell *et al.* (1973); Sprague (1955); Subandi and Compton (1974); Suwantaradon *et al.* (1975); Troyer and Brown (1972); Zuber *et al.* (1971).

PIM, population improvement, forage legumes

Anderson *et al.* (1972, 1974); Bingefors and Ellerström (1964); Bingham and Saunders (1974); Busbice (1968, 1969, 1970); Busbice *et al.* (1974); Dunbier and Bingham (1975); Dunbier *et al.* (1975); Elgin *et al.* (1970); Gorz and Haskins (1971); Haag and Hill (1974); Hanson *et al.* (1972); Hill *et al.* (1972); Hunt *et al.* (1971); Kehr (1976); Lowe *et al.* (1974); Obajami and Bingham (1973); Pedersen and Hill (1972); Posler *et al.* (1972); Rogers (1976); Simon *et al.* (1975); Thomas (1969); Tysdal and Crandall (1948); Tysdal *et al.* (1942).

PIM, population improvement, grasses

Breese (1969); Breese and Hayward (1972); Carlson (1971); England (1967, 1968); Foster (1971a, b, c, 1973a, b); Hanson and Carnahan (1956); McWilliam and Latter (1970); McWilliam *et al.* (1971); Thompson and Wright (1972).

PIM, population improvement, other and general

Comstock *et al.* (1949); Curtis and Hornsey (1972); Doggett (1972); Doggett and Majisu (1968);

Eenink (1974); Ehdaie and Gess (1973); England (1974); Hecker (1972); Hill (1971a, b); Hill and Haag (1974); Hornsey (1975); Jan-Orn *et al.* (1976); Johnston (1968); Khan (1973); Parlevliet (1973); Simmonds (1986); Swanson *et al.* (1974); Thurling (1974); Vaughan *et al.* (1976); Watts (1963, 1965, 1970); Wellensiek (1947, 1952); Wricke (1966); Wright (1973, 1974, 1977).

HYB maize

Chase (1969); Cornelius and Dudley (1974, 1976); Darrah and Hallauer (1972); Duvick (1974); Eberhart *et al.* (1967); Eberhart and Russell (1969); Efron and Everett (1969); Fleming (1971); Funk and Anderson (1964); Gama and Hallauer (1977); Geadelmann and Peterson (1976); Good and Horner (1974); Hallauer and Sears (1973); Hayes (1963); Hoegemeyer and Hallauer (1976); Horner *et al.* (1972); Mangelsdorf (1974); Otsuka *et al.* (1972); Rinke and Hayes (1964); Russell (1969, 1974); Russell and Eberhart (1975); Russell and Vega (1973); Schnell (1975); Sharma *et al.* (1967); Sprague (1955); Sprague and Tatum (1942); Stringfield (1964); Stuber (1994); Stuber *et al.* (1973); Suwantaradon and Eberhart (1974); Walejko and Russell (1977); Wallace and Brown (1956); Weatherspoon (1970).

HYB, other crops

Anderson *et al.* (1972, 1974); Briggle (1963); Davis (1974); Fedak and Fejer (1975); Hooks *et al.* (1971); Janossy and Lupton (1976); Johnson and Brown (1976); Johnson and Schmidt (1968); Lamberts *et al.* (1968); Mackay (1973); Matzinger *et al.* (1971); Meredith and Bridge (1972); Miller and Lucken (1976); Miller and Lee (1964); Netterick (1968); Pedersen and Hill (1972); Quinby (1963, 1962, 1973, 1975); Rosales and David (1976); Sage (1971, 1976); Shebeski (1966); Thompson (1964, 1966); White and Richmond (1963); Williams (1960); Zeven (1972).

CLO, clones

Brown *et al.* (1968, 1969, 1969, 1971, 1973); Copersucar (1987); Ferwerda and Wit (1969); Harris (1978); Haunold (1975); Hogarth (1968, 1971, 1977); Holden (1977); Howard (1961, 1962, 1970); James and Miller (1975); Krug *et al.* (1974); Maris (1966, 1969); Mendiburu and Peloquin (1977); Mendoza and Haynes (1974); Miller (1977); Miller and James (1975); Mok and Peloquin (1975); Plaisted and Peterson (1959); Plaisted *et al.* (1962); Ross and Brookson (1966); Rowe and Roberts (1996); Simmonds (1962b, 1966, 1969b, 1976b, c, 1994b); Spangelo *et al.* (1971); Swaminathan and Howard (1953); Tai (1974, 1975, 1976); Tarn and Tai (1977); Warner (1963).

Backcrossing

Briggs (1959); Duvick (1974); Knight (1945, 1956); Reddy and Comstock (1976); Thomas (1952).

Selection, including indices

Bos and Caligari (1985); Brim *et al.* (1959); Byth, Caldwell and Weber (1969); Byth, Weber and Caldwell (1969); Caldwell and Weber (1965); Dudley (1974); Eagles and Frey (1974); Elgin *et al.* (1970); Hanson and Johnson (1956); Lewers and Palmer (1997); Manning (1956); Pesek and Baker (1969, 1970); Robinson *et al.* (1951); Rosielle *et al.* (1976); Rosielle and Frey (1975); Siddig (1967); Simmonds (1985, 1991b, 1996); Singh and Bellman (1974); Smith (1936); Subandi *et al.* (1973); Thurling (1974); Walker (1960); Widstrom (1974); J.S. Williams (1962); Young (1961); Young and Weiler (1961).

Computers, numerical taxonomy

Rathjen and Lamacraft (1972); Sneath (1976); Whan and Buzza (1977).

Breeding strategy, genetic base

Allard and Bradshaw (1964); Allard and Hansche (1964); Banks (1976); Baker and Curnow (1969); Committee on Genetic Vulnerability (1972); Creech and Reitz (1971); Curnow (1961); Day (1973); Eberhart *et al.* (1967); Efron and Everett (1969); Federer (1963); Finlay and Wilkinson (1963); Finney (1958); Hammons (1976); Hanson *et al.* (1972); Paschal and Wilcox (1975); Simmonds (1961, 1962a, 1976b); Spoor and Simmonds (1993); Stevens (1948); Thorne and Fehr (1970); Walker (1969); Warner (1953); Webster (1976).

Chapter 6
Trials and Multiplication

6.1 Introduction

Between the identification by the breeder of a potential new variety among his stocks and its entry into farming practice there lie two essential steps: the decision that the stock is indeed to become a new named variety and the multiplication of seed or planting material up to a level at which it can be effectively marketed. The basis of decision is a series of trials which are intended to forecast the agricultural performance of the variety. Multiplication is aimed at the rapid production of an adequate supply of seed of good quality. Both processes, the subject matter of this chapter, are generally subject to an ever more complex series of official, legally based constraints. So, as time goes on, the trials-multiplication phase tends to become longer and more expensive, the time often, indeed, approaching that taken to breed the variety. As far as possible, therefore, trials and multiplication overlap in time (Fig. 6.1) even if this means, as sometimes it must, that multiplication effort is wasted if a variety fails to satisfy official requirements. Trials, then, are the logical place to begin (Section 6.2). We shall then go on to see how trials information of various kinds is used to reach decisions, official or other, on whether or not a new variety shall be marketed and we shall finally consider the principles (but not the practical details) of multiplication.

6.2 Trials

Objectives and kinds of trials

The object of any trial or set of trials is to predict something, nearly always the performance, in some agriculturally significant character, of a new genotype in relation to one or more standard varieties. By far the most important character is yield and, in effect, most trials are yield trials of new varieties against old; but other characters (maturity, lodging, composition, quality and so forth) are often also measured, or at least estimated, and enter the decision mechanism. A few trials, however, which we shall encounter recurrently in this chapter under the heading of DUS, attempt to do something rather different: they attempt to

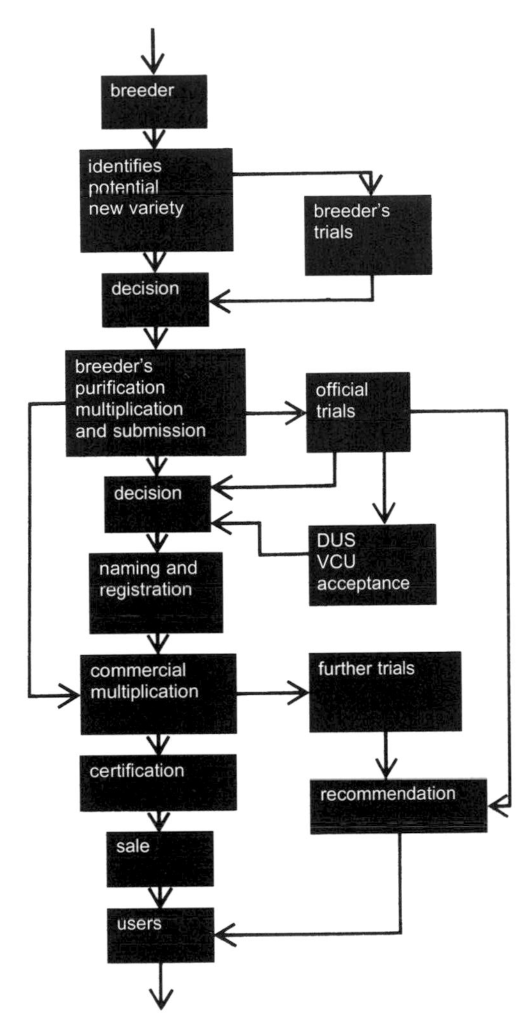

Fig. 6.1 The general course of multiplication and trials in a system subject to strict legal controls, such as obtains in western Europe.
Notes: Official acceptance on to the registered list of permitted varieties is conditional upon passing the appropriate DUS/VCU tests; sale of seeds is futher conditional upon commercial certification; recommendation is a quite separate element (which may or may not be practised) and has no legal force.

establish the nature of a new variety in absolute terms by posing the question: is this new variety distinct from all others currently in the system and is it sufficiently uniform and stable to meet the present arbitrary requirements in these respects? Or, briefly, is it different and potentially certifiable? We shall distinguish the two kinds as 'performance' and 'DUS' trials, when necessary. They differ in objectives (agricultural and legal, respectively), in structure (randomized field performance trials and otherwise) and in cost (performance trials are far more numerous and costly). We shall refer to DUS again below; performance trials are the subject of this section.

In some countries and some situations, the breeder's performance trials are the only ones done and they will extend over several years, from the time at which he first identifies the new variety as a potentially good one, until at least the moment of decision to market. In others (Fig. 6.1), breeders' trials are succeeded by various kinds of official performance trials designed to aid decisions on registration or recommendation. Who does performance trials however is immaterial. The principles are constant because all trials ask the same question: how does this new variety compare with the standard variety (or varieties)?

Design of trials

Until the twenties and thirties, virtually all variety trials, and other agricultural experiments too, were systematic; that is, plots were laid out on an arbitrarily determined plan – for example, by numerical or alphabetical order. On a very uniform site this might serve quite well but experimenters were well aware that, on many sites, soil variation could produce yield differences larger than those they expected to find between varieties. So neither variety mean performance nor errors attaching to the estimate of it could be properly determined. The solution lay in ensuring that soils effects were as evenly distributed over varieties as possible by application of the first two principles of trials design: replication and randomization. Several plots of each variety are grown in each trial and their positions are strictly randomized (under certain restraints which we shall come to in a moment). In the absence of near-perfect fore-knowledge of soil variation this is the nearest we can get to ensuring accurate and unbiassed estimates of means and errors.

The third principle of trials design is 'local control', the need for which arises from the pattern of soil variation. It is universal experience that neighbouring patches of soil are similar, distant patches different. The best comparison of two varieties therefore comes from neighbouring plots and, extending this principle, the best comparison of a set of varieties comes from a compact block of small plots, within which block soil variations likely to be minimal. Note that a long, narrow block would not be satisfactory because the two ends would tend to be more different than the two most distant points within a compact block; theoretically, in fact, the ideal block is square. The necessary replication and randomization referred to above as the first two principles of design are respectively between and within blocks. Average comparisons are thus as precise as possible and randomization ensures that local soil effects on varieties are unbiassed. The blocking provides 'local control'. We have thus arrived at the basic experimental design, the 'randomized complete block'. We shall see below that there are numerous variants of it; all observe the principles of replication and randomization while imposing constraints that enhance local control.

These principles are now universally accepted. They govern all experimentation, not only the agricultural kind, though they were developed in that context by the genius of the late R.A. Fisher (Finney *et al.*, 1964). They are inseparable from modern statistical theory, in particular, from the notion of analysis of variance, though they can be presented, as we have just seen, in commonsense terms.

The numerous variants of the basic complete randomized block design may be summarized as follows. First there are Latin squares (and several related designs) which impose restraints upon randomization within replicates in such a way that an extra share of the unavoidable soil variation is taken out; this reduces the residual error (Fig. 6.2). Latin squares are very effective designs for small numbers – say six varieties or fewer – but impracticable for large ones because the number of replicates must equal the number of varieties. It will be noted (Fig. 6.2) that the rows and columns can be (and are) separately randomized after the basic design has been (randomly) chosen. For middling numbers of varieties (say 5–20), complete randomized blocks, with fewer replicates than varieties, are fairly standard. It is always recommended that blocks should be as compact as possible, but they do not have to be square or even rectangular (Fig. 6.2). Strict randomization of varieties within blocks obtains; there are compelling statistical reasons for not adjusting what look like *a priori* unlikely juxtapositions. For larger numbers of varieties, a range of 'incomplete block' designs exist, collectively referred to as lattices (simple, triple, balanced, rectangular cubic). Details lie beyond our scope but can be found in the appropriate statistical texts, e.g. Cochran and Cox, 1957; Snedecor and Cochran, 1989. They all have in common the use of compact smallish blocks any one of which contains only a proportion of the total entries. Thus variety and block effects are confounded and variety means must be adjusted, but the local control so achieved keeps the error down and enhances precision of comparison. A serious drawback of the lattice designs is that there are always restraints as to the exact number of varieties that can be accommodated; thus, for simple lattices with n varieties it is a condition that n must be a perfect square, with $\sqrt{n}$ plots per block. Another is that randomization and analysis is more complex; but the general availability of computers now effectively negates this.

Relative efficiencies of trials designs can be estimated by the ratios of some measure of error, e.g. error variances or standard deviations (cf. Fig. 6.2). On this basis, randomized blocks have been found to be about 67 per cent as efficient as Latin squares and 50–78 per cent in relation to several lattice designs (Le Clerg, in Frey, 1966).

The reader whose statistical knowledge is limited would do well to study Fig. 6.2 and work the numerical examples it poses. It is not feasible to display a lattice design in the space available but the three principles emerge clearly enough.

Such, then, are the basic principles of trials design, founded on replication, randomization and local control. They leave unanswered the important ques-

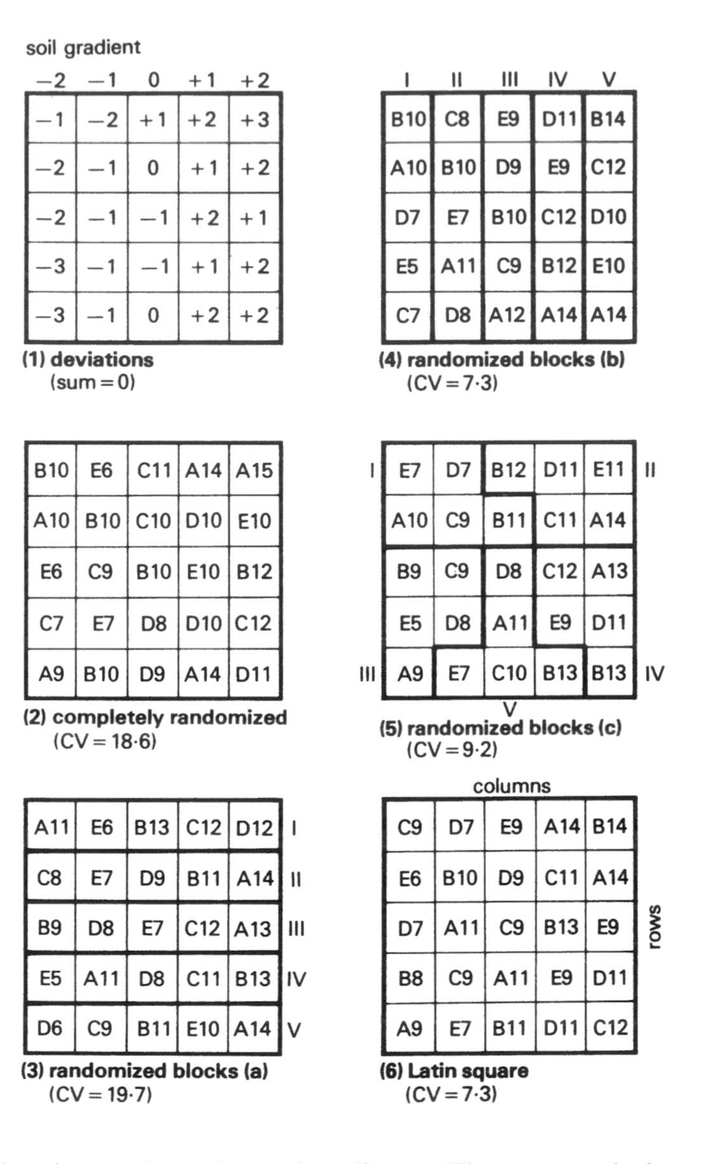

(1) deviations (sum = 0)

(2) completely randomized (CV = 18·6)

(3) randomized blocks (a) (CV = 19·7)

(4) randomized blocks (b) (CV = 7·3)

(5) randomized blocks (c) (CV = 9·2)

(6) Latin square (CV = 7·3)

tions of plot sizes and numbers of replicates. These can only be answered empirically, crop by crop and place by place. The larger a plot the more accurate it is, to a degree depending on soil heterogeneity; but large plots are expensive and they compromise, under fixed resources for the trial, the degree of replication and hence the control of error; on the other hand, large plots waste relatively less space than small ones in guard rows (see below). So a practical compromise emerges, based on joint analysis of past trials and aimed at least errors for low costs. In practice, replicates commonly lie in the range 3–6 and plot sizes vary enormously with the crop, from about 1 m^2, or even tiny 'hill-plots', in cereals to about 0.15 ha in rubber.

Fig. 6.2 Constructed illustration of the principles of trials design. Five varieties with true (unknown) means A(12), B(11), C(10), D(9), E(8) in various layouts on a site showing a strong fertility gradient. Soil effects (item (1), deviations) are composed of the fertility gradient effects *plus* small random additions and are assumed to be non-interactive with varieties. The remaining squares (item (2)–(6)) are five contrasted randomized designs; the relevant coefficient of variation (c.v., error standard deviation per cent of general mean) is shown in each case. The reader should work the examples described in the notes below

Notes: The reader should tabulate the data by varieties for the five squares, (2)–(6), and do analyses of variance, which will take the following skeleton forms:

(2)

	df
varieties	4
error	20
total	24

(3)–(5)

	df
varieties	4
blocks	4
error	16
total	24

(6)

	df
varieties	4
rows	4
columns	4
error	12
total	24

Note that, since soil effects sum to zero, the grand total for each square must be 250 and the grand mean 10. Compare the results for the randomized blocks (3), (4), (5); which is the most accurate and why? In practice, for want of reliable knowledge of soil fertility, something like (5) would invariably be adopted. Now compare the results for (2), (5) and (6). Calculate relative efficiencies from ratios of error variances and standard deviations. You should find (on standard deviations): Latin square 79, randomized blocks 100, completely randomized 202. For comparison, tabulate and graph estimates of variety means and their respective 5 per cent confidence limits. Differences lie in the progressive removal of soil effects from error; why? Note that strict randomization produced two odd-looking effects which were *not* adjusted (a line A–E in (2) and three adjoining A plots in (4)). For the purpose of this example, soil effects were chosen to be slightly larger than variety effects; verify this by examining appropriate variances and ranges. The coefficients of variation attaching to the best and worst designs were, respectively, (6) 7.3 and (2) 18.6; in practice, the first would be regarded as quite satisfactory; the second as rather bad.

The need for guard-rows springs from the fact that varieties differ widely in competitive ability. Since the object of a trial is to predict the performance of a variety in pure stand (i.e. competing with itself), guard rows are omitted only in exceptional circumstances, for example in rough 'sorting' trials of large numbers of closely related lines or when seed is in very short supply. But they are essential for any critical trial. They normally take the form of a simple extension of the plot in both dimensions so that one or two rows or a measured margin right round the plot are discarded, or disregarded, as the case may be. An alternative, but unsatisfactory, procedure is to grow a distinct variety in the guard rows; this is troublesome to do and will probably lead to biassed means.

The mechanics of trials work lie beyond our scope but a few basic principles are worth a short summary. First, the patch of land chosen must be as uniform as possible, both inherently and in terms of recent cropping history. Second, all the features of management should be as near as reasonably possible to current 'good farming practice'; since those are generally the conditions at which the new varieties being tested are aimed; time of sowing/planting, seed rate/crop density, spacing, fertilizing, weed control, irrigation, pest and disease control and harvest time would all come under this heading. Third, economy and efficiency alike demand that trials should be mechanized as far as possible; a considerable range of specialized equipment (especially seeders, harvesters, weighing machines) has been developed in recent years; sometimes, commercial equipment can be suitably modified for plot work. Fourth, great care will be taken to ensure accuracy of labelling of seed containers and of adherence to the planting plan; elaborate field labelling is not normally regarded as essential but some marking to assist location on the plan can be very helpful. Fifth, though yield is normally the basic observation of a trial it will often need to be supplemented by sample measurements of moisture content or chemical composition so that dry matter or economic yields can be estimated; smooth integration of field and laboratory operations is therefore indicated; some quality measurements, however, can, indeed often have to be, deferred to a less busy season of the year. Sixth, observations will be made during growth and may be very important in interpreting results; besides routine observations of emergence, stand, disease incidence, lodging, maturity and so forth, which all bear on varietal evaluation, accidents do happen and must be recorded as a basis for decision as to whether certain plots, or even a whole replicate, must be regarded as 'missing'. Seventh, though the field plan and field notebook are the traditional instruments of record for the plant breeder, there is a strong tendency to replace at least the latter by record sheets adapted to the computer; as a first step, the adoption of symbols or numbers directly transcribable to cards or tape is already common; ultimately, machine-readable records may well be generally adopted. The temptation to record more information than can be effectively used should, however, be resisted; a few critical observations made by someone who really knows the material will often be much more useful than a mass of routine data. The garbage-in-garbage-out principle remains a good one, however useful the computer may be in doing tedious sums quickly and accurately.

One final comment about trials. The majority (probably the great majority) of plant breeding trials investigate only one factor (i.e. varieties) keeping all other features constant; they are non-factorial. Sometimes, however, other treatments may have to be imposed and designs become factorial in nature. Examples might be: irrigation, fertilizer levels, harvest dates, cutting treatments. These will often pose very considerable problems of design and execution and the breeder would be wise to seek professional statistical advice before embarking on such trials for the first time, even though they may subsequently become routine.

Trials systems

A breeder will sometimes discard an unpromising line on the evidence of a single trial but will generally take important decisions as to whether or not to proceed with a potential new variety only on the basis of trials replicated over sites and seasons. Similarly, official performance trials always make provision for such replication. Sets of trials, or trials systems, are therefore decision systems based on many comparisons; decisions rest on the magnitude of differences between new varieties and standards rather than on mere statistical significance – which is rarely of much interest. Statistically, the problem is one of estimation of difference rather than its detection; this would be relatively easy were it not for GE effects.

A single variety trial says nothing about GE interactions because the varieties (V) × replicates (R) component, the highest order interaction, is defined and used as error. Repeated trials of the same varieties (V), replicated over sites (S) and/or years (Y), however, permit joint analysis and the separation of interactions (VS, VY, VSY) while still leaving VR items for the estimation of error. The statistical complications (heterogeneity of errors, non-orthogonal representation of varieties over trials and so forth) can be very considerable; they lie beyond our scope and should usually be referred to a professional statistician. Some generalizations are possible, however. Thus, it is common experience that years effects and VY interactions are larger than sites and VS (Table 6.1). This is simply an empirical rule that reflects the fact that a set of sites is generally chosen as much for administrative convenience as for agricultural diversity. A trials system that deliberately aimed to bring out GE effects would seek very diverse sites, and there must presumably be a level of diversity at which the common rule stated above would no longer work. So long as the rule does work, however, and the aim is to get good average comparisons of varieties, it would

Table 6.1 Analysis of variety and variety-interaction effects in sets of trials; components of variance as percentages of the sums of the three interactions; V varieties, Y years, S sites. The first four entries are means of two to four data sets each. Sources: US data on cotton, soybean and tobacco from Matzinger in Hanson and Robinson (1963); UK data on wheat, barley, swede and sugar beet from M. Talbot and F.J.W. England (personal communication).

Item	wheat	barley	cotton	soybean	tobacco	swede	sugar beet	means
V	19	38	94	127	448	43	106	125
VS	37	2	(4)	(9)	1	9	19	12
VY	30	26	(13)	40	22	91	51	39
VSY	(34)	73	83	52	77	(0)	30	50

Notes: Negative components are conventionally treated as zeros and their presence is indicated by bracketed entries in the table. Their occurrence should remind the reader that components of variance must be viewed cautiously as general orders of magnitude rather than as accurate estimates. The data are only illustrative; a much larger and random sample would be necessary to draw definitive conclusions. With these reservations, it appears (in agreement with general experience) that interactions involving years tend to be substantially larger than those involving sites. Data such as these could be made the basis for enquiries into the design of trials systems (see text).

be better to replicate trials over years than over sites within years. All trials systems accept that both kinds of replication are necessary but, inevitably, in the interests of speed, tend to compromise on the number of years.

If GE effects are small or negligible a mean difference between any two varieties, with confidence limits, will adequately summarize the information inherent in the trials but interpretation should be tempered by the reflection that a different choice of years or sites might have revealed interaction. If GE effects are substantial, a mean difference again summarizes the information (but not all of it) and interpretation becomes much harder. In particular, the mean difference will predict agricultural performance adequately only if the trials sites and seasons truly represent the universe of environments which the variety will meet in practice (which is unlikely).

There is no simple way out of this difficulty. In principle, it might be possible to recognize homogeneous sets of sites chosen so as to minimize site interactions and thus permit separate decisions for the different areas (e.g. Horner and Frey, 1957). But there is nothing that can be done about seasons, the more important effect. In practice, GE effects are usually virtually ignored in coming to trials decisions. No doubt some bad decisions are taken but the truly indifferent new variety and the excellent one are very likely to be recognized as such, whatever the interactions. Alternative approaches are outlined below.

Formally, interactions are items in analyses of variance. There is another way of looking at them which we shall now explore (Fig. 6.3). If, over a series of trials, the yield of one variety is compared with that of another, it commonly turns out (at least if there are substantial differences between trials) that the two are correlated. Obviously, a regression slope of $b = 1$ would indicate that both varieties were responding to environment in the same way and to the same degree; that is, the varieties would be non-interacting and any mean difference between them would be revealed by an intercept $a \neq 0$. By contrast, a regression slope greater or less than unity would indicate interaction. Regardless of the position of the mean, a low slope ($b < 1$) would indicate stability of performance (one variety less responsive to environmental effects that change the performance of the other) whereas a steep slope ($b > 1$) would indicate the reverse. In practice, positive correlations between varieties over trials, with regressions ranging below and above unity, are very generally found when sought (examples in Table 6.2). Correlations are often high (in our experience they average about $+0.8$), which means that quite a large part of any interaction is accounted for; but not all, because the residual deviations from the line could contain some interactions as well as error (Fig. 6.3).

So variety-on-variety (V-V) linear regressions often account for a surprisingly large part of the differences between varieties, including interaction effects of years and sites. This is just an empirical fact which deserves more exploration than it has yet had. There is no obvious biological reason for it, nor for supposing that the relation need always be linear. As an empirical observation, however, it has some interesting implications for the interpretation of variety

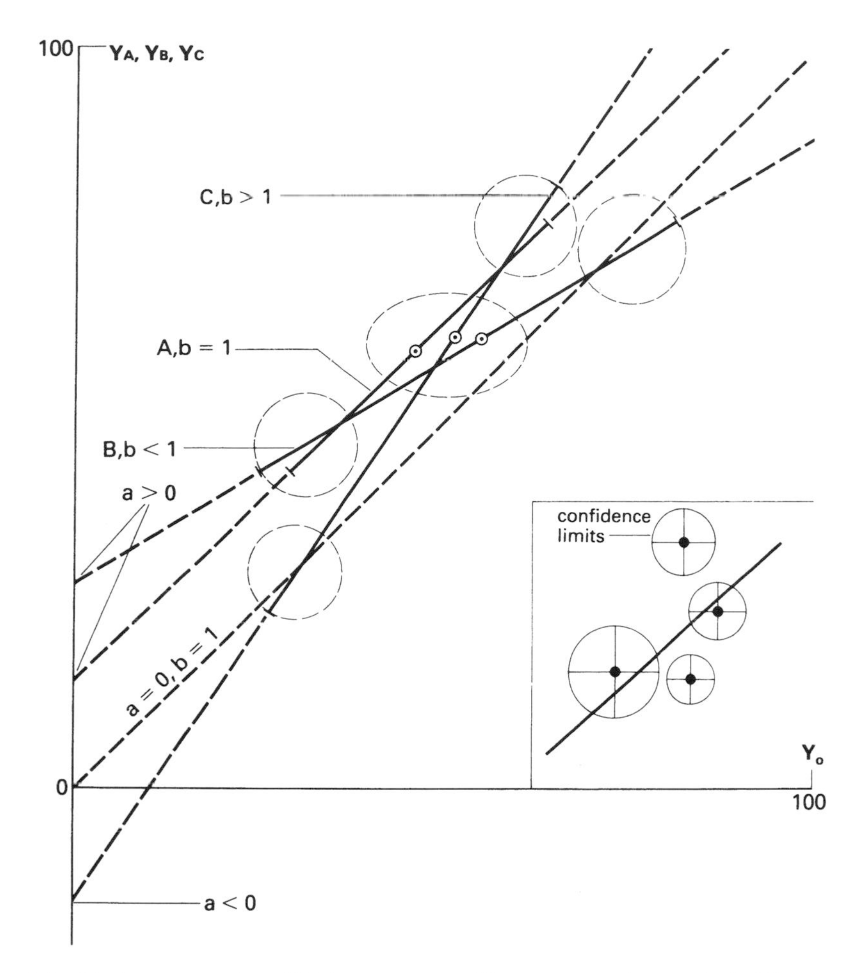

Fig. 6.3 Variety-on-variety (VV) regressions. Three new varieties (A, B, C) regressed, over trials, on the common standard (D). Inset shows that interactions may remain unaccounted for by regression and apparent as real deviations from the line.
Notes: In relation to D: A is non-interactive, b = 1 and the mean difference (= a) describes the relative worth of A over the observed range; B and C are interactive, slopes ≠ 1, means do not describe worth and B is more stable (less responsive) than C. The dashed areas enclosing parts of the lines show how deceptive a narrow environmental range can be in comparing means in the presence of GE effects.

trials, as follows. First one's view of the worth of a new variety (A, B, C) in relation to the standard (D) (Fig. 6.3) will, if interactions are present, depend upon the choice of sites and the chance of years; a narrow range of high yielding trials would tell a very different story from a similar range of low yielding ones; indeed if the ranges were narrow enough, no regressions might be apparent at all and decisions could be very bad indeed (Fig. 6.3). Second, only in two cases is it proper to compare varieties by a single number, namely when $b = 1$, $a \neq 1$, and $b \neq 1$, $a = 0$; in the first, the varieties differ by a constant amount of yield and in

Table 6.2 Variety-on-variety regressions in Guyana sugarcane. Mean yields of cane (t/ha) with regression parameters (a, b) of all six cultivars taken in pairs. The variety treated as the independent variate (Y) is on the left. In the right-hand column are the regressions of the six varieties on the means of the other five. The source is the same as Table 3.1 but this table also uses data that were inapplicable to Table 3.1.

Cultivar	Mean	A	B	C	D	E	F	Other five
A. D625	59.1	–	–52.6 1.638	–67.0 1.612	–102.0 1.920	–90.3 1.405	–96.6 1.336	–92.0 1.667
B Diamond 10	68.2	+32.1 0.610	–	–4.4 0.928	–31.4 1.187	–18.7 0.817	–17.5 0.735	–13.1 0.915
C Co 421	78.2	+41.5 0.620	+4.7 1.078	–	–29.9 1.288	–16.9 0.894	–17.3 0.819	–4.9 0.957
D. POJ 2878	83.9	+53.1 0.521	+26.5 0.843	+23.2 0.776	–	+10.3 0.693	+10.7 0.628	+18.3 0.766
E. B 34104	106.4	+64.3 0.712	+22.9 1.225	+18.9 1.119	–14.8 1.444	–	–2.2 0.932	+14.5 1.131
F. B 41227	116.6	+72.3 0.749	+23.8 1.360	+21.1 1.221	–17.0 1.592	+2.4 1.074	–	+18.6 1.238

Notes: Each regression is based upon 11 points and correlations were all in the range 0.71–0.94. The regression coefficients are of the kind that makes $b_{yx}.b_{xy} = 1$ (see text) so that any regression can be calculated from the complementary one; the reader should verify this for a few cases. Calculate the mean regressions across rows and compare the results with the regressions of varieties on trials means in the right-hand column; they are very similar (but not expected to be identical). The six cultivars are in approximate historical sequence and the three successive pairs (AB, CD, EF) represent cycles of sugarcane breeding. Plot regressions of B on A, C on B and so on to F on E and consider how GE effects might affect your judgements of varietal differences (see Fig. 6.3). Plot a scatter diagram of varietal means against b from the last column and try to decide whether breeding progress for yield was accompanied by any consistent trend in stability of performance (see Section 5.9). Note that no simple standard errors attach to these regression coefficients but you may take it that the extremes in both directions diverge significantly from unity.

the second by a percentage of yield; more generally, therefore, two numbers are required to compare two varieties, an a and a b (which may sometimes be, respectively, 0 and 1). The third point follows from the second: calculations based on percentage comparisons must often be wrong because they embody unjustified assumptions about interactions; they are meaningful only if the intercept is zero – or in the trivial case of a range so narrow that there is no regression. And, fourth, these ideas suggest a way of thinking about the economy of trials systems. Supposing the empirical observation of linear regressions to have been well established, economy would be promoted by concentrating fewer trials into high and low yielding environments; given the large effect of seasons, this might be difficult but not necessarily impossible (e.g. when, as with Barbados sugarcane, a dominant effect of rainfall is evident). Alternatively, it might be feasible to create contrasted environments by a suitable treatment (irrigation maybe?)

A statistical point about regressions should be made here. Ordinary regressions of Y on X and X on Y minimize the squares of deviations in Y and X respectively. Unless $r = 1$, one regression is not the reciprocal of the other. For some purposes the coefficient which minimizes squares vertical to the line and has the property $b_{yx}.b_{xy} = 1$ is useful (Table 6.2). The matter is not well covered in the standard elementary texts and the interested reader should consult a statistician; the choice of regression still presents problems.

The idea of thinking about V-V regressions rather than (or as well as) means as an approach to interpreting sets of trials data goes back to Yates and Cochran (1938) but is yet little applied in practice. Time will show if it is really useful. It grew out of studies of GE effects in which individual varieties were regressed on trials means (rather than on other varieties). Obviously, one set of data provides as many regressions as varieties and the regressions must average unity (Table 6.2). The idea is not useful in the context of trials decisions, where interest always centres on the new variety-standard comparisons, but has been distinctly valuable in the study of phenotypic stability, or otherwise, of genotypes (see Section 5.9). However, there is a hazardous assumption that interactions are repeatable.

So regressions provide one way of thinking about sets of trials results but the reader should appreciate that it is not free of statistical problems and that statisticians would probably not agree about how best to proceed; much more thinking and experience is wanted. As another approach, Patterson (1979) proposes the fitting of additive constants to varieties and trials as a method of coping with the problems of non-orthogonal data which arise when all varieties are not in all trials. This procedure is analogous to the fitting of general combining abilities (GCA) to breeding results. In effect, one fits to each datum an equation of the form: general mean *plus* variety effect *plus* trial effect *plus* residual (analogous to SCA). This yields variety means adjusted for environmental effects. In comparison with crude, arithmetical means, variety comparisons are much improved but, if results are to be generalized to a prediction of the relative performance of varieties in agriculture (which is the purpose of trials), the assumption must be made that the environments sampled are close to average.

The decisions that flow from a set of trials, whether taken by the breeder himself or by bodies officially charged with the task, rest essentially on comparing new varieties with standards in respect, primarily, of yield but taking other characters into account, too; so the intuitive selection index so characteristic of plant breeding reappears. We have seen above that the traditional emphasis upon crude trials means or percentages as the bases for decision is weak in the presence of substantial GE effects but that it can be effectively supplemented by emphasis either upon regression relationships – which subsume both means and GE effects – or upon statistical adjustment by fitting additive constants to varieties and trials. Two potentially important effects of transferring emphasis from means to regressions would be: (1) to sharpen decisions as to what sort of niche a new variety might be adapted to; and (2) to enforce closer attention to the choice of sites. However trials are to be interpreted, there is much merit in incorporating fairly numerous standards (especially agriculturally contrasted ones) and in maintaining some common standards over good runs of years. They will both sharpen contemporary comparisons and help to facilitate the re-analysis of past sets of trials.

Within the traditional framework, there have been a good many studies of trials design (e.g. Sprague and Federer, 1951; Patterson *et al.*, 1977; Kempton and Fox, 1997). Essentially, examination of variance components in past trials of the crop concerned permits the construction of model systems in which the effect of varying the numbers of replicates, sites and years can be tested. Thus minimal requirements for discriminating between varieties (at defined levels of difference and significance) can be estimated and some approach made to the problem of optimal allocation of effort. Several other studies of trials systems bear on a rather different question, namely that of how the breeder may most economically reduce rather large numbers of potential varieties to a few of the best of them; the conclusion is that the total trials effort should remain roughly constant over years, declining numbers of survivors being compensated by increasing intensity of assessment (Finney, 1958; Curnow, 1961).

Finally, there is a question that ought sometimes to be asked about trials systems but very rarely is. The purpose of trials is to predict, and they are satisfactory only in so far as they predict reasonably accurately. But any test of accuracy must depend upon the quantitative comparison of trials results with the performance of varieties in practical agricultural use. With the partial exception of some industrial crops, of which the produce is accurately weighed, good agricultural yield data by variety are rare and almost never available on a statistically satisfactory basis. If such data were available, tests would take the form of comparisons of mean differences between varieties in trials *vs.* agriculture or, if GE effects were present, of colinearity of regressions. The little evidence we have is conflicting. In rubber, there is a high correlation between trials and agricultural yields but rubber is a crop in which experiments are run on estates and on an estate scale; in wheat and sugarcane, in Australia on the other hand, Davidson (1962) thought that trials substantially overestimated the potential of new varieties. In the UK it has been noted (e.g. see Fiddian *et al.*, 1973) that national cereal yields were more or less static for a time despite the regular release of new varieties that were believed, on trials evidence, to be higher yielding. Collectively, the evidence is poor but this is a question well worth asking and answering. If trials were indeed found to predict well, then we could use the results with a confidence that can not at present be justified; if not, then we should need new kinds of trials.

The problem, it should be noted, is particularly acute in those crops in which agricultural worth is estimated indirectly, notably, the forage crops destined for grazing rather than cutting. Here, animal performance is implicitly predicted from arbitrary measurements of gross production qualified by chemical characteristics. It is here that trials procedures stand in greatest need of justification, especially as there is a substantial risk of distorting the proper objectives of plant breeding by substituting the satisfaction of arbitrary trials criteria for real agricultural performance as a measure of worth.

6.3 Requirements and conditions

Historical

Official, legally founded involvement in the sale of seeds of plant varieties has three main elements which are historically quite distinct. First, there is seed certification, which was rather widely developed in Europe and North America during the nineteenth century. The original object was to improve and maintain the quality of seed offered for sale by testing viability and cleanliness (freedom from dirt and weeds). The International Seed Testing Association (started, 1924) promulgated the now generally accepted international rules for seed testing. Seed testing became inextricably linked with seed certification when concern with varietal purity as a feature of seed quality was added to the list of requirements. So, nowadays, a seed certificate universally means that the seed crop was inspected and found true to varietal type *and* that the seed itself is of good quality in the more limited sense (sound, clean, viable and healthy). Historically, seed certification started on a permissive basis as a general encouragement to good standards; the universal tendency nowadays is to make it compulsory so that uncertified seed may not legally be sold.

The second element is recommendation, which has long been locally important (e.g. in Sweden and the UK since the 1920s) but is yet far from universally adopted. It sprang from the recognition that standards of varietal purity varied widely, that the naming of varieties was often chaotic and that some varieties were better than others. Farmers stood in need of objective advice as to which varieties and stocks to grow. A recommendation system therefore, such as that of the National Institute of Agricultural Botany (1919) in the UK, had to sort out varieties and test them in extensive trials in order to make effective 'recommended lists'. Certification and recommendation have many interests in common and each encourages the other but they are essentially different operations. Certification is fundamental to the maintenance of high standards of purity and condition; recommendation is a valuable but strictly inessential service to farmers which may, as we shall see below, be overtaken by other processes. On the NIAB, see Wellington and Silvey (1997).

The third and most recent element is plant variety rights. The idea developed in the 1930s in several European countries simultaneously; it is now nearly universal in Europe, making headway in North America and likely to become even more widely adopted. It sprang from the realization that, if commercial plant breeding were to be encouraged to the benefit of agriculture and society, then means had to be found to allow breeders to profit from their products. This they could only do if there were some legal restraint upon the multiplication and sale of seed. Previously, seeds of IBL and OPP or vegetative material of CLO, once on the market, were free for all to multiply and sell outwith the control of the breeder. HYB varieties, on the other hand, retained an inbuilt economic protection because the breeder alone had the parents; we saw above (Section 5.5) that this fact has influenced, perhaps even somewhat distorted, plant breeding methods and objectives.

Several methods of offering protection were tried; patents and the use of trademarks had little success and were, anyway, simply inappropriate. The method nearly universally adopted was to specify conditions for the registration of a new variety and then confer on the breeder a 'right and title' in that variety so that all seed sold could be produced by him and his agents and subject to royalty payments. Clearly, any such system depends upon a high degree of public control of the seed trade and, in particular, upon a rigorous certification procedure. Systems of this kind were well established in Europe before the late war but the key historical event (1961) was the Convention of Paris for the Protection of New Varieties of Plants. The eight European signatories, *plus* any other states joining later, formed UPOV (Union pour la Protection des Obtentions Végétales) with the objective of harmonizing domestic legislation and ensuring international protection among themselves for the products of their own breeders; so varieties bred in a UPOV country can be protected in all other UPOV countries but those bred outside cannot. In view of the increasing international movement of crop varieties, this will probably be a stimulus to the expansion of membership of UPOV.

The basic requirement of a plant variety rights (PVR) scheme is simply that a new variety and its name shall be registered and the breeder legally permitted to control its multiplication for his own reasonable profit. That is, the scheme could be permissive, allowing the registration and sale of all varieties offered that passed the basic DUS requirements. An extension, in the opinion of many an undesirable one, is that varieties shall be not only registrable but also proved to have some value for cultivation and use (VCU); the list of permitted varieties would thus be restricted to those thought to be of value. Several European countries have had such restrictions for many years and the practice is surely growing; following the UK Seeds Act (1964), DUS requirements were first established for a few crops; others were later added to the list and VCU requirements are steadily coming in, crop by crop. The trend towards variety lists restricted by VCU demands seems clear. We shall discuss some of the issues later and here merely note that plant variety rights require for effective operation only DUS tests and certification. VCU requirements are historically adventitious; it would be perfectly possible to have them without PVR (as a restrictive extension of a recommendation scheme for example) but, in fact, the idea has grown up alongside PVR.

Decisions upon exploitation

Our brief historical survey in the last section showed that there were two main kinds of official restraints upon the exploitation of new varieties, namely certification and plant variety rights, the former long established, the latter quite recent, but both growing. In this section we shall see how, in very broad terms, decisions upon the exploitation of new varieties are in fact taken against a varying social background of official constraints. There are three cases to be considered.

First, there may be no or only minor constraints and the breeder may therefore make his own decisions. In the absence of compulsory certification and of a PVR scheme, a breeder can handle the stocks and name the variety as he wishes and sell the seed for what he can get. This was the situation in most advanced temperate agricultures up to about seventy years ago. Voluntary certification, however, gained force and the (local) practice of recommendation undoubtedly tended to raise standards of varietal performance and purity and of seed quality; commercial competition no doubt helped to do the same. Given a discriminating population of farmer-customers and efficient and competitive breeders (as in North America generally) this is a perfectly effective way of doing things but it does deny the possibility of plant breeders' rights (PVR).

Another situation in which the breeder can make his own decisions without any official control occurs where the breeder himself is the user of the variety, as with plantation companies growing and breeding sugarcane, rubber, oil palm, etc. Cottons bred by the Cotton Corporation in tropical Africa would fall into the same category, though the growers are small farmers rather than company estates. The selling of such varieties to other companies would be a normal commercial transaction unrelated to certification or PVR (which simply do not exist in the circumstances of tropical plantation agriculture and, indeed, yet hardly exist in the tropics at all).

A variant of this situation is one in which the breeder's own views as to whether or not to exploit a new variety are tempered by reference to some voluntary-collective organization which tries to weigh up private and public interests in coming to decisions. Thus, local release of sugarcane varieties bred by some trade-cooperative or government-backed effort would normally be subject to the advice of a local committee of scientists, growers and administrators.

The second situation is one in which PVR exists but the list is open to all eligible varieties, without reference to any 'value for cultivation and use' (VCU) criteria. The requirements for registration (without which a variety may not be marketed at all) are relatively simple. The breeder must pay the appropriate fees, offer an acceptable name and provide the required batches of seed or planting materials for trials. The responsible officials will then carry out the DUS tests appropriate to the crop which are essentially tests of: (1) distinctness (is it different from all other varieties currently on the official list or in hand?); (2) uniformity (is it sufficiently uniform to meet the published arbitrary requirements in this respect?); (3) stability (does it breed true from year to year, having regard to any inherent reproductive peculiarities of the crop?). The observations made to determine distinctness and uniformity involve very minute description of the variety but, in the limit, description may not suffice, so side-by-side comparison of growing plants is sometimes necessary.

DUS tests establish the legal identity of the variety and, in addition, demonstrate that it is potentially certifiable at a very high level. Any doubts and queries having been resolved (a breeder may be required to resubmit), the variety is then either denied entry, on stated grounds, or admitted to the official list of permissible cultivars. Admission and publication constitute the breeder's 'right and title' to demand royalties upon sales; conversely, admission also involves the obligation to make commercial seed available if demanded. The legal term of rights is generally eighteen years but few cultivars, at least of annual crops, will last so long. All sales of seed during the term are subject to royalties, so a natural corollary of any PVR scheme is compulsory certification; farmer-to-farmer, over-the-hedge sales are banned (or, perhaps more realistically, discouraged).

All countries which operate PVR schemes are signatories of the Paris Convention and are members of UPOV; they agree to admit, on request (but with some qualifications), all cultivars registered and protected in other UPOV countries to their own national cultivar lists and afford them the same legal protection. In consequence, plant breeding is rapidly becoming more international in character as the scope becomes wider and the competition warmer.

This situation (PVR with the necessary DUS but no VCU) would be perfectly stable if left alone; in practice it is transient, or has been in Europe in recent years, to a much more restrictive situation the third in our list, in which the VCU criterion is used to shorten the list of permissible cultivars. VCU trials, as we saw above, are performance trials, over sites and years. The officials responsible for them, assisted by various committees, take decisions, on the basis of trials results, as to whether new varieties should or should not be admitted to the list of permissible cultivars. The VCU criteria are necessarily rather vague; at least some advantage, though maybe quite a small one, over established standards is implied; a definite yield advantage would presumably always be decisive but quality characters and disease reactions are taken into account. In fact, the intuitive selection index recurs. Results of VCU trials must be published and the breeder has the right of appeal against adverse decision. The existence of a VCU criterion (when it has worked its way through the system, at least) should mean that all cultivars on a list are good and some are excellent. This would imply a declining need for recommendation which was, historically, designed to sort out the chaos of varietal names and identify a reasonable range of the better cultivars for recommendation to farmers. Nowadays, certification and DUS testing assures an orderly list of varieties and VCU should eliminate the bad or indifferent ones.

Finally, perhaps, some assessment is wanted. If the objective is good varieties efficiently moved into agricultural production, all three situations can work very well. Competent breeding and management ensure that excellent cottons, sugarcanes and rubbers go into production without any need of legislation or certification to control the flow or guarantee the quality of planting material. The example of North America shows that a combination of critical farmers, competitive breeders (emphasizing hybrid varieties) and a good certification system can be highly effective. In Europe, PVR systems, with or without various levels of VCU, coupled with efficient certification, also work well and ensure economic returns to breeders (albeit at some cost in bureaucracy).

That the PVR concept in general is a good one can not, I think, be doubted. It manifestly fulfils its basic objective of encouraging commercial breeding that is competitive both within itself and with any governmental breeding systems that may exist; by increasing the total flow and by promoting the international movement of protected cultivars, agriculture and society at large must surely benefit. In the slightly longer run, it is hard to believe that any agriculturally advanced country can stand outside the PVR system. Any doubts one might have must, I think, centre on two points: first, the possibly stultifying effect of too rigid an application of DUS criteria (which are legal rather than agricultural in intent); and, second, the application of VCU principles (which are certainly unnecessary and expensive and can be regarded as potentially disadvantageous to farming). The second point is the weightier: officials have to take decisions on what may be rather thin trials evidence; conventional approaches to interpretation of trials are not above criticism, as we saw above, and bad decisions will undoubtedly sometimes be taken, as they have been in the past; farmers, with some guidance perhaps from recommendation, are perfectly capable of making their own choices, which do not necessarily agree with those of the officials. The VCU criterion seems to have come to stay, however; if interpreted broadly it may work quite well but the risks and disadvantages of narrow interpretation are real; unrestricted lists, even if they were longer, would probably have served farming better.

6.4 Multiplication

Breeder's multiplication

There are four requirements placed upon a breeder in developing stocks of a new variety for commercial multiplication. Three – purity, quality and health – are common to all multiplication systems and are basically the requirements of certification at a very high level; the fourth – uniformity – is a special requirement placed upon the breeder by custom and usage as well as, much more formally in recent years, by DUS testing.

Purity implies freedom from admixture by material of other varieties of the same species, freedom also from genetic contamination by foreign pollen and absence of weed seeds. It is achieved by scrupulous attention to cleanliness of equipment, to careful choice of land in relation to previous cropping history and to appropriate spatial isolation of multiplication plots. The need for spatial isolation stems from the hazard of cross-pollination. It is least in strict inbreeders, where, nevertheless, outer or windward rows are often discarded, greatest in crops liable to long-range pollination by wind (maize, sorghum) or by insects (brassicas, cottons, clovers). Quality implies well-developed, highly viable seed or planting material, to be achieved by good agronomic practice in a suitable environment.

Health implies freedom from seed-borne diseases and it is attained by a combination of choice of site for multiplication, chemical treatments with fungicides (or sometimes with insecticides intended to kill insect vectors) and roguing. In practice, health problems are generally greatest in clonal crops, for obvious reasons. Vegetatively transmitted diseases caused by fungi, bacteria and viruses – the last are perhaps especially troublesome – can all too easily be carried over in propagating material into commercial multiplication. Seed-propagated crops are sometimes troubled by seed-transmitted fungi, such as the cereal smuts and bunts, but they are, in effect, free of viruses. The health problems are trivial in comparison with those faced by the breeder of some clonal crops, such as potatoes, in which the maintenance of health during breeder's multiplication is a major preoccupation.

In meeting these requirements, of purity, quality and health, the breeder is simply doing what the commercial multiplier will later do but doing it at an extremely high level; the effects of a mistake or bad luck will inevitably increase down the line, so standards must be of the highest.

The problems posed by the uniformity requirement vary with the crop. A clone is a clone, so here the problems are generally nil or at most slight, though recurrent somatic mutations, such as bolters and wildings in potatoes, may have to be rogued out. The breeder of IBL is, at the limit, in much the same situation as the breeder of CLO but is always faced, as the CLO breeder is not, with the problem of purification before release. When his selections are submitted for trials at about F_6 there will nearly always still be visible segregation for minor morphological characters. Since DUS tests impose a requirement for near absolute uniformity (as, for legal reasons, they must) an elaborate system of successive ear-row examinations is usual (Fig. 6.4). These ear-rows are (e.g. in cereals) spaced rather widely, so that individual plants can be minutely examined; fertility is kept down to minimize lodging; and the general principle is to discard whole rows that show any off-types. Official tests, trials, purification and initial multiplication thus proceed concurrently. With regular, functionally diploid inbreeders such as wheat, barley, oats, many pulses, tobacco and others, problems rarely arise; when they do, they are usually attributable either to a higher-than-usual rate of out-pollination (consequent on more open-flowering) or to a persistent cytological peculiarity leading to apparent recurrent mutation, e.g. speltoids and fatuoids in wheats and oats.

While demands for apparent uniformity can reasonably be set at a very high level for CLO, IBL and, similarly, for HYB based on highly inbred pure lines, the same is not true of OPP (and SYN). These populations, by their very nature, can not be uniform; indeed, in so far as uniformity is related to inbreeding it would be positively undesirable from the point of view of agricultural performance to set the demands too high. Furthermore, there is no evident means of setting objective standards of uniformity in such material since the notion of 'offtypes' in a continuously variable population is virtually meaningless. There is no alternative but to rely on common sense and common experience of what is 'reasonable' in the circumstances. In practice, experienced breeders and officials responsible for DUS testing and certification would probably not disagree too widely as to what is 'reasonable' but it must be recognized that disagreements will tend to sharpen with time. Agriculturally, a good deal of variation, of non-uniformity, is inevitable and indeed desirable; legalistically, uniformity contributes to certainty of identification and therefore relates to the smooth operation of PVR. So there is a conflict and the inevitable compromise is an uneasy one. The breeder of CLO, IBL and HYB, whether he is working in a PVR context or not, has clearly defined uniformity standards to aim at; they may be high or low, according to circumstances, but they should be reasonably attainable; the OPP/SYN breeder is in an altogether more difficult position, in that true uniformity is unattainable and a 'reasonable' degree of it is objectively undefinable.

Whatever the nature of the crop, the breeder, in introducing a new variety to agriculture, incurs an obligation to maintain it in marketable form so as to be able to release, at intervals, supplies of authentic seed for commercial multiplication. Under any PVR scheme the obligation is, as we saw above, a legal one; the breeder *must* maintain his variety in good shape and make seed available for sale. The obligation lapses only with the withdrawal of the variety from the register or list. The technical requirements of such maintenance and release are exactly the same as for the first release and, for the maintenance of uniformity, purity, quality and health, the same methods are used.

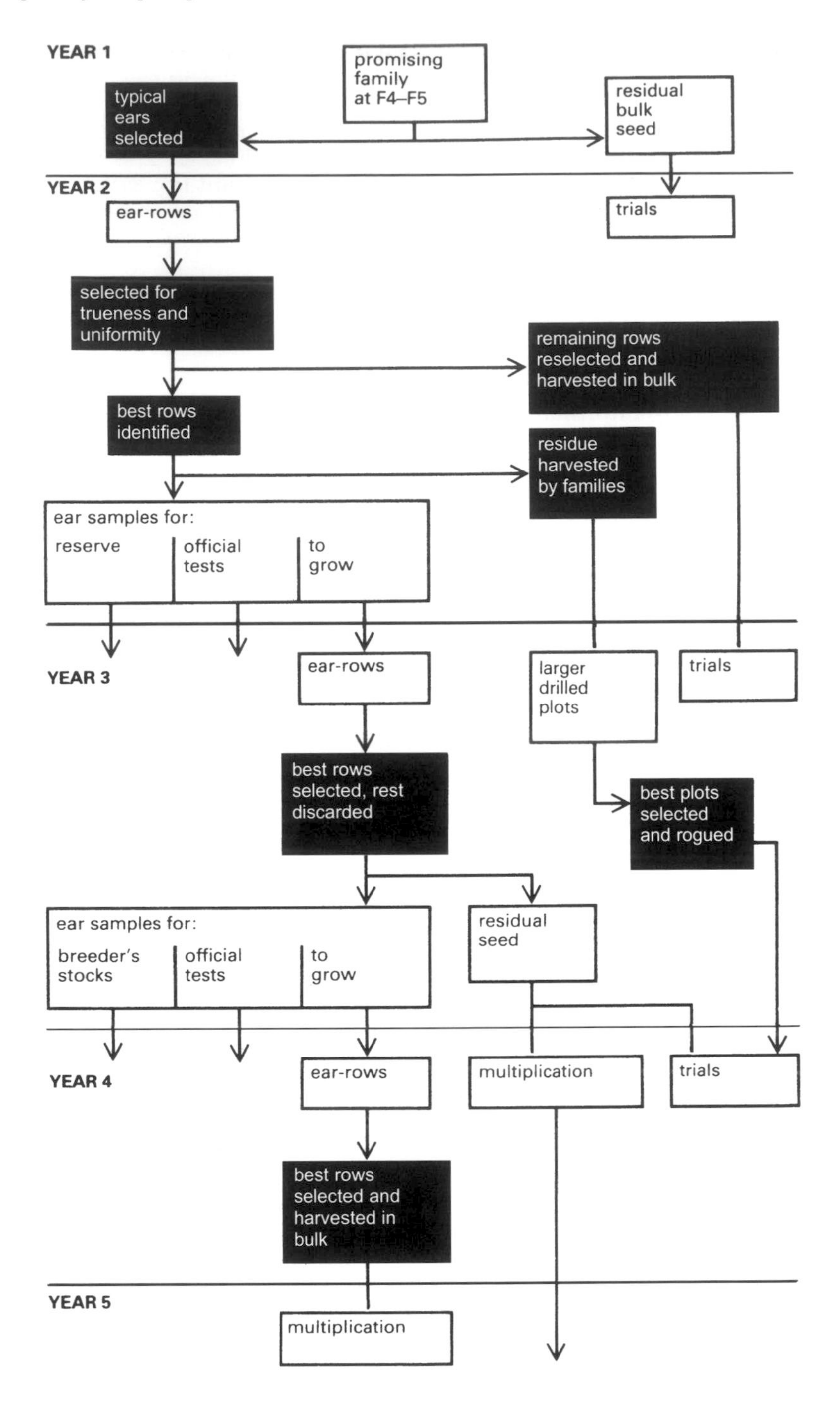
YEAR 1
promising family at F4–F5
typical ears selected
residual bulk seed
YEAR 2
ear-rows
trials
selected for trueness and uniformity
remaining rows reselected and harvested in bulk
best rows identified
residue harvested by families
ear samples for:
reserve
official tests
to grow
YEAR 3
ear-rows
larger drilled plots
trials
best rows selected, rest discarded
best plots selected and rogued
ear samples for:
breeder's stocks
official tests
to grow
residual seed
YEAR 4
ear-rows
multiplication
trials
best rows selected and harvested in bulk
YEAR 5
multiplication

Fig. 6.4 The trials-multiplication phase of a cereal IBL such as wheat, barley or oats. Notes: The diagram relates to a situation in which PVR exist and seed requirements for official tests and trials have to be met concurrently with the breeder's own trials. The diagram is quite generalized and probably describes no real programme exactly. Note that the purity of stocks in early trials need not be very high, that two main releases for commercial multiplication are provided for and that ear samples submitted for official tests represent the multiplication lines. The track at bottom centre is the start of recurrent breeder's–commercial multiplication cycle that will persist as long as the variety is grown.

Commercial multiplication

In technologically advanced agricultures, commercial multiplication of stocks, whether or not a PVR scheme is in operation, is dominated by the demands of certification. These cover both genetic and seed quality aspects. The leading schemes are those of the Organization for Economic Cooperation and Development (OECD, 1969) and, in North America, the Association of Official Seed Certifying Agencies (AOSCA, 1971). Historically, a seed line used to be maintained for as many generations as it could be certified for. Both schemes favour the idea of 'limited generations', which are, in practice, becoming generally adopted; this places a maximum (commonly 3–5) on the number of generations permitted to intervene between basic seed and the seed sold to farmers. Nomenclatures vary somewhat. The OECD scheme recognizes: Pre-basic, Basic (produced under the supervision of the breeder) and Certified generations 1, 2, 3 . . . n. AOSCA in the USA recognizes: Breeder's, Foundation, Registered, Certified; while Canada interjects a Select stage between the first two. If the uncertified product is permitted it would probably be called Commercial.

Whatever the scheme, principles are identical and practices similar (OECD, 1969; AOSCA, 1971; Feistritzer, 1975). Elaborate rules are laid down for: choice of land in relation to previous cropping; isolation from potential contamination (mechanical or genetic); agronomic practices; cleanliness and suitability of equipment and stores; labelling methods and packing; roguing. Growing crops are subject to inspection (by carefully prescribed methods) and seed samples tested against the usual criteria of soundness, viability, cleanliness and health. Standards are, as far as possible, objective and usually of the form 'not more than x per cent of . . .'. The nearer the start of the sequence the higher they are, being relaxed somewhat in the later generations. The outcome is an orderly flow of good quality seed, pure (within narrow practical limits) as to variety. The OECD scheme is internationally orientated and (like UPOV) aims to promote the traffic in seeds among member countries; seed sold under an OECD label ought, therefore, to attain an internationally agreed standard, and no doubt usually does so.

The foregoing summary applies generally to all crops, inbred and outbred, seed propagated and clonal. But there are two qualifications, relating to HYB and CLO respectively. The idea of 'limited generations' is, of course, inapplicable to hybrid varieties and to those SYN, such as grasses and clovers, in which generations are anyway limited by biological features (Sections 5.4, 5.5). This merely means that the early phases of multiplication, i.e. of parental clones or lines, must be on a scale sufficient to produce the necessary flow of certified seed at the end of the sequence. Principles and practice of certification remain unchanged. The methods of producing HYB seed on the commercial scale present many features of interest and are summarized separately in the next section.

As to clones, diseases, and especially virus diseases, sometimes impose specialized propagation methods as a feature of the earliest phases of multiplication. Thus, potatoes in Scotland go through a virus-tested-stem-cutting (VTSC) process between the breeder and the seed farmer, so there is an extra stage before cultivars enter the usual procedure of certification with limited generations. Similarly, in a number of clonal fruits (such as applies, cherries, raspberries, blackcurrants and strawberries in the UK), 'nuclear stock associations' have been set up, usually by grower-cooperatives with official backing; their object is to support the development and maintenance of virus-free stocks of established and new varieties and promote an orderly flow of certified planting material to growers.

Hybrid varieties

Hybrid varieties and those SYN populations which are based on controlled crossing between two parental populations, lines or clones (Section 5.5) survive only for one generation of agricultural production. Multiplication is therefore biologically, rather than legally, limited to one generation. In this section we shall be concerned with the many, and often ingenious, ways in which such varieties are multiplied for commercial use. The main points are summarized in Table 6.3. Five main methods may be distinguished, namely: mechanical, sex, nuclear male sterility (NMS), self-incompatibility (SI) and cytoplasmic male sterility (CMS). We shall deal with them seriatim. Some day, incidentally, it may be necessary to add a sixth category: chemical. Selective male gametocides have been sought for years but no really effective ones have yet been found.

Mechanical methods of emasculation, by pulling or cutting the tassels, dominated the maize scene until a CMS system became available (Fig. 6.5). This crop, with its copious wind-borne pollen and easy detasselling, is outstandingly well adapted to mechanical control. Otherwise, the method is possible only if high seed costs can be accepted; in the palms mentioned in Table 6.3, plant density is low and the plants are long-lived, so seed costs per hectare per year are low enough.

The next two methods, sex and NMS, have much in common. Both depend upon roguing pollen-bearing plants out of the functionally female parental line or population. Sex is of limited use because seed-propagated bisexual crops are

Table 6.3 Methods of controlled crossing to produce seed of HYB (and some SYN) varieties. Based largely on Duvick (in Frey, 1966).

Method	Leading features and examples
1. **MECHANICAL**	**Principle:** in hermaphrodites, emasculation followed by hand or natural pollination; in diclinous plants, simple protection and hand or natural pollination **Problems:** labour requirements; feasible only if relatively numerous seeds produced per operation **Examples:** maize detasselled and wind-polinated (Fig. 6.5); European tomatoes and US cucurbits produced by hand pollination; SYN populations in oil palm and coconut (diclinous) also produced by hand-pollination
2. **SEX**	**Principle:** mix two dioecious populations and rogue out, before flowering, males of one, females of other; progeny of surviving females will be population cross segregating 50 male : 50 female **Problems:** roguing is costly **Examples:** spinach in USA; in principle, could be used for hemp (*Cannabis*)
3. **NUCLEAR MALE STERILITY (NMS)**	**Principle:** found in all species, if sought; usually recessive (conventional symbols *Ms*-*ms*); mix female–hermaphrodite (*ms ms* + *Ms ms*) line with hermaphrodite pollinator population (*Ms Ms*); rogue out hermaphrodites (*Ms ms*) in former; in principle, facilitated by closely linked dominant marker (Fig. 6.6) **Problems:** roguing is costly and genetic expression not always satisfactory; good linked markers are very rare; natural pollination may be unreliable **Examples:** castor in USA; suggested but yet impractical for several other crops, e.g. cotton, carrot, tomato, sunflower, cucurbits, barley
4. **SELF-INCOMPATIBILITY**	**Principle:** interplant, either (a) two self-incompatible (SI) but cross-compatible clones or inbred lines and harvest all seed; or (b) an SI clone or line to act as female with an SC pollinator; several variants are possible, mostly designed to spare the costly SI material (Figs 6.7, 6.8) **Problems:** maintenance of SI lines by hand-pollination without automatically selecting for self-fertility and hence unwanted 'sibs'

Contd

Table 6.3 *Contd.*

Method	Leading features and examples
4. **SELF-INCOMPATIBILITY** (*contd*)	**Examples:** several vegetable brassicas, marrowstem kale (Fig. 6.7), sunflower, cacao (Section 5.4); other fodder brassicas (e.g. swedes, Fig. 6.8) are likely; 'rough' hybrids, with many selfs, often suggested and a few achieved in clovers, alfalfa and several grasses
5. **CYTOPLASMIC MALE STERILITY (CMS)**	**Principle:** CMS is of wide natural occurrence and can sometimes be constructed by backcrossing desired nucleus into foreign cytoplasm; male sterility results from nucleo-cytoplasmic interaction; one-several genes are involved, 'restorers' usually being dominant (conventionally *R–r*); male-sterile lines (cytoplasm *plus* non-restoring nucleus, *rr*) interplanted with hermaphrodite pollinator line and seed harvested only from the former; if product need not be seed-fertile (e.g. beet, onion), pollinator line need not carry restorer(s) (*R*), otherwise (e.g. maize, sorghum) it must. Several variant schemes are possible (Fig. 6.5). **Problems:** efficient sterility and restoration are rare and can be very laborious to construct; cytoplasm itself (always present in the final product) often has deleterious effects on vigour, yield, disease resistance, etc.; pollination difficulties except in naturally wind-pollinated crops **Examples:** outstanding successes in onions, sugar beets, maize and sorghum in the USA; used to a limited extent in a few minor crops (carrot, petunia); projected but so far unsuccessful in many others (notably wheat, cotton, tobacco)

Fig. 6.5 Production of HYB maize seed by mechanical (detasselling) and cytoplasmic male sterility (CMS) methods.
Notes: CMS cytoplasm is shaded, normal cytoplasm unshaded: *R* is the dominant restorer needed to suppress the CMS effect. Lines A, B, C, D are the constituents of a double-cross, with A related to B and C to D but (AB) relatively unrelated to (CD). ABCD (5), the double-cross made wholly by detasselling, is the traditional HYB maize, displaced by the partial or full CMS systems (ABCD (1) (2) (3)). AD (4) is the cytoplasmic single-cross which was approaching dominance in the USA when disrupted by the Southern Leaf Blight epidemic provoked by the susceptible T-cytoplasm (Chapter 7). AD (4) also represents the system that has been highly successful in sorghum; in onions and sugar beet (vegetative products, seed fertility in HYB immaterial), AB (*rr*) in CMS cytoplasm is acceptable.

very few. NMS is potentially of wider utility but its use must be limited by the necessity of costly roguing (Fig. 6.6).

There are several kinds of self-incompatibility systems in plants, some characterized by morphological differentiation of the mating types (heteromorphic), some not (homomorphic); examples are given in Table 1.5. In principle, any of these could be adapted to HYB production; in practice, only homomorphic systems have been seriously investigated in this connection and only the homomorphic–sporophytic system has actually found much practical use. Genetic details lie beyond our scope; Nettancourt (1977) gives a compre-

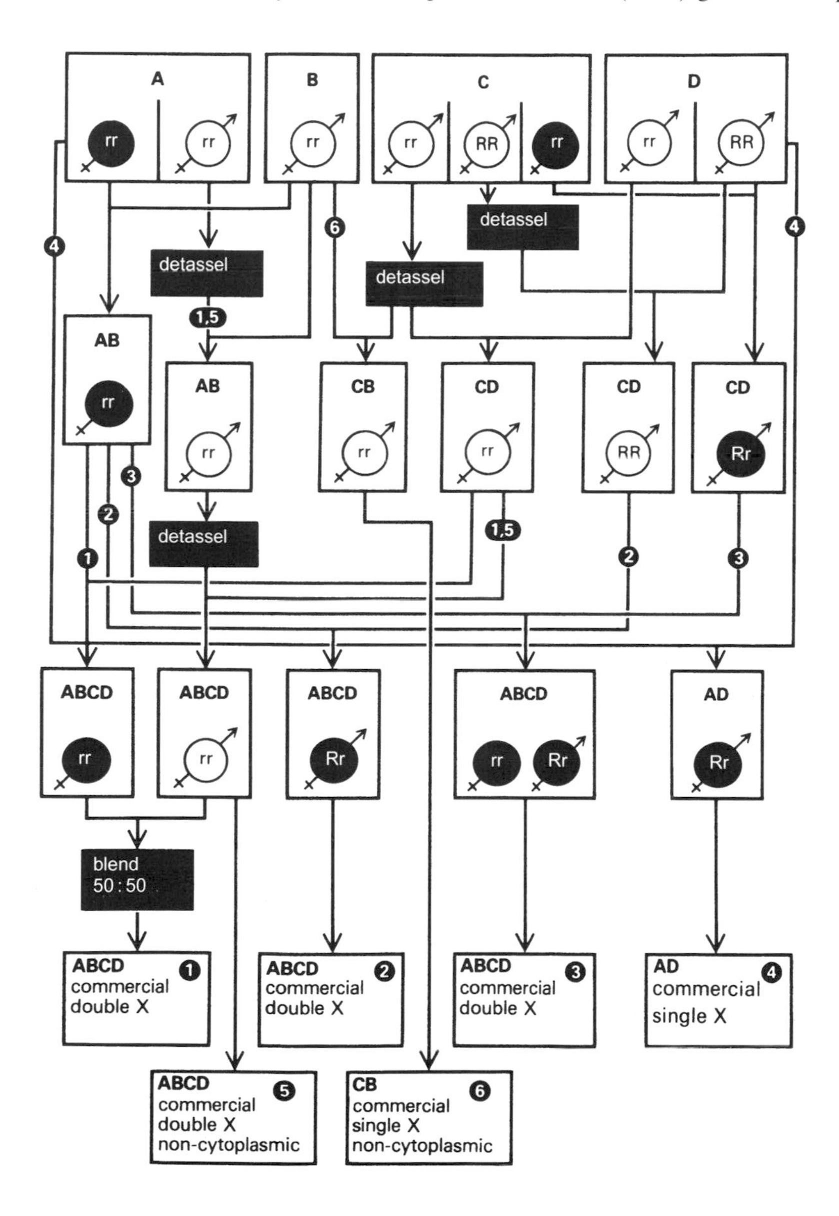

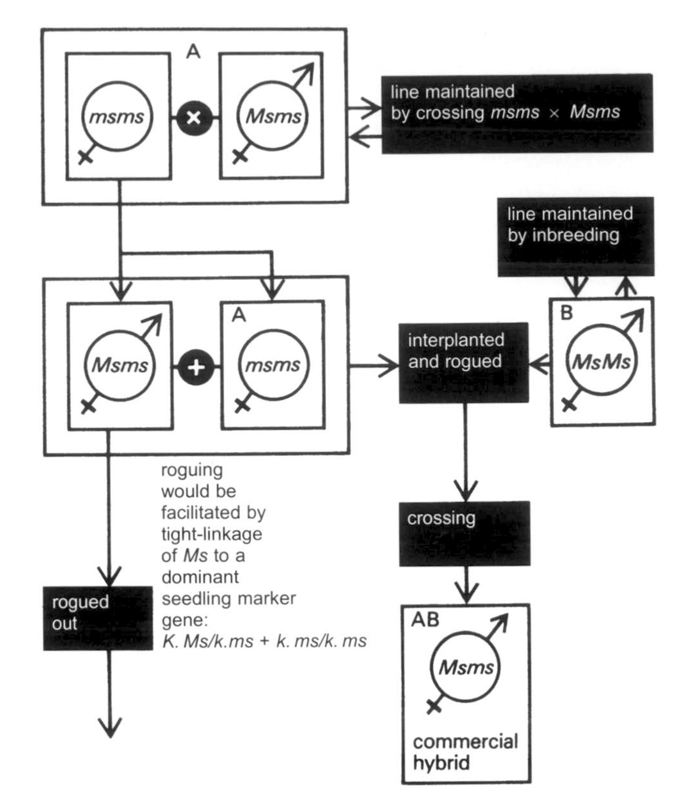

Fig. 6.6 Use of nuclear male sterility (NMS) to produce HYB varieties.
Notes: In practice, useful mutants are always recessive so *msms* is functionally female. Note that seed yields of the female-producing line are low because only half the population can be harvested. In the final phase, roguing out of hermaphrodites in one line is necessary, before they flower; genetic marking would facilitate this but, though simple in principle, has so far proved impossible in practice for want of good markers sufficiently closely linked. The system proposed for barley (Burnham in Frey, 1966) involves a cytogenetic mechanism on translocations to aid the roguing.

hensive review and outlines will be found in most genetics texts (e.g. Williams, 1964). Essentially, homomorphic systems depend upon the existence of several-many alleles at the *S* (incompatibility) locus; pollen-or tube-growth is inhibited by essential identity of genetic constitution as between the grain and the style (self-incompatibility, SI); self-fertilization is therefore generally prevented and unrelated plants may or may not be cross-compatible; however, inactive, self-compatibility (SC) alleles are known and many SI plants can be caused to set some seed by bud–pollination. In gametophytic systems (as in many Solanaceae, grasses and legumes) the genotype of the pollen determines its phenotype and therefore its reaction on a given style; in sporophytic systems (as in Cruciferae and Compositae), some alleles show series-dominance and the reactions of style and pollen grain depend upon zygotic constitution with respect to *S*.

All HYB-producing systems, both realized and proposed (Figs 6.7, 6.8) depend upon the use of at least one, and usually several, inbred lines homozygous for an appropriate *S*-allele; such lines are maintained by bud-selfing and are intrinsically infertile anyway, so seed is expensive. In principle, the lines should be SI when grown in the field and therefore reliable receptors for pollen from the complementary line. In practice, repeated bud-selfing automatically selects for self-fertility as a genetic background effect, regardless of *S*-allele constitution, so such lines often give trouble; they tend to produce within-line crosses (so-called 'sibs') in sufficient frequency to ruin an otherwise promising HYB variety. They have been especially troublesome in vegetables, such as Brussels sprouts, in which the demand for uniformity is high.

The proposals for making HYB varieties in *Brassica napus* (Fig. 6.8) follow the same principles as in *B. oleracea* (Fig. 6.7). Because *B. napus* is a self-fertile allotetraploid (AACC) and tolerant of inbreeding, the inherent vigour and fertility of inbred lines should pose no problems. The chief difficulty seems to be in constructing parental lines which are sufficiently SI to be reliably crossed. The available methods are: finding good *S*-alleles in *B. napus*, backcrossing them in from *B. campestris* or incorporating them in synthetic *B. napus* constructed *de novo* by crossing *oleracea* and *campestris* (Section 8.3). The same principle, of using various combinations of SI and SC lines, could readily be adapted to the production of the rather unusual HYB variety based on the interspecific combination, rape (*Brassica napus*, AACC) × turnip (*B. campestris*, AA); the cross is easily made and the triploid (sesquidiploid) products (AAC) have many attractions as rape-like forage plants (Mackay, 1973, 1977).

Turning now to CMS systems (Fig. 6.5), knowledge of the existence of maternally inherited male sterilities goes back to the early years of this century and Krug in 1932 made the first proposal for its use in producing HYB maize. The cytoplasm used by Rhoades in his classic study of CMS in maize, however, was not a useful one and was lost. It was not until the middle 1940s that the S and T cytoplasms were identified and it was some years before they (especially T) could be incorporated in breeding programmes. Meanwhile, highly successful CMS systems were developed in onions and sugar beets in the USA and their use, in fact, predated the development in maize. Sorghum, the fourth significant application of this technique, followed in the middle 1950s.

There are three requirements of an effective CMS system, namely: (1) an efficient cytoplasm lacking undesirable side-effects; (2) nuclear-genetic adaptation of the stock, especially in connection with restorer genes; (3) an efficient pollination mechanism in the field, whether by wind or insects. Good cytoplasms are not common and many have to be rejected. CMS has been found at low frequency in many crops and would probably be found, at least very rarely, in all if intensively sought. Alternatively, it can sometimes be constructed by backcrossing to a related species so that the desired nucleus is introduced to a 'foreign' cytoplasm. Examples are provided by cotton (*hirsutum* nucleus into *anomalum* or *arboreum* cytoplasm), tobacco (*tabacum* nucleus into *debneyi*,

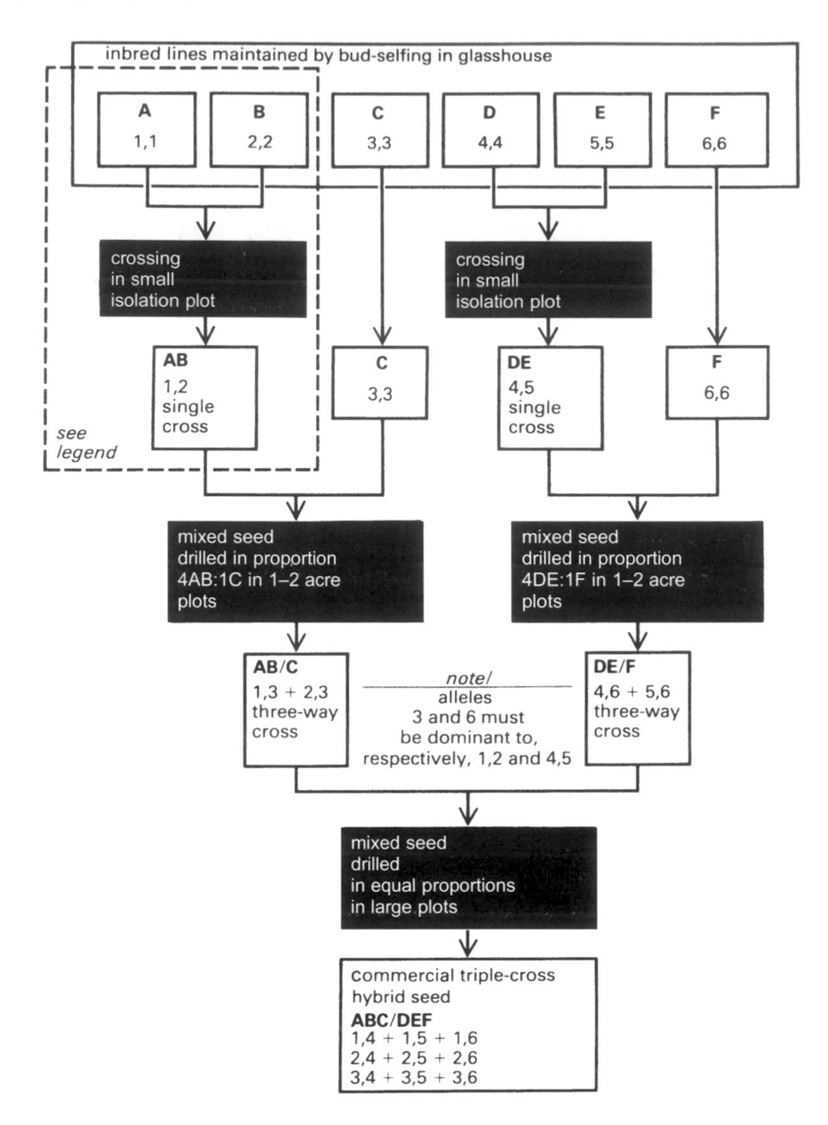

Fig. 6.7 Triple-cross kale produced by use of the self-incompatibility system characteristic of diploid species of *Brassica* (Thompson, 1964, 1966).
Notes: Six lines, each homozygous for different *S*-alleles and with dominance relationships indicated in the figure, are necessary. The assigned *S*-allele numbers are arbitrary. The area in the top left surrounded by a dashed line represents the simplest possible system, used in vegetable crops such as Brussels sprouts, in which uniformity is necessary and very high seed costs tolerable. The complicated system adopted in the kale is imposed by the necessity of economy in the use of weak and infertile inbreds to produce commercial seed cheaply enough; genetically, the product is perhaps nearer to SYN than to HYB.

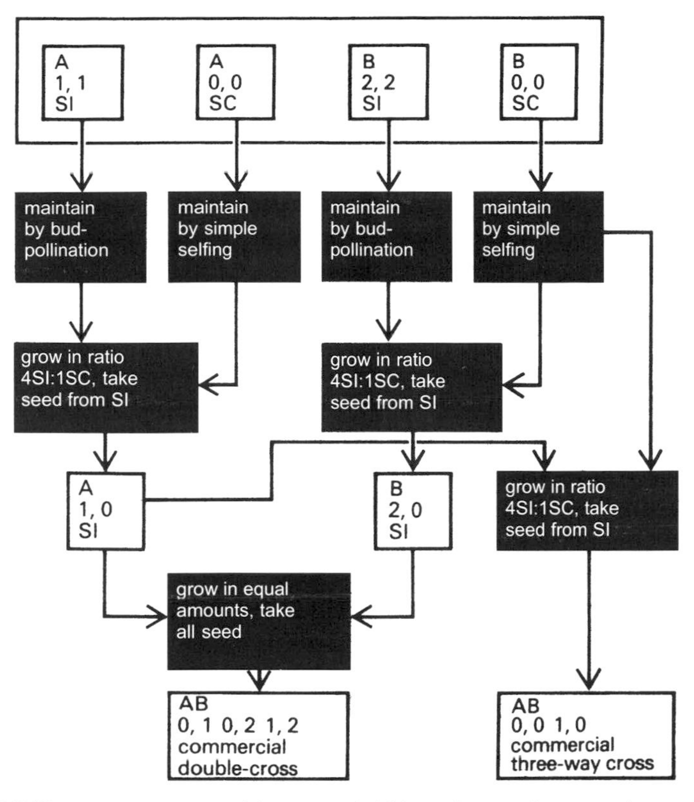

Fig. 6.8 HYB systems proposed for tetraploid brassicas such as swedes and rapes (Gowers, 1973).
Notes: *Brassica napus* (4 ×) is tolerant of inbreeding and generally highly self-fertile. Good *S* alleles are scarce and the systems proposed are designed to use them to best advantage and to economize on the use of lines which must be maintained (expensively) by bud-pollination.

megalosiphon or *bigelovii* cytoplasm) and wheat (*aestivum* nucleus into *timopheevii* or *Aegilops caudata* cytoplasm; *durum* nucleus into *Ae. ovata* cytoplasm). Incomplete male-sterility and undesirable side-effects are common. Thus floral abnormalities and poor growth occur with the tobacco cytoplasms mentioned above and even T in maize, by any standards a successful cytoplasm, slightly retards growth and plants carrying it turned out to be disastrously susceptible to Southern leaf blight (Table 6.3, Fig. 6.5, Chapter 7.2).

Nevertheless, workable cytoplasms can sometimes be developed. The nuclear–genetic requirement is basically for an effective restoration system when this is necessary (e.g. in maize and sorghum but not in onions or beets). Probably, several genes are always involved but, if some of them are common in parental populations, the need may only be apparent for one or two. In maize, the S and T cytoplasms require different restorers, one for S and two for T but, of the latter, Rf_2 is commonly present and homozygous in inbreds so only Rf_1 need usually be considered. Lack of restorers has sometimes seriously inhibited, even completely blocked (e.g. in cotton), the development of a CMS-based HYB programme. Even if a good restorer is available it may occur in an undesirable genetic background and give trouble in backcrossing. In maize, Rf_1 lies near the centromere on chromosome III, so restricted recombination leads to persistence of donor characters and the necessity of extensive testing of converted lines.

The third requirement, efficient cross-pollination in the field, is likely to be met by many outbred species whether wind-or insect-pollinated. One notes that sorghum, though tolerant of inbreeding, has some outbreeding features, including the copious production of wind-borne pollen. In cotton, if a CMS system could be developed, insect pollination would be necessary; fears have been expressed that this could be difficult to achieve in a crop that suffers worse from insect pathogens than most and must regularly be treated with insecticides. Inbred crops are, broadly, characterized by more or less closed flowering and relatively sparse pollen production. No such crop has yet been successfully adapted to a CMS system and it seems likely that inbreeders will always be beset by pollination problems. A likely CMS system in peppers (*Capsicum*) failed for want of sufficiently active insects. In wheat, the object of very active HYB research, good seed setting is rarely achieved and there is speculation that it may be necessary to breed more open-flowering lines; this, however, might encourage the spread of ovary-infecting diseases such as smut and ergot.

In summary, it looks as though outbreeders are not only better adapted genetically than inbreeders to HYB methods (Section 5.5) but are also better adapted morphologically to CMS technology. But, even in an outbreeder with a reliable pollination system, CMS, with its joint need for a good cytoplasm and correctly adapted nucleus, remains a formidable undertaking.

6.5 Summary

The objective of a trial is to test new varieties against standards in respect of agricultural performance characters, of which yield is by far the most important. Breeder's trials help the breeder to come to decisions on potential new varieties; official trials are the basis of registration under plant variety rights schemes (PVR) and of recommendation. All trials observe the statistical principles of replication, randomization and local control. Individual trials are aggregated into trials systems, so that important decisions are always taken on several-to-many comparisons of new variety with standards. If genotype-environment (GE) effects are important, simple mean comparisons become invalid but recourse may be had to statistical devices (regressions, additive constants) to aid interpretation.

There are three kinds of restraints upon the marketing of new varieties. Sometimes (e.g. in tropical agriculture generally) they are absent or but little developed; in technology-based agriculture, however, they are weighty and becoming more so. They are: certification (aimed at promoting the sale of seed which is pure as to variety and of good quality, viability and health); recommendation (aimed at identifying for farmers the best available varieties); and plant varieties rights (PVR, aimed at providing a legal framework whereby breeders may exact royalties on sales of seeds of their varieties). PVR is growing rapidly; basically, it confers a 'right and title' in a new variety that passes the prescribed tests for distinctness, uniformity and stability (DUS); additionally, the current tendency is to extend the scheme to include performance (value for cultivation and use, VCU) tests whereby the new variety is required to be not only acceptable but positively superior. It is doubtful whether VCU requirements are in the best interests of either breeders or farmers.

Once bred, tested and accepted, new varieties must be multiplied. The initial phase is in the hands of the breeder whose task is to present stocks to the commercial multiplier in a very high state of uniformity, purity and health. They can then undergo the limited number of generations of commercial multiplication prescribed by the prevailing certification scheme and reach farming in volume and at the appropriate level of purity. PVR and certification schemes (which are formally independent but practically linked) increasingly take on an international character. They set out the requirements for, and methods of attaining and testing, purity and seed quality in great detail.

Hybrid varieties (and some synthetics), though subject to the same sort of PVR and certification requirements, differ from other kinds in that their multiplication is limited biologically to a one-generation process. Methods of making hybrid seed may be classified as: mechanical, sex, nuclear male sterility (NMS), self-incompatibility, and cytoplasmic male sterility (CMS). The last, CMS, is the most important in practice; it has worked well in onions, beets, sorghum and maize but has so far failed, for various reasons, in several other crops. The difficulties of hybrid technology are considerable; either crossing systems or natural pollination may be limiting.

6.6 Literature

Trials and trials systems

Allard and Bradshaw (1964); Baker (1968); Campbell and Lafever (1977); Ceccarelli, Grando and Hamblin (1992); Cochran and Cox (1957); Cox (1958); Curnow (1961); Davidson (1962); Eberhart and Russell (1966, 1969); Federer (1963); Federer and Sprague (1947); Finlay and Wilkinson (1963); Finney (1958); Finney (1988); Finney *et al.* (1964); Horner and Frey (1957); Joppa *et al.* (1971); Kempton and Lockwood (1984); Keuls and Sieben (1955); Knight (1970); Moran-Val and Miller (1975); Nester (1996); Patterson (1979); Patterson and Simmonds (1989); Patterson *et al.* (1977); Pearce (1983); Rasmusson (1968); Rasmusson and Lambert (1961); Spoor and Simmonds (1993); Sprague and Federer (1951); Yates and Cochran (1938).

PVR and certification

AOSCA (1971); Byrne (1992, 1993a, b); FAO (1961); Feistritzer (1975); Hanson and Weiss (1964); OECD (1969); UPOV (1991); USDA (1961); Wellington (1974); Wellington and Silvey (1997).

Multiplication, general, including pollination

Bateman (1947a, b); Faegri and Pijl (1971); Feistritzer (1975); Frankel and Galun (1977); McGregor (1976); OECD (1969); USDA (1961).

Hybrid seed production

Duvick (1965); Frankel and Galun (1977); Duvick in Frey (1966); Meer and Nieuwhof (1968).

Incompatibility

Arasu (1968); Gowers (1973); Hermsen (1969); Heslop-Harrison and Lewis (1975); Mackay (1977); Nettancourt (1977); Rowlands (1964).

Male sterility

Duvick (1965); Edwardson (1956, 1970); Harvey *et al.* (1972); Jain (1959); Sage (1976).

Chapter 7
Disease Resistance

7.1 General

Kinds of diseases

For the purpose of this discussion, diseases are caused by pathogens whether virus, plant or animal in nature; physiological disorders, e.g. mineral deficiencies or toxicities, are of a different character, are better not regarded as diseases anyway and are thus excluded. As Table 7.1 shows, we can usefully recognize six major groups of causal agents, namely: air-borne fungi, soil-borne fungi, bacteria, viruses, eelworms and insects. These could be extensively subdivided if the pathological niceties were to be observed but, for our present purpose, the broad classification will suffice. A few other animal groups such as mites, slugs and birds might have been included but they are either unimportant or, in the case of birds such as pigeons and *Quelea*, pose wide biological problems which the plant breeder can do nothing about. Similarly, plant breeding is irrelevant to the control of elephants and of one of the more destructive insect groups, the locusts and also of the many pests of stored products.

Table 7.1 shows that all six groups of pathogens cause destructive diseases and that all have had attention from plant breeding, some of it successful. Crops differ markedly in the arrays of pathogens that attack them: thus the small grain cereals are conspicuously subject to air-borne fungal epidemics; most solanaceous crops are especially subject to viruses (the potatoes outstandingly so); cotton is damaged by a great variety of insect pests; sugarcane has significant pathogens in every one of the six groups of Table 7.1 but none seems predominant; and so on. No simple generalizations emerge. One might guess that severity of attack and economic loss has tended to provoke a roughly proportional effort at control (including breeding) and that, viewed very broadly, the overall significance of the six groups of diseases might go something like this:

$$A > B > C = E = F$$

Certainly, groups A and B, the fungi, must be economically as important as all the rest put together and they have surely taken well over half the total breeding

Table 7.1 The main features of crop diseases summarized from a plant-breeding viewpoint.

A. FUNGI, AIR-BORNE

Biology: specialized, aerial pathogens, often (smuts, rusts) with complex life cycles but not always (several imperfect fungi); pathotype-(race-) differentiation usual

Epidemics: recurrent, often seasonal, depending upon presence/absence of host and/or upon weather

Spread: wide and rapid, often intercontinental, new pathotypes rapidly disseminated

Damage: usually to biomass by destroying photosynthetic apparatus; some quality effects (e.g. fruit blemishes)

Examples: blight of potatoes; rusts of wheats, barley, oats, maize, coffee and other crops; powdery mildews of cereals, grapes and other crops; smuts of cereals, sugarcane and other crops; leaf spots of bananas; South American leaf blight of rubber; Southern leaf blight of maize; blast of rice

Observations: the most conspicuous and probably the most damaging group of diseases; have claimed by far the greatest part of plant-breeding attention; some successes but many failures (see text); grade into a great array of non-obligate and non-specific minor pathogens

B. FUNGI, SOIL-BORNE

Biology: obligate and specific (in varying degree), long persistent in soil; life cycles complex (e.g. wart, clubroot) or simple (Fusaria, Verticillia); some pathotype-(race-) differentiation

Epidemics: follow initial introduction and spread; thereafter chronic rather than epidemic in infected land

Spread: slow since not air-borne, usually man-assisted; new races generally restricted in distribution

Damage: usually to biomass by stunting plants or killing them, leading to poor growth and gappy stands; initiation of secondary storage rots in root crops; quality effects by blemishing root crops

Examples: potato wart, brassica club root, very many *Fusarium* and *Verticillium* wilts, take-all of cereals, several latent-tuber diseases of potatoes, 'root diseases' of tropical tree crops

Observations: range from obligate, highly specific pathogens showing pathotype-differentiation to semi-saprophytic and much less specific ones; can be devastating (e.g. Panama disease of bananas) but, on balance, probably less destructive than the air-borne group, A; have claimed a substantial breeding effort in total and some considerable successes

C. BACTERIA

Biology: usually species-specific, semi-obligate strains of wide-ranging organisms, but with little (or no) pathotype-differentiation as to crop genotype

Epidemics: chronic, rather than epidemic, after initial introduction/development

Spread: slow, usually man-assisted

Damage: various; often to biomass by stunting or killing plants (soil-borne vascular wilts) or by killing foliage (cotton blackarm)

Examples: potato ring rot, vascular wilts of many crops, leaf scald and ratoon stunting of sugarcane, cotton blackarm

Observations: rather few bacterial 'species' involved; relatively unimportant as a whole but very troublesome in specific instances; often have much in common with the soil-borne fungi; little breeding but some of it very successful (cotton blackarm)

Contd

Table 7.1 *Contd*

D. VIRUSES

Biology: always obligate pathogens but with varying levels of species-specificity; pathotype-differentiation common with respect to crop genotype

Epidemics: tend to be systemic and chronic in perennial crops rather than epidemic; in annuals, epidemic if transmission permits

Spread: in infected planting material (generally slow) or by insect vectors (which can be rapid and widespread), or mechanical

Damage: to biomass, by stunting host, rarely lethal

Examples: potato viruses X, Y, leaf roll and others, swollen shoot of cacao, bunchy top of bananas, sugarcane mosaic, bean mosaic, tobacco mosaic, sugar beet yellows, rice viruses, many viruses of clonal fruits

Observations: probably next in importance after the fungi (A, B) and a good deal of resistance breeding has been done, some of it very successful indeed; generally most damaging in perennial crops but can have annual epidemic aspects due to insect transmission

---

E. EELWORMS

Biology: specific, obligate pathogens few, some showing pathotype-differentiation with respect to crop genotype

Epidemics: persistent in soil and chronic once introduced, rather than epidemic

Spread: slow, since not air-borne, usually man-assisted

Damage: to biomass, by stunting growth through root and stem damage, sometimes accompanied by gall-formation; sometimes lethal, causing gappiness

Examples: potato cyst eelworms, cereal stem and root eelworms, sugar beet eelworms, banana eelworms, white tip of rice, root-knot of tobacco

Observations: a great many non-specific eelworms, often more or less saprophytic, also damage crops and a few are also significant virus vectors; the specific pathogens have much in common biologically with the soil-borne fungi (B); some breeding has been done, some of it (potato cyst eelworm) moderately successful

---

F. INSECTS

Biology: most pathogens have moderate host-range (sometimes polyphagous), non-specific to preferred crop; some show pathotype differentiation with respect to crop genotype (wheat hessian fly, several aphids, e.g. of raspberries)

Epidemics: characteristically seasonal in response to presence/absence of crop and/or weather but infestation can be chronic rather than epidemic in tropical perennials

Spread: seasonal, often limited to a few kilometres, but mass-dispersal also occurs (e.g. in aphids); wingless or weakly flying species often man-assisted

Damage: to biomass, by stunting growth (e.g. aphids), by cutting roots, by destroying leaf tissue (Colorado beetle, sugarcane froghopper, toxic mealybugs), by destroying stems (sugarcane and maize stem borers, twig-boring beetles); by direct destruction of agricultural product (fruit and tuber borers, grain seed infestation); some quality effects (fruit blemishes by scales, cotton stainers)

Examples: see under damage; insects which have had attention from plant breeding include wheat hessian fly, maize stem and ear borers, cotton jassids, legume and raspberry aphids

Observations: a great many other less specific and less damaging pests occur; some insects are important as vectors of viruses (especially aphids), occasionally of fungi (Dutch elm disease); principal damage by many borers is by way of secondary fungal/bacterial pathogens; relatively little resistance breeding has been done and few notable successes recorded

---

effort. Indeed there is sometimes a tendency to think of breeding for disease resistance as being practically synonymous with breeding for resistance to the air-borne fungal pathogens since these have attracted so much effort, are among the most conspicuous diseases and have provided both dramatic successes and striking failures. Table 7.1, however, should serve as a reminder that the field of resistance breeding is a wider one than it sometimes seems to be.

Effects of diseases

By far the most important effect of nearly all diseases is to reduce biomass and hence yield (Fig. 3.1). As Table 7.1 shows, this may happen in four ways: by the killing of plants, leaving gaps in the stand beyond the capacity of neighbours to compensate (e.g. vascular wilts, various other soil-borne fungi, some boring insects); by general stunting caused by metabolic disruption, nutrient drain or root damage (e.g. many viruses, aphids, eelworms); by killing of branches (e.g. many boring insects, some fungal diebacks); by destruction of leaf tissues (e.g. many rusts, mildews, blights, leaf spots and some 'burning' insects). These four routes by which diseases affect yield are, on the whole, distinct, though some intermediate situations can be found. Other effects of diseases are, overall, less important economically and have certainly received far less breeding attention. They include: damage to the crop product itself, initiated in the field but often more apparent after harvest (e.g. cereal smuts, various rots and borers of fruits and tubers); and quality effects (e.g. insect or fungal blemishes of fruits and tubers).

In summary, then, the diseases with which we are mostly concerned affect biomass and hence, presumptively, yield. Some caution is, however, needed here. It is certainly true that economic loss will be estimated by some function of intensity of infection; the question is, what function? Simple proportionality can rarely be assumed and general field experience, indeed, would often immediately suggest a curved relationship between resistance and economic worth (Fig. 5.10). Plant pathologists have put much effort into developing scales for the estimation, on an internationally agreed basis, of disease incidence; the use of such scales tells us about the disease but not about the damage it causes. Ideally, the plant breeder would like an economic scale or, at least, one that specifically related yield loss to disease incidence. Such scales are certainly rare and can be developed only by fairly massive experimentation. The results are rarely simple because they have to take into account varying seasonal patterns as well as overall attack. However, many experiments comparing yields in sprayed and unsprayed plots of susceptible varieties agree in showing that diseases do indeed cause yield losses, even if good yield-incidence equations can rarely be written. In practice, we can usually be confident that breeding resistance to conspicuous diseases will be economically worthwhile in terms of potential yield losses offset. Minor, or at least inconspicuous, diseases are another matter: they may sometimes be either much more or much less damaging than appears to first inspection. At all events, the plant breeder will want to be confident that the resistance he seeks will be genuinely worthwhile in economic terms.

Epidemics

An epidemic is an outbreak of a disease in which infection starts from a low level and progresses to a high one. Air-borne pathogens typically go through an annual epidemic cycle depending upon the presence or absence of the crop and/or upon seasonal weather. Persistent, soil-borne pathogens may show an epidemic phase while first becoming established, over several years, in a new locality but then generally settle down to a chronic phase (Table 7.1). In this section we shall be concerned with the elementary principles of epidemics in general but with the seasonable air-borne pathogens especially in view. The treatment is largely based on Van der Plank (1968).

Consider a planting of a susceptible variety early in the infection-season and suppose that there is a little infection (carried over from residues, wind-borne from outside, etc.). Any spores produced will land on fresh susceptible tissue and the rate of infection will be proportional to the infection level. As infection proceeds, however, some spores will land on already infected tissue so the infection rate will slow down and ultimately be related to the amount of *uninfected* tissue that remains. Thus we have in the initial ('geometrical'. 'logarithmic' or 'exponential') phase:

$$\frac{dD}{dt} = rD \tag{1}$$

$$D = D_0 \exp rt \tag{2}$$

Where D is infection at time, t, D_0 is initial infection and r is multiplication rate of the pathogen. If D is measured on a scale 0 to 1 (zero to total infection) equation (2) will normally hold pretty well up to about $D = 0.3$. The following equations, however, are more general and take into account the later slowing down of infection:

$$\frac{dD}{dt} = rD(1 - D) \tag{3}$$

$$D = \frac{1}{1 + C\exp -rt} \tag{4}$$

where D, t and r are as before and C is a constant that depends upon D_0 (but is not equal to it). If (4) is to be fitted to data the following version of it may conveniently be used:

$$\log_e\left(\frac{1}{1 - D}\right) = rt + \log_e(1/C) \tag{5}$$

Where the regression coefficient estimates r and the intercept $\log_e(1/C)$.

Comparing (1) and (3) it will be seen that they will be similar when D is small so (2) is a special, initial, case of (4).

Equation (4) is one version (out of several) of the general 'logistic equation' used to describe a great many biological growth processes. As Van der Plank shows, it describes several fungal epidemics of crops quite well, though it must sometimes be modified to take account of latency of infection and the fit can be affected by variation in r during the season in response to changes in weather. However, for our purposes we can ignore these (and other) complexities and concentrate upon seeing what the equation signifies for the plant breeder (Fig. 7.1) First, the constant $C(=(1 - D_0)/D_0)$ is a large number which tends to infinity as D_0 tends to zero; so (an obvious result) no infection, no epidemic. Second, even if D_0 is very small, the decrease of the exponential term in (4) with rising t is so potent that epidemics regularly emerge from virtually imperceptible sources; initiators of potato blight epidemics have been estimated as one infector plant per square kilometre. Third, given that D_0, though small, is greater than zero, the course of the epidemic depends upon r (Fig. 7.1); if $r = 0$ (multiplication of the fungus completely inhibited) $D \simeq 0$ and there will be no

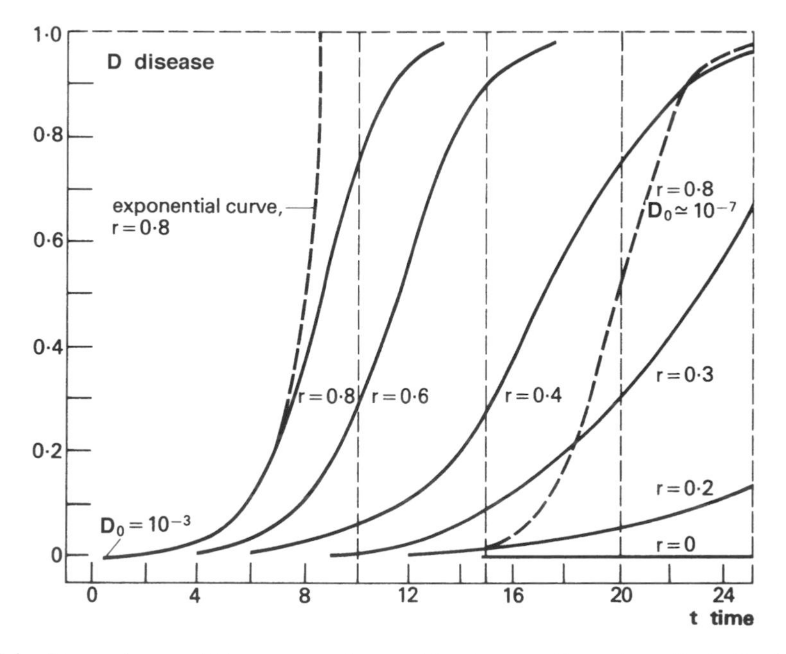

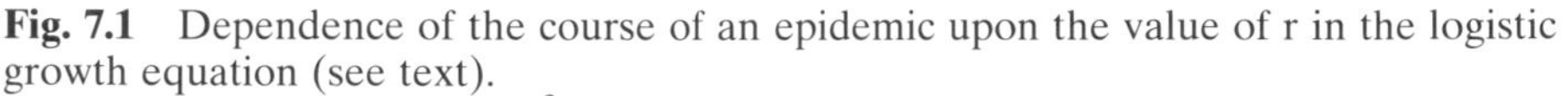

Fig. 7.1 Dependence of the course of an epidemic upon the value of r in the logistic growth equation (see text).
Notes: D_0 was taken to be 10^{-3} (0.1 per cent) and the time scale was chosen so that r lay between 0 and 1: the experimental curve on the left for $r = 0.8$ agrees with the logistic up to $D \simeq 0.3$. The vertical dashed lines represent maturity, to illustrate how terminal incidence varies with maturity and r. The dashed curve, to the right, illustrates the delaying effect of vertical resistance when D_0 is very small (here about 1.13×10^{-7}) even if r is large (0.8).

epidemic; as r increases, so do the earliness, severity and rate of development of the epidemic also increase. And, fourth, the final severity of attack will depend not only upon the pattern of the epidemic (i.e. upon r) but also upon when the crop matures (Fig. 7.1).

The plant breeder's task is thus clear and it emerges from the last two points just made. It is to reduce r to a level at which the final attack, at the time of crop maturity, has an acceptably small effect upon yield. It is good if r can be made equal to zero (i.e. immunity) but inessential; much successful plant breeding (as well as most chemical treatments) depend simply upon reducing r below the level found in an unprotected, susceptible crop. Obviously, the actual magnitude of r depends upon the time scale chosen because the exponent in (4) is rt; a convenient approach (adopted in Fig. 7.1) is to put $r_s = 1$ for a highly susceptible variety and choose the time scale accordingly; the breeder's objective, then, is to reduce r in the range 1 to 0.

7.2 Disease resistance

Kinds of resistance

With the exception of total resistance (i.e. immunity), which is a clearly defined condition of 'no disease', resistance is always manifested on a scale relative to something else that we arbitrarily label 'susceptible'. In plant breeding situations, a practical standard of susceptibility is usually apparent (e.g. the currently most susceptible standard variety) but the choice is arbitrary and reflects the genetic history of the material. Thus the worker on potato blight who was accustomed to north temperate cultivars would have to adjust his scale if he came to deal with unselected South American Andigena Group materials; the worst of them extend the scale, revealing that what had seemed highly susceptible in the cultivars was really somewhat resistant. So susceptibility-resistance scales are arbitrary, even though practically workable. Resistance simply means 'less disease' on whatever scale is adopted.

Two main manifestations are apparent, namely: resistances due to inhibition of infection and inhibition of subsequent growth of the pathogen. Infection inhibition has several aspects, as follows. By far the most numerous examples are provided by the hypersensitivities evoked by a great many specific fungal air-borne pathogens and some viruses (e.g. Table 6.3). Infection evokes a rapid, localized reaction by the host; a patch of host cells surrounding the infection point die and the pathogen also dies (or at least stops growing); a necrotic fleck marks the spot and, typically, immunity is the result. An almost instant mutual recognition, at the molecular level, of pathogen and host, is implied but not yet understood. Sometimes (perhaps frequently) a phytoalexin is involved. Phytoalexins are specific chemicals of which about thirty (often of polyphenolic or terpenoid character) have been identified in various plants, conspicuously in

Leguminosae. They are produced in response to attempted infection by the evoking pathogen and, being themselves fungicidal or fungistatic, are at least a component of some hypersensitive inhibitions of infection. They are often evoked by completely foreign fungi which would never invade the host in question (e.g. the potato blight fungus on broad bean pods); and by pathotypes of a potentially pathogenic fungus to which the particular host genotype is hypersensitive. We shall return to genetic questions in the next section. From the plant breeding point of view, hypersensitivities are superficially attractive because of the immunity ($r = 0$) that they commonly seem to offer; in practice, they have often been disappointing, as we shall see later.

Other examples of infection inhibition are less numerous and less well defined than the hypersensitivities – which are many and clear even if poorly understood physiologically. Mechanical inhibition can be recognized in relation to several smuts, for example, the closed flowering habit of wheats and barleys which excludes the spores of ovary-infecting fungi. Tight husks in maize help to exclude several ear-infecting insects and hairy leaves in cotton deter invasion by jassids. A number of examples of resistance to insect pests are thought to have an antibiotic (i.e. toxic chemical) basis but it is often unclear whether infection- or growth-inhibition is implied. Furthermore, the dividing line between this situation and insect infection resistance due to 'non-preference' is far from sharp. Entomologists are usually quite uncritical.

Growth inhibition is apparent in fungal pathogens by restriction of lesion development and/or of sporulation; in viruses and animal pests by restriction of multiplication of the pathogen population. That there are differences between host genotypes in these respects has been demonstrated on innumerable occasions but the mechanisms, though surely broadly nutritional in character, are essentially unknown. We can assert that potatoes resistant to blight, wheats resistant to rusts and barleys resistant to mildew show various combinations of slow lesion growth and reduced sporulation but we have no real idea why. Chemical inhibitions (e.g. by polyphenolic materials) have sometimes been suggested but very rarely substantiated. Among insect pests, chemical deterrence, whether by nutritional unattractiveness ('non-preference') or toxicity ('antibiosis') (the distinction in practice is quite unclear), has often been invoked. In maize, resistance to first-(but not to second-) brood European corn borers is connected with the presence in the plant of specific substances, DIMBOA and its precursor 6-MBOA; in cotton, glandless forms, deficient in the normal polyphenolic constituent, gossypol, are rather resistant to the boll weevil but, unfortunately, more than usually susceptible to bollworm and other insect pests that were previously unimportant.

For the plant breeder, the conclusions to be drawn are perhaps simpler than this inevitably rather confusing discussion would suggest (Fig. 7.2). First, potent infection inhibitions, nearly (not quite) always due to specific hypersensitivities, offer the possibility of putting $r = 0$ and are therefore superficially very attractive. Second, the very various resistances (but with $r > 0$) which turn up when-

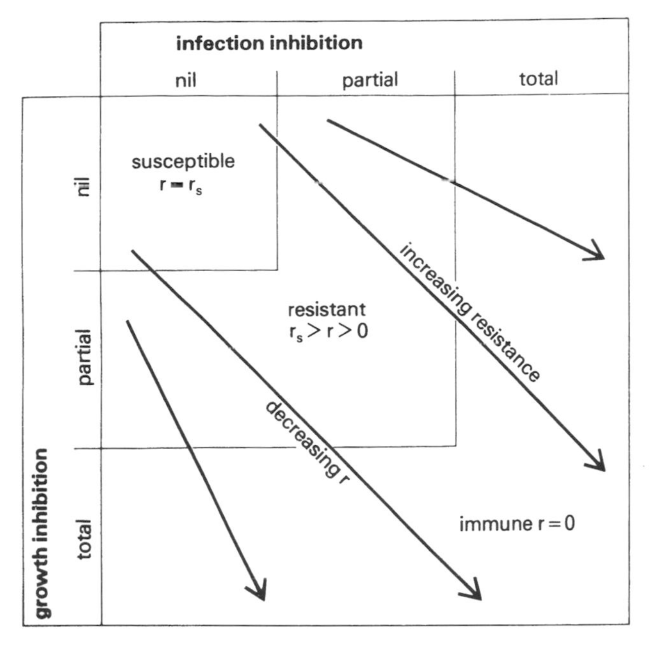

Fig. 7.2 Inhibition of infection and growth of pathogen as components of resistance. Notes: Increasing resistance reflects a declining epidemic parameter, r (Fig. 7.1). The common hypersensitive immunity to infection lies in the right-hand column of the diagram and may be superimposed upon any degree of (potential, unexpressed) growth inhibition.

ever they are sought, may have infection and/or growth-inhibition components and may be variously mediated by (usually little understood) nutritional, specific chemical or mechanical factors. Better understanding would be good to have but, fortunately, the plant breeder does not need to understand resistance in order to use it; it is enough to recognize a reduction of r (i.e. $r < r_s$) and to be able to select for it. At an empirical level, selection for low attack in relation to an appropriate crop maturity works pretty well.

Since 'tolerance' is sometimes mentioned in the literature as an aspect of resistance, it deserves mention. If the word means anything in this context, it must mean that a variety is attacked by a pathogen but does not, contrary to expectation, suffer damage. *A priori* this seems an unlikely situation: all experience says that visible attack damages the host, even if the amount of damage and consequent loss of yield is unknown. The word is clearly often misused and should be avoided unless it can be shown to be appropriate (as occasionally it is).

Genetics of reaction

As a preliminary generalization (to be qualified later), most resistances are fairly clearly classifiable as due either to one or a few major genes (with or without modifiers) or to polygenes, with the obvious corollaries of continuous variation and biometrical, rather than Mendelian, analysis. Tables 7.2 and 7.3 give two examples, out of the many hundreds recorded in the literature, of major gene resistances. One relates to an airborne fungal pathogen and gives evidence of a genetic background effect on the expression of the resistance gene. Even in the most favourable genetic background (that of the resistant parent line itself) the gene does not confer immunity: r is low but not zero. The other example, that of the potato viruses, is uncomplicated by modifiers and is concerned with genes which effectively confer immunity (r = 0). It also illustrates an important feature, not shown by the tobacco example, namely, pathotype-specific reactions. Thus A and C are strains of virus Y; genes *Na* and *Nc* confer hypersensitive immunity to each independently but not to both; *Ny* is resistant to A, C and all other strains of Y and so is *Ry* (a 'resistance' rather than a necrotic gene, though the difference does not matter in the present context). *Nx* and *Rx* are analogous.

Table 7.2 Segregation of a major gene for resistance to blue mould (*Peronospora tabacina*) in tobacco. Frequencies as percentages. Main classes in bold face. Adapted from the data of Marani, Fishler and Amirav (1971, Table 1) by permission of Dr Marani and the publishers (*Euphytica*).

Percentage (parental inbreds R and S)	Resistance score						Plants
	1	2	3	4	5	6	
	Susceptible					*Resistant*	
R parent	0.2	0.8	0.2	0.6	4.2	**94.0**	518
RxS–F_1	0.0	0.6	1.1	7.8	**46.8**	**43.8**	528
RxS–F_2	**8.9**	**18.9**	2.7	4.8	**23.7**	**42.0**	1047
(RS)xS–BC	**13.4**	**37.2**	3.8	4.6	**23.5**	**17.4**	1043
(RS)xR–BC	0.4	1.2	1.6	2.3	**21.7**	**72.9**	516
F_3 lines	**6.9**	**14.6**	1.9	3.5	**21.5**	**51.6**	801

Estimates of resistance scores of resistant segregates in relation to background genetic constitution:

Constitution	R	(RS)R	RS	(RS)S
Per cent R	100	75	50	25
Scores	5.9	5.6	5.6, 5.5, 5.3	5.2

Notes: The data provide clear evidence of a dominant resistance modified in expression by environmental and/or polygenic background effects (see also Table 4.1, Figs 4.1 and 4.2). The reader should plot frequency diagrams of the six lines in the top table and estimate (by inspection) the distributions of susceptible and resistant classes (which obviously overlap). Verify that Mendelian expectations are approximately met (though formal ratio testing is hardly justified). Calculate resistance scores, as in the bottom part of the table, of resistant and susceptible classes; the latter are all about 1.7–1.8; the former vary in a systematic fashion with the background genetic constitution. Plot scores against per cent parent R in the background and estimate the score when R = 0. Interpret.

Table 7.3 Major-gene resistance to viruses in potatoes. Adapted from several tables in Cockerham (1970) by permission of the Editor of *Heredity*.

1. Joint segregations (simplex × recessive) for pairs of dominant *N* (hypersensitivity) genes conferring resistance to specific viruses.

Viruses	Genes	Progeny hypersensitive to				Remarks
		both	one	other	neither	
X, B	*Nx Nb*	60	64	56	59	1:1:1:1
X, A	*Nx Na*	116	3	11	103	Coupling linkage
X, C	*Nx Nc*	66	68	72	56	1:1:1:1
B, A	*Nb Na*	63	44	61	59	1:1:1:1
B, C	*Nb Nc*	60	57	67	55	1:1:1:1
A, C	*Na Nc*	66	59	56	49	1:1:1:1

2. Segregations (simplex × recessive and simplex × simplex) for dominant *N* (necrosis, hypersensitivity) and *R* (resistance) genes conferring resistance to viruses.

Viruses	Genes	Combination	Progeny			Remarks
			Rx	*Nx*	neither	
X	*Rx, Nx*	*Rx* × rec.	306	0	306	1:1
		Rx × *Rx*	144	0	46	3:1
		RxNx × rec.	119	80	66	2:1:1
			Ry	*Ny*	neither	
Y	*Ry, Ny*	*Ry* × rec.	46	0	50	1:1
		Ry × *Ry*	49	0	20	3:1
		RyNy × rec.	121	69	78	2:1:1

Notes: Remember that potatoes are autotetraploid and that the simplex genotype is of the type *Aaaa*. These data are a small abstract of a huge body of information. Virus B is a strain of X, and A and C are strains of Y. The necrotic (N, hypersensitivity) genes are strain-specific (*Na*, *Nb*, *Nc*) or comprehensive (*Nx*, *Ny*); the resistance genes (*Rx*, *Ry*) are both comprehensive and therefore effective against all strains. The coupling linkage shown is between genes active against different viruses. In total, some nineteeen genes are known, from both cultivars and wild species (backcrossed into Tuberosum background); seven (at four or five loci) are active against X viruses, twelve (at three loci) against Y viruses. As an exercise the reader should abstract the seven single-gene backcross segregations and test agreement with the expected 1:1 ratios and homogeneities.

Pathotype-specificity is so common a feature of major-gene resistance that it can be regarded as normal and expected to emerge from search or experience even when initially undetected. Historically, the situation has repeatedly been revealed by (1) discovery of a gene conferring resistance to the prevalent pathotype; (2) evolution (or emergence to attention) of a new pathotype that will attack the hitherto resistant genotype; (3) discovery of another resistance gene; (4) evolution of yet another pathotype; and so on. Pathotypes that can attack genotypes bearing a specific resistance gene or genes are said to be 'virulent' towards them; single or multiple specificities in host and pathogen are matched in one-to-one correspondence.

In the cases of potato blight and several cereal rusts and mildews, which have been the most extensively studied, the signs are that the numbers of specificities known to investigators are limited only by the time and effort put into the work: lists can, indeed, become formidable (e.g. Wiberg, 1974). Given any number or combination of host resistances, there will be matching pathotypes. In general, if there are n specificities, there will be 2^n host genotypes and 2^n pathotypes, ranging from the simplest (host r, pathogen o) to the most complex (host R_1, $R_2 \ldots R_n$, pathogen P (1, 2, ... n)). In potatoes, 12 R-genes are known, $2^{12} = 4096$ and the most complex pathotype (P (1 ... 12)) has long been known. It will be noted that: (1) a resistance gene can only be identified with the aid of a non-virulent pathotype; thus *Na* or R_1 in potatoes can only be detected by the use of, respectively, virus A (C is no use) or a blight pathotype lacking specificity 1 (no matter what else it carries); and, contrariwise, (2) pathotype specificity can only be identified within the limits of the genes available for testing; thus a blight pathotype might be P (1, 2, 3, 4) but, if only R_1 and R_2 were available, we could say nothing about specificities 3 and 4.

The facts of reciprocal host-pathogen specificities led Flor, in his notable work on flax rust in the USA, to the now widely accepted 'gene-for-gene' hypothesis upon which host and pathogen genes are expected to occur in matched pairs. Details lie beyond our scope but the subject has been thoroughly reviewed by Day (1974). The 'gene-for-gene' idea is well supported experimentally for several rusts, smuts and mildews and for the wheat hessian fly (*Mayetiola destructor*); It can reasonably be suspected to hold also in a great many other instances, indeed whenever host–pathogen specificities are known to occur. The genetic features of pathogen adaptation lie beyond our scope; this subject, too, has been well reviewed by Day (1974). We will simply note here that mutation *plus* sexual (i.e. normal meiotic) recombination is, predictably, the basic method but that some fungi (e.g. the potato blight fungus) seem to adapt remarkably efficiently without the benefit of conventional sexual mechanisms.

Major-gene resistances are, then, generally characterized by pathotype-specificity with all that this implies. But there are exceptions, that is, resistance genes that the pathogen apparently cannot match. *Ry* and *Rx* in potatoes (Table 7.3) are examples and others are provided by the dominant resistance in sorghum to Milo disease (*Periconia circinata*), by the recessive resistance in oats to Victoria blight (*Helminthosporium victoriae*); the resistance in maize to southern leaf blight (*Helminthosporium maydis*) might be thought of in the same category, since resistance/susceptibility here seems to be of a unitary nature, though cytoplasmic rather than nuclear in determination (see below Table 7.5, Fig. 7.4). A few more examples could be adduced but they are not numerous; pathotype specificity in relation to major-gene resistance must be regarded as the norm.

Most resistances are dominant, but dominants are easier to detect and study than recessives so the sample may be biassed. Multiple allelism is not uncommon, the extreme case being the *Rp* locus in maize, with at least fifteen alleles controlling response to rust, *Puccinia sorghi.* Many, probably a majority, of resistances to the air-borne fungi and to viruses are hypersensitivities; contrariwise, the occurrence of a hypersensitivity immediately provides *a priori* grounds for suspecting a pathotype-specific resistance. Major-gene resistances are often absolute (i.e. confer immunity, $r = 0$) against the appropriate pathotypes but need not be so and there are plenty of examples of resistance well short of immunity ($0 < r < r_s$) (e.g. Table 7.2 and Simmonds, 1991a, Fig. 15).

Pathotype-specificities are especially characteristics, as we have seen, of the specific, obligate air-borne pathogens. But examples occur in all the other five major groups of pathogens recognized in Table 7.1, for example: flax wilt (*Fusarium*), potato wart (*Synchytrium*), brassica club root (*Plasmodiophora*); several bacterial wilts (*Pseudomonas*); many viruses (e.g. Y in potatoes and other Solanaceae such as *Capsicum* peppers); potato cyst eelworm; wheat hessian fly and several aphids (e.g. of *Rubus*). Gene-for-gene situations have been proved in several and may reasonably be inferred in all.

On the face of it, the facts of pathotype specificity seem discouraging from the plant breeding viewpoint and sometimes, indeed, they are. We shall see later, however, that, though major gene resistances have commonly failed against the air-borne fungi, they have not been completely unsuccessful even there and have had some very striking successes against other classes of pathogen.

Polygenically-controlled disease resistance has been reviewed by Simons (1972) citing examples from potatoes, maize, small grain cereals, tobacco and cotton. By now the list has been considerably extended and it is clear that appropriate study of any crop in relation to any of its diseases reveals polygenic variation in reaction (Simmonds, 1991a). An example is provided in Table 7.4. Resistance in quantitatively inherited – there is no sign of Mendelian segregation – but not simply additive (on the chosen scale); it is, however, fairly highly heritable, as shown by a good response to three generations of selection. Other examples reveal the same sort of pattern, often with high heritabilities. In potatoes, the clonal repeatability of response to the blight fungus is quite high (i.e. grades of resistance-susceptibility are readily recognized) but the specific component of combining ability (SCA) is large; hence, the resistance of progeny is not well predicted either by parental phenotype or GCA (Killick and Malcolmson, 1973) but it can be quite effectively selected for.

In all such examples of polygenic control of response, resistance varies continuously over varieties; it rarely approaches immunity ($r = 0$) but r is frequently sufficiently smaller than r_s to provide economically acceptable control of the disease. The mechanisms are generally unknown in any detail but, predictably, include infection resistance, slow lesion growth and reduced sporulation. In general, there is no evidence of pathotype specificity and often, as when complex pathotypes are deliberately used for testing, definite evidence against it. However, it can never, in the limit, be strictly excluded because (obviously) a weakly expressed specific hypersensitivity ($r > 0$) would tend simply to disappear into the continuous polygenic segregation. As always in studies of

Table 7.4 Polygenic control of resistance to *Helminthosporium turcicum*, causing northern leaf blight in maize. Mean resistance scores (R) of parental inbred lines and F_1 progenies (n in the range 52–120) on a scale 0 (very susceptible) to 10 (highly resistant): the second figure in each entry (σ) is the phenotypic standard deviation within families. Adapted from the data of Jenkins, Robert and Findley (1954, Table 1) by permission of the publishers of the *Agronomy Journal* (the Crop Science Society of America).

Parents	A and B		A and C		A and D		Means	
Generation	R	σ	R	σ	R	σ	R	σ
P_1 (= B, C, D)	1.24	1.22	0.02	0.10	0.54	1.18	0.60	0.83
P_2 (= A)	9.30	0.68	8.74	1.22	8.98	0.66	9.01	0.85
F_1	8.16	1.42	7.34	1.36	7.70	1.22	7.73	1.33
$F_1 \times P1 = BC$	4.46	1.74	3.22	2.04	4.10	1.98	3.93	1.92
RS_1	6.40	1.88	4.60	2.64	2.64	1.92	4.55	2.15
RS_2	8.10	1.20	6.94	1.78	4.52	2.00	6.52	1.66
RS_3	8.36	1.26	7.44	1.74	6.22	2.22	7.34	1.74

Notes: Results are based on crosses of the resistant line A by the three susceptible lines B, C and D. Three cycles of recurrent selection (RS_1, RS_2, RS_3) followed a backcross of the F_1 to the more susceptible parent in each case. RS was, in effect, mass selection starting from BC and based on controlled crossing among the 10 per cent (approx.) more resistant plants in each family. Calculate expectations for F_1 and BC in each case, on the basis of simple additivity; compare with observed figures and interpret. Using mean σ in the last column and a table of ordinates of the normal distribution, fit distributions to R for P_1, P_2, F_1 and BC and plot the results as frequency diagrams. Examine and interpret apparent trends in σ; are the parental inbreds really less variable than the F_1? Examine responses to selection and (assuming $i = 1.7$) calculate realized heritabilities. Do all three populations respond and in a consistent fashion? Would you expect continued response if selection were carried on into RS_4?

polygenic systems, the number of genes involved is strictly unknown. It does not have to be large. The data of Table 7.4 suggest few genes showing some dominance; in cotton, the distinction between more or less additive major-genes and polygenes in relation to blackarm (bacterial blight) resistance is sometimes quite unclear. In the limit, of course, this merely means that we could, given good environmental control of infection and sufficient study, break most polygenic systems down into specific oligogenic components; polygenic systems are categories of experimental convenience (or necessity).

To summarize this section, we may say that two broad categories of resistance are apparent: (1) that due to one or few major genes (large effects, often dominant, often hypersensitive, usually pathotype-specific); and (2) that due to polygenic systems (several/many genes, small individual effects, large environmental component, general inhibition rather than hypersensitivity, not pathotype-specific). The distinction is historically and operationally a sound one, even though various qualifications of detail need to be entered.

We should perhaps add a third category of genetic control of resistance to the two already discussed, though there is but one good example of it extant: cytoplasmic control. This became apparent when uniformity for a single cytoplasm (T) in US maize induced extreme susceptibility to a previously unimportant disease, southern leaf blight (*Helminthosporium maydis*). The epidemic pathotype was found to be T-specific. Conversely, it also became apparent that all the common, non-T, cytoplasms conferred an excellent resistance to the disease. Thus an apparently highly reliable ('durable'), pathotype-non-specific resistance was replaced by a pathotype-specific susceptibility. The analogy with major (nuclear) gene resistance is obvious but nothing is known of the number or nature of the cytoplasmic determinants. Substantial cytoplasmic effects on disease reaction would usually be easily detected. Their rarity probably means that they are genuinely rare or genetically fixed and therefore undetectable. Southern leaf blight offers something of a cautionary tale for plant breeding and we shall return to it later (Fig. 7.4, Table 7.5).

Vertical and horizontal resistance

Vertical and horizontal resistance (VR, HR, respectively) are a pair of terms introduced by Van der Plank (1963, 1968) and now so widely used that they merit special discussion (Fig. 7.3). VR is, in effect, synonymous with a major-gene pathotype-specific resistance in relation to which non-virulent pathotypes (and only they) show r near or equal to zero. If $D_0 = 0$ (virulent pathotype absent) resistance is absolute or nearly so. If D_0 is very small (virulent pathotype very rare indeed) C in equation (4) is very large and any epidemic is so delayed that attack at crop maturity is nil or small; see the dashed curve to the right-hand side of Fig. 7.1. If D_0 stands at a level at or approaching 'normal' for the disease, the epidemic takes its course and the outcome depends upon the size of r determined by the genetic background, not upon D_0. In short, VR can determine anything from immunity to disaster, depending upon the initial frequency of the virulent pathotype, the size of r determined by the major gene itself and the size of r determined by the genetic background. Horizontal resistance (HR) is not pathotype-specific; it is commonly polygenic (but need not be); so long as there is some initial inoculum ($D_0 > 0$), r is the principal determinant of the course of the epidemic and therefore of the state of the crop at maturity (Fig. 7.1). Thus HR characteristically slows the progress of an epidemic without inhibiting its initiation, whereas VR typically delays the start. When the pathotype virulent to a VR becomes prevalent, VR itself disappears and the outcome is determined by the HR in the genetic background (i.e. by the size of r).

Formally, as Van der Plank points out, VR displays a host–pathogen interaction (a special kind of GE effect) when tested by analysis of variance of a suitable experiment; HR does not. Thus a test of two potatoes (r and R_1) against two blight isolates (P(0) and P(1)) would show just such an interaction. This is true but not operationally very useful in distinguishing VR from HR. If the virulent pathotype were lacking, there would be no interaction and, if it were present, an analysis of variance would not usually be needed to demonstrate the fact.

Table 7.5 A diallel cross of maize made in order to investigate the inheritance of resistance to Southern leaf blight, due to *Helminthosporium maydis* (Lim, 1975, reproduced by permission of the Editor of *Phytopathology*).

	A	B	C	D	E	F	G	H
A	–	50	84	59	59	69	55	64
B	93	–	77	52	50	55	52	55
C	99	98	–	76	77	83	81	79
D	97	93	98	–	44	63	53	54
E	94	91	98	91	–	62	52	61
F	98	96	99	98	96	–	54	61
G	97	93	99	89	91	98	–	57
H	97	93	99	93	91	97	92	–
IBL/T	64	30	73	34	39	66	42	52
IBL/N	97	83	99	91	85	97	82	94
G_T	+0.9	–7.2	+20.4	–5.6	–4.9	+2.1	–5.1	–0.6
G_N	+1.3	–1.7	+3.8	–1.3	–2.5	+2.5	–1.3	–0.8

Notes: The cross was an incomplete half diallel but the eight parental inbreds (IBL) were recorded and are here analysed separately. In the table, the top RHS refers to materials with T cytoplasm (susceptible) and the bottom off-diagonal to the N-cytoplasm (resistant). Below are the relevant parental IBL and the calculated GCA values (G_T and G_N, according to cytoplasm). All data in the body of the table are resistances (R) derived from the disease scores (D) recorded by Lim by the relation R = 100 – D (see section 7.4, end, on scales). The data have been partially summarized by Simmonds (1991a, Fig. 7).

As an exercise, which will teach a great deal about both HR and the analysis of plant breeding data, the reader should do the following calculations. Compute all expected R (i.e. R_e) from the general means and given values of G, keeping the data for the two cytoplasms strictly separate, of course. Compute also the mid-parent values for the two sets of 28 numbers, plot graphs and calculate OP regressions. As a general measure of goodness of fit, calculate r^2 values for all sets, whence: T-cytoplasm, GCA fits 0.93, OP regression 0.65; N-cytoplasm, GCA fits 0.82, OP regression 0.34. All but the last are rather close; what points are discrepant? Calculate also expectations for the two sets of IBL, using their own GCA values, plot graphs and calculate r^2 (T 0.70, N 0.72); absolute numbers are very different but agreement is quite close. Why? (Consider expression of HR genes in different cytoplasmic backgrounds). Finally, compare the two sets of GCA estimates at the bottom of the table; plot a graph and calculate r^2 (=0.76). What are the leading conclusions for the plant breeder? Could the maize breeder have coped with the disease even without the confusing effect of cytoplasm? The answer is almost certainly yes, immediately suggested by the presence of values greater than 80 in the upper off-diagonal of the table. The great leaf-blight scare was an artefact that need not have caused panic.

Van der Plank (1968) draws a distinction between 'strong' and 'weak' VR genes which has sometimes been misunderstood and which deserves comment. The words do *not* mean what might be expected, namely genes determining low and high r, that is strongly and weakly inhibitory of pathogen development, respectively. They apply, not to gene expression at all, but to the variability of the relevant virulent pathotype. A 'strong' gene counters a pathotype that is of poor viability when set in competition with other pathotypes in the absence of strong selection for its particular specificity; a 'weak' gene is specific to a pathotype of high competitive viability. There is little doubt that the distinction is biologically a real one; Van der Plank's examples, drawn from the cereal rusts

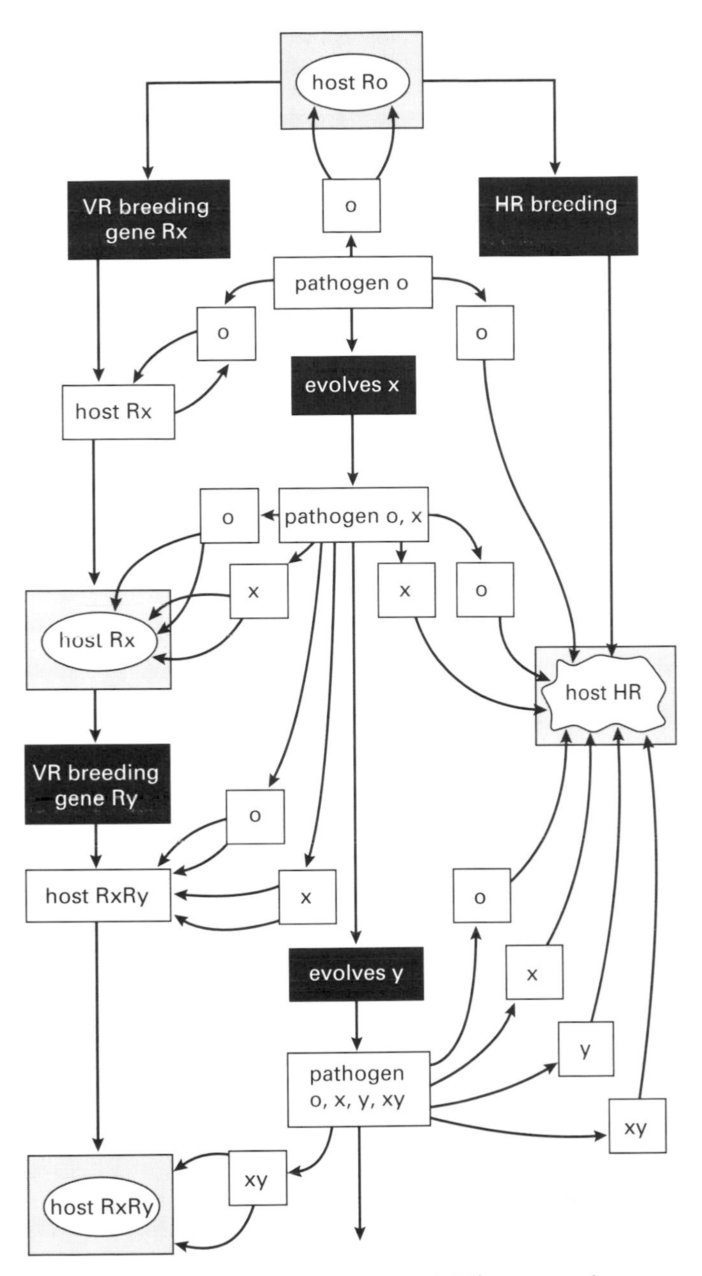

Fig. 7.3 Vertical (VR) and horizontal resistance (HR) compared.
Notes: Degree of susceptibility to prevalent pathogen indicated by size of stippled zone in host 'boxes'. The pathogen evolves specificities (O→X→XY) to match successive host immunities. The HR host is resistant to all but immune to none. 'Boom-and-bust cycles' are indicated on the left (see also Fig. 7.4) and the 'vertifolia effect' (Fig. 7.5) is implicit.

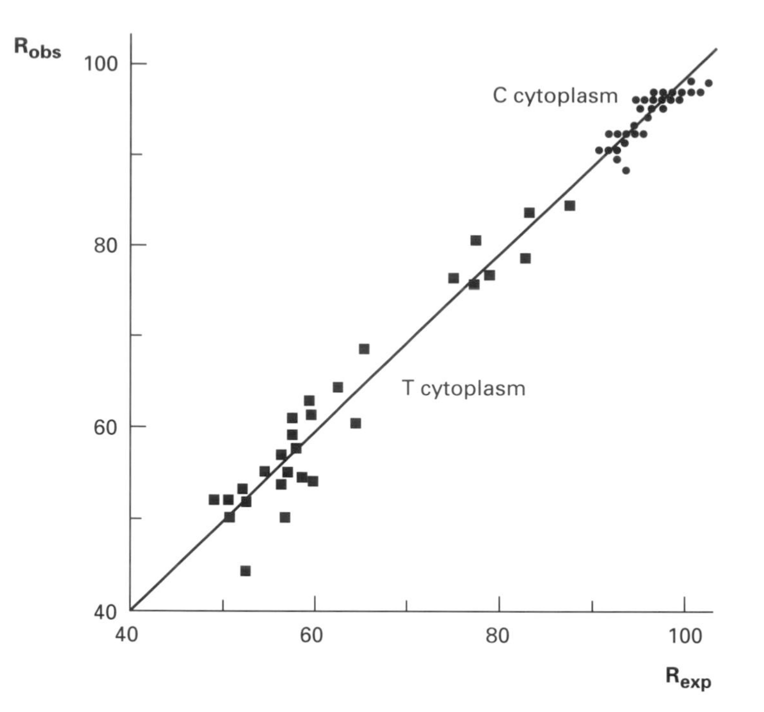

Fig. 7.4 Results of study of HR to Southern leaf blight of maize in the USA. Combining ability analysis of a diallel cross fully described in notes to Table 7.5 R is a measure of resistance, so the higher the score, the more resistant the cross. All combinations in C cytoplasm are nearly immune but the best crosses in T cytoplasm are fairly highly resistant, indicating that breeding would not be difficult without recourse to cytoplasm. Data of Lim (1975) analysed and plotted by Simmonds (1991a).

and potato blight, show that specific pathotypes do indeed show differential rates of survival in the absence of positive selection by host genotype; thus there are, in a sense, 'weak' and 'strong' pathotypes but this terminology is, unfortunately, precisely the opposite of that which follows from Van der Plank's 'weak' and 'strong' gene distinction. 'Weak' genes relate to 'strong' pathotypes and *vice versa.* The terms should be avoided, especially as breeders tend to use them in the obvious genetical sense as applying to effects upon r, i.e. amount of infection.

There is also another term introduced by Van der Plank, which has been widely enough used in the literature to merit comment, namely: 'vertifolia effect'. This derives from a German potato variety carrying two pathotype-specific hypersensitivity genes (R_3 R_4); this variety was immune to the initially-prevalent blight population but very susceptible indeed (r small) to the virulent pathotype P(3, 4) when it emerged. Evidently, the presence of a potent hypersensitivity (or indeed any other major gene effect that makes $r = 0$) precludes selection for HR in the genetic background. Hence, HR will tend to be, on average, low in varieties bred for VR and the failure of VR, when it happens, will be correspondingly dramatic. This is the 'vertifolia effect', repeatedly observed over the past fifty years or so – but only fairly recently interpreted – in potatoes, small grain cereals and other crops. It is of great importance for plant breeding strategy and we shall revert to it later.

Terminology

The terminology of host–pathogen relations in a plant-breeding context bears the accretions of years and uniformity has been attained neither of definition nor of usage. In this chapter we try to use a restricted and consistent terminology, confining mention of various other words and phrases which may be encountered to this section alone.

First consider the pathogen. Like all living organisms, pathogens vary. Strains which differ in general vigour or pathogenic effectiveness but are undifferentiated in relation to host genotype are said to vary in 'aggressiveness'; strains which differ in performance on specific host genotypes are said to differ in 'virulence', ineffective strains being described as 'non-virulent' or 'avirulent'. A 'pathotype' (or 'biotype') is a strain which has more or less defined virulence characteristics – defined, that is, within the limits of the diagnostic genes available. A 'race' ('physiologic race') is not quite the same. Traditionally, 'races' of cereal rusts and mildews were, and still mostly are, defined by an arbitrary letter-number system on the basis of virulence towards an arbitrary series of 'differential hosts' of unknown genetic constitution; if finer discrimination is required, as it often is, more differentials are added and the labelling system extended. Races are therefore pathotypes, or, often, populations of pathotypes, the virulence spectrum of which is more or less defined but not in terms of specific host genes. Strains of potato-blight, by contrast, have always been identified by reference to host *R*-genes (rather than to arbitrarily chosen potato varieties). Similarly, there will be situations when it is more useful to think of pathotype P(6) of wheat stem rust virulent to hosts carrying the gene *Sr*6 than of 'race XYZ 123' (or whatever it may be). Certainly, for the plant breeder, it is specific virulences that count. The signs are of a general move towards replacing the race by pathotype classifications (Day, 1974); any such classifications are, of course, operational, not taxonomic.

Second, consider host response in terms of host–pathogen relations. This approach gave rise to Van der Plank's terms 'vertical' and 'horizontal' resistance (VR and HR) which have become widely, though not universally, accepted. VR is the same as 'race-specific' or 'pathotype-specific' resistance or even just 'specific' resistance. HR has also been called 'race-non-specific', 'pathotype-non-specific', 'partial', 'general' and 'field' resistance, even, on occasion, 'tolerance'. 'Slow rusting' and 'adult-plant resistance' are specialized terms used by workers on wheat stem rust for the same thing. 'Field resistance' has been especially favoured by potato workers (who have done more than most over many years to establish the existence and importance of HR) but is not now kindly regarded. As terms, VR and HR have their critics (e.g. Day, 1974, prefers 'specific' and 'general') but they are reasonably descriptive and widely used.

Third, a purely genetical terminology is possible. Thus we have 'major-gene' or 'oligogenic' resistance contrasted with 'polygenic'; the former implies one or a few Mendelian determinants, the latter few-to-many genes individually indistinguishable in segregation. Some writers treat 'major-gene resistance' as effectively synonymous with VR, 'polygenic resistance' with HR. This is nearly accurate but not quite and it should be avoided; as we have seen, some major-genes confer HR and polygenic resistance may have VR components. Thus, genetical terminology will sometimes be essential but should not be allowed to assume overtones as to host–pathogen relationships.

Fourth, the word 'durable' is finding favour as a neutral word that refers to the hoped-for end-point of breeding activity without implications as to biological mechanisms. An antonym does not seem to have been suggested: 'transient', perhaps?

Finally, the terms 'weak' and 'strong', as applied to genes and pathotypes, and 'vertifolia effect' have been discussed in preceding paragraphs; so has 'tolerance', which should be avoided unless precisely used.

7.3 Epidemics, populations and the genetic base

Epidemics and agriculture

Wild plants rarely have epidemic diseases. Pathogens, plant and animal, are common enough; they take their toll but rarely kill or even seriously damage their hosts. For this state of affairs it is generally agreed that the genetically heterogeneous nature of virtually all wild populations, their dispersed, discontinuous character and natural biological control (e.g. hyperparasitism) are responsible. Hosts and pathogens survive together in complex equilibria in which neither dominates. It is virtually universal experience that the pathogens of crops can be found (if sought) on the wild relatives, but in a non-epidemic state. One has the impression that interplant competition and grazing or other biotic factors pose far more potent selective pressures than pathogens, though it would admittedly be impossible to give any quantitative substance to this statement.

Agriculture introduces three important changes. It narrows the genetic base, since, normally, only a selected fraction of a wild population can be taken into cultivation. It generates relatively large, more or less continuously distributed populations. And it changes whole ecosystems more or less fundamentally, creating habitats profoundly altered for both hosts and pathogens. It by no means follows that all pathogens become epidemic: they obviously do not and, indeed, the majority remain what they were before, unimportant. But some will increase, favoured by a rare combination of changed environment, host susceptibility and contiguity in large populations.

Primitive agricultures are characterized by genetic diversity and spatial complexity. Genetically pure stands are rare and plantings are in discrete backyards, plots, patches or small fields rather than in tens or hundreds of hectares. Diseases are always present but rarely epidemic. Natural selection in highly heterogeneous populations ensures a modest resistance to the leading pathogens and conservative methods, both of agriculture and exploitation of the local ecosystem, damp down potential epidemics. Peasant agricultures are often of rather low productivity on a per-hectare basis but they are essentially stable, unless perturbed by demographic or other irresistible social pressures. Even the more intensive systems, such as rice in Southeast Asia and bananas in upland East Africa, were remarkably free of the epidemics which are typical of technologically advanced agricultures of the same crops. Perhaps the greatest disease hazard is from pathogens introduced from outside. Lacking previous adaptation, the crop may suffer until, in time, selection has its effect and the disease declines. For example: maize in West Africa was severely attacked in the 1950s by the introduced rust, *P polysora*, a common but unimportant disease in tropical America; according to Van der Plank, VR breeding failed, for the normal reasons, but the pathogen, in a few years, evoked an effective HR in the (outbred) local maizes and the disease declined naturally. Again, potato blight is a Central American disease to which South American potatoes were wholly unadapted when the fungus spread there in the 1940s; an epidemic ensued and the availability of HR (e.g. in Colombia – Thurston, Heidrick and Guzman, 1962) soon became apparent and there is no doubt that the South American potato population has the capacity for adaptation to the disease (though the response, in a clonal crop, could not be as rapid as in maize). No doubt, many of the equilibrium situations that we see now in peasant agricultures have emerged historically from local adaptation to newly introduced diseases. However, the temptation to suppose that such epidemics were always followed by genetic adaptation in the hosts should, perhaps, be resisted; who knows what crops may have been destroyed and replaced by something else because they could not adapt quite quickly enough?

Technologically more advanced agricultures (and especially the monocultures) present large contiguous areas to pathogens and a relatively, or even very, narrow genetic base as well. It is here that conditions are most favourable for the development of diseases. The vast majority of the epidemics which have engaged the attentions of pathologists and breeders have had, as a causal element in their history, either a chance susceptibility associated with a narrow genetic base or a positive disadaptation to a disease resulting from preceding plant breeding. Examples of both are numerous. Chance susceptibility is evident in: the north temperate potatoes devastated by blight in the 1840s (it took about forty years of breeding to develop a useful degree of HR); the European grape clones attacked by the American root aphid, *Phylloxera*, around 1865 (later the subject of an outstanding breeding success); the Asian bananas (in effect only two clones) attacked in tropical America and elsewhere by Panama disease (banana wilt – *Fusarium*) and two leaf spots (*Mycosphaerella*); the very narrowly based Arabica coffees devastated by leaf rust (*Hemileia*) in the 1870s; and, a current example, Dutch elm disease (*Ceratocystis ulmi*) which has almost destroyed the (narrowly based) English elm. All these examples (and many more could be collected) relate to serious diseases with considerable, even profound, social consequences.

The distinction between the preceding examples and diseases encouraged by 'disadaptive' breeding is a fairly narrow one for both are connected with the genetic base and the exposure to infection of populations not known in advance to be susceptible to unforeseen diseases. Two categories are obvious: first, and perhaps commonest, are the 'boom-and-bust' cycles generated by the operation of Van der Plank's 'vertifolia effect' mentioned above; second, and perhaps more dramatic, because less predictable, are the trivial (or even unknown) diseases which suddenly spring to prominence because new host genotypes evoke them.

'Boom-and-bust' cycles – the term is due to Suneson (1960) – follow from the cyclical failure of VR genes to deter newly evolved pathotypes (Figs 7.4 and 7.5). Given that an effective VR gene (e.g. a potent hypersensitivity) forbids the recognition of any underlying HR, it follows that VR varieties will, on average, have the mean HR of the breeding population. If this is low, as it usually is, then the VR 'boom' is followed by the 'bust' due to the 'vertifolia effect' (Fig. 7.6). The 'breakdown' (so-called; the term is obviously not a good one and should be

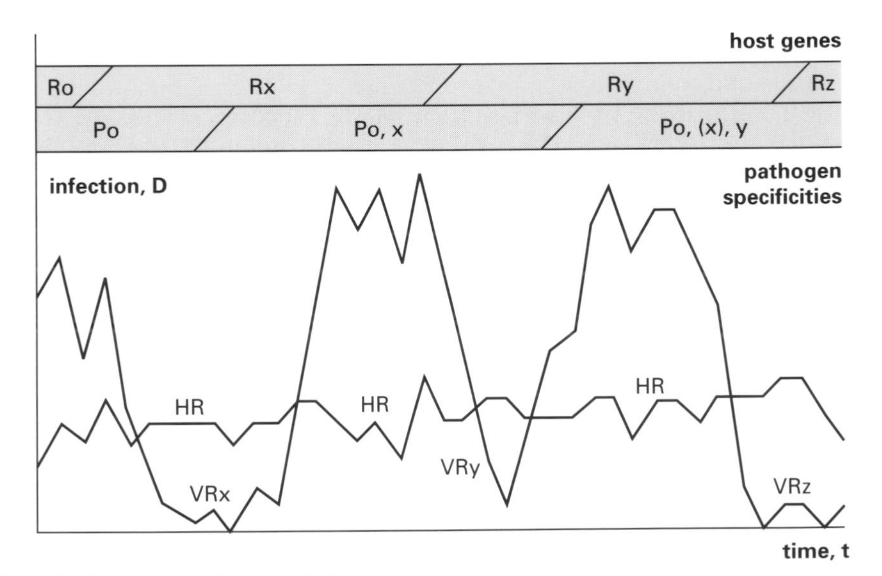

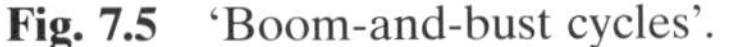

Fig. 7.5 'Boom-and-bust cycles'.
Notes: The periodic dys-synchrony between host genes and pathogen specificities that produces the cycles is evident. The bracketed P (x) indicates a specificity that is selectively eliminated in the absence of the host gene (R_x) that evoked it; R_x is a 'strong' gene in the sense used by Van der Plank.

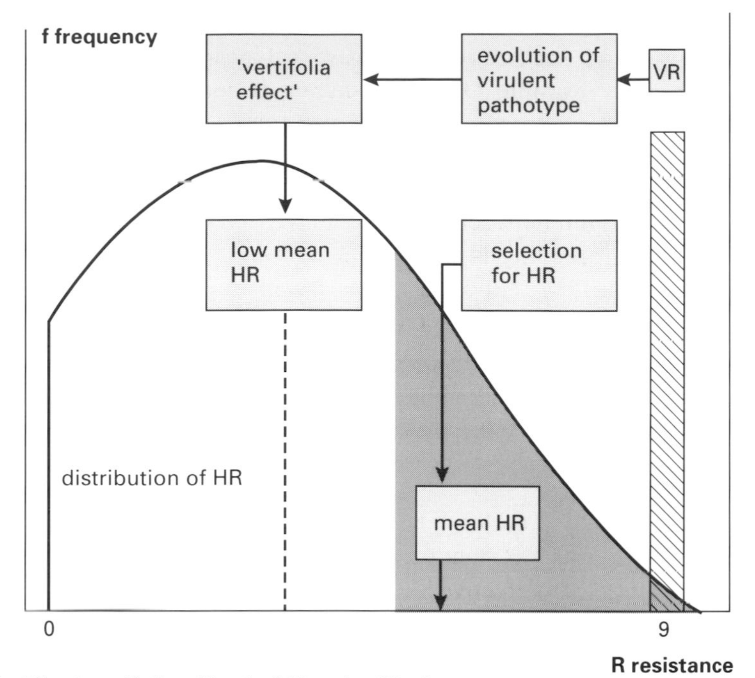

Fig. 7.6 The 'vertifolia effect' of Van der Plank.
Notes: Since HR cannot be selected for in the presence of a potent VR (column on RHS), the HR genotype of a variety in which VR has failed is likely to be about average (i.e. inferior); conversely, good HR is only likely if selected for *per se* in the absence of VR.

avoided) of VR became apparent within a decade of starting VR breeding against cereal rusts in the early years of this century. Since then, VR methods have recurrently and generally failed as a stable technique of control of the air-borne fungal pathogens of inbred cereals, potatoes, tobacco, coffee, rubber and many other crops. The period of usefulness of a VR in the inbred cereals has normally been put at about five years but a few have lasted longer and many others have failed in the breeder's plots, before the potentially resistant variety could even be marketed. Multiple VR genes (e.g. R_1 R_2 R_3 in the potatoes) have proved but little more effective or lasting; to pyramid them is useless.

The outcome of a VR failure is the exposure of whatever HR there is in the variety, as determined by the genetic background. If the HR is too low, the variety fails, as many have failed; if it is high enough for practical purposes (e.g. in combination with chemical control) the variety will survive at a level determined by its overall economic worth; many useful potatoes carry useless *R*-genes but those that proved too low in HR have long since disappeared.

Vertical resistance, therefore, provides a striking example of disadaptive breeding and one that has been recurrent over the past fifty to sixty years in many crops. It has not, however, been by any means a total failure. First, even against the air-borne fungal pathogens, there are (rare) circumstances in which it can be made to work pretty well and we shall return to these later in considering breeding strategy. Furthermore, even when VR breeding has ultimately failed, it has often provided a period of several years of resistance (the 'boom' part of the cycle) before the succeeding epidemic. Economically, the gains must have been considerable, though no one has, I think, ever seriously tried to calculate them. Second, against the less mobile pathogens, such as soil fungi and nematodes, VR has sometimes been very effective indeed, as against potato wart, potato cyst eelworm and flax wilt. Given that (as in these examples) pathotypes are of restricted distribution, there is no evident reason why a combination of VR, sensible crop rotation and phyto-sanitary good sense should not maintain effective control indefinitely. Such diseases have a potential for generating 'boom-and-bust' cycles but these should, in practice, be avoidable.

The second kind of disadaptive breeding to be considered relates to the generation of unpredicted susceptibility to minor or even unknown diseases. The two most striking examples come from the USA and we have touched upon them earlier. The Victoria oat and its close relatives, bred for crown rust resistance, were, by a genetic freak (see Day, 1974), homozygous for a dominant susceptibility to a previously unknown disease, Victoria blight, *Helminthosporium victoriae.* When the recessive resistance (evidently an HR) was restored, the disease disappeared. The other example, that of southern leaf blight of maize, is closely comparable except that resistance-susceptibility is cytoplasmically controlled and the disease was a known minor one rather than unknown. In both cases, epidemics followed the unwitting breeding and wide dissemination of narrowly based, genetically uniform stocks; in both cases too, the remedy was simply to restore the previously unrecognized but highly effective HR (Fig. 7.4, Table 7.5).

These are simply the two most dramatic examples available of what is, in practice, a common phenomenon. Any major change in varietal usage nearly always calls forth some change in disease spectrum: barley mildew, long thought to be unimportant in Scotland, became a serious disease when Ymer (polygenic HR) was replaced by Golden Promise (susceptible, very little HR). Similarly, international trials of varieties or breeding stocks very commonly reveal unpredicted, often unpredictable, disease reactions. Thus maizes which have acceptable polygenic HR to *Helminthosporium turcicum* and *Puccinia sorghi* in the USA turn out to be too susceptible for conditions in Kenya (Harrison in Leakey, 1970); and similar examples could be multiplied.

Collectively, these examples show that we have a great deal of highly effective HR in our crops, the existence of which only becomes obvious when it is either lost or more severely tested than usual. Conversely, we might almost define minor diseases as those for which a potent HR is locally present; on the same basis, major diseases are those which have not evoked an effective HR in the host, a process which is often, as we have seen, actually obstructed by transient VR. This is perhaps slightly to overstate the situation. But there is a substantial truth in the idea that we have the diseases to which we breed susceptibility.

Populations, diversity and the genetic base

The idea that genetic diversity offers some protection against diseases is not a new one but it has only received systematic recognition in fairly recent years (e.g. Simmonds, 1962a; Committee on Genetic Vulnerability, 1972). As long ago as the 1940s, Stevens (1948) pointed out that estimates of losses due to disease in US crops went in the order: clonal crops $\simeq$ inbreeders > outbred seed propagated crops. At the time he wrote, maize had only just made the transition from OPP to HYB and it was still remarkably healthy, in effect free of major epidemic diseases. Indeed, it was the southern leaf blight outbreak of 1970, a direct consequence of HYB breeding techniques, that provoked the Committee on Genetic Vulnerability to its activities (Fig. 7.4, Table 7.5). That committee concluded that the genetic diversity of US crops was indeed too narrow for safety. The same could probably be said of other advanced agricultures in which, as we have seen (Section 5.9), there is fairly widespread evidence of a narrow genetic base.

Three levels of diversity are relevant: within varieties, between varieties and in breeding reserves. The example of maize shows how powerful the first can be. OPP maizes, as we have seen, have diseases but not (or only transiently) epidemics: such populations adapt remarkably fluently, as the example of rust in Africa shows. By contrast, maize inbreds predictably exhibit numerous susceptibilities and southern leaf blight provides an example of one disease that, so to speak, got away. Significantly, the many known VR genes that confer resistances to the two rusts of the crop have hardly been used as such by breeders; they have been obviated by very effective polygenic resistance (HR but maybe with a VR component) in both OPP and HYB. One might speculate that widely based landrace type populations of inbred annuals would be, if not as flexible and responsive as maize, at least capable of considerable adaptation to diseases. They probably are but the evidence is hardly critical. It is at least likely that IBL that emerged from the landrace cereals (or were bred from them but without the incorporation of VR) often carried a modest to good HR selected more or less unconsciously (e.g. Lupton and Johnson, 1970, on UK wheats in relation to yellow rust). It thus seems that the intravarietal diversity of outbred OPP in maize, rye, sugar beet and forages, and probably IBL landraces as well, does indeed offer some protection against diseases, not primarily because the populations are variable but rather because they are adaptively responsive.

The second level of diversity is between varieties. A spatial mosaic of unrelated genotypes, whatever the breeding-propagation system, is very unlikely to be uniformly susceptible to any disease. Potential epidemics would therefore tend to be damped down. VR genes which would be useless in pure form would, against a heterogeneous pathogen, contribute to the buffering process. Over a number of cycles, such populations could show adaptive adjustment to diseases, by change of composition, even without genetic recombination. The idea of deliberate varietal diversification in advanced agricultures has been gaining ground in recent years, for example in the USA and Europe; the practical difficulty, as the Committee on Genetic Vulnerability recognized, is that the necessary range of good but unrelated varieties is usually not available. That the idea may have much merit is suggested by recent experience with wheat rusts in Kenya; the diseases have long been very damaging but have declined in recent years in apparent response to a deliberate policy of maintaining a long list (about thirty) of diverse varieties from international sources (Guthrie and Pinto in Leakey, 1970). The idea of varietal diversification obviously has especial force when the varieties themselves are highly uniform (i.e. CLO and IBL); Stevens, as we saw, found that disease losses in these were greater than in the intrinsically more variable OPP. To the extent that narrowly based HYB approach CLO and IBL in uniformity they will need to be treated like them rather than like the more adaptable OPP from which they came.

The limiting situation is one in which the varietal mixture is at the plant-to-plant rather than at the field-to-field level. Peasant clonal agricultures of bananas and root crops provide examples. In advanced agriculture, 'multilines' and deliberate mixtures of IBL varieties compounded with specific disease resistance in view are relevant and will be referred to again below.

The third level of diversity, that in the actual breeding population of the crop (as distinct from the producing population), we have discussed in general terms previously (Section 5.9). Disease resistance is only one aspect of the matter, one reason among many for maintaining adequate genetic reserves and a strong flow of new parents into a programme. The need can not become less and might, indeed, increase if a policy of deliberately enhancing varietal diversity at the producing level were adopted.

7.4 Breeding

Strategy

Resistance to a disease is but one character among the many of which the plant breeder must take account (Section 5.8). If resistance were all that mattered, resistance breeding would often be easy; the difficulties all lie in combining sufficient resistance with the other characters that go to make a satisfactory variety. Very high levels of HR to blight in potatoes have been available, and in breeders' use, for years but currently successful potato varieties are no more than moderately resistant to the disease. No doubt the highly resistant varieties that will surely emerge in time will be somewhat susceptible to something else and, if seriously so, will fail, whatever their excellence in other respects. In short we must remind ourselves that a disease resistance is but one character among the many that enter the intuitive selection index. We should also recall (Fig. 5.9) that the economic worth of resistance will rarely, probably indeed never, be linearly related to expression, a principle which is further illustrated in Fig. 7.7.

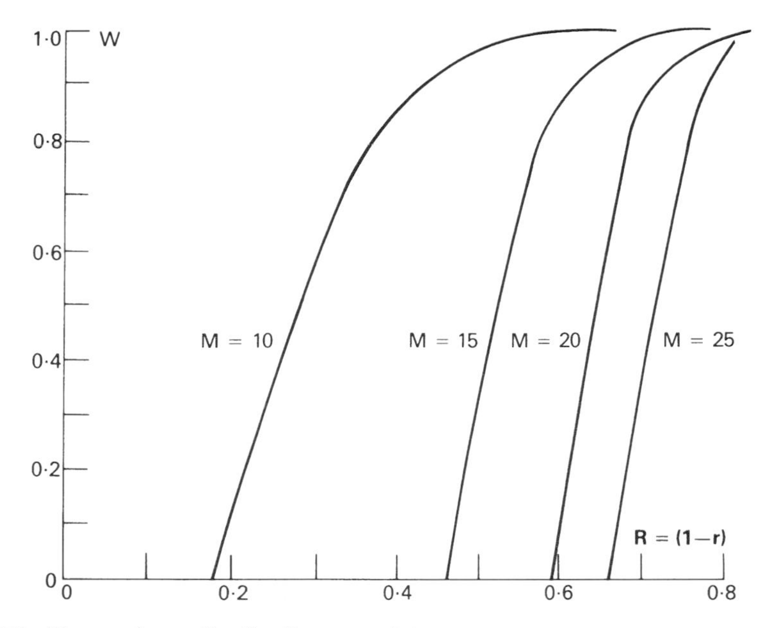

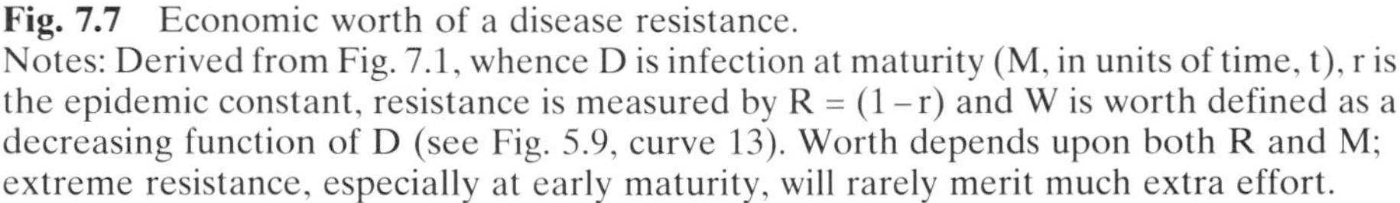

Fig. 7.7 Economic worth of a disease resistance.
Notes: Derived from Fig. 7.1, whence D is infection at maturity (M, in units of time, t), r is the epidemic constant, resistance is measured by R = (1 – r) and W is worth defined as a decreasing function of D (see Fig. 5.9, curve 13). Worth depends upon both R and M; extreme resistance, especially at early maturity, will rarely merit much extra effort.

To put disease resistance in its proper place in breeding strategy then, we note that: (1) a little resistance to several diseases is usually necessary because extreme susceptibility to what is normally thought of as a minor disease can kill an otherwise excellent variety; (2) high resistance, approaching immunity, is good to have if it can be got without compromising other characters; but (3) the example of many successful varieties shows us that enough is enough – middling resistances wisely used can be agriculturally very satisfactory. And a complementary conclusion follows (4): resistance breeding *per se* is a mistaken strategy unless it takes the form of a preliminary working up of stocks; it is all too easy to breed useless resisters.

Turning now to a consideration of the place of different kinds of resistance in breeding programmes, there is no doubt that polygenic HR has always been and remains by far the most important because it accounts for the multitude of middling resistances, far short of immunity but quite good enough for practical purposes, that maintain a fair control of the minor diseases and often some of the major ones too. A continual modest effort, amounting to little more than the discard of lines or clones that seem to be more than usually susceptible, suffices: in short, a strategy of watchful neglect.

Something more positive is needed for the major diseases, those that merit a substantial share of the effort of a programme. The choice lies, broadly, between HR and VR and the main issues, explored earlier in this chapter, are summarized in Fig. 7.8. First, there may not be a choice, if VR is lacking and the pathogen is not pathotype-differentiated. In that case, one of the rare HR-conferring major genes would be very attractive, if available; if not, polygenic HR is the alternative, and by far the commonest, recourse. Some resistance can nearly always be constructed thus, sometimes high resistance; breeding can be laborious but the product is likely to be stable and the method is applicable, in principle, to any crop and any pathogen; since resistance will normally be less than absolute, modest losses and/or some continuation of agronomic control measures will normally have to be accepted.

Second, if VR is available, a choice (which may be a difficult one) must be made; whether to use it or not. The correct strategy surely is to use it only if the

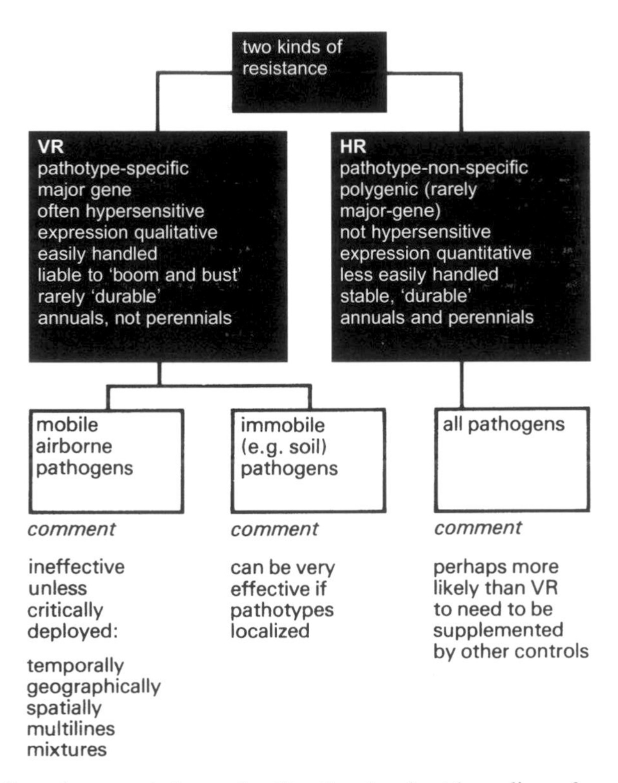

Fig. 7.8 Leading characteristics and utilization in plant breeding of vertical and horizontal resistance (VR, HR) to pathogens.

way is clear to its critical deployment. The simplest case is that of the immobile soil pathogens against which VR, as we have seen, can be highly effective, so long as virulent pathotypes are spatially localized. In practice, pathotype distribution cannot be known before tester-genotypes are available, so the breeding and the mapping are likely to develop together. The more difficult case is that of VR in relation to the mobile air-borne pathogens, especially the epidemic fungi of annual crops. (We will return to perennials later.) Here we have to face the fact that single VR genes and combinations of several of them have virtually always failed sooner or later and usually sooner; there can surely be no justification for the continued use of VR genes against such pathogens unless they can be used in such a way as to give reasonable confidence of stability of protection. Simply to put out new varieties for wide cultivation, confident only that resistance will be transient and underlying HR probably low, is no longer good enough.

The possible methods are listed in Fig. 7.6. To be effective they all depend, in principle, upon a factor which we have so far only just touched upon (Section 7.2). Van der Plank's distinction between 'strong' and 'weak' VR genes, we saw, was based upon the character of the corresponding pathotype, not upon the gene itself: 'strong' genes correspond with weak pathotypes, those which are of poor competitive ability in the absence of the corresponding gene and so tend to die out; 'weak' genes correspond with strong pathotypes, those that can persist in competition. The evidence on which the distinction is founded is the observation of the frequency of a specificity in the absence of the relevant gene. On this basis, Van der Plank thought that rather few genes were 'strong': perhaps one, three and none respectively of those known relating to stem rust of wheat, blight of potato and crown rust of oats. In principle, 'weak' genes are ultimately useless, alone or in any combination or under any method of deployment, because the pathogen will be freely adaptable, unhindered by counter-selection. 'Strong' genes, by contrast, could, in principle, be used in the five following ways (Fig. 7.8).

First, they could be deployed temporally, one succeeding another (cyclically if necessary) so that the pathogen population was never wholly adapted to the prevailing host genotype: the result should be rather like a 'boom-and-bust' cycle (Fig. 7.5) with reduced amplitude, approximating, in the most successful limit, to a good HR. This approach would imply a high degree of control and discipline and also operation on the regional, perhaps continental, scale. It has not been tried and probably never will be: the threshold between 'boom-and-bust' and 'damped boom-and-bust' cannot be a wide one.

Second, VR genes could be deployed geographically in the special circumstances in which one crop population protects another by acting as a filter, so to speak, for the pathogen. Van der Plank described a beautiful case in North American wheat. The gene Sr_6, one of the most effective available against stem rust, worked well in Australia for about fifteen years, then failed. In North America it continued to protect the northern spring wheat area because the annual northwards migration of the rust spore population is filtered through the susceptible winter wheat area to the south. The gene Sr_6 has fortunately not been introduced into the prevalent winter wheat varieties. Pathotype P(6) is of poor viability so it is scarce on the winter wheat and the spores arriving at the spring wheat (Sr_6) are predominantly avirulent; so D_0 is very low and the annual epidemic is acceptably delayed (Fig. 7.1). The spring crop is protected by denying the use of the gene to the winter crop; if Sr_6 were backcrossed into the winter wheat varieties, the result would be (as shown by the Australian experience) annual epidemics in both. The example is striking but, as a general method of using VR genes, geographical deployment would seem to be of little general utility because it depends upon very special circumstances.

Third, spatial deployment of VR genes at the between-varieties, field-to-field level has often been written about and recommended but rarely systematically practised. We saw above that a recent low level of attack by wheat rusts in Kenya has been attributed to a programme of varietal diversification; time will show whether or not this was simply a happy accident. In principle, a mosaic of fields carrying two or more VR genes (so long as they are 'strong' in Van der Plank's sense) should have a damping effect on epidemics.

The fourth and fifth methods of using VR are similar to each other: a 'multiline' is an agronomically homogeneous population formed by backcrossing two or more VR genes into a single genetic background and mixing the products. A mixture, in this context, is a mixture of distinct varieties having complementary VR genes; the components must be homogeneous as to maturity (and, perhaps, quality) but may otherwise be quite diverse. Biologically, both are comparable with the spatially differentiated population of varieties mentioned above but the scale of variation is finer (plant-to-plant rather than field-to-field). Multilines were first proposed for wheat in Mexico over forty years ago but have not been widely tried; practical examples include a Colombian wheat, Miramar, and an oat multiline in Iowa (Browning and Frey, 1969). The success of the oat is interesting because Van der Plank's observation that all known VR genes in oats were, in his sense, 'weak' would not encourage a belief in the utility of mixtures or multilines in the crop; perhaps the oat genes were 'stronger' than he thought? Barley mixtures are currently being tried in the UK and gave some promising indications (Wolfe, 1977) but no real achievements. Multilines and mixtures (and spatially differentiated populations, too), it should be noted, cannot offer the immunity which has long been sought from VR genes; the best that can be achieved is a damping effect on epidemics which would look the same in the field as a HR that reduced the epidemic constant, r.

More generally, all five methods of deploying VR have in common the features that they deliberately diversify crop populations (in time or place) and they aim to produce not an immunity, but a simulated HR. This being so, it is hard to see that they offer anything that a true HR does not do just as well and much more reliably. No doubt they will have specialized local uses – perhaps valuable ones – as opportunities offer but one cannot foresee an indefinite continuation of what is, after all, a rescue operation, an attempt to salvage something from programmes that have essentially gone wrong.

The suggestion is sometimes made that VR and HR should be combined, producing VR without the 'vertifolia effect', so to speak. The suggestion fails on the fact that selecting for HR in the presence of VR is impossible; if the selector has the special pathotype, the VR is already useless and might as well not be there and, if he has not, he cannot select.

We conclude, therefore, that a reasonable breeding strategy for tackling the mobile air-borne pathogens of annual crops is to concentrate upon HR as the basic approach, using such VR as are available *only* if they can be critically deployed to simulate HR. In the longer run it is hard to believe that VR genes will not come to be generally regarded as potato breeders have regarded them for forty odd years, as a nuisance. The same argument applies, but with even greater force, to perennial crops, especially long-lived trees where mistakes are lasting in effect and slow and expensive to put right; coffee leaf rust and South American leaf blight of rubber provide cautionary examples but, significantly, examination of a number of rubber clones in a field epidemic immediately provided visual evidence of a useful level of HR in a few of them.

A general move away from VR (save in special circumstances) towards HR therefore seems to be indicated and is, in fact, surely taking place. The undertaking seems formidable because HR has been considered to be laborious to work up in breeding stocks. However, there are two encouraging points, the first being that it should be easy at least to improve on the very low level of HR that normally underlies a failed VR; the second is that quite modest levels of HR may be, in practice, all that is required. This second point is suggested by the frequent observation that apparent resistance depends upon the amount of infection offered to the plant; a clone, line or population that is rather resistant in a large pure stand may score susceptible if closely surrounded by infectors generating a spore-rain greater than the pure stand itself would generate; 'a mildew'd ear blasting his wholesome brother' perhaps? The Scottish barleys, Ymer and Golden Promise, mentioned earlier, illustrate the point: Ymer, in pure stand, has a useful HR to mildew but scores susceptible in a severe test or even on the field scale as a neighbour of Golden Promise.

Methods

The methods used for assessing disease resistance are extremely various but may conveniently be grouped in three categories for the purpose of summary. Details are beyond our scope.

First, experimental infection in glasshouse or laboratory is often used, especially in relation to VR, where an all-or-nothing reaction is expected. Thus hypersensitivities to pathotypes of the air-borne fungi are commonly detected on seedlings in the glasshouse or on detached leaves or on leaf-discs in the laboratory; resistance to viruses, whether VR or HR, is usually tested in the glasshouse by mechanical inoculation or, if necessary, by standard infestation with viruliferous insects; resistance to potato cyst eelworm, both VR and HR, is estimated by standard inoculation of pots of soil with the appropriate pathotype(s), followed by counting of cysts on the exposed root system; soil-borne fungi that produce visually estimable effects can be handled similarly (e.g. potato wart, brassica club root); aphid resistance can sometimes be estimated by scoring the behaviour of a standard sample placed on a leaf. And so on. The methods in general demand rather precise pathological control but they are neat and accurate and give quick, clear results.

Second, field testing using artificial infection or, perhaps more usually, augmented natural infection is widely used. The objective is, basically, to ensure that D_0 is not limiting so that differences in r between genotypes can be the more reliably identified. There can be no exact control of pathotype, of course, but, where HR is concerned, this does not matter. A still-effective VR can, of normally, be readily recognized. Artificial infection may be achieved by inoculating seed or planting material before planting, by interplanting 'spreader' rows of plants of very susceptible varieties, by spray or other infection of the crop, by soil inoculation with eelworms or fungi, in short by a variety of methods suggested by the biology of the pathogen. If HR is being sought and there is a known risk that natural pathotypes will be avirulent to VR genes in the material, then particular care will be taken to see to it that any added inoculum carries the appropriate virulence. Uneven infection is nearly always a problem, especially perhaps with some of the naturally 'patchy' soil pathogens; hence, if important selection decisions are to be made, replicated controls (susceptible and resistant tester stocks) to provide a scale will be essential.

Reliance upon natural field infection is but a simple form of the preceding. Given a good site, it can be very effective indeed; probably, the choice by the Rockefeller Foundation workers of the Toluca valley site in Mexico for testing HR to potato blight did more than any other single thing to promote understanding of the potential of HR in plant breeding. Not all sites, even good ones, will be equally effective every year because seasons vary. But most programmes make at least some use of off-station sites selected for disease-producing potential. Controls are again essential, of course, and the breeder will be continually wary of any seeming immunity as probably reflecting an unsuspected VR. Chosen sites are often called 'hot-spots'.

Finally, we must consider the question of scales. No, or few, problems arise where major-gene effects are concerned, whether VR or (rarely) HR; classification on a yes/no, diseased/not or some such basis generally suffices. Polygenic HR, with a continuous distribution of expression that shifts with time, however, presents problems. Whatever the basis of assessment (per cent leaf area infected, aphids per stem etc.) controls are vital to provide an objective basis for a scale that is liable to vary both between and within seasons (Fig. 7.1). As to the scales themselves, pathologists (presumably for obvious psychological reasons) seem drawn to scales that run from 0 (no disease) to, say, 5 (much disease).

Recently there has been a tendency to recommend 0–9 scales on the grounds that these are convenient for computing and can easily accommodate the shorter scales that are often preferred in practice. To the plant breeder, however, accustomed to the notion that the higher the score for a character the better, the pathologists' habit seems to be the reverse of sensible; he naturally prefers to estimate resistance rather than amount of disease and would need to reverse disease scales if diagrams such as Figs 5.10–5.12 were to make visual sense (see also Fig. 7.7 where it was necessary to write $R = (1 - r)$ and to connect W to D by a negative relationship). Similarly, disease scales were reversed, to make resistance scales, in the preparation of Tables 7.2 and 7.4. We hope that such pathologists as may read this chapter will take note. The potato breeders have long used resistance scales (0–9).

7.5 Summary

Diseases of crops are caused by a variety of pathogens, including fungi, bacteria, viruses, eelworms and insects. All can be, at least to some extent, combated by plant breeding and all provide examples of both successes and failures. Probably, the fungi, especially the obligate, airborne pathogens, those which cause many of the most dramatic and damaging diseases, have had the most attention. Most of the damage done by disease is to yield, by reducing biomass of the crop but blemishing of the product is also sometimes important. Epidemics depend upon initial infection/infestation, D_0, and upon the epidemic constant r (Fig. 7.1). The plant breeder's task is to reduce r to zero if feasible (immunity), but, more usually, to a level at which losses are reduced to an economically acceptable level.

Manifestations of resistance to a pathogen are very various: they include inhibition of infection/infestation and of growth and reproduction of the pathogen, all processes which tend to reduce r, the epidemic constant. Genetic control of resistance may be by major-genes (often controlling an infection-resistance) or by polygenes. Empirically, major-gene resistances (often, not always, dominant) are very often found to be pathotype-specific; that is, they are potent against some specific genotypes of the pathogen, but not against others; contrariwise, polygenic resistances are pathotype-non-specific. This generates a useful working distinction between vertical resistance (VR) and horizontal resistance (HR); the former is pathotype-specific and major-gene determined, the latter pathotype-non-specific and (nearly, not quite, always) polygenic. Some writers prefer other terminology (e.g. specific and general resistance) for the two classes. Because pathogens are living organisms capable of evolutionary response, VR has very frequently selected out new pathotypes virulent to previously resistant hosts; the evidence strongly suggests a one-to-one, gene-for-gene, correspondence of specificity between host and pathogen. Much VR breeding, especially against the mobile air-borne pathogens such as rust, mildews, smuts and potato blight, has therefore failed but it has served very well against several less mobile, soil-borne pathogens.

In comparison with wild populations of plants and primitive agricultures, advanced, technology-based agricultures tend to narrow the genetic base of producing populations and thereby positively encourage epidemic disease; there are several striking examples of disadaptive breeding, of epidemics actually promoted by plant breeding practices. 'Boom-and-bust' cycles, generated by the recurrent failure of VR, are one example; southern corn leaf blight, generated by the nearly universal use in the USA of an unsuspected 'susceptible cytoplasm', is another. A general need for reasonable genetic diversity as an insurance against such hazards is apparent.

Breeding strategy is increasingly being directed towards the development of HR (already very effective against a great many diseases). VR has its uses against immobile soil pathogens known to be spatially localized and may yet find wider use than hitherto against the air-borne pathogens in producing a sort of simulated HR in populations heterogeneous as to VR in time or place. A disease resistance is but one character of the many that enter the plant breeder's intuitive selection index and should have as much weight as is economically justifiable but no more. In practice, quite moderate HR is often good enough, even if more would be better. Methods of handling disease resistance in a breeding programme are very various and depend upon the biology of the pathogen. Fairly refined glasshouse and laboratory methods and rougher, less well controlled, but still perfectly effective field techniques are used; the latter, perhaps, will tend to gain currency as HR displaces VR as a breeding objective. Scales used in such work should relate to resistance not to infection (or susceptibility); in estimating resistance in tests, controls are essential.

7.6 Literature

General, fungi, bacteria, viruses

BEPP (1973); Bjorling (1966); Brian and Garrett (1972); Buzzell and Haas (1972); Cockerham (1970); Day (1974); Flor (1971); Gallegly (1968); Good and Homer (1974); Holmes (1954, 1965); Holton *et al.* (1959); Horsfall and Dimond (1959, 1960); Ingham (1972, 1973); Johnson and Taylor (1976); Kim and Brewbaker (1977); Klinkowski (1970); Leppik (1970); Macer (1963); Marani *et al.* (1971); Person *et al.* (1976); Robinson (1969, 1971, 1973); Rodrigues *et al.* (1975); Saari and Wilcoxon (1974); Sidhu (1975); Simmonds (1991a, 1995); Stevens (1948); Tatum (1971); Ullstrup (1972); Van der Plank (1963, 1968, 1975); Walker (1953); Wiberg (1974).

General, animal pests

Gallun *et al.* (1975); Maxwell *et al.* (1972); Painter (1951, 1958); Rhode (1972); Widstrom (1974); Zuber *et al.* (1971).

Epidemics

Kranz (1974); Mode (1958); Van der Plank (1963, 1968, 1975); Zadoks (1972).

Breeding methods and strategy

Borlaug (1959); Browning and Frey (1969); Caldwell in Finlay and Shepherd (1968); Committee on Genetic Vulnerability (1972); Crill (1977); Day (1973, 1974); Eenink (1976); Fiddian *et al.* (1973); Frey *et al.* (1973); Fry (1975); Hanson *et al.* (1972); Harlan (1972, 1976); Howard *et al.* (1969); Jenkins *et al.* (1954); Killick and Malcolmson (1973); Knight (1945, 1956); Knott and Dvorak (1976); Lim (1975); Lupton and Johnson (1970); Nelson (1973); Parlevliet and Zadoks (1977); Roane (1973); Russell (1978); Simons (1972); Storey (1958); Thurston *et al.* (1962); Umaerus (1969, 1970); Van der Plank (1963, 1968); Wilcox (1983); Wolfe (1977).

Disease losses, economics

Carlson and Main (1976); James (1974); James *et al.* (1972); Ordish and Dufour (1969); Slootmaker and Essen (1969).

Chapter 8
Special Techniques

8.1 Introduction

The great bulk of plant breeding endeavour is devoted to the manipulation, at a fixed level of ploidy, of genetic variability that already exists within the confines of a single cultivated species. Within these limits, the breeder is concerned to change gene frequencies, under more or less regular sexual reproduction, to economic advantage, whether he is breeding, IBL, OPP, HYB, or CLO. The result is enhanced local adaptation in terms of yield and quality of new varieties. In this chapter and the next, we shall be concerned with an array of special techniques which are essentially accessory to this 'mainstream' plant breeding of which some can be considered under the umbrella term 'biotechnology'. They are quite diverse in origin and in scientific nature but all have in common the feature that they complement the more conventional techniques. Many are ingenious, even elegant, and have had valuable applications. At the other extreme, a few touch upon the frontiers of biological knowledge or technique and have yet to be significantly realized in practice, although progress is being achieved in what may be called 'genetic engineering' but at a slower rate than its advocates had hoped.

The principal techniques to be considered are summarized in Fig. 8.1. They are: polyploidy (auto-and allo-, within and between species); wide (inter-specific or inter-generic) crossing which is sometimes a preliminary to allo-polyploidy, sometimes a start to a backcrossing programme; haploidy, as a means of making 'instant inbreds' or of expediting breeding at a higher level of ploidy; mutagenesis and transformation as means of generating variability that is not naturally available; and *in vitro* methods (covered in Chapter 9), which touch upon a remarkably wide range of mainstream plant breeding techniques from the multiplication of clonal stocks to 'genetic engineering'.

8.2 Polyploidy

Introduction

We start by reminding ourselves that our crops are naturally, and without any intervention from plant breeders, somewhat unequally divided (Table 1.1)

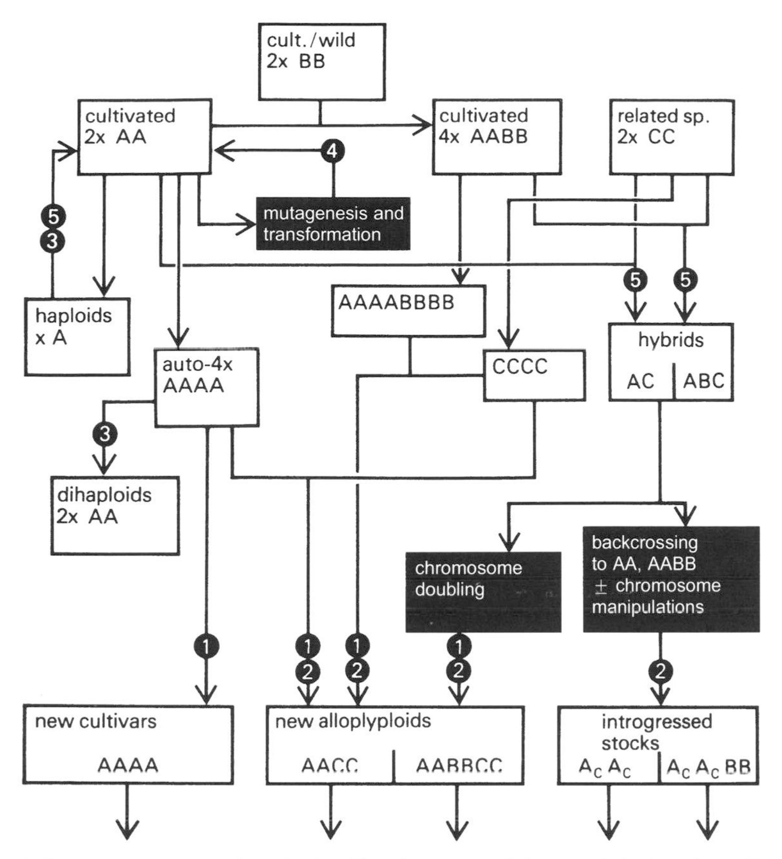

Fig. 8.1 Diagram to summarize the leading features of the special plant breeding techniques discussed in this chapter.
Notes: The numbers refer as follows: (1) the use of polyploidy to produce new autopolyploid cultivars from normally diploid species and new allopolyploid crop species from inter-specific hybrids; (2) the use of wide inter-specific crosses to produce (see (1)) new allopolyploids and, by backcrossing techniques, cultivar genotypes introgressed by pieces of foreign chromosomes; (3) the use of haploids and dihaploids to facilitate breeding; (4) the use of mutagenesis to produce desired genetic changes (mostly at the diploid level); (5) the use of *in vitro* techniques to facilitate the production of haploids, transgenics and difficult hybrids.

between the more numerous diploids (e.g. maize, rice, barley, sugar beet, faba bean and most other grain legumes, rubber, cacao) and the allopolyploids (e.g. wheat, oats, tobacco, sugarcane, cottons, *arabica* coffee) *plus* a few, more or less autopolyploids such as potatoes and lucerne. Several groups contain both diploids and polyploids in the crop itself (e.g. potatoes, cottons, coffees, wheats, oats) or in the crop and its wild relatives (e.g. potatoes, tobacco). It is generally agreed that the allopolyploids owe their success to the same factors that have contributed to the success of wild allopolyploids, namely: the vegetative vigour ('gigas' characters) typical of many polyploids and, probably much more important, the permanent hybridity, genetic buffering and evolutionary flexibility conferred by a multiplicity of genomes. Since there is always an upper limit to the degree of polyploidy tolerated by a group of plants, the possibilities open to the breeder effectively resolve themselves into two: the exploitation of autotetraploidy in diploid crops and of allopolyploidy in both diploids and existing polyploids.

A word about techniques will not be amiss. Chromosome doubling occurs naturally in all plants at low frequency as a result of mitotic failure. Most products are lost but occasional doubled cells situated in critical places in meristems give rise to polyploid sectors or (rarely) even whole plants. The frequency of such events is raised by various external factors such as heat shocks or forced regeneration of shoots from calluses but such techniques are, at best, laborious and unreliable. In practice, chemical treatments are always used: there are several active chemicals which inhibit anaphase separation at mitosis, of which by far the most reliable and widely used is colchicine. This substance is extracted from the autumn crocus, *Colchicum autumnale* (itself, perhaps surprisingly, a diploid); it is rather toxic to man but is a useful drug for the treatment of gout (and has been known as such for 2000 years).

At appropriate concentrations in aqueous solution, colchicine inhibits mitotic anaphase, so the treatment of a meristem with it for a few hours causes an accumulation of polyploid cells. Subsequent growth results in polyploid sectors, chimeras or even whole meristems. Young zygotes (i.e. seed embryos or small seedlings) respond best; doubling of whole vegetative meristems is difficult, often very difficult indeed (mixoploid chimeras are the usual outcome). The practical point that emerges is, therefore, that breeding plans should be so arranged that any colchicine treatment is, if possible, applied to seeds or small seedlings that are abundant and easy to produce, rather than to rare or precious plants or to vegetative meristems.

Autopolyploidy

There is probably no diploid crop in which at least a few autotetraploids, whether of natural or experimental origin, have not been examined. In the 1930s, autotetraploid lines of maize and tomatoes were important objects of cytogenetic research and, for a time in the late 1940s and 1950s, a good deal of plant breeding effort was devoted to autotetraploidy but the total practical outcome has been small.

In comparison with diploids, autotetraploids are often (not always) larger and lusher in growth and, because of multivalent formation at meiosis, always initially less seed-fertile. This might suggest that autotetraploidy would find its principal use in crops grown for a vegetative product; and so it does. Autotetraploid red clovers and ryegrasses are the only recorded successes and they have by no means supplanted the diploids. They have vegetative vigour, digestibility and palatibility to stock and the clovers have an unpredicted bonus in the form of some eelworm resistance; they are sufficiently seed-fertile for practical purposes. Among other crops grown for vegetative products it has generally turned out that the 'gigas' feature of autotetraploids are compensated by lower dry matter contents; hence tetraploid kales and turnips (Fig. 8.3) were no more productive than diploids and triploid sugar beet synthetics had but a passing vogue (Hornsey, 1975).

Autotetraploid bananas fall in quite a different category. They have no attractions *per se* over the established autotriploids; indeed they have the positive disadvantages of weaker foliage and enhanced fertility. They are bred because they offer the only available chance of adding a disease resistant genome, A′, to an agronomically acceptable genetic background, AAA. Their production rests, not upon the use of colchicine, but upon the possibility of fertilizing rare unreduced eggs (AAA) by haploid pollen from a disease resistant male parent. Autotetraploid bananas have been produced thus in large numbers but have yet had no substantial practical success; they still represent, however, the only evident practical approach to banana breeding – which is surely one of the odder and most specialized corners of the whole field of plant breeding.

Experience (of these and many other crops) thus does little to support the common expectation of thirty years ago that the 'gigas' characters of autotetraploids would prove advantageous in crops grown for vegetative products. The natural autotetraploidy of the potato therefore provokes the question as to why it was selected. The short answer is that we do not know. In the Andes, diploid cultivars derived from wild diploids were largely supplanted by the autotetraploid Andigena Group but the basis of selection has never been determined. In north temperate countries, the first introductions were, presumably, all autotetraploid and local evolution produced the Tuberosum Group, our modern potatoes, from them. But breeding experiments with diploids and of crosses between diploids and dihaploids (out of autotetraploids – see below, Fig. 8.4) suggest that diploids are intrinsically no less productive than autotetraploids. So our potatoes may be autotetraploid by accident. There is no decisive evidence for superiority over diploidy (though the history of the crop in the Andes is suggestive).

There are a few crops in which autotriploidy may be valuable, though not yet widely adopted. These are hops, citrus and pyrethrum. Clonal propagation is, of course, implied. The inevitable seed-sterility is a positive attraction in the first two crops, as it was also in the evolution of the triploid bananas. In pyrethrum, seed fertility/sterility does not matter because the flower, not the fruit, is the crop product.

Turning now to autopolyploidy (in effect tetraploidy) in plants grown for a seed product, many have been studied and none has yet been successful. The invariable initial sterility can be reduced by selection but has not so far been eliminated. Indeed it would perhaps be surprising if it could be, since success must depend upon regular 2-2 disjunction of quadrivalents or upon near-perfect bivalent pairing by chromosomes which are, by origin, undifferentiated genetically. The greatest progress (after many years of work) has perhaps been made with rye (Kranz, 1973), barley (Gaul, 1976) and sorghum (Doggett and Majisu, 1972). Larger grains, enhanced protein contents and higher yields are the aim. Whether these objectives can be achieved, if and when fertility is pushed up to a level at which the ears are well enough filled, is yet uncertain. It is at least possible that negative correlated responses will reduce the grain size and protein content as yields rise (see Section 5.8).

In summary, autotetraploidy has not fulfilled the hopes that were once entertained for it; indeed its direct uses are probably exceeded in value by indirect uses in the manipulation of allopolyploids and wide inter-specific crosses, as we shall see below.

Allopolyploidy

The existence of many allopolyploid crops (Table 1.1), mostly derived from allopolyploid wild ancestors (such as the tetraploid cottons and wheats) but some from hybridization in cultivation (hexaploid wheats, tetraploid brassicas and triploid hybrid bananas), might suggest wide opportunities for the plant breeder to imitate nature in the production of new allopolyploids. Indeed, scores, perhaps hundreds, have been made experimentally, often as steps in introgression programmes. Only one, however, has yet attained the status of a new crop species but others may follow. The two leaders are triticale and raphanobrassica (Figs 8.2 and 8.3). Among the herbages, allopolyploid clovers and *Festuca-Lolium* hybrids may also find a place and there are also interesting possibilities in other crops as diverse as jute and *Rubus* fruits.

The history of triticale goes back to the late nineteenth century when wheat–rye hybrids were first examined. Serious breeding did not start until the 1930s (in Sweden) and it is only in the last few years that triticale has emerged into practical agriculture (in Canada) as a new crop with the hardiness of rye and the yield and nutritional quality of wheat. Triticale was by no means an 'instant species'; the delay was due mainly to the need for prolonged selection for fertility and for exploration of the many possible genome combinations. As it turned out, the obvious choice (the octoploid AABBDDRR) was not the most adaptable (but may yet be successful). Other possibilities are obvious and it can hardly be doubted that we are still only at the beginning of exploitation of complex wheat–rye combinations. As Fig. 8.2 suggests, the rye genome takes its place naturally in the *Triticum–Aegilops* complex, so it seems an unnecessary nomenclatural elaboration to give triticale a new generic name.

The other example, Raphanobrassica, is analogous to triticale in three respects, namely: (1) in being a cross between two cultivated species (and having some good characters from both); (2) in the infertility which had to be overcome in the early generations; and (3) in the evident possibilities of producing other new and useful allopolyploids (Fig. 8.3). And, another parallel, the

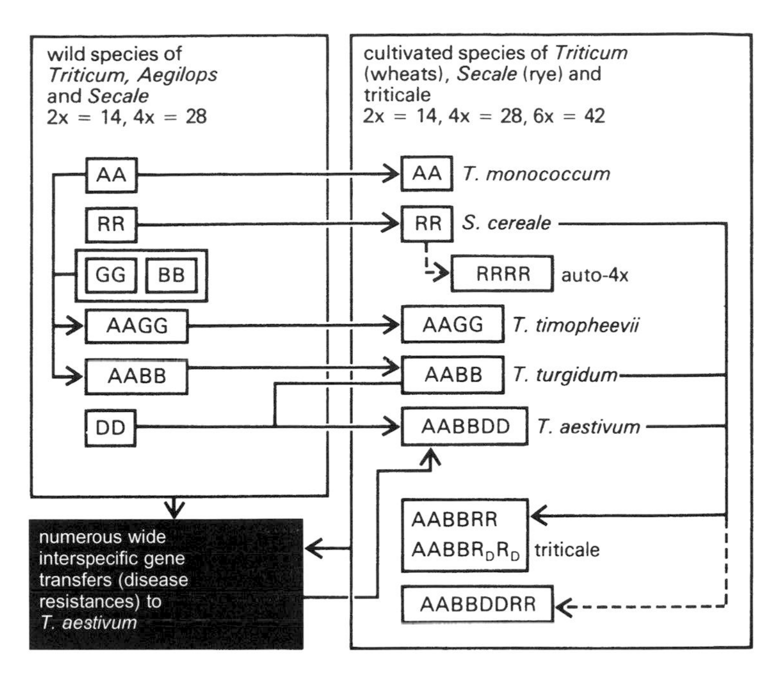

Fig. 8.2 Wheats, rye and triticale, a summary of relationships.
Notes: Genome constitutions and ploidies are indicated by the appropriate capital letters. The cultivated wheats had four origins and contain four different genomes (A, B, C, D). The breeding of triticale has added another genome (R), so far with success only at the hexaploid level; the R_D genomes in the second kind of triticale listed are rye (R) with some D material and they derive from a heptaploid stage (AABBRRD) in the breeding of it. The routes to autotetraploid rye and octoploid triticale (AABBDDRR) are shown with dashed lines to indicate that they have not yet been successful. Examples of interspecific transfer to *T. aestivum* are given elsewhere in this chapter. Based on Feldman and on Larter in Simmonds (1976a), considerably simplified.

Raphanus genome takes its place in *Brassica* and a new generic name seems equally inadvisable. Raphanobrassica varieties may yet emerge into agriculture in the mid-future, providing a useful combination of the hardiness of the kales with the quick growth and disease resistance of the fodder radish. Though the first products are leafy, rape-like plants, there is no evident reason why later allopolyploids should not be bulb-formers reminiscent of the turnips and swedes.

Figure 8.3 illustrates several other useful points about polyploids. First, autotetraploids had little or no success in comparison with diploids but have been very useful in making difficult allotetraploids (both *B. napus* and Raphanobrassica). Second, two rather unusual kinds of allopolyploids are represented in the bottom left-hand corner of the diagram; the hexaploids (of which AAAACC was best explored) were attractive, but were as unsuccessful as the autotetraploids (and for the same reasons), whereas the triploid (AAC) types may yet find a place as a very special case of HYB breeding (see Section

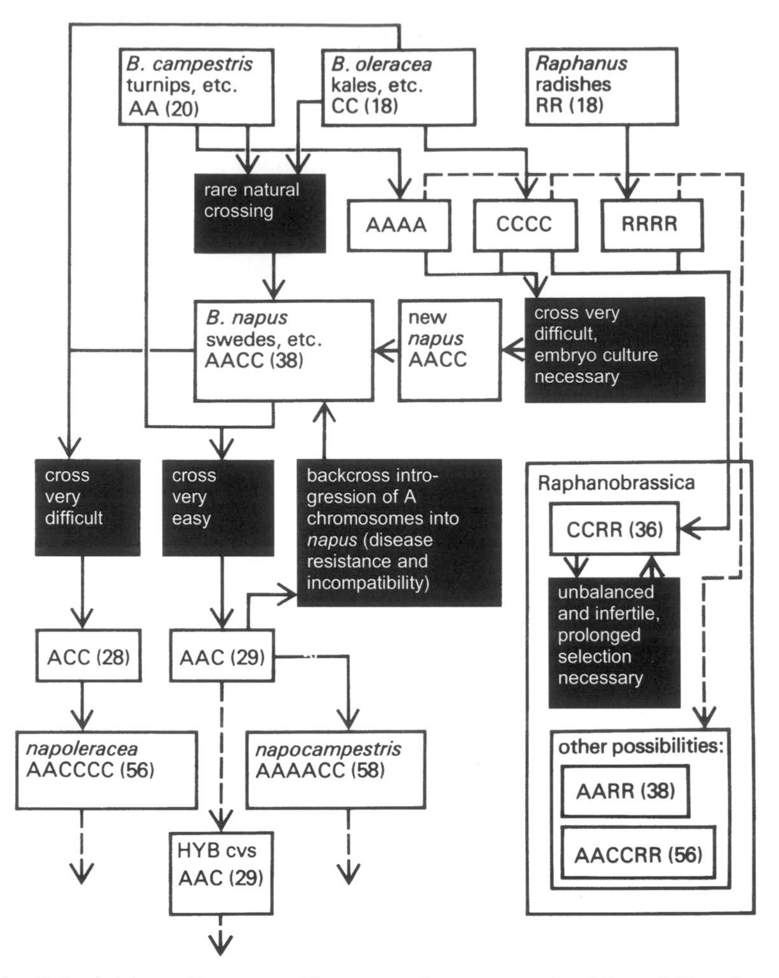

Fig. 8.3 Polyploids and inter-specific crosses in a group of cultivated Brassicas. Notes: Only three *Brassica* spp. are included in the diagram. The inclusion of *B. nigra* (black mustard, NN, 2x = 16) would extend the diagram by adding two more natural allotetraploids (*B. carinata*, NNCC, 4x = 34 and *B. juncea*, NNAA, 4x = 36); the three diploids and three allotetraploids constitute the 'Triangle of U'. This diagram shows: (1) how breeders are reconstructing *B. napus* from its source diploids, using autotetraploids and embryo culture methods; (2) introgression of A chromosomes into *napus*; (3) attempted exploitation of new A + C combinations as HYB varieties and allohexaploids; and (4) attempted exploitation of intergeneric allopolyploids (Raphanobrassica). Based on McNaughton in Simmonds (1976a); McNaughton and Ross (1978).

5.5). Third, the artificial *napus* stocks, constructed from parental diploids by way of autotetraploids, provide a neat example, not of a new allopolyploid, but of an old one in which there was a plain necessity to widen a narrow genetic base (Section 5.9); *campestris* (A) chromosomes can easily be backcrossed into *napus*, as the diagram shows, but artificial *napus* is the best way of enhancing the C contribution and of widening *napus* in general.

We have seen that, if a cross is difficult, it will usually be economical (and may also be easier) to make it at the tetraploid level, as shown in Fig. 8.3. In addition, there is much to be said for the principle of using highly heterozygous autotetraploids as parents because: (*a*) they will generate variability at the allotetraploid level and thus open up possibilities for selection even if only a few crosses are available; and (*b*) there is a chance that crossability might have an element of nuclear genetic control and so be favoured by gametic segregation.

Conclusions about the use of polyploidy are as follows. Autopolyploidy is of little direct practical value but is of great importance in making allopolyploids and other wide crosses. Evidently, 'gigas' luxuriance contributes more to moisture content than to biomass and, for seed-producing crops, infertility is a severe barrier. Allopolyploidy has far greater potential, by reason of the permanent hybridity and possibilities of new character combinations that it offers. There must be a significant potential for the construction of new crops (e.g. triticale) and the reconstruction of old ones (e.g. *B. napus*). In the longer term, these may well turn out to be rather important steps in the continued evolution of polyploid crop complexes.

8.3 Wide crossing

General

Wide crossing as a feature of programmes designed to generate allopolyploids we have discussed in the preceding section. Here we are concerned with the transfer of genes or of relatively small blocks of foreign chromosome across inter-specific boundaries so that the recurrent parent genomes are left substantially unaltered. Whatever the complications (and they can be many), the process is always essentially a backcross system. At the simplest level (Fig. 8.4), the recurrent and donor parents may be near enough genetically that, though distinct species, backcrossing goes as easily as with an intra-specific transfer. This can be so at both the diploid and polyploid levels. *Hirsutum-barbadense* transfers in cotton provide the classical example (Section 5.7). Often, uneven polyploidy, 3x, 5x, 7x, etc.) presents no special obstacles, though there may be a good deal of sterility, erratic chromosome numbers and some very peculiar segregates in early generations. All the routes shown on the right-hand side of Fig. 8.4 have been used in potatoes for the transfer of disease resistances (to viruses, blight and eelworms) from diploids and wild allopolyploids to autotetraploid cultivars. Examples could be multiplied endlessly but it would be profitless to do so. Hundreds of inter-specific transfers have been made, at various levels of ploidy, with varying degrees of difficulty and with or without various tricks such as the use of embryo cultures, grafting techniques, autotetraploid parental lines and bridging crosses. All such trans-

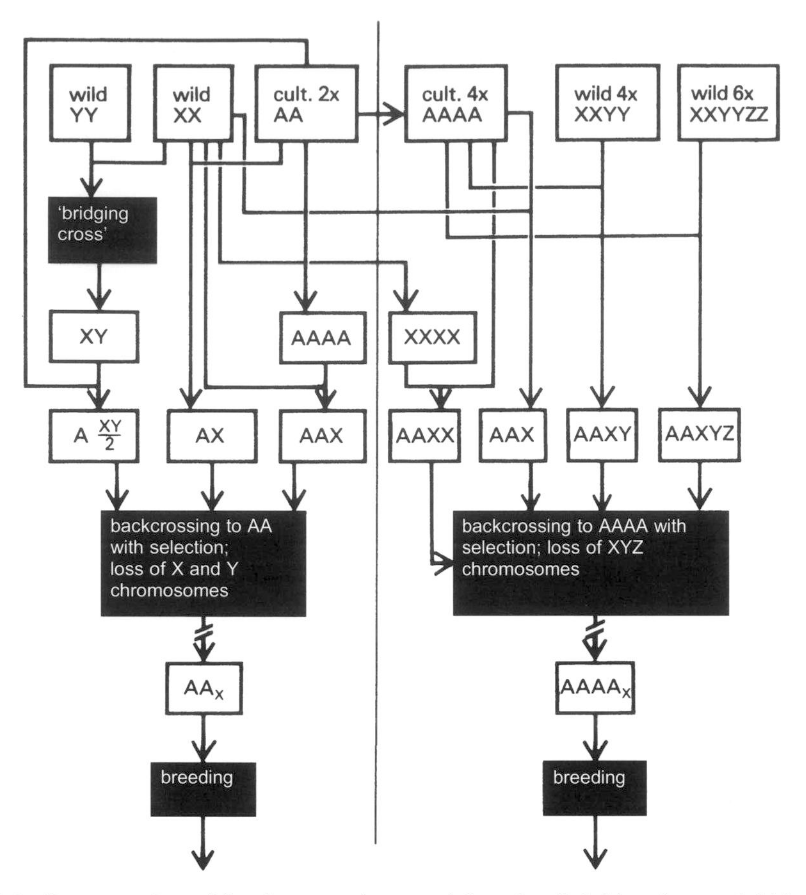

Fig. 8.4 Introgression of foreign genetic material at the diploid and tetraploid levels. Notes: All routes depend upon backcrossing, with selection for the desired foreign character. Simple crossing and backcrossing works where the cultivated and wild genomes are little differentiated. Induced tetraploidy on one side of the cross may be useful. The 'bridging cross' approach applies where a wild genome other than that of the desired donor is nearer genetically to the cultivated genome; then, XY may be more easily crossed to AA than XX itself; the principle applies equally at the polyploid level (e.g. in wheats and tobaccos). The tetraploid routes on the RHS of the diagram can all be exemplified in the potatoes.

fers depend upon there being sufficient homology between the recurrent cultivated genome (A) and the foreign donor genome (X in Fig. 8.4) so that the desired gene can be incorporated without unacceptable genetic disturbance of other characters – which, however, sometimes occurs and some crosses fail because of it.

The genes transferred have nearly always been disease resistances but Fig. 8.3 provides an example of the transfer of self-incompatibility (needed for the development of HYB cultivars in *Brassica napus* – see Section 6.4). Cotton provides the example of the fibre strength of *G. thurberi* introgressed into *G.*

hirsutum by way of an *arboreum-thurberi* synthetic allotetraploid; and tobacco (Wernsman *et al.*, 1976) the example of general adaptation improved by the introgression of *sylvestris* chromosomes into *tabacum* by way of a synthetic allopentaploid.

Chromosome manipulation

When the foreign chromosome is but distantly related to its equivalent in the recurrent parent, ordinary pairing and genetic recombination cannot be relied upon to effect the transfer. In that case, special techniques, referred to collectively as 'chromosome manipulation', must be used to incorporate the foreign material. The starting point is a backcross programme which results in either an 'addition line' or a 'substitution line'. The former is aneuploid, with one foreign chromosome added (e.g. AABB + x), the latter euploid but unbalanced, with the foreign chromosome substituted for one of the recurrent chromosomes (e.g. AABB – a + x). Lines such as these are effectively viable only in polyploids so chromosome manipulations are, in practice, virtually confined to the few allopolyploid crops, such as wheat and tobacco, for which there is an excellent background cytogenetic understanding and control. Given these circumstances, the incorporation of genetic material from addition or substitution lines depends either upon the management of homoeologous pairing or upon the restructuring of chromosomes by translocation. We shall now consider these two methods.

Chromosome pairing in regular allopolyploids is only between homologous chromosomes within genomes (e.g. in wheat, 2A with 2A, 2B with 2B but not 2A with 2B). But the constituent genomes of an allopolyploid carry extensive homologies residual from earlier evolutionary identity. Genomes and individual chromosomes that are so related are said to be 'homologous' (e.g. wheat 2A with 2B). It is a condition for evolutionary success of an allopolyploid that homoeologous pairing is suppressed, so that homologous bivalents alone are formed; extensive homoeologous pairing leads to multivalent formation, sterility and genetic disruption. The functional diploidy of all well established allopolyploids shows that some sort of homoeologous pairing suppression is regularly evolved; in hexaploid wheat, it is known to be due to a dominant gene in chromosome 5B (Riley and Kimber, 1965) supplemented by others located elsewhere (Driscoll, 1972). In other allopolyploids such as tobacco, cotton and oats, control may be confidently inferred but its genetic nature is yet unknown. The sterilities characteristic of the early generations of new allopolyploids (see comments above on triticale and Raphanobrassica) are no doubt, partly at least, due to homoeologous pairing; *per contra*, the evident improvement of fertility under selection must imply enhanced homologous bivalent pairing and the suppression of homoeologous combinations.

The use of translocations depends upon the facts, well known from many cytogenetic studies of diverse organisms, that any energetic radiation (e.g. X-rays) breaks chromosomes more or less randomly and that broken ends then either heal or rejoin or fuse in new combinations. For the plant breeder the last is the interesting possibility because it includes the formation of new chromosomes carrying translocated bits of the foreign chromosome; if those bits contain the desired gene but confer no undesirable genetic background effects, the way is open to incorporation of the new chromosome, along with its resistance, in the genome of the recurrent parent. Technically, this is not easy because, though translocations are readily enough produced, most of them are unsatisfactory and a great deal of screening, with much progeny testing, is usually necessary.

These, then are the two available methods. In practice, translocations have been more widely used and there are good examples in tobacco (transfer of tobacco mosaic virus resistance from *N. glutinosa* to *N. tabacum* is the classic case) and wheat (rust resistances from rye and wild diploids – see Fig. 8.2). Only in wheat is the management of homoeologous pairing control good enough to permit the use of the alternative method and an elegant example is given in Fig. 8.5. Homoeologous pairing, and hence incorporation of the desired m segment, was made possible by the use of the 5B-suppressing effect of *Ae. speltoides*; an alternative approach is to use lines deficient in the critical segment of 5B, thus eliminating homoeologous pairing control rather than suppressing it.

Finally, to try to place chromosome manipulations in context, we should recall that the great majority of inter-specific gene transfers are accomplished without recourse to them. They are usually either unnecessary or demand a higher degree of cytogenetic control than is available. However, their use could increase as knowledge improves.

Sugarcane, a special case

The situations we have been discussing so far (allopolyploidy and the inter-specific transfer of small chromosome segments) cover the great majority of cases of the use of wide crossing in plant breeding. Sugarcane is exceptional in not fitting into either of the above categories. In this crop, the long-established 'noble' canes (*Saccharum officinarum*, 2n = 4x (?8x) = 80, presumptively allopolyploid from allopolyploid wild ancestors) were universally displaced about sixty years ago by complex, highly polyploid, hybrid derivatives of noble canes crossed by wild forms of *S. spontaneum.* The latter has variable chromosome numbers in the range 2n = 40 – 128. Backcrossing to nobles yielded populations with 2n = 100–125, 5–10 per cent of wild chromosomes and much aneuploidy and gametic sterility. Modern cultivars are all of this nature and have proved outstandingly successful (Tables 3.1 and 6.2). Sugarcane breeding seems to be unique in respect of its successful use of hybrid aneupolyploids. This irregular ploidy has been no obstacle to breeding (Table 3.1, Fig. 11.7). Sugarcane clones cover huge areas and about 20 serious diseases are well-controlled by HR (Section 7.4) with no recourse to spraying.

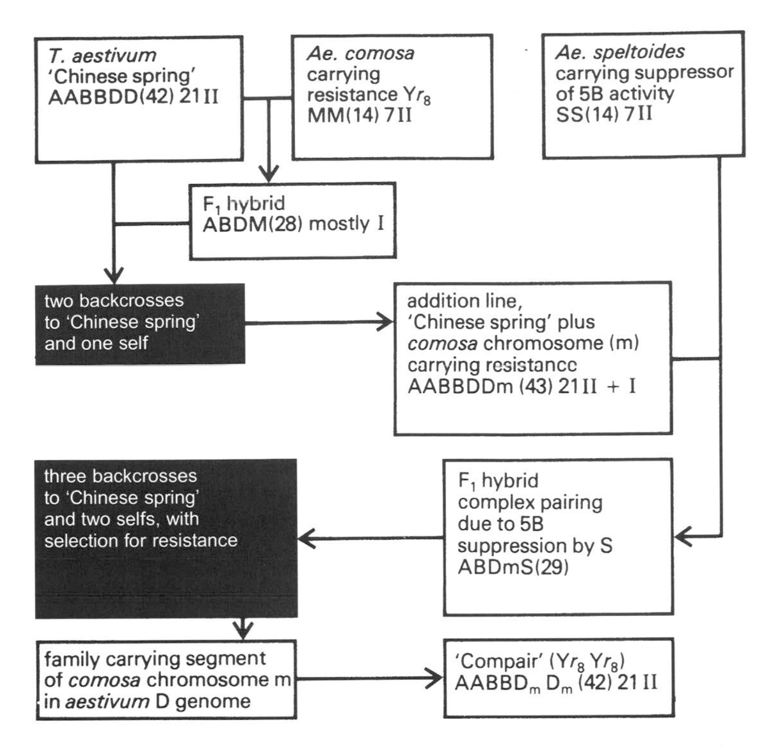

Fig. 8.5 Introgression of resistance to yellow rust from *Aegilops comosa* into wheat, using the 5B-suppression system of *Ae. Speltoides* (based on Riley, Chapman and Johnson, 1968).
Notes: I indicates univalent chromosomes, II bivalents. There were two backcross programmes in succession. The first added a single chromosome (m), carrying the desired gene Yr_8, to an *aestivum* genotype, forming an 'addition line'; the second used the 5B-suppressing activity of *Ae. speltoides* to encourage homoeologous pairing and the consequent insertion of a segment of m into the wheat genome D. The *comosa* chromosome was 2M, known to be homoeologous with 2D. In principle, a similar end result might have been achieved (though less elegantly) by irradiation of the addition line, followed by a search for (rare) translocation products in which the necessary bit of m had been inserted into the wheat genome but without concomitant deleterious effects.

8.4 Haploidy

Nature and production

Haploidy with respect to the sporophytic chromosome set is characteristic of the gametophytic phase of the life cycle in all angiosperms that show normal sexual reproduction. Occasionally, in circumstances which we shall shortly outline, haploid cells can divide and grow along a sporophytic development path, thus producing whole plants with the gametophytic chromosome number of the parents. Haploids, with reduced numbers, are therefore complementary to polyploids, with increased numbers. Haploids out of diploids are often called 'monoploids', especially in the maize literature, a term which brings out the contrast with polyploidy. Haploids out of polyploid parents are often simply called 'haploids' without ambiguity but, in practice, a distinction is sometimes useful, so those derived from tetraploids are 'dihaploids' and the words 'trihaploid' and 'tetrahaploid' could be used if needed; collectively, all such are 'polyhaploid'.

Haploids, of all kinds, are occasionally produced in all plants, normally by female parthenogenesis whereby an unfertilized egg cell develops; much more rarely, male parthenogenesis, wherein the male gamete alone replaces the zygote, produces a haploid plant derived from the paternal parent. Haploids sometimes appear as components of twin embryos/seedlings but the connection is not close and many twin pairs are diploid, the products of developmental accidents such as twin embryo-sacs or divided zygotes. In flax, however, Thompson (1977) suggests that haploids in twins occur in sufficiently high frequency (about 6×10^{-3}) to be useful in plant breeding, and are easily doubled with colchicine. With this exception (and, perhaps, others) natural haploids are normally too rare to be useful so, if wanted, they must be deliberately made, nearly always by female parthenogenesis (Fig. 8.6). The principle is to pollinate the female parents from which haploids are desired by pollen carrying a distinctive dominant seedling marker. The rare unmarked progeny are then (presumptively) maternal haploids. This method works well in maize where, as we have seen (Section 5.5), it has contributed usefully to the production of parental inbred lines. Well controlled, easy cross-pollination and a good seedling marker are essential; the method is obviously not applicable to inbreeders which are laborious to cross.

Dihaploids out of autotetraploid potatoes have been produced in some quantity (now several thousands) by essentially the same method but with help from a fortunate cytogenetic accident. The cross tetraploid female by diploid male would be expected to produce many triploids but it does not; total progeny are few and predominantly tetraploid, the products of selection of occasional unreduced, i.e. diploid, pollen grains; the rest are triploids and maternal dihaploids. A dominant (colour) marker in the pollen parent helps to sort out the dihaploids but, even if this is unavailable or unworkable, the relatively small numbers of progeny can be screened on phenotype and chromosome counting. Thus the very infertility of the cross (caused by the block to triploid production) is an advantage. Pollination is laborious but screening is relatively simple, though not so elegant as in maize.

The same genetical principle, of using a seedling marker in the pollen parent, has been exploited in a slightly different form in *Brassica* and cotton. Any matromorphic plants from a wide cross, that would normally produce few, visibly hybrid, progeny, will be, presumptively, haploid or, perhaps, doubled haploid. In practice, diploid matromorphs in *Brassica*, once hoped to be a source of 'instant inbreds', always turn out to be heterozygous, the parthenogenetic products, presumably, of partial meiotic failure (Eenink, 1974).

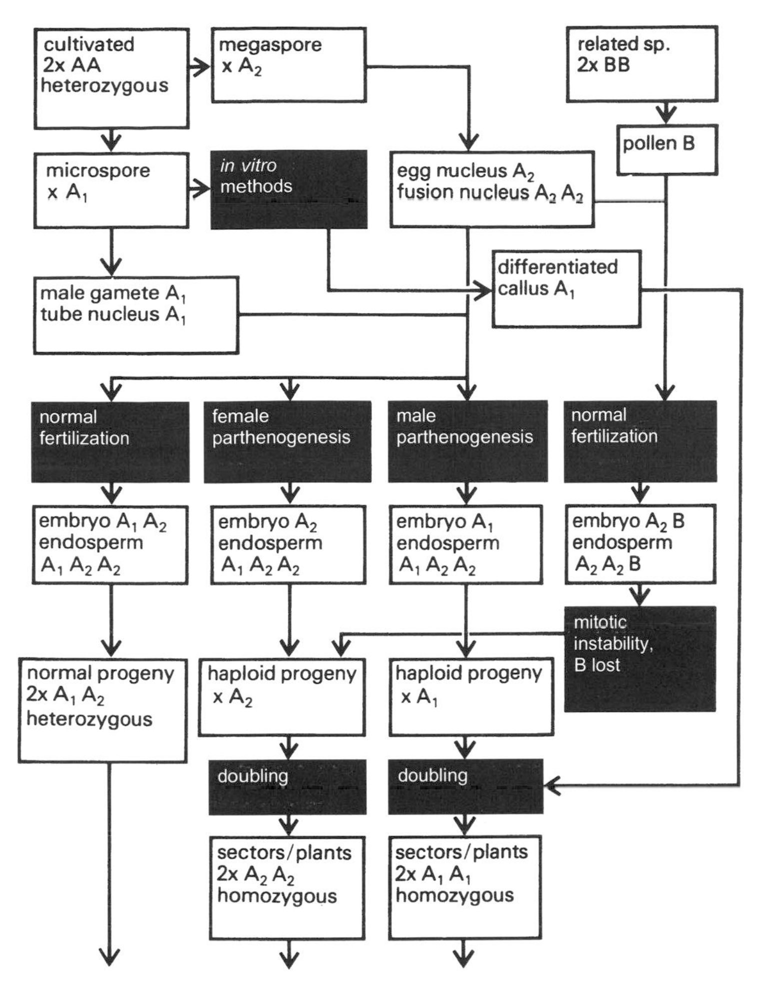

Fig. 8.6 Sources of haploids in plant breeding.
Notes: There are five routes, of which female parthenogenesis has been the most important (especially in maize – see Section 5.5). Efficient exploitation depends upon genetic marking of male gametes so that diploid plants (A_1A_2) can be distinguished from haploids (A_2) in the seedling stage. Female parthenogenesis is often associated with twin embryo-seedling formation and more or less obscure embryology. Male parthenogenesis is much rarer; haploids derived from pollen culture *in vitro* (Figure 8.8) are genetically equivalent, except for cytoplasm. The route shown at the RHS of the diagram is known in barley (*Hordeum sativum* × *H. bulbosum*); it depends upon dis-synchronous mitosis between the two genomes, such that B chromosomes are lost in development. For simplicity, the diagram is confined to haploids (= 'monoploids') out of diploid parents. Haploids out of tetraploids are functionally diploid and are called dihaploids; they are well known in potatoes, where they are of female-parthenogenetic origin (see text).

Three other methods of haploid production shown in Fig. 8.6 are, so far, less important. Male parthenogenesis has been exploited a little, at very low frequency, in maize (Section 5.5). It has the interesting and potentially useful feature of moving a haploid genome into a foreign cytoplasm at one step. *In vitro* cultivation of pollen as a callus/cell culture, followed by differentiation of new meristems and hence whole plants from single cells could, in principle, provide a massive source of haploids; technical problems have so far proved insuperable but the method may yet be useful (see Section 9.2). Haploids produced thus would be genetically equivalent to male parthenotes but without change of cytoplasm. Finally, barley haploids have been produced by the route shown at the right-hand side of Fig. 8.6. The rather wide cross *H. sativum* × *H. bulbosum* produces F_1 plants in which, as a result of dis-synchronous mitosis, *bulbosum* chromosomes are lost in development, leaving *sativum* haploids, genetically equivalent to the products of female parthenogenesis. The method is thought to be more promising in barley than the *in vitro* approach (Dale, 1976) but is clearly of very narrow applicability.

Uses of haploids

We will consider haploids, in the limited sense, and polyhaploids separately. Haploids as such are of little or no use because they are weak and sterile. Any practical interest they may have centres upon the use of the homozygous diploids that derive from them by chromosome doubling. Somatic doubling occurs spontaneously in many haploids, leading to haplo-diploid chimeras and hence to the initiation of pure lines, by selfing in diploid sectors; the process (which is fairly frequent but not very reliable) could presumably be assisted by the use of colchicine but, as we saw above, this is technically unsatisfactory when applied to vegetative meristems. Supposing that diploid sectors can be generated, haploids then clearly provide a neat and potentially rapid means of producing homozygous lines. They have been used thus in maize (Section 5.5) with some success but not very extensively. They have been tried, but with no significant practical success to date, in several inbreeders such as barley. In these, and any other crops in which homozygous seed-propagated lines were desired, haploids would have considerable attractions, but only if they could be produced cheaply and in large numbers. A population of doubled haploid lines from the F_1 of an IBL is nearly equivalent to an array of late-generation lines such as emerges from a single-seed-descent programme (Section 5.3). To offer an effective substitute, haploids would have to be produced in several, even many, thousands per F_1, in order to sample the cross properly and to allow for inevitable losses due to failure of chromosome doubling; in maize (Section 5.5) these losses are large, of the order of 90 per cent.

It must be concluded that any large-scale use of haploids to produce pure lines quickly must await a revolution in the techniques of producing them. In highly outbred diploid crops in which homozygotes are either inviable or are unpropagable because of self-sterility, they are unlikely to be of any practical value at all.

Potatoes are the only crop in which dihaploids have been studied in quantity but a fair number have also been examined in alfalfa. In potatoes, being functionally diploid (2n = 2x = 24) out of outbred autotetraploids (2n = 4x = 48), they are often weak and infertile from the exposure of deleterious recessives. The few vigorous and fertile ones often generate remarkably productive families from crosses among themselves or with true diploid cultivar material; this observation has bearing upon our uncertainty as to what, if any, is the selective advantage of autotetraploidy over diploidy in the potato (see above, Section 8.2). At all events, it is established that potatoes can be reduced to and bred at the diploid level; this should offer simpler genetics and the possibility of constructing multiple disease resistances more easily than at the tetraploid level. For exploitation, selected diploid stocks can easily be got into a good tetraploid background by crossing as males to tetraploids when, as we have seen, most of the progeny are tetraploid, generated by diploid (non-reduced) pollen grains. A diploid (or dihaploid) homozygous for a desired dominant thus generates duplex tetraploids, with advantage in subsequent crosses. Potato dihaploids are likely to find modest uses along these lines; in the longer run, diploid cultivars are not inconceivable. There has recently been some effort to convert potatoes back from clones to seed-propagated progenies using some dihaploids for the purpose (Simmonds, 1997).

8.5 Induced mutations

General

The existing variability in all living organisms, including our crop plants, has been generated by mutation and subsequent recombination, accompanied often by some structural rearrangements of the chromosomes. Most natural mutants are recessive and they occur at low frequency, of the order of 10^{-6} per generation. At the molecular level, mutations are alterations of base-sequence in the DNA, in situations that lead to altered polypeptides and hence to phenotypic effects. The broad operational distinction of classical genetics between the 'point mutation' of a gene (when genes were 'beads on a string') and the chromosome structural change (e.g. duplication, inversion, translocation) remains good in the plant breeding context. The formal genetics of the past fifty years, especially of *Drosophila* and micro-organisms, has of course pushed the level of analysis to genetic and molecular levels far beyond those available, or even conceivable, in crop plants. That some, perhaps many, 'point mutations' are fine structural alterations rather than base substitutions need not trouble the plant breeder; his working distinction is between the visible phenotypic change that obeys Mendelian rules and gross chromosome alteration apparent under the microscope.

The mutagenic effect of X-rays was discovered by Muller in the 1920s. Later, any short-wave radiation up to the UV level and energetic particles were also found to be effective; since the early 1950s, a great array of chemicals (see below) have also been found to be active. No treatment is truly selective (though there are differences in the spectrum of changes produced). Hence, any treatment produces a more or less random array of point mutations and structural changes from which the experimenter must sort out those of interest for his purpose. A substantial part of the marvellous refinement of current genetical analysis in micro-organisms is attributable to the development of selective media and methods that separate the interesting mutants from the rubbish. The plant breeder faces, as we shall see, exactly the same problem of recognizing and isolating what he wants.

The mutagenic chemicals are all alkylating agents of which seven groups have been recognized: sulphur mustards, nitrogen mustards, epoxides, ethylene-imines, sulphates and sulphonates, diazoalkanes and nitroso-compounds. They vary in solubility, volatility and convenience in use; all are toxic, some very toxic. Ethyl-methane-sulphonate (EMS) and ethylene-imine (EI) have been the most widely used.

The experimental details are beyond our scope but some general principles are apparent. To a first approximation, results are proportional to dose (= intensity × time). But all mutagenic treatments are damaging, so an overdose kills too many cells (or seedlings) for practical convenience and an underdose produces too few mutants. Somewhere, there will be an optimum (Ehrenberg, 1960) but it will rarely be precisely defined. The optimum will depend upon the exact forms of the dose-response relationships, which are rarely well known and, anyway, are liable to vary with dose-rate and external conditions (such as oxygen tension and moisture). In practice, preliminary experiments are used to establish a dose and method of treatment that neither kills too much of the material nor cripples the survivors.

Dose is adjusted by adjusting intensity (dose-rate) or time or both, according to convenience, intensity by varying the radiation source itself (if it is variable) or by choosing the appropriate distance from source (e.g. an active isotope) or by varying the concentration of a chemical mutagen applied in solution. Treatments by penetrating radiation are adaptable to any material, pollen, seeds, seedlings, buds or whole plants; chemical treatments in solution are unsuitable for pollen but are well adapted to ungerminated seeds. The first step in a mutation-induction programme, then, is to establish methods of treatment appropriate to the plant material and the chosen agent and to estimate a suitable dose. In practice, the treatment is the easy part; nearly all the hard work lies in the subsequent sorting out.

Applications

Applications are summarized in Fig. 8.7. For simplicity, we first consider only diploids. In seed-propagated species treatment of pollen produces a mixture of

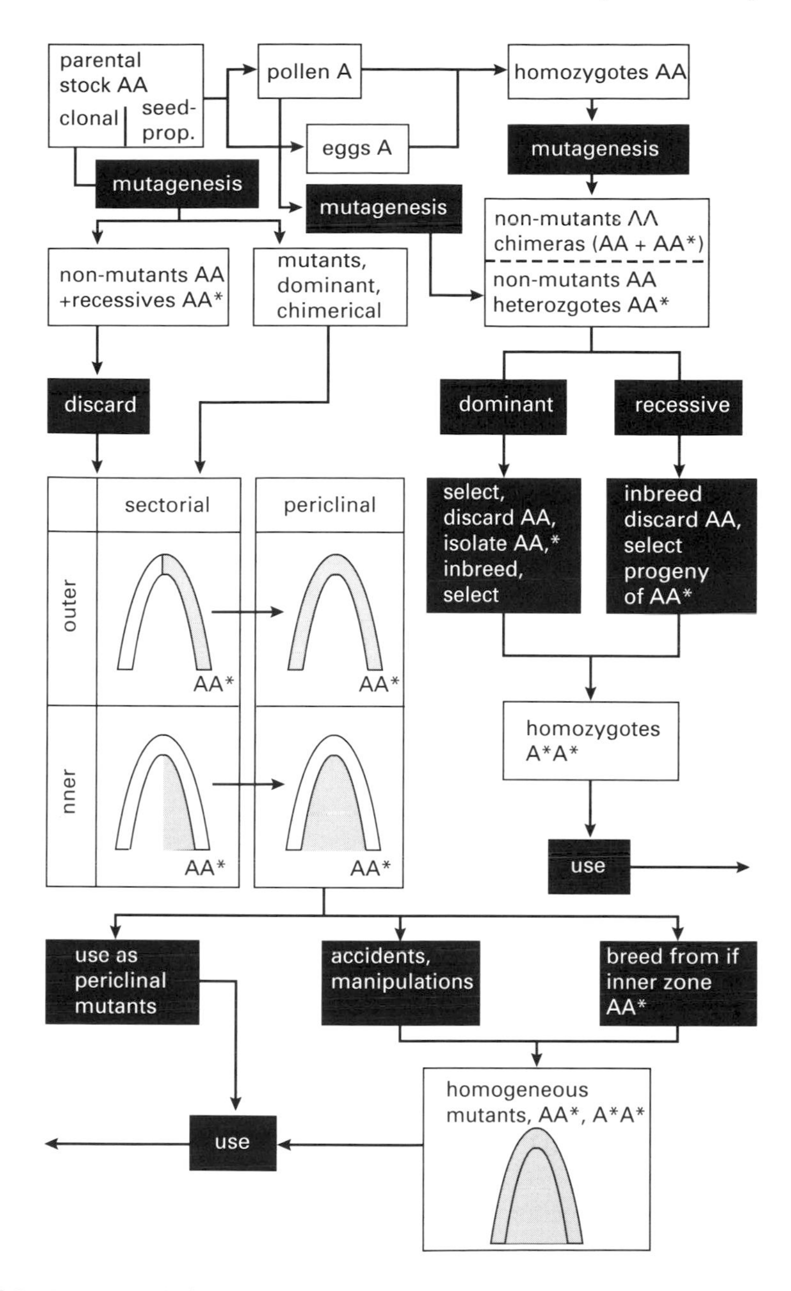

Fig. 8.7 Mutagenesis in plant breeding.

Notes: The genome carrying the desired mutation is indicated by A*. Observe the broad distinction between induced mutants in clonal plants (which must be dominant and are always primarily chimerical) and in seed-propagated plants in which treatment can be applied to the haplophase (in effect to pollen); in the latter case, a generation of selfing must intervene before selection for a recessive mutant can begin. Mutagenic treatment of zygotes of seed propagated plants (e.g. barley grain) presents the same initial problems of chimerism as clones.

non-mutants (in respect of the desired mutation) plus a small minority of mutant heterozygotes. If the mutant is dominant, which will be rare, it should be immediately recognizable and isolable by selfing; if it is recessive, it will be detectable only after inbreeding. Treatment of the diplophase can only, since mutation is a single-nucleus event, lead to plants chimerical for mutant-heterozygote and unchanged tissue. This implies that phenotypic selection for a dominant mutant must pay attention to likely sectors rather than to whole plants, while the search for a recessive after a round of selfing is likely to reveal it in less than the simple Mendelian expectation of one-quarter; for example, an ear of barley that was one-quarter sectorially heterozygous in the appropriate cell zone for a recessive mutant might be expected to produce homozygous mutants only at the rate of $\frac{1}{16}$; smaller sectors could well go undetected and mutations in cell layers that do not contribute to sexual reproduction can never be recovered. Not surprisingly, meaningful statements about mutation rates after treatment of the diplophase can hardly ever be made; for practical purposes, it is enough that desired mutants, whether dominant or recessive, can sometimes be detected as whole plants after a generation of selfing.

From the foregoing, it will be evident that treatment of the haplophase (in effect, of pollen) is an altogether neater and more manageable system than treatment of the diplophase (i.e. of seed or seedlings). Unfortunately, few crops lend themselves to it because few produce a sufficient bulk of pollen or are reliably crossable on the scale required; maize is ideal but the inbred cereals and legumes must, in effect, always be treated in the diplophase. In practice, most programmes rely on seed treatment followed by empirical sorting out of the products after selfing.

Turning now to clonal crops, the aim is normally to induce a specific change in an existing clone. In effect, recessives are excluded except perhaps in the rare event (none come to mind) in which the clone in question is already heterozygous at the locus. All mutants are chimerical, initially sectorial, later periclinal (Fig. 8.7). Details vary with the structure of the growing point in relation to morphogenesis; two functional layers (as in Fig. 8.7) are common but three occur in some plants. The outer layer (I) generates the epidermis *plus* a variable amount of leaf mesophyll; the inner layer (II) generates the rest, including reproductive cells; if there is a third, innermost, layer (III), it generates pith and adventitious roots.

Sectorial patterns depend upon the number of initial cells in layers in the growing tip; they may be quite complex and are nearly always unstable, giving rise to periclinal structures. The first step, then, after mutagenic treatment of young buds or shoots of any clone, is to grow a sequence of clonal generations, selecting simultaneously for the desired mutant characteristic and for vegetative stability; the latter characteristic will be achieved only when uniformly periclinal lines of sub-clones have been isolated and it may take several propagations to do so. Thereafter, any promising periclinal (i.e. stable) lines that survive may be useful and used as such; alternatively homogeneous, non-chimerical plants may be generated, either by breeding from the mutant clone (though only if the mutation is in layer II) or by horticultural manipulations that produce new adventitious meristems from the mutant layer (Fig. 8.7).

Mutagens are undiscriminating agents that always produce a complex mixture of mutations, chromosome structural changes and non-genetic (or at least non-nuclear) aberrations. So, if 50 per cent of treated plants die, the rest normally look pretty sick. Mutations are never found at first in a 'clean', stable genetic background: in seed-propagated material, there will nearly always be some undesirable mutants and infertile segregants caused by structural chromosome changes to be discarded in the early generations after treatment; in clones, there will be erratic phenotypic effects as well. It follows that mutation-induction programmes should be big enough to provide (presumptively) for the production of several, even of many, of the desired mutants, to cover both the possibility of allelic differences and the likelihood that some will be unusable because of the ineradicable side-effects.

It will now be apparent that mutagenesis as a plant breeding technique has its limitations. What are they and what are the circumstances in which the technique is likely to be useful? There are three points to make. First, mutagenesis is indiscriminate and generates large volumes of undesirable variants which must be laboriously sorted out and discarded; any serious programme must therefore be large and it must extend over several generations of seed or clonal propagation (as the case may be); if desired mutants are identified they are likely to be associated with undesirable side-effects (other mutations, translocations, sterility, etc.), necessitating 'cleaning up' of the genetic background, perhaps even needing an outcross in order to do so. Second, desirable mutants are at best uncommon, often rare, even after treatment; if good ones, in acceptable genetic backgrounds, are to be isolated, an easy method of screening is necessary; readily visible changes in stature, maturity, flower colour and so forth are favourable and so are sharply defined changes in resistance to a disease or pest which can be revealed by a laboratory/glasshouse test or a field epidemic; more subtle changes, however, that can only be detected by refined chemical methods or replicated field trials, are liable to be prohibitively expensive to work with – though not necessarily impossible. Third, and last, the isolation of recessive mutants (the commonest class, after all) may be, in effect, forbidden by the genetic nature of the crop; only diploid seed-propagated crops which can be easily inbred are favourable for their isolation; all others – polyploids, clones, highly self-incompatible diploids – are impossible subjects or nearly so and functionally dominant mutants alone are a feasible objective in them.

To summarize, therefore, a mutation-induction programme must: (a) be on a scale sufficient (in numbers of plants and generations) to 'clean up' the inevitable adverse background effects; (b) have at its disposal an effective means of recognizing the rare desirable mutants if and when they occur; and

(c) be genetically reasonable, seeking recessives only if they are, in fact, detectable.

Results and prospects

There was a time, in the 1950s, when the protagonists of 'mutation-breeding' gave the impression that the then fairly new method was about to revolutionize plant breeding. This has not happened, nor is it likely to do so. Instead, the technique has simply taken its place as one more addition to the plant breeder's repertoire, occasionally very useful but usually irrelevant; recombination, far more often than the need for specific alleles, limits plant breeding progress. One lesson which emerged from the early, over-optimistic days was the necessity of clear, limited and biologically reasonable objectives; merely to treat material in the hope that something useful would turn up (and this is not an unkind description of a good deal of work) was, it appeared, simply a waste of resources. Another important lesson was that, even if objectives were reasonable and mutants obtained, the effort required to sort out the desired products in usable form was often comparable with that demanded by a conventional breeding programme.

What, then, have been the actual achievements and what are the prospects? In the late 1960s, Sigurbjörnsson and Micke (in IAEA, 1969) identified 77 cultivars which owed their origins directly to mutation-induction programmes. Eleven were in barley, 28 in ornamentals, 38 in other crops. The list would be longer now. Many such cultivars however have failed to make a mark and notable successes have been few; perhaps the most striking success is the great series of *erectoides* (dwarf, short-strawed, lodging-resistant) mutants in barley pioneered by Swedish breeders. Several of these, recessive, alleles are now widespread in the current genetic base of European two-row barleys, supplementing the naturally occurring Abed dwarfing gene, morphologically a different kind of mutant. Other mutant characters listed include plant habit and morphology, maturity time, fruit shattering, major-gene disease resistances, some biochemical quality features and flower and foliage characters in ornamentals.

Looking to the future, one can foresee a continuing need for new mutants – dominant or dominant and recessive, according to genetical circumstances – affecting the following groups of characters: (a) favourable changes in gross morphology, for example, dwarfs in tree crops (some are already known in apples), determinate growth habits in a variety of crops, non-flowering habits in crops grown for a vegetative product (such as have been produced in sugarcane); (b) disease resistances, in circumstances in which monogenic resistance is expected to be of lasting value (Section 7.4); and (c) biochemical or quality characters (if effective screening methods can be devised). Mutagenesis surely has a secure place in the plant breeder's repertoire of techniques, even if a much more restricted one than was once supposed; it is probably easier to find a gene than to make it.

8.6 Summary

This chapter is concerned with four groups of specialized techniques which all have places in plant breeding but are broadly peripheral to the main stream of work. Polyploidy occurs naturally in many crops and crop–wild plant complexes, allopolyploids far exceeding autopolyploids in numbers and importance. Autopolyploids of diploid crops have been extensively tried but have had little practical use. Induced allopolyploids include triticale and raphanobrassica (which are, in effect, new crop species) and the resynthesis of existing allopolyploids (e.g. *Brassica napus*) is sometimes advantageous. Allopolyploidy seems to have much greater breeding potential, offering, as it does, the opportunity of constructing permanent hybridity between parents with complementary characteristics.

Wide (inter-specific, sometimes inter-generic) crossing (other than that undertaken with allopolyploids in view) is virtually always succeeded by a backcrossing programme aimed at incorporating a small piece of alien chromosome in the recurrent genome. Disease resistance is usually the objective. The subjects chosen are more often polyploid than diploid and the incorporation of the alien chromatin is sometimes aided by 'chromosome manipulations' designed to exploit translocations or chromosome pairing control.

Haploids, out of diploid parents, offer an attractive possibility of generating homozygous lines at one step and so of short-circuiting breeding programmes in appropriate crops (especially IBL). They are usually produced by female parthenogenesis but several other routes, including pollen culture *in vitro*, are known. Methods of production are yet laborious and/or unreliable and haploids have had little practical impact; but the principle remains attractive.

Genetic changes – gene mutations, chromosome structural alterations, others undefined – are readily induced, in quantity but unselectively, by a variety of physical and chemical methods. Desirable changes so produced must be sorted out from the mass of undesirable, deleterious changes by selection over several generations, sexual or vegetative, as the case may be. Most mutants are initially chimerical; recessives are usually sought only in species which can be inbred. Mutagenesis is an established addition to the plant breeder's repertoire of techniques and is of definite though limited utility. Most successes recorded to date have been in barley (e.g. dwarf mutants) and ornamentals.

8.7 Literature

Polyploidy

Bingefors and Ellerström (1964); Bingham and Saunders (1974); Doggett and Majisu (1972); Dudley and Alexander (1969); Eigsti and Dustin (1955); Gerstel (1966); Hecker (1972); IAEA (1974); McNaughton (1973); McNaughton and Ross (1978); Mendiburu and Peloquin (1976);

Smartt and Haq (1972); Somaroo and Grant (1972); Sybenga (1969, 1973); Thomas (1969); Zillinsky (1974).

Wide crossing, cytogenetic manipulation

Driscoll (1972); Ellerström and Zagorcheva (1977); Gerstel (1945); Knott and Dvořak (1976); Law and Worland (1973); McNaughton (1973); McNaughton and Ross (1978); Mayer (1973, 1974); Riley and Kimber (1965); Riley and Lewis (1966); Sears (1972); Wernsman *et al.* (1976); Zillinsky (1974).

Haploidy

Bingham (1971); Bingham and Saunders (1974); Chase (1964, 1969); Choo, Reinbergs and Kasha (1985); Collins (1977); Eenink (1974); Feaster and Turcott (1973); Kasha (1974); Kasperbauer and Collins (1974); Kimber and Riley (1963); Melchers (1972); Park *et al.* (1976); Raina (1997); Rajhathy (1976); Reinbergs *et al.* (1976); Riggs and Snape (1977); Snape (1976); Thompson (1977); Walsh *et al.* (1976).

Mutagenesis

Auerbach (1976); Bird and Neuffer (1987); Ehrenberg (1960); Gardner (1961); Gustafsson *et al.* (1971); IAEA (1964, 1969, 1970a, b, 1972, 1973, 1974); Konzak, Kleinhofs and Ullrich (1984); Rédei (1974); Yonezawa and Yamagata (1977); Y. Yoshida (1962, 1964).

Chimeras

Dermen (1960); Neilson-Jones (1969).

Chapter 9
Biotechnology and Crop Improvement

S. Millam[1] and **W. Spoor**[2]

9.1 Introduction

Classical breeding techniques have been largely responsible for the major gains in plant yield and quality achieved over the last century. However, such factors as, increased consumer demand for food or plant-derived products, and the potential for deriving additional industrial products from crops may necessitate the introduction of new crop lines in a more rapid and precise manner than possible from conventional breeding alone. This will be further exacerbated by the predicted rise in world population, along with problems precluding increases in crop area, such as erosion, rising salt levels, pest and diseases, climate change and political pressures. Considering all these factors, it is clear that methods of crop improvement beyond the present capabilities of classical breeding may be required to meet future demands.

The rapidly advancing portfolio of techniques categorized under the general heading of biotechnology has stimulated great interest and debate, not all of a positive nature. However, in this chapter describing the context, methodologies, targets, current status and future prospects of biotechnology the reader will be given an informed and balanced view of this topic and potential importance. The ethics and implications of this rapidly moving science will also be considered, adding to the current public debate on the use of transgenics.

In the context of plant improvement, plant biotechnology revolves around two key areas. Plant cell and tissue culture, which has a number of highly useful applications to agriculture in its own right, is linked with some of the techniques of molecular biology, a discipline that has seen rapid advances in recent years, predominantly driven by medical research.

9.2 Plant cell and tissue culture

The foundations of this science were laid as early as 1838 when Schwann and Schleiden outlined the theory of totipotency. This concept underpins all plant

[1] Research Leader in Crop Genetics, Scottish Crop Research Institute, Invergowrie, Dundee
[2] Head of Biotechnology, SAC, West Mains Road, Edinburgh EH9 3JG

cell and tissue culture manipulation and is the capacity (common in plants) of a single cell to regenerate a complete organism. Though a great deal of research was carried out in the late nineteenth century, the science evolved relatively slowly. In 1902 Haberlandt kept single cells of *Tradescantia* alive for 27 days, but no actual cell divisions were observed. In 1927 Went discovered a substance known as auxin which was found to promote cell division. During the 1930s scientists such as White, Gautheret and Nobecourt advanced the techniques so that unlimited growth of roots could be obtained, and cell growth stimulated by auxin was observed. The next step in assembling the foundations was made by Skoog in 1948 who made the important discovery that adenine (a purine) enhanced cell proliferation and resulted in buds forming on previously undifferentiated tissue. This work was amplified by Miller who in 1955 discovered that cytokinins were found to promote cell division. Skoog and Miller (1957) combined the use of the plant growth regulators auxin and cytokinin to explore their effects on root and shoot formation in plant cell culture and laid down the blueprint for all plant tissue culture techniques used today.

Since these foundations, the technology has evolved rapidly, not only underpinning research tools, but stimulating development of a number of important industries and some useful applications to agriculture, listed below. All techniques are based on the capacity to grow plant material on defined media aseptically, under controlled environmental conditions.

Applications of plant cell and tissue culture to crop improvement

- Micropropagation and meristem culture
- Microspore culture
- Embryo culture and somatic embryogenesis
- Protoplasts and somatic hybridisation
- *In vitro* germplasm conservation and cryopreservation
- Gene transfer technology

Micropropagation and meristem culture

Though micropropagation is relatively simple, essentially a scaled-down version of conventional vegetative propagation by cuttings, it is a powerful tool. By culturing meristem shoots or buds in a sterile environment, under optimum conditions for plant growth and development, rapid clonal multiplication can occur in many crop plant species. The subculturing period and efficiency varies greatly between species: taking a potato as an example, a single shoot after three weeks culture could be subcultured into five or six microcuttings. Each of these, after a further three weeks culture, would produce five or six plants each and within one year over one million clonal plants could be derived from the original. This has applications for the rapid introduction of improved lines and is a key tool for many species, particularly soft fruits and ornamentals. In the

Netherlands alone annual production of plants via micropropagation exceeds 53 million plants per year (Pierik & Ruibing, 1997). However, this method is also used extensively for the rapid production of plantation crops. Meristem culture has been a key element in the production of virus-free potato planting stock which has eliminated the paracrinkle virus from clones of the King Edward cultivar. Furthermore, this technique, being non-seasonal and highly efficient in usage of space is being applied to a number of accelerated conventional breeding programmes.

Microspore culture

This *in vitro* technique is designed to produce routinely haploid cells, which possess the gametic rather than the somatic chromosome number. Anthers containing pollen are removed from the plant, surface-sterilized and placed on culture medium. Embryos are induced to form and may subsequently develop into plants. The principal use in plant breeding is as an intermediate in the production of homozygous plants, produced by chromosome doubling, i.e. an alternative method to repeated cycles of inbreeding in self-pollinated crops (Section 8.4). Haploid plants have been generated in more than 50 species, mostly in the Gramineae, Solanaceae and Cruciferae. Unlike many of the other techniques of cell and tissue culture, anther culture is particularly applicable to monocotyledonous crops. Anther culture has a number of related applications, notably the use of doubled-haploids in quantitative genetics and molecular mapping (Becker *et al.*, 1995).

Embryo culture and somatic embryogenesis

Embryo culture has been a standard adjunct of wide crossing programmes in the form of 'Embryo rescue'. The techniques developed for this purpose have been used and further developed for the production of plants from embryos procured by non-sexual methods.

The process of embryo initiation and development from vegetative, or non-gametic cells (compared with anther culture which includes gametic embryogenesis) can, in some species, result in the production of somatic embryos from cell, tissue or organ cultures. This development can proceed directly, or indirectly via a callus phase, but is highly dependent on cultural conditions. A generalized strategy is that the application of auxins to competent cultures induces embryos, and the removal of the auxins results in the development of normal plants from the embryos (Komamine *et al.*, 1990). In theory somatic embryogenesis could provide a system for a high level of production of uniform plants with potential for 'artificial seeds' and technology take-up in low-tech situations (Litz & Gray, 1995). There has been progress in some horticultural species, notably carrot, and also in some tree species (Hammat, 1992).

Protoplasts and somatic hybridisation

A protoplast is defined as a cell from which the entire cell wall has been removed. First isolated manually, the advent of the use of cellulase and pectinase to digest cell wall material reported by Cocking in 1960 facilitated the isolation of protoplasts from a wide range of species. However, until 1971 no practical example of regeneration into an intact plant was made (Takebe, 1971), but this work involving tobacco was successful in 1972 when a plant was regenerated out of a fusion between two species (Carlson *et al.*, 1972). This methodology, now known as somatic hybridisation, initiated a great deal of interest and resulted in speculation suggesting that wide crosses and the creation of a vast range of fusion material could play a significant part in agricultural improvement. This has not been the case. Though scientifically an elegant tool, regeneration from protoplasts has been restricted, in practice confined to only a few species from the Solanaceae and Cruciferae. There are a very limited number of practical examples arising from somatic hybridization and the technique has been largely superseded by advances in transformation technology. The two most prominent successes using this technology are the products of Cytoplasmic Male Sterile Brassica lines at Cornell University, USA (Christey *et al.*, 1991), and disease-resistant commercial tobacco cultivars grown on substantial acreages in Ontario, Canada (Brown & Thorpe, 1995). However, these plants were not direct fusion products and had undergone three backcross generations. Protoplast fusion in theory does allow the transfer of more than a single gene. However, the inherent randomness of the approach implies that a large effort in the selection of material has to be undertaken. Also, the uncontrollable risk of other negative traits being transferred and the problem of genetic stability of such hybrid material has precluded a wider uptake of this theoretically elegant, but technically demanding, technology.

In vitro germplasm conservation and cryopreservation

The use of *in vitro* techniques for germplasm conservation has been advocated, particularly as a conservation strategy for specific genotypes, or where alternatives (e.g. seed storage) are not appropriate. Storage of plant germplasm may include, as a relatively short-term measure, introduction into axenic culture. This facilitates a non-seasonal, easily maintained and transportable source of clonal plant material with great potential for facilitating international exchange of genetic material. However, for medium-term (more than 6 months) storage the use of growth inhibitors (plant growth regulators or osmotic control) may be required for certain species, to alleviate the economic and technical problems of repeated subculturing. The long term effect of maintaining such material for extended periods has been relatively poorly investigated. In many cases however, the transfer of material maintained on such media to normal growth media results in a phenotypically normal status being restored, though this may take a small number of subculturing cycles.

A related *in vitro* approach, using cold-storage, at temperatures of 2–10°C offers only a partial solution, with related technical problems. Possibly the most interesting methodology involves the technique of cryopreservation. This involves the storage, in liquid nitrogen at –196°C of viable tissues in either the vitrified (ice-free) or crystalline (ice+) status. Such techniques are a long term storage option and have been investigated in a number of crops. The topic has been recently reviewed by Dumet *et al.* (1998). Cryopreservation is, in general, an accessible technology, and can be applied to both pilot scale and substantial collections of material.

9.3 Transgenic plant technology

Background molecular principles

Probably the most important tool in plant biotechnology is that of transgenic plant production. As previously stated, this is strongly linked to plant cell culture techniques, but some background molecular biological principles need to be emphasized to establish a clearer understanding of the principles and applications.

- In basic terms, most organisms are composed of cells which contain DNA in the chromosomes.
- The chemical structure of DNA molecules contains information used by cells as the code for the organism.
- The 'words' for the DNA recipe (genes) are derived from a 4 letter alphabet – comprising adenine, guanine, cytosine & thymine.
- The entire recipe, the genome, may contain 3 million to 4 billion letters.
- DNA from any organism is chemically and physically uniform across organisms except for the sequence and number of letters in each recipe.
- One of the great discoveries of biotechnology is that DNA from *one* organism can function in *another.* DNA 'sharing' has been exploited in breeding programmes for many years by way of sexual reproduction.

With the knowledge that foreign DNA was capable of functioning in other organisms, for many years attempts were made to 'feed' DNA into plants, with no success. However, a combination of factors, not least of which was the rapid development of DNA technology in the late 1970s, combined to facilitate the transfer of alien DNA into plants.

Agrobacterium

One of the principal agents now involved in DNA transfer had been known since the beginning of the twentieth century, when the naturally-occurring soil bacterium (*Agrobacterium tumefaciens*) was found to incite crown gall disease.

This organism was extensively studied and findings on the tumour-inducing mechanisms and their physiological basis summarized by Braun (1958). Crown galls were found to synthesise auxins in large amounts, causing cell proliferation in the plants. However, at this stage the actual mechanism by which additional auxin was synthesized was unknown. Significant progress was made in the 1970s when modern DNA technologies provided evidence that virulent strains of *Agrobacterium tumefaciens* contained large extra-chromosomal element, harbouring genes involved in crown gall production. Further work found that this element was a very large (>200 kb) plasmid which, because of its role in plant tumour induction, was designated the Ti (tumour inducing) plasmid. It was shown that infection and subsequent oncogenesis in the plant was caused by the Ti plasmid from the bacterium becoming inserted into the *plant* genome. Zambryski *et al.* (1983) developed a significant advance in vector development by deleting, using restriction enzymes to cut DNA at known sites, the oncogenicity genes from the T-DNA and replacing these sequences using DNA ligation with DNA sequences derived from *E. coli* cloning vectors. Thus the capacity for transferring gene sequences from *E. coli* to *Agrobacterium* and then, potentially, into the plant genome was developed. The first report of successful transformation (the permanent uptake of new genetic information) of a plant was by Herrera-Estrella *et al.* (1983) who inserted a chimeric gene, with the nos promoter linked to the neomycin phosphotransferase II (nptII) gene, into the T-DNA; crown gall cells were produced which expressed resistance to the antibiotic kanamycin sulphate. The system was further enhanced by the excision of the phytohormone biosynthetic genes from a T-DNA region, thus eliminating the ability of *Agrobacterium* to induce aberrant cell division. In 1985, Horsch *et al.* (1985) developed a simple approach to the transformation of plants, using plant tissue regeneration methodology and the presence of a selectable marker (the nptII gene conferring resistance to the antibiotic kanamycin – still in common use today) which form the basis of many transformation projects in widespread use all over the world.

Agrobacterium transformation is now a core research tool and a practical aid to cultivar improvement, with methods for the transformation of over 120 plant species published. Developments of the technique include vastly more efficient vectors, with a range of selectable or scorable markers for rapid identification of transgenic lines. Though antibiotic resistance genes are efficient for selection, their presence in a commercial product, though probably harmless, is undesirable. Alternative markers are in use, as well as methods for eliminating the antibiotic resistance genes.

An example of this methodology is given below.

Case study – transformation of potato using *Agrobacterium tumefaciens*

- Initiate plant tissue cultures of potato, and establish stock of micropropagated plants (6 weeks)

- Day 1 – set up cultures of *Agrobacterium tumefaciens*
- Day 2 – prepare explants (leaf strips/petioles/internodal stem sections)
  – incubate explants with *Agrobacterium* culture for 45 minutes
  – plate explants onto media, co-cultivate for about 48 hours
- Day 4 – explants replated onto regeneration medium, supplemented with antibiotics, to kill off *Agrobacterium*, and to select transformed material
- Day 18 – replate on fresh media to maintain antibiotic selection pressure
- Day 32 – excise regenerating shoots, maintain selection pressure
- Day 46 – selected shoots tested using PCR for presence of novel gene
- Day 60 – positive plants grown (under controlled conditions)

Agrobacterium also exists in another form, *A. rhizogenes* which causes 'hairy root disease'. This is caused by a similar mechanism to the Ti plasmid, in this instance the causative agent is termed the Ri plasmid (root inducing). This species of *Agrobacteria* has also been used for transforming plants, but is not as widely understood or utilized as *A. tumefaciens.*

Alternative transformation systems

There are limitations to the range of species that can be transformed using *Agrobacterium.* Many monocotyledonous plants were, until very recently, considered difficult to transform using *Agrobacteria*; and woody plant species and legumes are two other groups of plants generally recalcitrant to *Agrobacterium*-mediated transformation, though exceptions do exist. For these groups a number of alternative systems for gene transfer have been developed as follows.

Alternative transformation systems

- Biolistics
- Silicon fibre
- Electroporation/PEG treatment of protoplasts
- *In-planta* transformation
- Sonication

Biolistics

A number of novel approaches to solving the problems of gene delivery into intact tissues have been investigated. The most successful and widely applicable system is that originally developed by Klein *et al.* (1987) at Cornell, USA. Biolistics can be defined as the introduction of substances into intact cells and tissue through the use of high-velocity microprojectiles. It is primarily a mechanism for breaching cell walls and membranes which are the principal barriers to DNA delivery. A microprojectile can be any small, discrete particle capable of being accelerated so that it penetrates cells or tissues, and should be small enough to enter the plant cell or tissue without adverse effects and also capable of carrying DNA on its surface or interior. Such microprojectiles are typically a high density metal such as gold or tungsten.

The prototype system utilized a chemical explosion from a modified .22 calibre rifle to accelerate tungsten particles coated with DNA and, while it was successful in delivering DNA, many target cells were killed. Systems have been considerably refined since then, and are now based on either helium or electrical discharge systems. There are a number of perceived advantages of biolistics over the *Agrobacterium*-mediated methodologies. Non-hosts of *Agrobacterium* such as cereals can be transformed and plasmid construction is simplified since DNA sequences for T-DNA replication and transfer necessary for *Agrobacterium* protocols are not necessary. Also, once the equipment has been purchased, licensed or constructed, the preparation is simple and the procedure rapid.

Plant material used for particle bombardment is an important factor in the success of such transformation systems. Direct gene delivery to an explant subsequently used to initiate embryogenic cultures is a commonly used approach. Immature embryo explants have been frequently used with success in barley, pearl millet, papaya, maize and rice (Dunder *et al.*, 1995). An interesting alternative is the use of microspores as a target tissue, and such an approach has been used successfully in barley (Wan & Lemaux, 1995). Seed tissue of soybean has been used for the recovery of numerous independently-derived transgenic soybean plants expressing herbicide resistance, tested in large scale field trials and subsequently successfully marketed in the USA.

Silicon carbide whisker mediated transformation

A very simple method for transferring DNA into plant cells was derived from a method used for the injection of macromolecules into insect embryos and was first applied to plants by Kaepller *et al.* (1990). The protocol employs silicon carbide fibres, plasmid DNA and plant cell walls which are vortexed together briefly and then cultured. The fibres appear to penetrate the plant cell walls, the DNA adheres to the fibre surface and is propelled through the cell wall by collisions between cell aggregates while vortexing. This technique requires very little preparation, and is rapid and inexpensive. However, a plant regeneration system is a pre-requisite, and the relative crudeness of approach implies that it is unlikely to result in high frequency DNA delivery into organized cells without further technological development. There are also safety implications regarding the use of the silicon fibres. However, maize plants have been produced showing expression and transmission of single and multiple copies of introduced herbicide resistance by this method.

Electroporation/PEG treatment of protoplasts

Protoplasts are potentially useful experimental systems for gene transfer, due to their physiological status, i.e. a freely accessible plasmalemma. Furthermore, no vector for transformation is necessary and DNA uptake could be considered a relatively straightforward physical process, thus circumventing any potential host range limitations. Chemical (usually polyethylene glycol, PEG) or electrical systems have been used, or combinations of both (Shillito *et al.*, 1985), and though dependent on a successful system of regenerating plants from protoplasts (see protoplast section above), can result in fertile transgenic plants. Major species transformed by such methodology include rice and maize. However, the technical demand of protoplast regeneration and the increase in prominence of more straightforward systems have precluded the widespread adoption of such methodology for routine crop plant transformation.

In planta *transformation*

The rationale for developing this system was to overcome problems associated with the regeneration of plants from cultured explants and subsequent re-differentiation. In fact the system developed for the widely-used model species *Arabidopsis thaliana* does not involve any tissue culture stage. Simply, developing plants are taken at the stage where the primary inflorescence is 10–15 cm high and soaked in a plastic tray containing *Agrobacterium* suspension and placed in a vacuum chamber at 10^4Pa for 20 minutes. Following vacuum treatment the infiltrated plants are transferred to soil and kept in humid conditions to recover. After maturity (4–6 weeks) the seeds are harvested, and sown under appropriate selection pressure to obtain resistant plants, at rates quoted as up to ten seedlings per initial plant treated (Bechtold & Bouchez, 1995). Obviously, many aspects of the *Arabidopsis* phenotype(e.g. short life cycle, small plant and very small genome) are significant factors in the success of this simple system. However, attempts to modify this methodology for plants used for agricultural purposes are current in a number of laboratories.

Sonication

A new and efficient *Agrobacterium*-based transformation method suggested capable of overcoming limited host specificity and based on sonication has recently been reported (Trick & Finer, 1997). The target host plant tissue is subjected to brief periods of ultrasound in the presence of *Agrobacterium*. Stable transformants of soya bean were obtained using this method. Histological analysis revealed that the treatment produced small and uniform fissures and channels in the plant tissue possibly allowing better *Agrobacterium* penetration. Preliminary results suggest a significant increase in potential transformation efficiencies and this methodology offers great potential for increasing both range and efficiency of conventional transformation.

Certainly, improvements to existing transformation technologies are expected in the near future. However, the existing methodologies, notably *Agrobacterium* and biolistics, are in widespread use in research organisations and increasingly perceived as a valuable part of the portfolio of methods available to commercial breeding companies.

9.4 Target traits for transformation

Until recently, it was possible only to transfer single gene traits into plants by transformation; however, recent developments in gene cloning and vector construction may allow the packaging of several genes in one cassette. Even the introduction (or blocking) of single gene traits has enormous significance for the production of improved crop plants (as below).

Examples of single gene targets for transformation

- herbicide resistance
- insect resistance
- virus resistance
- fungal and bacterial resistance
- quality traits

Herbicide resistance

The mode of action of most modern herbicides is the disruption of a specific metabolic process. Consequently, manipulation of the plant to detoxify the herbicide or modify the enzyme target site would confer a degree of herbicide resistance. This would allow the use of non-selective herbicide treatments. Many herbicides act on single enzymes enabling a transformation strategy for resistance to be established (as set out in the case study below). The benefit of such a system, in theory, is that smaller applications of herbicides may be required to effect weed control.

Case Study: Transgenic tobacco plants expressing a bacterial detoxifying enzyme are resistant to 2,4-D
(from Streber and Willmitzer, 1989).

- The *tfd* gene was found to encode the first enzyme involved in the 2,4-D degradative pathway. It was found to be a single polypeptide that was able to remove the acetate side chain from 2,4-D, known as 2,4, dichlorophenoxy acetate monoxygenase (DPAM).
- The *tfd* gene was found in the bacterium *Alcaligenes eutrophus*, and was cloned and sequenced.
- Plasmids were constructed and used to transform tobacco.
- Resultant shoots were selected on kanamycin and analysed by Northern Blotting (a system for detecting specific RNA molecules) for the synthesis of DPAM-RNA.
- Transgenic plants were sprayed with 2,4-D.
- Seeds from selfed transgenic lines were germinated on sterile medium con-

taining kanamycin to test for the inheritance of the introduced trait. In a good approximation to the expected 3:1 ratio, 258 were found to be resistant and 83 sensitive.

- Southern blots (a system for determining the stable introduction of DNA sequences) revealed at least three copies of the DPAM gene in some lines.

Insect resistance

The creation of insecticidal transgenic plants has been a major objective of many research teams. If resistance is engineered into the crop there would be considerable positive environmental benefits in that pesticide applications would be substantially reduced. Considerable advances have been made in determining the value of a broad range of insecticidal proteins incorporated into transgenic crops. Much work in this area involves a strain of bacterium *Bacillus thuringiensis* (Bt). Within minutes of ingestion by an insect the toxic proteins produced by Bt destroy the insect gut. These toxins are highly specific, none being toxic to mammals and many very specific to narrow ranges of insects (Brousseau, 1997). Key objectives in establishing a programme of transgenic Bt plants included isolation of the specific Bt gene, and correct expression mechanisms (i.e. in the right part of the plant, at the right time) which have resulted in such elegant strategies as the development of wound-inducible promoters.

Another natural plant defence mechanism against insect damage is production of protease inhibitors, which often accumulate in seeds and storage organs of plants. These inhibitors bind to the active site of proteases enzymes. Protease enzymes cleave the peptide bonds to initiate protein breakdown. The isolation of genes that encode plant protease inhibitors and their transfer to over expression of these genes in crop plants has been widely investigated (Gatehouse & Gatehouse, 1998). Other novel mechanisms include the insecticidal protein cholesterol oxidase (Corbin *et al.*, 1998).

The long term viability of these approaches alone, as effective means of insect control, has to be questioned. Resistance to the Bt toxin has been found in a number of populations of diamond back moth, and already pyramiding of insecticidal genes has been suggested as a way of prolonging the effective life of such genes (Zhao *et al.*, 1997). Thus VR problems recur (Chapter 7).

Virus resistance

Viruses can be responsible for major crop losses and the establishment of a transgenic virus resistance mechanism for major crops has had enormous commercial significance. The first virus-resistant transgenic plants were produced using a transgene derived from a viral pathogen. This novel approach, using the pathogen genome itself as a source of resistance, was first formally proposed by Sandford & Johnson (1985). The existing phenomenon of cross-protection, in which plants infected with one strain of a virus are found to be immune to infection by another virus, is an already existing method used in horticulture. This concept offers great opportunities in conferring virus resistance in major crops. Coat-protein mediated resistance has been widely applied and demonstrated to be durable under field conditions.

There are several different approaches to imparting virus resistance in plants. Non-coat protein transgenes have also been investigated, including genes coding for functions such as cell-to-cell movement and viral proteases. Replicase-mediated resistance has been reported to different viruses, e.g. potato virus X, potato virus Y and cucumber mosaic virus. Strategies include introducing the entire wild-type viral replicase genes or truncated versions in which deletions have been made. Though the mechanisms of the resistance are unclear, virtual immunity to infection can occur. Movement protein-mediated resistance has been suggested as a means of engineering broad spectrum resistance in plants. RNA-based strategies or the use of satellite RNA also have potential. Inoculation of transgenic plants with the parent viruses has been shown to result in the amplification of satellite RNA transcripts, leading to conferred protection against more severe effects. The down-regulation of expression of plant nuclear genes based on antisense RNA has resulted in resistance to the geminivirus tomato golden mosaic and tobacco mosaic virus.

Plant-derived genes, often related to hypersensitive response, have also been suggested as possible targets for transformation. Plant genes encoding anti-viral proteins offer potential. Ribosome inactivating proteins, possessing broad-spectrum antiviral activity, have been identified in a number of plant species, although neither the virus particle nor the viral genome are thought to be the direct target of this activity (Kavanagh & Spillane, 1995). Non-plant, non-pathogen derived transgenes have also been postulated as virus-resistance mechanisms, with the use of antibodies (Tavladoraki *et al.*, 1993) or interferon-based systems (Kavanagh & Spillane, 1985).

Considerable effort has been placed in the molecular mapping of virus resistance genes, notably in the potato, and this approach may be applied in future crop improvement programmes.

Fungal and bacterial resistance

Research into the creation of transgenic plants conferring fungal and bacterial resistance has been undertaken using a wide range of strategies (Lamb *et al.*, 1992). These areas of research would appear to be more difficult than other transgenic pathogenic resistance approaches, due to the more complex nature of the fungal or bacterial pathogens, and of the mechanisms of resistance. However, some reports of successful transgenics exist. The effects of chitinase or beta glucanase modifications to crop plants have been tested in a number of species. The enzyme stilbene synthase gene derived from *Vitis vinifera* has also been utilised to express specific phytoalexins in transformed plants, conferring additional resistance to *Botrytis* under experimental conditions. Progress in this area needs to be carefully thought through as previous efforts to reduce the economic importance of many fungal and bacterial pathogens using genetic resistance based on single gene characteristics have not proved to be particularly successful for many plant pathogen combinations (VR, Chapter 7).

Quality traits

Many key biosynthetic pathways in plants are controlled by a succession of single gene steps. Notable examples are the sucrose-starch pathway in potato tubers and the fatty acid biosynthetic pathway in oilseed crops such as *Brassica napus.* The first high profile transgenic commodity available to the public was the well-publicized FLAV-SAVR™ tomato in 1995, modified to reduce softening of the fruit (for review of biochemical mechanisms involved see Grierson & Schuch, 1993). Transgenic potatoes modified to produce industrial starch are being grown in Holland on a commercial scale. The fatty acid profile of oilseeds is a particularly interesting case. Fatty acids, though of crucial importance in human and animal nutrition, have a number of industrial applications, with approximately 15 per cent of the world production employed for industrial purposes such as lubricants, inks and plasticizers. The qualitative manipulation of seed oils involves modifications of the fatty acid composition. The selection of target fatty acids for genetic manipulation has been largely determined by the availability or ease of isolation of the appropriate gene clones (Table 9.1). Oilseed rape is particularly amenable to gene transfer technology due to the ease of transformation and relatedness to *Arabidopsis* (model system for molecular genetics). Murphy (1996) described the use of ESTs (expressed sequence tags) generated by rapid-throughput automated cDNA sequencing for isolation of genes involved in the biosynthetic pathway of fatty acids.

Table 9.1 Examples of seed oil modification by genetic engineering in *Brassica napus* (see also Murphy, 1996).

fatty acid	uses	gene	modification
Stearic acid (C:18 saturated)	margarines, cocoa butter substitute	V9 stearoyl-ACP desaturase	increase from <5% to 30%
Lauric acid (C12 saturated)	detergents, soaps	lauroyl-ACP thioesterase	increase from 0% to 25%
Erucic acid (C22 mono-unsaturated)	chemical	various	>70%

Current status of transgenic crops in world agriculture

It is illuminating to bear in mind that the first report of transformation of a crop plant was as recent as 1983. Since then, the technology has developed to such an extent that it has been predicted that 40% of the USA crop area will be transgenic by the year 2000 (USDA statistics). The rapid uptake of transgenic crops can be seen in Table 9.2. In 10 years of field release of transgenic varieties in the USA, 1120 cases were related to herbicide resistance, 911 to insect resistance, 808 quality traits and 383 were virus resistance trials. The major crops were maize (1476 trials), tomato (378) and potato (363) (Source: USDA, statistics). The recent status of research in the EU is presented in Table 9.3. For a review of R&D regulation and field-trialling of transgenic crops see Dale (1995).

Table 9.2 Transgenic crop areas in USA 1996–1998 (million acres).

	1996	1997	1998
cotton			
transgenic	1.8	3.2	7.0
non-transgenic	12.8	10.6	6.2
soybean			
transgenic	1.0	9.0	20.0
non-transgenic	63.2	61.9	52.0
maize			
transgenic	1.0	6.0	17.0
non-transgenic	78.0	74.0	64.0

Table 9.3 Numbers of EU field release and product development projects (1996).

Crop	Product development	Field release	Total
Oilseed rape	22	32	54
Maize	11	31	42
Potato	18	23	41
Sugar beet	12	26	38
Other vegs	22	7	29
Ornamentals	23	1	24
Tomato	5	14	19
Wheat	13	2	15
Sunflower	8	4	12
Tobacco	3	8	11

Source: European Crop Biotechnology – A Strategic Review & Directory. *BioBusiness*, April 1996.

Future prospects

The first generation of transgenic crops, for example maize, cotton, potatoes, squash, soybeans and tomatoes engineered for herbicide tolerance, insect resistance, delayed ripening or virus resistance have already made a significant commercial impact (see above). The future prospects are based around technology uptake and development. Two clear areas are the transformation of minor crops using existing technology and the further advancement of gene constructs and engineering for multiple resistance. A list of possible projects is presented below.

Proposals for field trials in USA (1997) – examples

- squash engineered for resistance to four viruses
- virus resistant tomato
- potatoes resistant to Colorado potato beetle and potato leaf roll virus
- beet resistant to beet necrotic yellow vein virus
- grapes engineered for sulfonylurea tolerance
- strawberry resistant to fungal infection
- walnuts with multiple resistance to virus and coleopteran insects
- herbicide tolerant lettuce
- blackberries engineered for altered fruit ripening
- petunias resistant to both bacterial and fungal infection
- fungus resistant creeping bentgrass

Gene stacking is a concept likely to achieve substantial input in the near future. Indeed, commercial companies are keen to add on extra traits to their existing genetically modified lines, e.g. virus plus insect resistance, or resistance plus quality traits.

Ethics and implications

The release of transgenic plants into the environment has been a topic of hot debate and some controversy in Europe, though much less of an issue in the USA. It should be stressed that all releases in USA and EU are subject to rigorous approval schemes involving risk assessments and public notification. In the UK, ACRE (Advisory Committee for Release into the Environment) has responsibility for policy and operation of the legislation controlling the release of genetically modified organisms (GMOs). This committee also gives advice to the Secretary of State on human and environmental concerns regarding the releases of GMOs and non-native organisms into the environment.

The marketing of GMO products has proved to be a contentious area within the EU. Some products, e.g. tomato purée, have been marketed in a very positive fashion, with excellent point-of-sale information. Other products have, however, been marketed with great insensitivity to opponents of the technology. Opposition to transgenic technology has been very high profile, with field releases disrupted in Europe. There needs to be a fully informed, balanced debate with the general public over the utilization of this technology, by all concerned.

However, the views of the general public can be represented by several factors; the success of the genetically modified tomato puree in the UK with over 750 000 cans sold in 1997, and the referendum held in Switzerland in 1998 which came out against banning transgenic research.

The enormous investments by biotechnological companies, with a need to see some return, have created a strong market push for the acceptance of transgenic crops. This has focused on the potential favourable environmental impact of transgenic crops. Pest and herbicide resistant crops require less chemical input, with consequently less environmental damage, and some traits (e.g. starch/oils) result in reduced processing costs and wastage. Conversely, the risk of monoculture, loss of biodiversity, risk of traits escaping into the wild and unknown effects on the ecosystem are all cited as reasons against transgenics. All these factors require investigation on a case-by-case basis subjected to scientific scrutiny rather than emotive outcry.

9.5 Molecular applications to crop improvement

Historical background

Breeders have traditionally improved plant varieties by selecting on the basis of phenotype. However, the phenotype of a plant is determined not only by its genotype but also by the environment in which it is grown. To circumvent such problems plant breeding techniques based on detailed statistical inference have been developed. The theoretical application of genetic markers and maps to increase the speed and precision of plant breeding have long been proposed. In 1923 Sax proposed identifying and selecting for minor genes of interest by linkage with major genes which could be scored more easily. This concept has been extended by many workers, and early maps were produced using structural and other types of mutants by following a segregation of two or more traits in a particular cross. If traits separated from each other in the offspring it was concluded that they were not linked, whereas tightly-linked traits would be frequently inherited together in progeny. As only a few genes were segregating for any one cross, many crosses were required to produce even the simplest genetic map. Also, most morphological markers cause major effects on phenotype and are undesirable in breeding programmes.

It has also been stated (Tanksley *et al.*, 1989) that these markers can mask the effect of minor linked genes making it impossible to identify desirable linkages for selection. Isozyme markers were developed as nearly-neutral genetic markers, although they were not sufficiently discriminating for many applications in plant breeding. The advent of gene cloning technology from the early 1970s onwards has resulted in fundamental advances in marker technology.

DNA-based markers

The genetic information which lives in the genes of higher plants is stored in the DNA sequence of the nuclear chromosomes and the organelle genomes. Replication of DNA is highly accurate and rapid in plants, but there are many mechanisms which can result in changes. Simple base pair alterations, or larger scale changes caused by translocations, deletions, or the effect of transposons may occur.

Natural variation in DNA sequence can be detected in a number of ways. The first major molecular marker system to be developed was that of restriction fragment length polymorphisms (RFLPs) which were initially developed in 1980 by researchers in human genetics. The technique uses restriction enzymes to cut DNA into small fragments of differing lengths. When two or more individuals are compared using the same RFLP probes there can be a difference in the length of the fragments, i.e. a polymorphism. Mapping with RFLPs involves the application of molecular technology to the basic concepts of transmission genetics. The method involves cloning a single copy DNA sequence from a species of interest and using this as a probe to follow the segregation of homologous regions of the genome in individuals from segregating populations. Due to the fact that many single copy clones can be analysed, genetic linkage maps can be constructed that contain a large number of markers at relatively close intervals. Notably, the RFLP markers, which are screened directly at the DNA level, behave as co-dominants, also the level of allelic variation for RFLPs in a population is greater than that for morphological markers.

Much effort was expended in the late 1980s to construct RFLP maps of major crop species such as potato (Tanksley *et al.*, 1989). Though RFLP technology was a major advance, the system requires relatively large amounts of plant material (1–10 g) and can be time consuming. However, rapid advances in molecular analysis followed, facilitated largely by the introduction of PCR (Polymerase Chain Reaction) – a method for rapidly amplifying specific DNA sequences.

The advent of RAPD markers (Randomly Amplified Polymorphic DNA) combined PCR technology with non-specific DNA primers to amplify random DNA fragments. The fragments can be polymorphic due to being present in one sample and absent in a second. As previously stated RFLPs produce co-dominant markers, whereas RAPDs yield dominant markers. RAPDs are quick and relatively cheap to use, and can be undertaken with as little as 10 mg of plant tissue, but problems persist due to lack of repeatability.

Simple Sequence Repeats (SSRs) and microsatellite-based markers are short segments of DNA consisting of small numbers of repeated nucleotide sequences. These marker systems, again based on PCR, have benefits in that they reveal greater polymorphisms than RFLPs as they exhibit co-dominant inheritance and are multi-allelic. The first stage in the process is the development of genomic libraries enriched at the pre-cloning stage for an appropriate microsatellite repeat. The second stage is to sequence SSR-containing clones in order to design suitable PCR-primers that amplify a particular SSR sequence. AFLP (Anchored Fragment Length Polymorphism) is a commercial system (licensed by Keygene) in widespread use by plant biotechnology and breeding companies; it uses a technique that results in the amplification of a random subset of the restriction fragments present.

Bioinformatics

An often overlooked facet of molecular marker strategies is the importance of bioinformatics. Contemporary molecular technologies generate huge amounts of data, with significant chances of introducing errors at many of the stages. Correct accreditation systems and methods for identifying inconsistent data are of great importance. Also of increasing relevance are the establishment and maintenance of plant genomic databases accessible through the WorldWide-Web.

Marker assisted selection

Sax in 1923 is generally credited with the idea of using marker genes linked to those difficult or impossible to select directly. This took some sixty years to take shape as a realistic option. The major reason was the shortage of suitable marker genes combined with the inadequacy or non existence of genetic maps in most crop species. Arás and Moreno-González (1993) set out the essential qualities of good markers: these should be polymorphic and the action of alleles should be easily recognized in all combinations whether heterozygous or homozygous. They should be recognizable early in plant development, and not interact with the marked locus nor have any direct effect on the morphology of the species in question. Prior to the development of molecular markers, very few potential markers were of value in selection. Molecular markers have themselves many, if not most, of the desirable qualities listed. Both protein and DNA markers are useful but the use of isozymes as markers is likely to diminish as the range of DNA markers increases.

Practical usefulness of markers will undoubtedly be greatest when they are closely linked to loci controlling simply determined traits such as disease resistance, self-incompatibility and morphological features such as shape, colour and size of plant parts. The disease resistance case is a particularly interesting one: the necessity for inoculation is obviated and selection can be carried out in an environment free of the disease. Selection efficiency is obviously greater the more closely linked the marker is to the target gene. If suitable markers flank it then less closely linked markers can still be highly effective. It is also possible to mix or match markers in that both could be DNA or isozyme markers or one each of the two types.

Tanksley and co-workers (1989) have developed applications in backcross breeding programmes where effective selection for the recurrent parental background genotype can be accelerated. The use of flanking markers can also reduce 'linkage drag', that is the amount of undesirable genetic material linked with the selected gene from the non-recurrent parent.

Quantitative trait loci

Perhaps the most interesting use of molecular markers is in the more efficient selection which they make possible for polygenic characters controlled by what are now generally called quantitative trait loci or QTLs. The use of molecular markers such as isozymes and RFLPs (Stuber, 1995) has made it possible to differentiate the quantitative effects of different loci on the same character. This effectively belies the simplifying assumption made in developing quantitative genetics models that the effects of different loci affecting the same character were small and approximately equal. The discovery that this was not so and that genes with substantial effects could be identified made it feasible to select those with the greater effects. Interesting work on QTLs has been carried out by Tanksley and co-workers (1982) on tomatoes, Stuber *et al.* (1987) in maize, McCouch and Doerge (1995) on rice and Bradshaw *et al.* (1998) on potatoes. A very considerable effort has gone into research and subsequent development of these ideas in a practical context with indications of very substantial returns. The levels of investment in research however are very high and it is only in highly commercialized crops such as maize that satisfactory returns on this investment could be expected in the short or medium term. Considerable concern has been expressed at the inefficiency of traditional selection methods and the strong possibility, even probability, that the best genotypes get away. The identification of the important QTL alleles and their fixation in commercial varieties could have economic consequences of the very greatest importance. Immediate application can be envisaged in the hybrid maize industry where initial production of inbreds and subsequent improvement could be facilitated by the use of QTL technology.

Particularly significant applications could be expected in the improvement of disease resistance, especially in cereals. Durable resistance (the HR of Van der Plank) is thought to be polygenic and probably amenable to QTL manipulation. There is even the possibility that use of markers could enable vertical (race-specific) resistance to be combined with horizontal resistance.

Future prospects

The rapid pace of change and innovation in the field of molecular biology will undoubtedly result in faster, more efficient and more informative methods applicable to improving the speed and precision of plant breeding. Specific examples include automated fluorescent sequencing equipment, with parallel data scoring and analysis with appropriate software. Further factors include automation, DNA chip technology, and an increasing user-friendliness of marker systems. Though large investments are required for such technology, the rate of progress expected justifies the cost.

9.6 Summary

This overview attempts to cover a broad and rapidly expanding area of science. The development of more accurate methods of gene detection and isolation in plants is proceeding at great pace, and this, coupled with increasingly efficient protocols for the uptake of genes into plants, offers great prospects for future crop technology. Certainly, the range of crop plants that can be genetically enhanced by such technologies is increasing rapidly, with hitherto 'recalcitrant' species now proving amenable to transformation. It is undoubtedly true to state that transgenic plant technology is still in its infancy, despite some notable successes and their impact on agriculture. The first report of a transgenic crop plant was as recent as 1983. In the last fifteen years useful progress has been achieved. The future, with such challenges as the uptake and expression of multiple genes, the elimination of selectable markers from plants and gene-switching mechanisms in plants, will be a scientifically stimulating era.

9.7 Literature

Plant cell and tissue culture

Becker, J. *et al.* (1995); Brown, D.C.W. and Thorpe, T.A. (1995); Carlson, P.S. *et al.* (1972); Christey, M.C. *et al.* (1991); Dumet, D.J. *et al.* (1998); Hammat, N. (1992); Komamine, A. *et al.* (1990); Litz, R.E. and Gray, D.J. (1995); Pierik, R.L.M. and Ruibing, M.A. (1997); Skoog, F. and Miller, C.O. (1957); Takebe, I. *et al.* (1971).

Transgenic plant technology

Bechtold and Bouchez (1995); Braun (1958), Dunder *et al.* (1995); Herrera-Estrella *et al.* (1983); Horsch *et al.* (1985); Kaepller *et al.* (1990); Klein *et al.* (1987); Shillito *et al.* (1985); Trick and Finer (1997); Wan and Lemaux (1995); Zambryski *et al.* (1983).

Target traits for transformation

Brousseau (1997); Corbin *et al.* (1998); Dale (1995); Gatehouse, and Gatehouse (1998); Grierson and Schuch (1993); Kavanagh and Spillane (1995); Lamb *et al.* (1992); Murphy (1996); Sandford and Johnson (1985); Streber and Willmitzer (1989); Tavladoraki *et al.* (1993); Zhao *et al.* (1997).

Molecular application to crop improvement

Arás and Moreno-González (1993); Bradshaw *et al.* (1998); McCough and Doerge (1995); Stuber, *et al.* (1987); Stuber (1995); Tanksley *et al.* (1982); Tanksley *et al.* (1989).

Chapter 10
New Crops and Genetic Conservation

10.1 Introduction

Most plant breeding takes place by a stepwise improvement of crops long established in a place, using a store of genetic variability, in the form of breeding stocks and working collections, that is large enough for immediate needs. Occasionally, however, the breeder faces the task of adapting an unadapted foreign crop to his local conditions or even of developing a new crop from wild or primitive materials. Success in any such venture depends upon the acquisition of a range of genetic stocks adequate to the purpose. The need for genetic variability brought in from somewhere else is never more apparent. Even in old, well-established programmes, the long-term need for a sustained supply of new variability is no less, as our earlier discussion of the genetic base (Section 5.9) will have shown. But this is a time of decay of genetic variability and it has become clear, over the past forty years or so, that we must accept the responsibility deliberately to conserve it for the sake of future progress of our crops, old, new and yet unconceived.

10.2 New crops

Introduction

We saw in Chapter 1 that our crops collectively represent a continuous history of domestication, geographical spread, evolution and, sometimes, extinction. These processes continue today, probably more rapidly than ever in the past. They are promoted by far-seeing agricultural scientists (including plant breeders) who perceive needs and possibilities and by farmers and technologists, with perception of economic potential; they have been supported, directly or indirectly, by governments who alone can assure the necessary long-term resources. We might here recall that governments historically, in the eighteenth to nineteenth centuries, supported crop introduction by means of voyages (such as that of the 'Bounty', 1787–90) and botanic gardens which, in those days, had economic as well as scientific and recreational functions (Purseglove, 1957, 1968). Unfortunately in more recent times the approach of some previously supportive governments has been one of indifference and reduced support, not only for national efforts in the field but also of the international agricultural research institutes which are well placed to further development of new crops.

At first sight, several distinct categories of new crops seem obvious; on closer inspection they turn out to overlap but are still categories of convenience. In the obvious sequence, suggested by crop evolutionary ideas, they are: (1) domestication of new crops from wild plants; (2) adaptation of old crops to new places; (3) construction of new crops from old by way of polyploidy; (4) a second cycle of domestication of previously discarded crop species.

New domestication

The outstanding example of recent domestication is *Hevea* rubber which unambiguously made the transition from being a wild forest species in Amazonian South America to being a clonally propagated cultigen in Southeast Asia within the last century. Other examples of recent domestications are numerous though less dramatic and clear-cut than that of rubber: oil palm, pyrethrum, the coffees and quinine, among others. In all these there is some question as to the state of domestication at the time at which deliberate large-scale cultivation began and the evolution of the crop may fairly be said to have started. Certainly, all were at least a little cultivated (but hardly more than gathered) until the last one or two hundred years. All have been developed as crops in places remote from their wild origins and all have become dependent upon systematic plant breeding for future progress, indeed probably for very survival as crops.

The line between new domestication and new adaptation in an existing crop (our second category) is thus not very sharp. That new domestication has been occurring in the recent past and continues vigorously now is, however, clear. Perhaps the best current example is provided by Australian efforts, over the past fifty years, to develop new forage grasses and legumes for the country. The background is that Australia has a mainly stock-based agriculture in warm, but often dry, climates and few useful native grazing plants. Stock-rearing in similar climates elsewhere is mostly based on ranching of native vegetation and there was, fifty years ago, no substantial body of developed warm-country grass and legume varieties upon which Australia could found a local development. Accordingly, a fairly massive programme of exploration in likely environments in the Old and New Worlds, followed by introduction, evaluation and breeding, was mounted and is now paying off. The programme is summarized in Fig. 10.1. The fifteen genera mentioned in the figure come variously from Africa and America; Asia so far seems to have had little to offer. All are warm-country plants and none would have a place in temperate agriculture. Which ones will have places in Australia is becoming clearer, there are certainly several likely candidates.

It will be noted in Fig. 10.1 that the programme as a whole is based in the

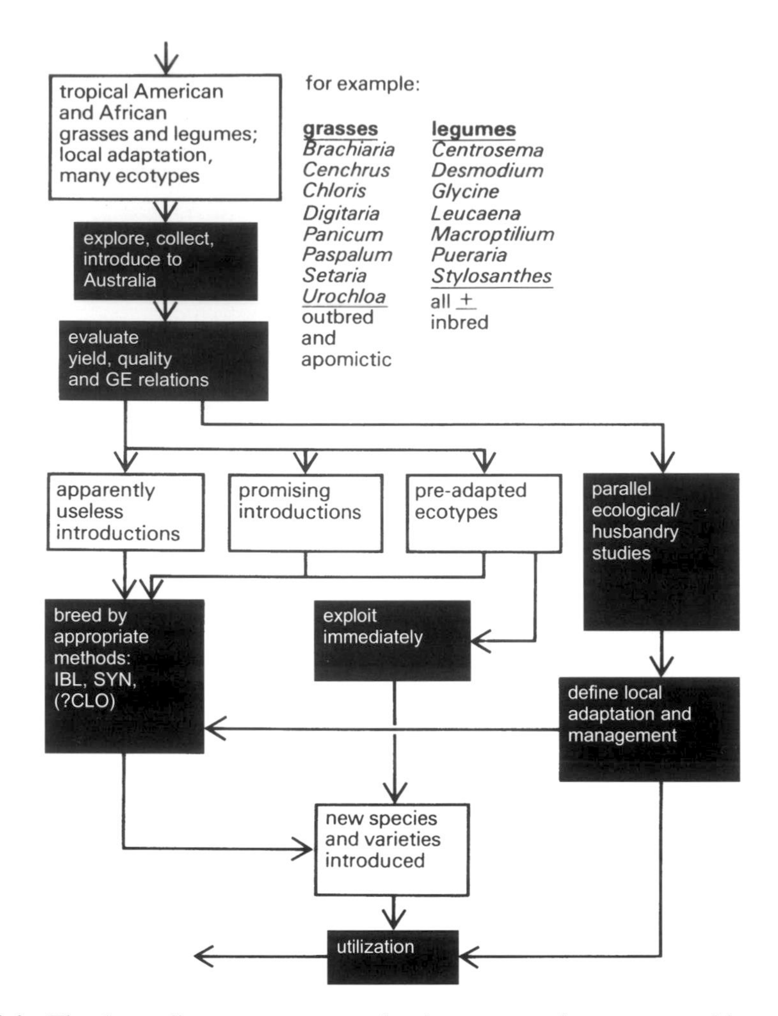

Fig. 10.1 The Australian programme to develop new grazing grasses and legumes. Based on Shaw and Bryan (1976), Hutton (1976).

coordinated activities of botanists, collectors, agronomists and animal scientists as well as of plant breeders; plant breeding, indeed, only becomes involved in the final phases of adaptation when it has already become clear, from initial screening, that introductions have potential. Rare introductions, indeed (e.g. an early one of *Stylosanthes*), are found to be 'pre-adapted' and can be multiplied straight away for agricultural use. Here, as elsewhere, grazing crops are, by their very nature, but little removed from the wild; there is no scope for altered partition, since the crop product is the whole aerial growth, and the breeder's task is effectively met by the discovery or production of locally adapted ecotypes. In the slightly longer run, it looks as though the Australian initiative is likely to provide, not only new grazing species for Australia, but also grasses and legumes for the more recently developing stock agricultures of tropical Africa and America (Section 11.4).

Other examples of new domestications might be sought among the timber trees. Tree breeding is only a few decades old but already substantial areas (especially in Scandinavia and the USA) are being planted with bred populations, locally even with selected clones (as with New Zealand *Pinus radiata*). Natural forests and natural regeneration are already inadequate, in many areas of the world, to meet the demand for timber. Economic pressures will increasingly ensure that, if trees have to be planted, productive genotypes are chosen for the purpose. So forestry is inevitably, though perhaps not very rapidly, becoming a branch of agriculture; ploughing, fertilizing and weed control are already quite widely practised; it will be surprising if bred, domesticated varieties do not very largely displace the traditional more or less random seedlings in the course of the next two or three crop rotations (say 100–200 years). Tree breeding lies beyond our scope but it is worth observing that it has something in common with breeding of grazing crops in that the cultivated tree is not too far removed from the wild one; the first objective is simply the adapted ecotype which can realize the biomass-potential of the local environment; but, in addition, the tree breeder may have some opportunity to exploit the potential of his species for improved partition (between stem and branch) and wood quality (straightness and density).

New domestication is therefore taking place, and will surely continue, in a variety of grazing crops and timbers. The genetic changes implied are not profound, for the basic requirements are met by the discovery or production of locally adapted ecotypes. Greater changes will no doubt follow as the full range of plant breeding techniques are brought to bear: perhaps new population structures or induced polyploids will be useful?

Another group of plants which may be expected to provide at least occasional new cultigens are those grown to provide complex chemicals such as insecticides and drugs. Often, chemical industry can provide the product, or a substitute for it, more cheaply; so indigo disappeared as a crop of the last century, displaced by aniline dyes. But pyrethrum and quinine persist successfully and natural rubber thrives because it is superior to synthetics for certain purposes and cheaper than its only direct competitor, synthetic cis-polyisoprene. The ever-increasing chemical sophistication of the pharmaceutical industry has led in recent years to hundreds, perhaps thousands, of investigations of plant sources of physiologically active materials, usually as starting points for industrial synthesis. The steroidal saponins of yams provide an example; production has hardly got beyond the stage of destructive harvest of wild plants but can clearly not be sustained indefinitely on this basis. Sooner or later, if the pharmaceutical industry wants a continual supply of specific natural steroids, a domesticated source (whether of a yam or of some other different plant, for related steroids are widespread) will have to be developed. Opportunities of this kind must be recurrent but would often have to be viewed rather cautiously as prospects for plant breeding: the drug itself might be superseded or a cheaper starting material found; again, cell/tissue cultures might conceivably provide an alternative source which would by-pass agricultural production of the plant. This may yet be slightly science-fictional but the industrial production *in vitro* of precious metabolites from plant cells treated as though they were micro-organisms is an active field of research and an undoubted middle-long-term practical possibility.

The prospects for new crops that have to be profoundly different from their wild ancestors before they become really useful is less evident. Could one imagine inventing wheat or maize *de novo* given the appropriate wild grasses? Progressive modification, though sometimes of a fairly fundamental kind (as with allopolyploid triticale and raphanobrassica), seems more likely. An example of a radical development is, however, currently available and serves to show that the possibilities, though perhaps rare, are not negligible.

Various lupins have been minor forages and grain in both the Old and New Worlds for a long time but have hardly made the transition to being proper cultivated plants. The group is botanically complex and ill-understood cyto-taxonomically; even chromosome numbers are yet incompletely known. At least four more or less cultivated species are concerned and there have been sporadic efforts over the past fifty years to complete the process of domestication. In Germany in 1928–29, von Sengbusch sought and found a few alkaloid-free ('sweet') mutants in two species and others have been discovered since then. In the past fifty years extensive efforts by breeders in Russia, Europe, the USA and Australia have brought a good deal nearer the goal of alkaloid-free, non-shattering, thoroughly domesticated lupins as fodder and grain crops for sandy soils. A striking feature of the Australian work, with *Lupinus angustifolius* now becoming a significant crop in Western Australia (Gladstones, 1975), has been the evident importance of major-gene mutants; most are recessive; one was induced by a chemical mutagen and the rest were revealed by careful search. They include genes for 'sweetness' (alkaloid content reduced fifty-fold), permeable seed coat, white flowers and seeds (useful as a marker in sweet lines), delayed and reduced pod shattering, early flowering (reduced vernalization demands), and several disease resistances. It now seems clear that it wants only time and effort to add several lupins to the list of crop species. Since there are 300–400 species of *Lupinus* (yet very poorly understood) the possibilities for new domestication, inter-specific introgression and perhaps allopolyploidy must be substantial (Hill, 1995).

The lupins are well on the way to domestication. Two less advanced, more speculative, examples are worth mention, just to indicate the possibilities. The buffalo gourd, *Cucurbita foetidissima*, native in dry places in the southwestern USA and Mexico, has been a 'camp-follower' plant of American Indians for millennia (Bemis *et al.*, 1978); the Indian people gathered and used it in a variety of ways, though the high content of bitter cucurbitacins was always an obstacle to eating the seeds or fruit. Since the 1940s there has been some interest in the plant as a possible domesticate adapted to dry places; it is a strong, sprawling vine with a huge, water-storing root. Plants are perennial and easily clonally or seed-propagated; they are monoecious and gynodioecious, so hybrid seed production could, in principle, be manipulated. The products are seed (protein 31 per cent, crude fat 33 per cent), starch from the roots and leafy fodder; yields of, maybe, 3 t/ha of seed and 13 t/ha of starch are projected but much will depend upon the way that breeding adjusts the partition between parts of the plant. The oil is of good quality, but the breeders will probably have to mimic the evolutionary history of other cucurbits and reduce the cucurbitacin content. A second example is even more speculative but suggests an interesting historical parallel. The babassú palm of Brazil (*Orbygnia speciosa*) is a sort of New World analogue of the oil palm (*Elaeis*) of West Africa (Markley, 1971). It grows wild over large areas and is extensively gathered, as an oil plant, mostly for local use. This is exactly the situation of *Elaeis* 100 years ago. The fruit-shell of *Orbygnia* is very hard, so one of the first preoccupations of the babassú breeder would have to be to find (or make) the analogue of the *tenera-pisifera* (thin-shelled) mutant of *Elaeis*. *Orbygnia* is tolerant of flooded land, has a good quality oil and there is plenty of variability, both in *O. speciosa* itself and in related species. As a third example, the Amazonian peach palm pejibaye (*Bactris gasipaes*) is already an important food plant but wholly neglected by plant breeding (Clementin, Smartt and Simmonds, 1995).

New adaptation

The geographical spread of crops to new places is one of the dominant processes in the history of agriculture, indeed in the history of mankind. Nearly every major production is distant, often very far distant, from the area of origin of the crop. The process is yet incomplete and old crops are being adapted to new places probably faster now than ever they have been in the past. Sometimes, crops appear to be 'pre-adapted' to homologous environments elsewhere and the move is easily made, as with rubber and oil palms in Southeast Asia, bananas in Africa and tropical America, cacao in West Africa, arabica coffee in tropical America, sugarcanes throughout the tropics, US cornbelt maize in the Danube basin and so on. Sometimes, again, the new environment is sufficiently different to forbid large-scale pre-adaptation; then, success depends upon happy accidents or, more often perhaps, upon purposeful breeding. Wheat owed its beginning as a major crop in North America and Australia to the identification of the rare combinations that provided the essential local adaptation: in the hard red spring wheat area of the USA and Canada, Marquis (1907) came from the cross Red Fife (eastern Europe) × Hard Red Calcutta (India) and Marquis descendants have dominated the area ever since.

The northward extension of wheat in North America and northern Eurasia during the past fifty years depended upon the deliberate breeding of cold-tolerant, early maturing types and the same applies to maize, which has effectively appeared as a crop in northern Europe only in the past forty years. The adaptation of potatoes to north temperate climates and of soybeans to the USA depended upon the adjustment by breeding of the day-length responses that determine maturity. Diseases and technological requirements dominated the adaptation of cotton in tropical Africa (perhaps one of the best examples extant of the development of a new crop in response to a deliberate policy decision rather than to existing economic forces).

The list could be greatly extended. Curiously, there exists no comprehensive genetic history of crop adaptation to new places, which is a pity because it could be remarkably interesting. Some generalizations seem possible however. First, the easy, potentially pre-adapted introductions have no doubt mostly been made, so future progress must increasingly depend upon introduction *plus* systematic plant breeding; this will certainly be true, for example, of the tropical soybean developments now under way (e.g. in Brazil and India). Second, the blocks to adaptation will be, in the future, as they have been in the past, quite various, including physiological factors (temperature, day-length and moisture responses), diseases and pests, and technological features. And, third, we should perhaps remind ourselves that all adaptation involves genetic erosion, so that the well-founded programme will make deliberate provision for the maintenance of a reasonably wide genetic base; here we touch both upon questions discussed above (Section 5.9) and upon genetic conservation, treated later in this chapter.

New polyploid species

In our survey of the use of polyploidy as a plant breeding technique (Section 8.2), we saw that its principal value was to ease difficult interspecific hybridizations rather than to produce new crop genotypes directly. However, autotetraploid crops grown for a seed product may yet be more valuable than they now seem to be and there must surely be opportunities for more new allopolyploids than triticale and raphanobrassica. The latter are certainly new species within any definition of the word. Autotetraploids should probably also be regarded as functionally distinct, being at least semi-isolated from their diploid ancestors; there will be no need for new genera or Latin names, however. Experience now shows that polyploids of both kinds are far harder and slower to develop than used to be expected but it seems certain that occasional enrichment of our crop resources will occur by this route. Probably, the easing of inter-specific crossing and base-broadening work in existing polyploids (e.g. *Brassica napus* – Section 8.2) will remain the principal use.

10.3 Genetic conservation

The need for conservation

Before the Second World War, very few people foresaw the explosive growth of world population that was then imminent; or the swing towards technology-based agriculture that was an unavoidable response; or yet, as a more remote consequence, the pressure on both natural ecosystems and the genetic variability of domestic plants and animals. From small beginnings in the 1940s, some general awareness of the magnitude and importance of the problems that face us has only emerged since the late 1950s. The Food and Agriculture Organization (FAO) of the United Nations has done more than any other body to promote understanding and a sense of urgency; it has achieved this partly by meetings and publications (FAO, 1967; Frankel and Bennett, 1970; Frankel and Hawkes, 1975), partly by helping to initiate the International Board for Plant Genetic Resources (IBPGR, 1975, 1976), to which further reference will be made below. The need – an urgent one in many crops – for genetic conservation is now generally agreed; how did it arise and what are the dimensions of the problem?

Successful plant breeding tends to narrow the genetic base of a crop in rough proportion to its success; in all advanced, technology-based agricultures, a few excellent varieties, themselves often inter-related, tend to cover large areas of land to the exclusion of all else. The closer the adaptation, the narrower the base and the effect is often reinforced by the choice of breeding methods (e.g. the use of HYB in crops traditionally bred as OPP). The general consequence is, as we have already seen (Section 5.9), a situation in which fifty to seventy years of highly successful breeding in many crops is liable to be succeeded by an uncomfortable check to progress plus unadaptability to unforeseen crises. As a matter of breeding strategy, the remedy is plainly to ensure a steady flow of parents from well-adapted reserves (Fig. 5.16), a principle which is being adopted in some programmes. The genetic variability in those reserves has to come from somewhere. Furthermore, the need to maintain reserves will never become any less; there will be a requirement in perpetuity. So long as plant breeding is practised, so long will new stocks be needed to improve local adaptation, to push crops into new environments, to meet crises (e.g. new diseases) and to satisfy changing social demands upon the product.

Until recently, agriculture in the poorer countries, mostly in warm climates at lower latitudes, presented quite a different picture. The great plantation crops, it is true, were (and are) intensively and efficiently bred; food crops, however, were largely neglected and great stores of genetic variability persisted. Plant breeders seeking variability only had to look in the right places to find it in plenty. In the past forty-odd years, these stores have declined as the methods of temperate, technology-based agriculture have penetrated into the more conservative systems; high-yielding cultivars, often bred thousands of miles away, displace the heterogeneous local varieties and mixtures; adaptation improves as variability declines. The wheats and rices of the 'green revolution' (Section 11.4) are but the most dramatic example of this trend, which is shared by many crops, if less conspicuously. In so far as a few high-yielding varieties produce more food than a multitude of less productive ones, the trend is clearly good and, anyway, inevitable; it will continue, as technology-based methods penetrate into more places and touch more crops. So genetic variability residing in actual producing populations of crops is inevitably on the decline and can only continue thus. Hence the need for genetic conservation: to compensate for declining native variability by deliberate maintenance.

It would be good to be able to give some quantitative substance to a description of decline by genetic statements about the probability of survival of alleles (for example) or even, though more crudely, by estimates of changing numbers of distinct phenotypes. Unfortunately this is impossible. The evidence of decline is purely anecdotal, though cumulatively powerful and wholly persuasive; travellers' tales, from experienced and perceptive travellers, at least, must be taken seriously. The decline however is certainly uneven and the urgency of the situation varies rather widely as between crops (a point to which we shall return later). In the longer run, it can not be doubted that a conservation effort will be required for all.

There are, of course, important wider issues under the general heading of 'genetic conservation'; they lie beyond our scope but deserve a brief comment. Many, perhaps all, natural and semi-natural ecosystems are at least at some risk from human activities. Man has a sorry record of persecuting animal species to extinction and it is of course, birds, tigers, whales and such like which tend most readily to come to the public mind in the general context of 'conservation'. The idea that a great many plant species are similarly threatened with extinction is now accepted, workers at the Royal Botanic Gardens, Kew, reckon that about 20 000 plants (roughly 5 per cent of angiosperms) are at hazard (Heslop-Harrison, 1974). Subsequent extensive destruction of tropical rainforest has probably exacerbated the situation considerably. The idea of protecting wild animal and plant populations by the conservation of large areas of natural and semi-natural ecosystems is agreed to be excellent in principle (though maybe, alas, expensive and insecure in practice). For the plant breeder two points arise. First, there may well be potential economic plants among the 20 000 threatened species; the prospect of trying to identify and preserve them specifically is daunting indeed but should at least be seriously thought about. And, second, it has occasionally been proposed that the ends of crop–genetic conservation would be served by the preservation of primitive agricultural ecosystems, crops, stock and presumably people included. In practice, of course, this would be socially inconceivable; crop variability will have to be conserved in collections designed and managed for the purpose.

Strategy

In the long run, genetic conservation measures will have to be applied to all crops; the need will be quite general. We now assume that the need is agreed and consider the consequent strategy of conservation, leaving methods to the next section. The strategic components to be considered in deciding just how to set up and maintain a major collection are (see Fig. 10.2): (1) contents and scope of collection; (2) places to be explored and (3) place of maintenance; (4) basic methods to be adopted; (5) choice of information system; and (6) utilization.

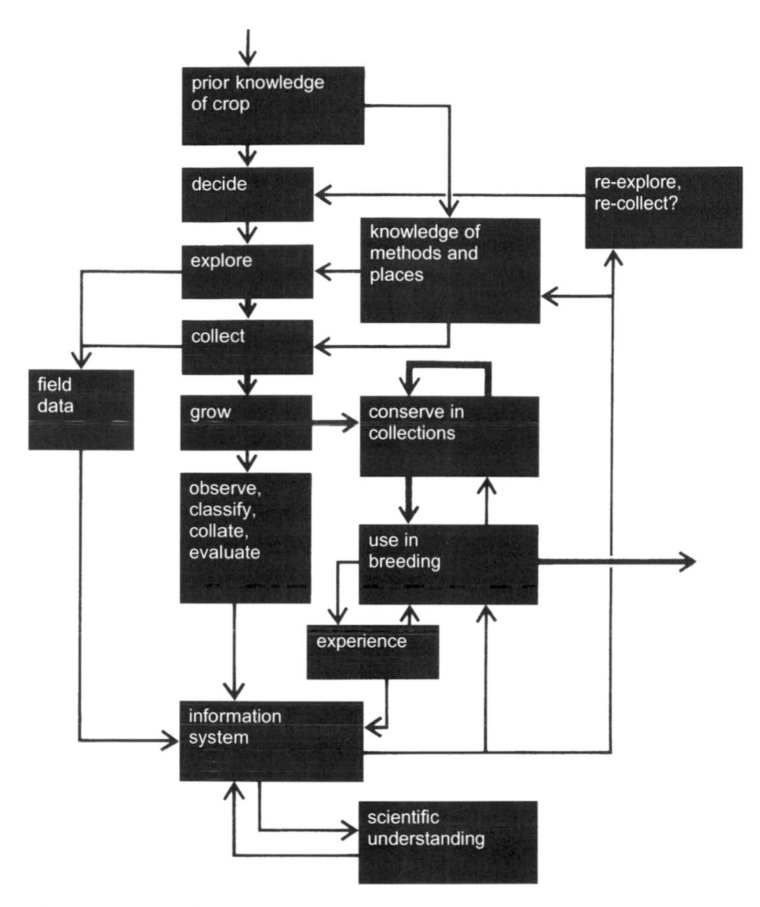

Fig. 10.2 The pattern of genetic conservation.

On (1), contents and scope, there will usually be little choice because the proper answer commonly should be: the whole crop and its cultivated and wild relatives of known or potential use in breeding. This does not mean that the objective of every collecting expedition must be all-inclusive or that every component collection should cover the entire range of material; many expeditions will naturally be selective and it will sometimes be sensible to concentrate different kinds of material in different sub-collections, as when a crop is widely differentiated ecologically. But the overall aim must be completeness as to the crop itself and useful relatives.

The places to be explored (2) will depend upon knowledge of the crop: where the variability is; what has already been collected; what areas have been searched and what have not. Sometimes, actual collecting will be unnecessary, as when a letter requesting known material will suffice. As to choice of site for maintenance (3), this will often be predetermined by political or administrative considerations. Obviously, any site chosen must be biologically appropriate to the material that is prospectively to be grown and maintained there. Scientific experience, communications, local climate, diseases and quarantine arrangements will all be relevant. Recalling that the intention is conservation in perpetuity, the probability of exceptional natural hazards (such as earthquake, eruption, flood or storm) may need to be considered. The question then arises: since rare events do occur, if only rarely, should not the collection be duplicated in the interests of long-term security? The answer surely is that it certainly should be, hard as it might be to convince administrators of the wisdom of the idea. Anyone who has had working experience of a collection knows that there is nearly always a trickle of unavoidable losses (Fig. 10.3) and will be acutely aware of the still greater hazards due to disaster, bad luck or carelessness.

Next, there are strategic aspects of methods to be applied to the collecting itself and to subsequent maintenance (4). The former has been considered from genetical first principles by Marshall and Brown (in Frankel and Hawkes, 1975); they conclude that a frequent best strategy will be to collect large numbers of more or less random small samples but that prior genetical knowledge might

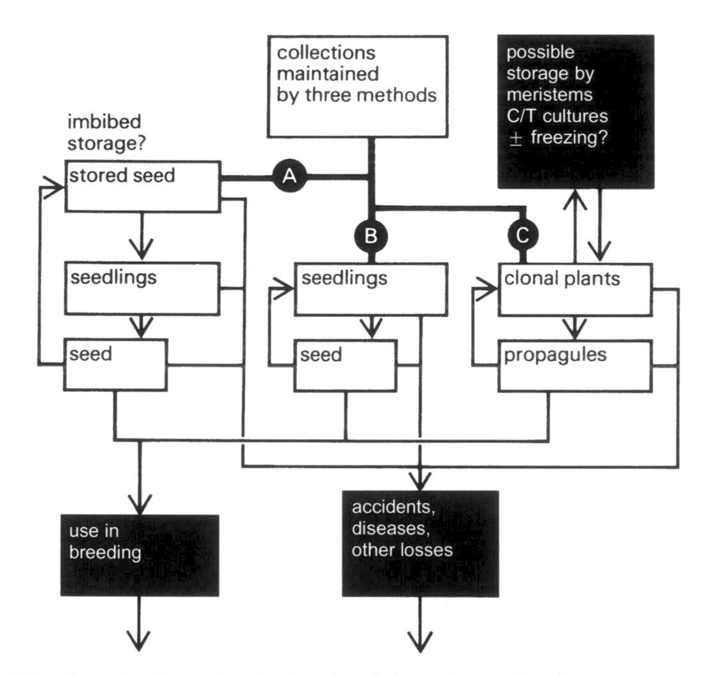

Fig. 10.3 The three basic methods of maintaining plant collections: A, as seed lines with seed storage; B, as seed lines without seed storage; C, as clones. See Fig. 10.4 and text.

sometimes suggest a different plan. Whatever the sampling strategy adopted, the collector will clearly need to time his activities rather carefully in relation to seasonal weather and crop cycle and will be alert to the possibility of exploiting rural markets as sources. Once collected, methods of maintenance and of long-term conservation, depend essentially upon the biology of the crop but, even so, there will be strategic decisions to be taken, for example: methods of propagation, storage conditions, size of samples, frequency of renewal and so forth. We return to these questions below.

Information (5) is an aspect that has received increasing attention in recent years. The current tendency seems to be to develop very elaborate botanical and agronomic descriptions (including adaptation, quality, disease reactions) of all accessions in a collection and store them in a readily-retrievable, computerized form. This system, it is argued should help the ultimate user, the plant breeder who can specify precisely what he wants; and so it should, if precise specification were possible. In practice, phenotype (especially if determined elsewhere and referring to a poorly adapted foreign stock) can be a bad guide to breeding potential; the breeder will often, therefore, have to do his own searching, evaluation and test-crossing. Some basic information about stocks is essential but there does seem to be a risk of over-emphasizing this aspect, at least in respect of utilization, though perhaps not of more general scientific understanding (Fig. 10.2). It will be obvious that, if an information system is to be internationally useful, then agreed methods of recording and description will have to be adopted and that these will have to be coupled to compatible, if not identical, computing systems. Progress along these lines is being made but there does seem to be some risk of over-elaboration; a 1970s 'descriptor' list for barley contained about 140 'characters' each with 2-several 'states'. And how to describe variable outbreeders?

Finally (6), utilization is the end-point, the very *raison d'être* of genetic conservation. Over the years, there should be a steady outward flow of stocks in response to requests for: particular lines (accession number 12345, known to be resistant to this or that pathogen); or for groups of lines having one or more desired characters in common ('dwarf, early-maturing types with large seeds'); or even for substantial blocks of material for developing a base-broadening programme or for biosystematic research (distribution of isozymes, perhaps). Those responsible for the collection will need to have some policy in regard to number of seeds or propagules in each sample distributed and the outflow will have to be smoothly compensated by the routine regeneration and maintenance processes. Finally, we note that, as a major collection approaches completeness or there is nothing left to collect, growth by exploring and collecting will cease and all new accessions will come from the breeders, incorporating often enough, no doubt, genetic material derived from the collection years before. This state of affairs is yet far off but, given vigorous collecting and the continued decay of variability, it must arrive some day.

Methods of maintenance

Methods of maintenance of collections are summarized in Figs. 10.3 and 10.4. Basically, there are three and they depend upon the biology of the crop. First (A in Fig. 10.3), there are the seed-propagated plants (roughly, all annual and biennial crops *plus* a few perennials) which must be maintained as seed lines (IBL or OPP according to mating system), assisted by more or less prolonged

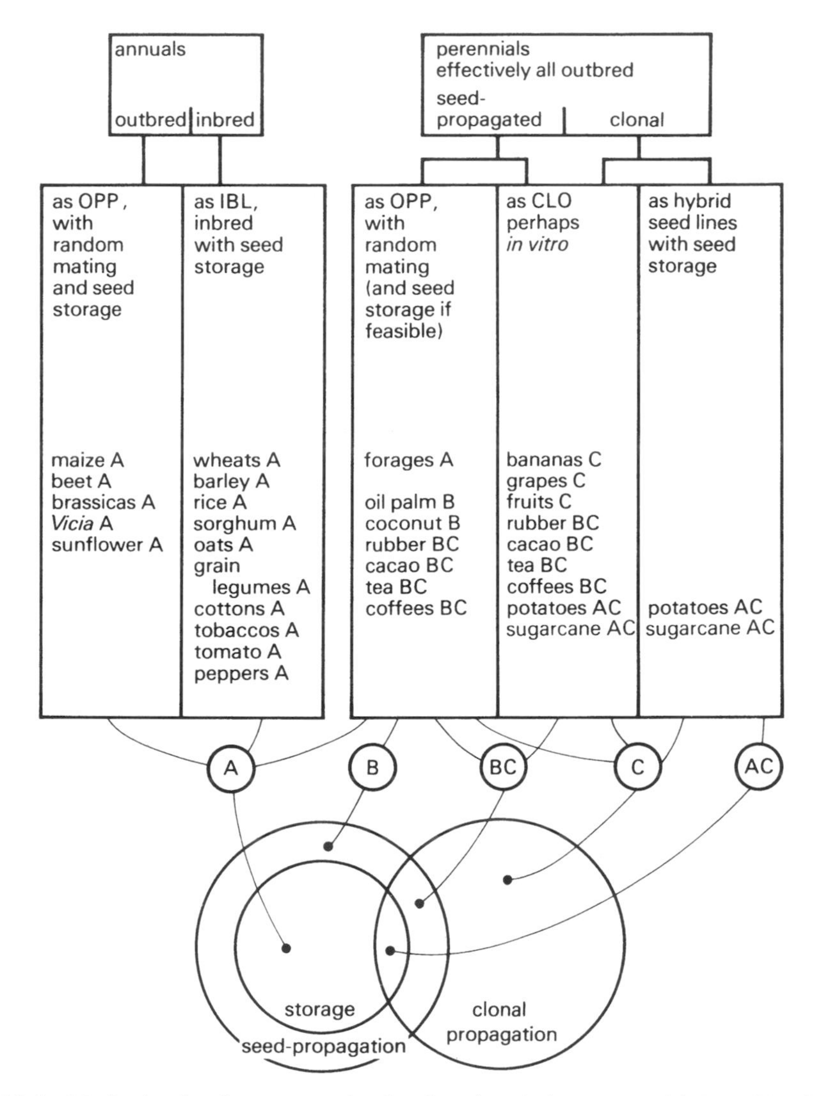

Fig. 10.4 Methods of maintenance of collections in relation to crop biology. For A, B, C see Fig. 10.3. The bottom part of the diagram shows that the three basic methods generate five groups when the possibility (or need) to maintain some crops by two methods is taken into account.

seed storage. Second (B), there are the seed-propagated perennials, the seeds of which cannot, on present knowledge, be effectively stored; they can only be maintained as long-lived plantations. And, third (C), there are the perennials which must either be maintained as clones or, at least, are most effectively treated thus. Examples are given in Fig. 10.4. Since many perennials can be handled in two ways, as clones and as seed, five logical categories emerge, AC and BC being added to the basic A, B and C. At one extreme, breadfruit, bananas, grapes and many other fruits could conceivably be maintained only as clones; at the other, coconuts and oil palms (indeed, nearly all cultivated palms) are not readily vegetatively propagable, have short-lived seeds and seem unlikely to yield to seed-storage technology; in between, as in potatoes, sugarcane and other tropical perennials, a judicious mixture of methods, by clones and/or seedlings, with or without seed storage, will be appropriate.

Potatoes deserve a special comment because there is long experience of maintaining collections and they represent the major practical example of category AC yet available (Fig. 10.4). Early attempts to maintain clones collected in the Andes failed; many came in already infected with viruses and subsequent spread in the glasshouse simply could not be checked. The only escape, for Andean materials wild and cultivated, is seed (though even this is not secure against potato spindle tuber virus, PSTV). However, there is nothing sacrosanct about Andean clones – they are probably short-lived in their native area anyway – so a judicious mixture of sib-mating and outcrossing to related, or presumptively related, materials was adopted in both the major collections, in the USA and UK; fertile Andigena tetraploids will stand a little inbreeding (e.g. occasional selfing) and several wild species are natural inbreeders and can be maintained as IBL. Thus, a judicious mixture of selfing, sibbing and outcrossing, according to breeding system and history of a stock, serves to maintain descendants from a collection. It is well-established that the seeds store well in cool, dry conditions. There are losses (cultivated forms can be very infertile) but they are far fewer than under clonal propagation. Named cultivars, which are not yet systematically conserved anywhere, I think, should start free of viruses and could be maintained in the appropriate isolation, supplemented by occasional meristem culture when necessary to eliminate viruses. In the longer run, maintenance by meristem culture (Section 8.6, see below) may assume greater importance, in which case the need for intermittent seed propagation would be greatly reduced, even eliminated. There may well be other crops in which a combination of clonal and stored-seed maintenance turns out to be appropriate (sugarcane certainly, cassava?).

In the course of time, new technology (see below) may suggest new methods for some crops; meanwhile, it will be prudent to assume that we are committed to a formidable total effort, using diverse methods of maintenance. Whatever the method adopted, the long-term integrity of a collection, which should, ideally, be permanent, is threatened in two ways: genetically random and genetically selective. The former includes mutation which is unavoidable and can be minimized in some crops by choice of conditions for seed storage (see below). It also includes genetic drift and random hybridization, especially in seed-propagated outbreeders; thus, in contrast to IBL which can be safely and easily maintained in pure form, OPP must pose considerable practical problems of adequate population size and security of pollination at each regeneration. Finally, it also includes purely random losses due to bad luck or carelessness. Genetically selective changes are potentially inherent in any growing of plants – and may well occur in seed storage, too. Some differential survival of clones, lines or of components of OPP in response to the environment provided for regeneration or growth is almost inevitable; diseases, which will occur whatever protection methods are practised, must also tend to be selective in effect. In short, absolute integrity in perpetuity can never be guaranteed. What can be done is to adopt the most conservative methods that are technologically feasible, replicate collections over diverse, well-chosen sites and exercise great practical care. Even so, crops will differ; categories B and C (Figs. 10.3 and 10.4) must be at greater risk than category A; IBL under long seed storage are the safest and easiest crops to handle, obligate clones the most risky.

We now turn to a short review of the technology of conservation, starting with seed storage and going on to some rather more speculative possibilities. Species vary widely in the aptitude of their seeds for storage in the dry condition. Some (e.g. many tropical perennials such as those listed in Fig. 10.4) can only be dried partially and briefly or not at all; they must in general germinate and grow as soon as the fruit that bears them is mature. Others (e.g. most, probably all, annuals, temperate and tropical) will withstand from a few months up to several years of ordinary air-dry storage, as, of course, a vast body of agricultural and horticultural experience goes to testify. It has been known for decades that viability in such seeds can be prolonged by dryness, low temperature and low oxygen tension but only quite recently have there been systematic studies to determine the possible practical limits of storage (review by Roberts in Frankel and Hawkes, 1975). With the proviso that very few species have been investigated, it begins to look as though very dry, very cold storage will preserve cereals seeds for many decades – maybe up to about 100 years – and other, less favourable, subjects (legumes, many grasses, vegetables) for several years, maybe a decade or two. These estimates are, of course, extrapolations. There seems to be some relationship between composition and longevity, starchy seeds being potentially longer-lived than oily ones. Whether there is truly a discontinuity between the tropical species referred to above, which must on present, imperfect knowledge grow at once or die, and the annuals, which will withstand at least some dry storage, is not known. The former have been described as 'recalcitrant' the latter as 'orthodox'. This curiously anthropocentric terminology clearly reflects the seed technologist's predilections rather than the biological facts; it should surely be dropped and replaced, if needed, by an appropriate terminology when the biology of the situation has become better understood.

At present therefore it looks as though some seeds can not be stored, some can be stored for a few years and some for many. For those that can be stored at all, Roberts suggests that hermetically sealed cans or laminated plastic-foil packets maintained at –18°C, with the seed predried to 5 ± 1 per cent moisture would generally offer a workable system. Oxygen tension can be ignored because it quickly tends to zero in sealed storage. If higher temperatures and/or moisture contents were adopted for technical or economic reasons, then the consequently shorter viability period would necessitate more frequent regeneration – with all the attendant costs and hazards.

Whatever conditions were chosen, periodical germination tests would be necessary and the aim would be to renew stocks when viability had fallen to about 90–95 per cent of initial. This seemingly rather high figure is justified by the observation that declining viability is closely accompanied by mutations, chromosome structural changes and other (undefined) aberrations; these are few at 90 per cent but very substantial at 50 per cent viability or less.

A workable technology for prolonged seed storage is thus emerging. Roberts *et al.* (1976) discuss practical details and give tentative costings. The technology, however, is unlikely to stand still. As our knowledge of seed biology (which is still remarkably weak and virtually confined to a few temperate species) improves, so too should the technology. Already there are several interesting possibilities in view, even within practical reach. For example, imbibed seed storage, modelled on the ancient observation that some seeds survive for years in damp soil, has only recently been recognized as a possibility (Villiers in Frankel and Hawkes, 1975); relatively high temperatures and good control of storage atmosphere are implied: could the usually short-lived tropical perennials be kept thus, perhaps? By contrast, pollen storage at extremely low temperatures (maybe approaching 0°K) seems to be at least a possibility for some species; genes and genetic variability could be stored thus, if not plant collections *per se*; one recalls in this context that semen storage is the only evident method of long-term genetic conservation available for domestic animals. Again, since at least some higher plant tissues can be deeply frozen without killing them, perhaps sterile meristems or regenerable tissue/cell cultures *in vitro* could be treated thus? A really reliable frozen-meristem technology could transform the problems of genetic conservation of clonal crops; frozen cell/tissue cultures for the purpose are clearly far off (see Section 9.2) and offer no evident attractions over meristems. Much closer to practical reality – and deserving of vigorous study – is the idea of maintaining growing clones as meristems *in vitro* rather than in the glasshouse or field (Section 9.2). Morel (in Frankel and Hawkes, 1975) found that grape meristems could be maintained indefinitely *in vitro* with one annual transfer. He calculated that 800 clones, replicated six times could thus be maintained in 2 m^2 of laboratory space, as against the hectare or thereabouts required in the field. If required, a five-fold monthly multiplication would be possible. The method seems to offer economy, security and good control of health; at the distribution phase it might ease the passage of material through quarantine, a major obstacle to the exploitation of foreign stocks in any clonal crop.

The preceding discussion is all based on the assumption that genetic conservation must be founded upon the maintenance of collections in perpetuity. It has been argued (e.g. by Simmonds, 1962a) that, in seed-propagated crops, widely based composite crosses would offer a cheaper alternative, or at least complementary, method. It is known that such populations, even of inbreeders, retain much variability, even though they show strong local adaptation (Sections 5.3, 5.9). Replication over sites would ensure the maintenance of some variability but the question is how much? Marshall and Brown (in Frankel and Hawkes, 1975) argue, that on recent evidence there is considerable genetic erosion associated with the improved adaptation that is generally observed. Very extensive replication would reduce the loss but would be expensive. On balance, it now look as as though composite crosses are not a substitute for collections; rather they are a cheap means of developing local breeding reserves, a convenient intermediate stage between the collection and formal exploitation in a breeding programme.

The state of the art

This is a time of rapid change in the genetic conservation scene, of a growing awareness and a growing sense of urgency. Throughout the world there must be, in total, hundreds of plant collections ranging in size from many small lots of local cultivars to a few gigantic accumulation of many crops. Information is very scattered (there is no single, coherent survey of the whole effort); correlation of the contents of different collections is, in general impossible; we do not yet know, I think, whether any major crop can be considered thoroughly collected (many certainly are not); and, as we have seen above, the technology of maintenance is developing fast enough perhaps to suggest new overall strategies. Any account must necessarily be rather vague but I believe that one important development is apparent: the tendency for the effort to become more international in character.

The earlier collections were all nationally developed and aimed at providing material for breeders and scientific study rather than at long-term genetic conservation. The great collections of the former Soviet Union All-Union Institute of Plant Industry (VIR, centred at St. Petersburg go back to the 1920s; they owe their origins to the vision of N.I. Vavilov, are still regularly being added to and recently contained about 200 000 accessions (Brezhnev, in Frankel and Hawkes, 1975). In the USA, the USDA has a history of systematic plant exploration and introduction going back to 1898 but it was not until relatively recently (1947–52) that the effect was concentrated in four Regional Plant Introduction Stations and the Inter-Regional Potato Introduction Station; more recently still (1958), responsibility for long-term conservation was placed on the US National Seed Storage Laboratory, Fort Collins, with about 83 000 accessions (Hyland, in Frankel and Hawkes, 1975). These are the historic collections; they both cover many crops and both have adopted long-term conservation as a leading objective. In addition, there have long been a multitude of small to fairly large collections of a great variety of crops alongside plant breeding programmes in many countries. Some survive pretty well but many have been lost or at least severely eroded in the course of time. The lesson seems clear, that long-term security, especially if it involves expensive storage facilities, is best promoted by large centralized laboratories with assured resources. National seed storage centres have been set up in Japan, Germany and the Nordic countries. The UK is disgracefully backward.

National seed-storage centres in temperate countries, however, only touch half the problem: what of clonal and tropical crops? Even for temperate, seed-propagated crops there is an evident risk that several major collections could jointly achieve less than satisfactory coverage if activities were uncoordinated. The problems of genetic conservation are truly international in character and are increasingly seen to be so. The trend to international action goes back to an FAO initiative of 1963, perhaps the point from which recent awareness primarily springs. Under a joint stimulus from FAO, the World Bank and the United Nations Development Programme (UNDP), the Consultative Group for International Agricultural Research (CGIAR) was formed in 1971. CGIAR, supported, with substantial funds, by the sponsors, governments, development banks, foundations and others, promoted the development of a group of agricultural research institutes as its primary task but also took on the job of genetic conservation; to do this, it founded, in 1974, the International Board for Plant Genetic Resources, with its own funds and a general remit to promote, in any way feasible, the conservation of genetic resources for agriculture and forestry (IBPGR, 1975). At present it supports: three regional programmes (Near East, Mediterranean, Southeast Asia); several crop-specific local programmes on a selective *ad hoc* basis; collecting activities, especially where there is a risk of genetic erosion; the development of information systems (Genetic Resources Communication, Information and Documentation System, GR/CIDS, at Boulder, Colorado, USA – see Rogers in Frankel and Hawkes, 1975); a variety of 'pump-priming' operations and some educational activities. Recognizing that resources are limited in relation to the task, it has started to define priorities for work by regions and by crops (IBPGR, 1976). It promotes international cooperation and, indeed, makes free exchange of material a condition of support. In general it proceeds by what might be described as enlightened empiricism. At all events, genetic conservation must, of its nature, become increasingly international in character and if the IBPGR did not already exist it would have to be invented. Administrative structure was recently amended and the name changed to International Plant Genetic Resources Institute (IPGRI).

10.4 Use of collections: base broadening

Several references have been made in the preceding text to the need to sustain genetic variance in the long term if progress is to be maintained. For this, intelligent use of collections is essential. Relevant points emerge in Fig. 2.1 and at pp. 322–3. However, concrete examples are wanting so some are briefly presented here in the natural context, namely of genetic resource conservation, the essential subject of the chapter.

The two uses of a collection

In the 1970s and 1980s there was widespread confusion abut the function of curators of collections, who might themselves both characterize (i.e., identify) and evaluate (i.e., determine the utility of) items in their care. It took a long time for the genetic resources community to realize that this idea was unworkable. In the longer run the point was taken that evaluation can only be for plant breeders working locally, because evaluation is beset by genotype × environment problems, which cannot possibly be resolved, even understood, at a single (usually distant) place. It is now universally agreed that collections are best accumulated, cared for, maintained and characterized at chosen central sites but that utility/usefulness can only be determined by plant breeders working in distant places, in the sites to which adaptation is desired.

The foregoing does not mean, however, that collections went completely unused in earlier times. From the 1920s there were many examples of what we may call *introgression* (borrowing a genecological word); that is, items from collections were chosen as having some desirable character (nearly always a disease resistance) and crossed and backcrossed to breeding stocks in the hope of transferring the desired character. This sometimes worked but sometimes did not. All too often, the desired character was either polygenic or was a vertical resistance (i.e. VR, Chapter 7), due to a major gene, readily nullified by evolution of a new pathogenicity in the pathogen. The wheat rusts, potato late blight, and leaf diseases of rice, barley and grain legumes provide scores of examples. Generally, introgression has been widely practised but has only very rarely been useful. Thus, at least 11 R-genes have been backcrossed from *Solanum demissum* into the cultivated potatoes, to no useful effect. To the plant breeder they are merely nuisances.

Recognition of the need to go beyond introgression to widen the genetic bases of diverse crops, either narrow at the start or seriously narrowed by subsequent selection, came much later and even now the concept has not been fully assimilated. The first plain statement was that of Simmonds (1962) which coincided with the initiation of the Andigena-Neotuberosum potato project described later. The principle was simply to recognise that the large scale loss of genetic variability due to response to selection had to be repaired somehow and that the best and most economical approach was to adopt a very widely founded base-broadening attack. This principle is now slowly becoming accepted. The approach aims at wide incorporation of genetic variability which, of its very nature, cannot be preselected. It generally turns out that the earlier evolutionary history of the crop must be reconstructed if good new parents are to be synthesized.

In effect then we recognize two basic ways of using a crop collection, namely: (1) *introgression*, adopting a pairwise crossing and backcrossing technique to introduce specific genes and (presumptively useful) characters; and (2) *incorporation*, which attempts a wide-scale incorporation of genetic variability, the value of which cannot be known until the incorporation has been nearly achieved. In contrast to the preceding, *incorporation* must go forward on a wide genetic base, must in a sense repeat crop evolution and must usually be regarded as a continuing process (Fig. 10.5).

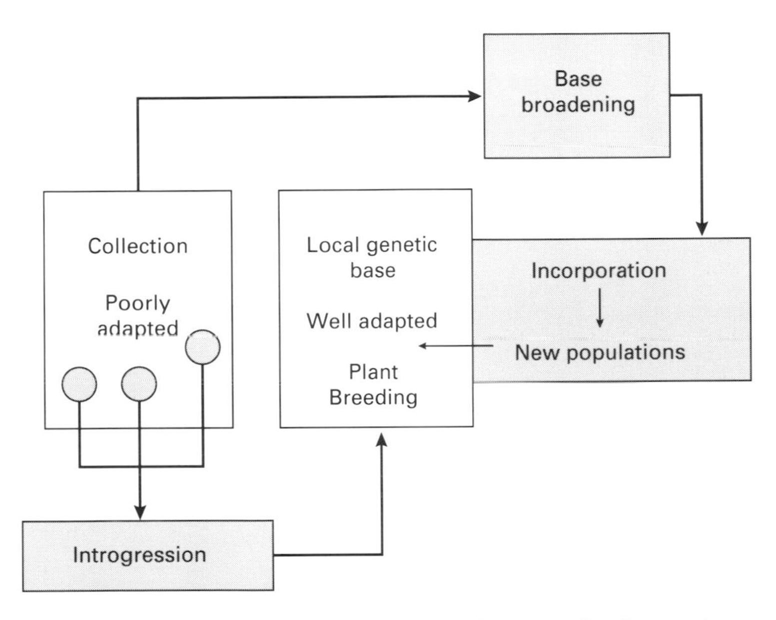

Fig. 10.5 The general pattern of introgression and incorporation in crop improvement. From Simmonds (1993) by kind permission of the Editor of *Biological Reviews of the Cambridge Philosophical Society*.

Functions and genetic principles

Introgression will have its uses but it will commonly continue to fail in future, as it has failed in the past. Either, the desired character is not simply monogenic (and therefore not readily backcrossable) or it is a VR susceptible to the emergence of a new pathotype. It may be that major-gene virus resistances will prove to be more useful and 'durable' than others because the agents may be less liable to generate new pathotypes. But this is quite speculative and far from certain as a generalisation. By contrast, *incorporation* has indefinitely wide applications and has yet been but little explored. The need sometimes arises from a narrow base *ab initio*, sometimes from a base narrowed by prolonged selection. A good deal is known about its application to potatoes, sugarcane and maize (Simmonds, 1993). Its use for barleys is plainly foreshadowed by the great Californian Composite Cross experiments; and its application to several tropical tree crops has been outlined in, for example, rubber, cacao and oil palm. But the great array of other crops, both temperate and tropical that would benefit from *incorporation* remain untouched, both experimentally and even conceptually. There has been very little awareness and thinking; the small grain cereals such as wheats, rices and millets and the inbred grain legumes are conspicuous examples of neglect. There have, however, been some signs of interest in soybeans.

Genetic principles of incorporation

The genetic principles underlying *introgression* are simple and obvious. Those underlying *incorporation* are not. The latter may be listed as follows:

(1) Large scale of operation is essential if a wide range of variability is to be caught in the net; a narrow start simply ensures an even narrower finish.

(2) Non-adaptation of the introduced materials will usually have to be assumed, even if the assumption is sometimes mistaken. The 'primitive' potatoes and sugarcanes looked useless but many 'primitive'-looking cereals and tree crops have attractions. The cereal pot trials described by Spoor and Simmonds (1993) provided some surprises, examples of adaptation of tropical land races to Scotland.

(3) The process must be complementary to conventional breeding; test-crosses may sometimes be undertaken but any general crossing with adapted materials is not only inappropriate but also foolish because it must corrupt the process of adaptation itself and tend to corrupt, even destroy, the base-broadening effort.

(4) The methods and breeding patterns to be adopted will depend on the biology of the crop, its breeding system and reproductive behaviour. The general rule must be 'quick and dirty' but always within the bounds of security and reliability. The crops already at least partly understood, all demanded different methods (Simmonds, 1993). An attractive possibility if it could be proved to be useful, would be to exploit pot tests as an initial stage (e.g. barley – Spoor and Simmonds, 1993).

(5) It is prudent to assume little inherent adaptation in the material and therefore to use large populations, to maximize recombination (by cyclical crossing if necessary) and to maintain only weak selection. Rigorous selection can only narrow the base unnecessarily, but usually without concomitant gain.

(6) Utilization of such broadly based populations will require considerable development of local adaptation before true potential can be recognized. To minimize undue narrowing of the base and further genetic loss, replication over several diverse sites is highly desirable.

(7) An important feature of local genetic adaptation is horizontal resistance (HR) to disease. Most important crops have 5–10 (or more) more or less serious diseases and new ones regularly appear (as in sugarcane). Usually, it will be sufficient to grow populations in disease 'hot-spots' and let natural selection do the work; sometimes, however (as in potatoes and sugarcane), some intervention by the plant breeder is needed. The profound importance of HR (generally, but especially for the small farmer) has been strongly emphasized by FAO (1986, 1988) and genetical features have been reviewed by Simmonds (1991).

(8) Progress may well be remarkably rapid on evolutionary time-scales (it was in potatoes and sugarcanes) but seemingly rather slow on human/plant breeding time-scales. This is unavoidable but it implies a long-term commitment and assured continuity in relevant programmes. But, even in years, time scales are not always long and both potatoes and canes showed major advances in a decade, a mere moment in crop evolutionary time. Strictly annual crops could well be quicker.

(9) The outcome of an effective base-broadening programme will be enhanced genetic variance in economic characters and either good materials *per se* or good parents for crossing into established programmes; both outcomes are known and both are to be expected.

A closing comment is justified because it is often a source of difficulty. The term 'pre-breeding' is used in various senses but should usually be avoided. It may refer to the development of potential parents either from adapted stocks or from what might be regarded as a base-broadening programme. Only confusion follows from loose usage.

An example from potatoes

Frequent reference has been made above to base-broadening of potatoes, perhaps the leading example of *incorporation* because it was the earliest and arguably the most successful.

The work was summarized, with references, in Simmonds (1993), and see also Smartt and Simmonds (1995, Chapter 93).

The potatoes are ancient Andean cultivars which started as diploids (2x = 24) from diploid ancestors; they are all outbred and self-incompatible (SI). In the Andes, at great altitudes, autotetraploid derivatives (4x = 48), the Andigena (ADG) Group, came to dominate and still do. They are adapted to short tropical days but to cool weather. A few ADG cvs were taken to Europe in the

16th century and descendants adapted to temperate climates though they tended to be 'late' (because of short-day adaptation) and were very susceptible to diseases. In Europe, and also in North America, local adaptation emerged in the Tuberosum (TBR) Group, which is effectively Andigena with minor morphological differences and adaptation to long summer days. The importance of day-length adaptation cannot be over-emphasized. Very few ADG were brought to the north and the crop made good genetic progress on a very narrow genetic base up to the middle 19th century when it was catastrophically attacked by the newly introduced late-blight fungus, *Phytophthora infestans*, introduced from Mexico to Europe and North America in the 1840s. By then, TBR was a major food crop. Blight still further narrowed the base, though the crop responded well to selection. Pedigree information showed that millions of hectares of a few clones all went back to half a dozen or so introductions. Attempts to use ADG for direct crossing to TBR were useless; products were sometimes vigorous but had many bad features, including disease susceptibility.

The only solution, and the method in fact adopted, was to recreate the TBR Group from ADG stocks, of which a large sample was available in the Commonwealth Potato Collection at the John Innes Institute, Hertford, England. An initial sample (1959) of about 3300 seedlings was grown, selected roughly and the selections carried on vegetatively in the next year. Thereafter seedling and tuber propagation alternated; seed was collected from isolated tuber plots in the hope that it was mostly crossed, even though the ADG potatoes are fairly self-fertile. This was a mistake, later remedied by making crosses, when it was found that selfing predominated. Little conscious selection was practised, though plots were deliberately grown every year in a blight 'hot spot'. Responses were very largely to natural selection for survival, growth and yield (i.e. tolerance of long days), with minimal interference by the breeder.

Day-length adaptation and disease resistance improved very rapidly and in three two-year generations (i.e., before 1970), the better materials rivalled TBR in performance, had a great range of tuber types, cooking qualities and resistances in them and showed some truly spectacular heterosis in crosses with TBR. So good were they that the name Neotuberosum (NTB) was invented for them. In the 1970s about 60 000 plants were grown, so the demands, though non-trivial were not great.

The work was carried on in Scotland in later years and many crosses made into a conventional breeding programme. Stocks were widely distributed round the world. A few current commercial varieties in the UK, Europe and North America are half NTB in constitution and many more are in prospect. The material is, of course, freely available to bona fide potato breeders and, if wisely used, can hardly fail to revolutionize potato breeding, temperate and tropical. It may be that the greatest potential of all will be in the tropics, for Third World agriculture. Diploid selections can readily be introduced into the programme if required (Simmonds, 1997).

10.5 Summary

New crops come to agriculture by fresh domestication from the wild, by adaptation of stocks from other places and by the construction of new cultivated allopolyploids. Fresh domestication has occurred in the recent past (rubber), continues now (African and American forage plants being developed as crops in Australia) and may be expected to go on indefinitely (with pharmaceutical plants and timber trees among the most obvious possibilities). Geographical spread coupled with new adaptation is, of course, as old as agriculture; it continues vigorously today. New allopolyploids are yet few but the possibilities can not have been exhausted. In short, there is no reason to think that the spectrum of crops grown in a place is static or need become so.

Genetic conservation is the process of preserving the genetic variability of our crops for the future use of plant breeding. The need stems from the recent general decay of variability consequent upon the spread of technology-based agricultural methods to erstwhile 'primitive' areas. The need, now generally agreed to be urgent, has only become widely apparent and accepted in the past thirty five to forty years. FAO has done most to promote general consciousness of the matter; collaborative international action has been essential and is supported by the International Board for Plant Genetic Resources (IBPGR) and its successor IPGRI (International Plant Genetic Resources Institute).

The strategy of genetic conservation starts from the view that, in the not-so-long run, living materials representative of all crops and their relatives will have to be maintained in substantial collections in perpetuity. A need for extensive collecting, classification and evaluation activities is implied. Once assembled, collections must be protected from genetic erosion and other losses, due to diseases and accidents; replication in diverse sites is implied. Methods of maintenance of collections depend upon the biology of the crop concerned. The three basic methods are: as seed-propagated lines, with or without seed storage and as clones (but some crops can be handled jointly by two of these methods). Methods for the long-term storage of some seeds in very dry, very cold conditions are available and will be valuable for some species; others can not yet be stored effectively and more research is needed. There are several prospective new technologies which should assist maintenance when fully developed, notably: imbibed seed storage, frozen pollen storage, and *in vitro* meristem maintenance (for clones).

Viewed broadly, the task is a dauntingly large one but we are at least more conscious of the need and the urgency now than we were even a few years ago and new technology may ease some of the more formidable practical problems.

The use of collections is summarized as *introgression* of single genes (often disease resistances) and *incorporation* of genetic variance on a larger scale. The former has been widely practised, but to little useful effect; the latter is now (belatedly) becoming understood and an example from potatoes is presented.

10.6 Literature

New crops

Bemis *et al.* (1978); Gladstones (1970, 1975); Hernandez Bermejo and Léon (1994); Hutton (1970, 1976); Janick (1996); Janick and Simon (1990, 1993); Markley (1971); National Research Council (1989); Purseglove (1957); Shaw and Bryan (1976); Smartt and Haq (1997); Stern and Roche (1974); White *et al.* (1971); Williamson (1975).

Genetic conservation

Brown *et al.* (1989); Burley and Styles (1976); Burton (1976); Committee on Genetic Vulnerability (1972); Creech and Reitz (1971); FAO (1967); Frankel and Bennett (1970); Frankel and Hawkes (1975); Grubben (1977); Harlan (1956, 1972, 1976); Hawkes and Lange (1973); Hawkes *et al.* (1976); Heslop-Harrison (1974); Hodgson (1961); Holcomb *et al.* (1977); Hoyt (1988); IBPGR (1975, 1976); James (1967); Matsuo (1975a); Parmand Hallauer (1997); Roberts *et al.* (1976); Simmonds (1993); Stern and Roche (1974); Wilkes and Wilkes (1972); Zhukovsky (1975).

Use of collections

FAO (1986, 1988); Simmonds (1962, 1991, 1993); Smartt and Simmonds (1995).

Chapter 11
The Social Context

11.1 Introduction

Plant breeding is an applied science that is devoted to changing nature rather than to understanding her – though understanding is never amiss and is sometimes essential. Over the world as a whole, we spend an unknown, but certainly quite large, amount of money on plant breeding. That money represents resources which might have been devoted to other purposes which might imply a consensus that it is well spent, that the products are, in some sense, economically advantageous. But advantageous to what extent, how and to whom? In trying to answer these questions we shall see that the general route by which economic 'benefit' is diffused through society is broadly discernible but that the counting of it is difficult; we can however be fairly confident that the advantage is often substantial, even if economists do yet disagree on how it should be measured. These ideas lead on to the notions of cost–benefit analysis (including discounted cash flow) and the identification of specifically economic criteria in breeding programmes. First, however, we shall try to place plant breeding economics in context by reminding ourselves that breeding is but one phase of agriculture research, the products of which are sometimes, perhaps often, interactive in effect. And we shall return to this theme at the end of the chapter in outlining the main points about the 'green revolution'; here, the technological issues are fairly clear but the social consequences far from simple.

The social context of animal breeding is reviewed by Lerner and Donald (1966). The animal is better explored than the plant field but both would benefit from further study.

11.2 Agricultural research and development

Research and development (R & D) in any industry is carried out in the expectation that the results, when applied in practice, will lead to a cheaper product, an improved product, a new product or some economically favourable combination of these changes. Thus the objective of applied R & D is advantage for someone, sometimes but not necessarily the innovator himself. Fundamental research is sometimes held to be justified along the same lines; however, the connection with applied outcomes is often remote and tenuous and it is probably better thought of as a substratum for applied research and in a cultural/educational context rather than as an economic activity *per se.* In this chapter R & D means applied R & D directed to (more or less) defined economic objectives. No agricultural research is ever pure.

Agricultural R & D as a whole aims to enhance the economic effectiveness of agriculture, either by improving the environments in which crops and stock are grown or by improving the genotypes or, more generally by improving both jointly. Thus husbandry research is the E component and breeding the G component; there may also be a G × E effect whereby new varieties are specifically adapted to changed husbandry. GE effects may be numerous but may often go unrecognized; any secular trend in the husbandry environment must tend to evoke them and, contrariwise, new varieties may alter husbandry. Dwarf cereals adapted to high fertility (Sections 3.5 and 10.5; Hutchinson, 1971b) are the most striking example and no doubt the adaptation of chickens to intensive indoor husbandry had a GE component too.

The technology-based crop agricultures, temperate and tropical alike, have mostly shown strong and sustained yield advances over the past thirty to forty years. Trends are always confounded with external influences such as weather, disease and changeable economic effects; they can seldom be confidently described by straight or even simply curved lines. In principle, if we had good enough knowledge, it should be possible to distinguish G, E and GE components of a trend. In practice knowledge is never adequate to do so; we can only observe that, in situations in which new husbandry and new varieties have as a matter of historical fact been adopted and yields have risen, then all three components are, presumptively, implicated (Fig. 11.1). And we might add that any finer analysis which sought to identify components of husbandry advances (say fertilizers, machinery, weed control, plant protection, etc.) should also, in principle at least, recognize interactions among these components too.

Although we have become accustomed in recent years to the idea of erratically rising yield trends as evidence of the effective exploitation of agricultural R & D, we should note that there is no necessary connection between the two. Yields can rise because weather is good or prices make generous husbandry worthwhile and they can fall for opposite reasons; well-exploited R & D may then merely improve a position that could have been worse (Fig. 11.1). In short, speculation about the causal factors underlying yield trends is fascinating but usually unproductive. We very rarely have good enough information to make a confident analysis, even though we can be sure that G, E (and certainly GE components too) will nearly always be present. So no simple shares can be allocated to G and E, even in principle, if a G × E component exists. That new varieties provide a substantial part of the benefits of agricultural R & D is however not in doubt and, as we shall see below, their effects are sometimes open to economic analysis, even if the long-term G component of yield trends is uncertain.

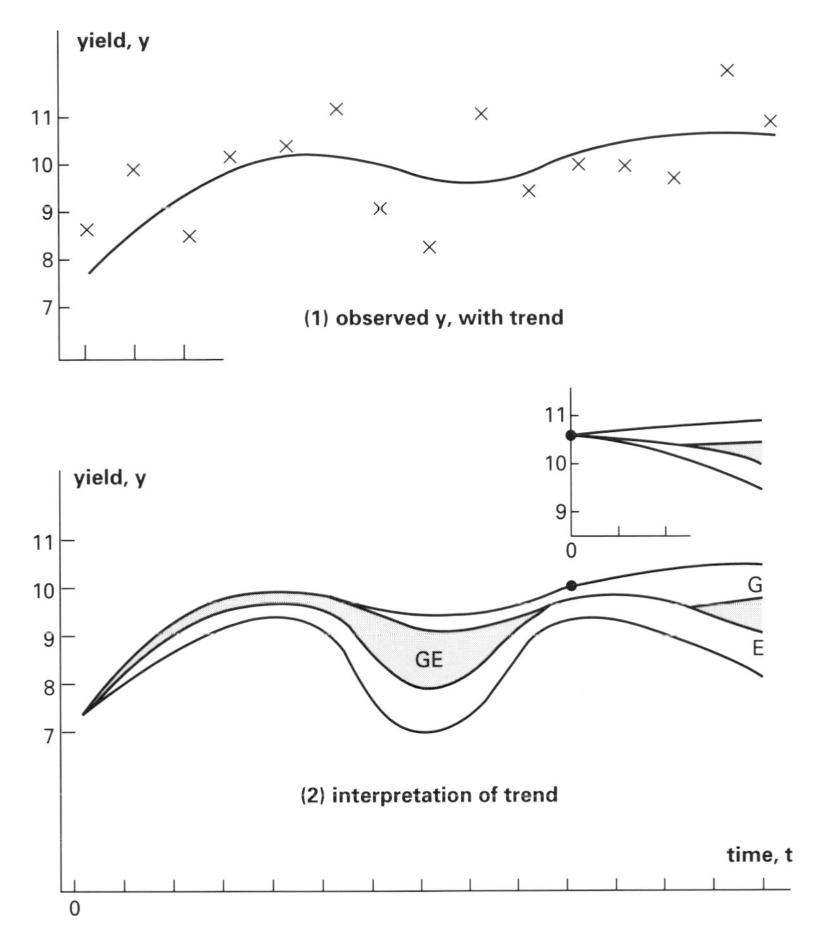

Fig. 11.1 G, E and GE components of yield trends.
Notes: Consider a 'true' underlying yield trend shown by the curved line in the top part of the figure. The mean is 10.0 and the 'observations' (crosses) were constructed by assuming a random deviate (sd = 1) about the trend; the mean deviation is near zero. An upward movement is clear enough but in practice such data would be described by a straight line. The lower part of the diagram 'interprets' the trend in terms of new technology, with varying G, E and GE components, applied from time 0. In the absence of new technology, the trend would presumably have followed the base line. The effect of assuming a new base and a new t = 0 is shown at the right centre of the diagram.

11.3 Profits, costs and benefits

The innovating farmer

We start from the familiar observation that the supply of a product, the consumers' demand for it and the price are interrelated: greater supply or reduced demand and the price falls, smaller supply or increased demand and it rises; conversely, supply and demand are responsive to price. Arithmetically, price-supply and price-demand relations are described by what the economists call 'elasticities'; numerically these vary widely between products and societies and even between sections of one society. In a free, competitive market there will be a tendency to equilibrium, producers trying to produce just as much of each product as they can sell at acceptable prices, consumers allocating their money between products according to their personal desires and prevailing prices. Profit, in a competitive market at equilibrium, is adjusted in the same sort of way as prices; it is the producer's margin between costs and receipts and it is set, by the collective judgement of the market, at a level which gives him a 'fair' return on the work, skill and money he invests in the enterprise. Producers, however, are unequal; the lazy, incompetent or unlucky ones fail, the rest prosper in varying degree. In short, the market tends to keep prices as low as the circumstances of production will permit.

Equilibria, however, are never long maintained. Supply (and therefore prices) fluctuate more or less randomly according to seasonal weather or other extraneous features; and fashions among consumers alter demand. Again, any departure from free competition – as by a monopoly or cartel or by government or other social intervention in the market – must tend to alter the equilibrium. Such disturbances are all but universal. Superimposed upon them, there are always, too, secular trends generated, on the one hand by changes in society at large (e.g. increasing or decreasing general prosperity) and on the other by improved (i.e. cheaper) methods of production; the first will alter patterns of demand, and hence supplies of products, while the second will reduce prices. The leading source of such price reduction is new technology.

Consider Fig. 11.2. A farmer, A, is producing a crop for a market. His receipts, R, exceed his costs by an acceptable margin, P, and he stays in business. His fellow farmers occupy different positions on the diagram (Fig. 11.2, inset). Now suppose that a new variety becomes available. Farmer A, an innovator, is alert to the possibilities and grows it; for a year or two his yields, receipts and profits are all increased. Thereafter, his neighbours, less inclined to innovation than he, look over the hedge, take the point and adopt the new technology themselves. Supply increases, prices fall, the market mechanism adjusts supply or price or both and a new equilibrium is approached. At this point, all the farmers, A included, once more make a 'normal' profit (the 'fair return'). Only farmer A and his innovatory colleagues made an extra profit (transiently at that) out of the new variety, so wherein lies the economic benefit?

Before trying to answer this question we must note that, as explained in the notes to Fig. 11.2, there are many possible routes to economic advantage. They are, variously, dependent upon or independent of other technology, may or may not incur altered costs of production per hectare but all have in common the feature that they reduce costs per unit of supply. Clearly, then, the net advantage of the new variety emerges as lower prices and it is consumers as a whole who reap the economic benefit. If this seems unfair to farmers (and breeders) one should reflect that they themselves are consumers of many different goods and

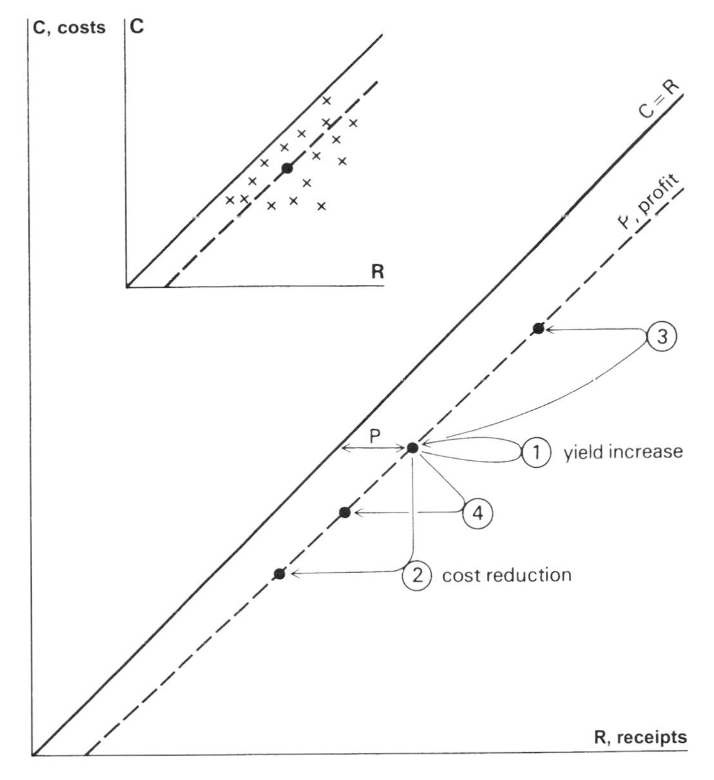

Fig. 11.2 New varieties and the innovating farmer.
Notes: C and R are in money per unit area. The inset shows a population of farmers, all more or less successfully producing the crop in question; all lie to the RHS of the C = R line. A few will be innovators, the majority not. The consequences of new varieties for an innovator are shown in the main diagram: (1) a new variety which increases yield under standard husbandry will increase R without affecting C; (2) the new variety may effect a simple cost reduction if it yields the same as the old one but is more cheaply harvested, does not need to be sprayed against a disease etc; (3) the new variety needs extra husbandry costs (e.g. more fertilizer) to achieve its full potential (i.e. it shows a GE effect in terms of Fig. 11.1); (4) is intermediate between (1) and (2) (affects both R and C); many more intermediate situations could easily be constructed. In all cases, the transient extra profits earned by the innovator are returned to a 'fair return' level by a shift of the scale of R and a rightward movement of the C = R line as the innovation spreads. Farmers who fail to adopt the new technology are likely to find themselves on the wrong side of the shifted C = R line and so be forced out of the enterprise.

that the consumers of their products are producers of other commodities subject to the same sorts of rules. The market tends to ensure, though imperfectly, that the benefits of all new technology are widely diffused through society as a whole.

The idea of benefit

We have used the word 'benefit' several times in preceding paragraphs as meaning the 'social profit' of an innovation. That benefits do exist and that plant breeding is one way of generating them is not in doubt. But we meet extreme, so far insuperable, economic difficulties in arriving at generally agreed means of estimating them. The difficulties are of two kinds, social and technical. The former arise from considering *who* benefits. At one extreme some economists would say that it doesn't matter, that it is enough to estimate a total cash benefit and assume that the market and other social mechanisms will distribute it equitably throughout society. At the other extreme, other economists would argue that the mechanisms are inadequate, that benefits are never equitably diffused and that these 'welfare' or 'distributional' effects are all-important. Proponents of the latter view then have to face the fact that neither information about nor the technical means of handling distributional problems is adequate.

Even if distribution is, for a start at least, ignored, the problem of estimating benefits is great. Essentially, there are two approaches. One appeals to the 'economic surplus' concept and the apparatus of classical economic theory. Very roughly speaking, it estimates the extra cash accruing to consumers from reduced prices. The other is to start instead at the producer's end and consider the resources saved by the use of the new technology; thus a new high-yielding variety saves land, fertilizers, machinery, labour etc., if produced at constant crop on a smaller area or it saves cash otherwise needed for the purchase of extra product. This method, in effect, is based upon a sort of resource balance-sheet. Both approaches suffer from the deficiencies of real-life economic data and both offer formidable difficulties of interpretation. An indication of the kind of approach used in consumer-surplus calculations is given in Fig. 11.3. Details lie beyond our scope: the interested reader should consult Wise (1975, 1978) and Lindner and Jarrett (1977) for discussion in an agricultural research context (and see also general references at the end of this chapter).

Private and public breeding

The plant breeding which supports technology-based agricultures is part-private, part-public, in mixtures which depend largely upon historical factors. In continental western Europe, long established plant-varieties rights schemes (PVR, Section 6.3) have swung the balance far towards private breeding, with state organizations undertaking supporting and background research and generally discouraged, by the private sector, from undertaking breeding *per se.* In Britain, with a relatively recent PVR scheme, the balance was more nearly equal swinging towards the commercial. In the USA the long-established division has been between those crops (notably hybrid maize) which offer inbuilt economic protection and are privately bred (see Section 5.5) and the rest (the IBL, OPP, SYN and CLO) which are traditionally bred by state or federal agencies. The object of PVR legislation is to encourage private breeding in the belief that a vigorous and productive domestic plant breeding system is economically favourable to agriculture and to society. Presumably, private breeding will grow as PVR schemes spread and develop. The tendency, already

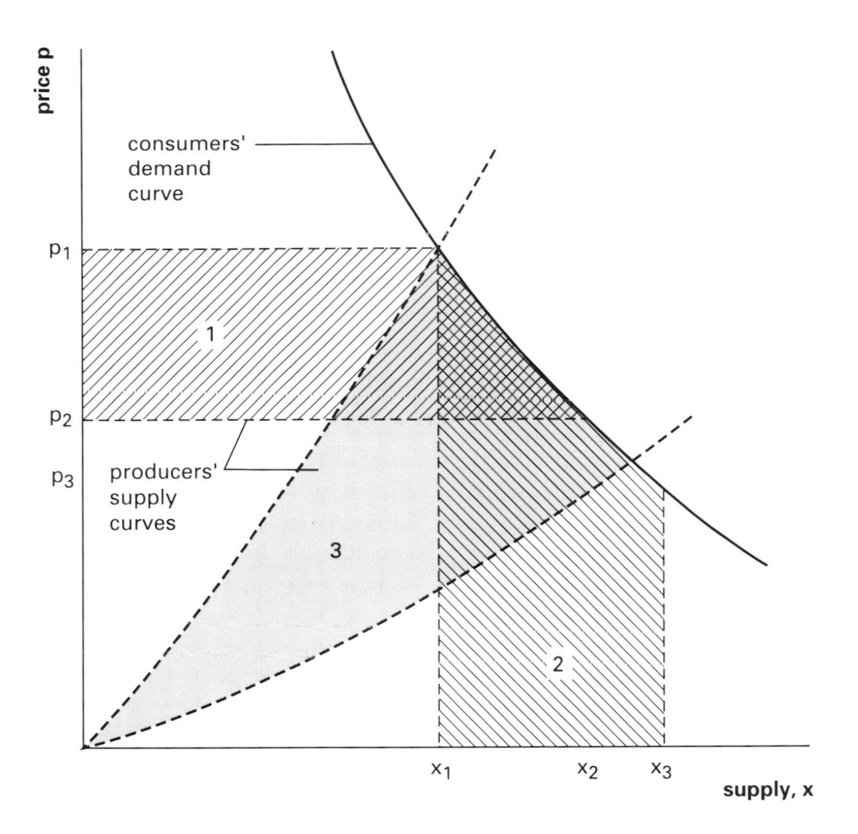

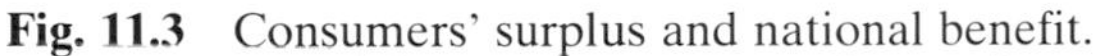

Fig. 11.3 Consumers' surplus and national benefit.
Notes: As price (p) falls in response to innovation and cheaper unit production, supply (x) increases towards a new equilibrium. Any area under the demand curve is a product of supply × price and therefore represents a sum of money. The shaded areas lying between pairs of supply curves are net social benefits. Three cases are distinguished. Under (1), production is 'elastic', producers neither gain nor lose and the social benefit is equal to the consumers' surplus; under (2), production is inelastic, over-production forces prices far down and producers may suffer a disbenefit (to the advantage of consumers). Case (3) is one of many possible intermediates. The reader is warned of very considerable economic complexities in constructing and interpreting diagrams of this kind; Wise (1978) suggests a practical simplification ('technical assessment') that relates benefit simply to the rectangle bounded by the lines p_1, p_2 and x_1.

plainly apparent, will generally be for the private sector to press for the reduction of state-supported breeding on the grounds that it offers 'unfair' competition; the state institutes, it is argued, should confine themselves to research that supports commercial breeding and, perhaps, to breeding a few crops that offer insufficient promise of profit to interest the private breeders. Thus private and public activities would be complementary. This is, superficially, an attractive argument but it leaves out of account the possibility that commercial pressures, especially if competition were less than perfect, could distort objectives. Thus, unnecessary HYB breeding could be encouraged if

HYB varieties were fashionable and seed were expensive; or disease resistance breeding could be discouraged if resistant varieties were likely (as with potatoes) to be 'grown on' by farmers and thus need lesser supplies of fresh seed. So there is a rather strong argument for maintaining, in the public interest, and in effective competition with the private sector, state-supported programmes that are free to work towards different objectives.

Given a PVR scheme, new varieties earn royalties in rough proportion to their success. Royalties, *plus* any unusually high returns on early seed sales while a new variety still has novelty value, constitute the economic *raison d'être* of private breeding (Fig. 11.4). New varieties bred by state-supported institutes also earn royalties which, in Britain at least, have been returned to the public purse and are not accounted for in the context of institute expenditure. All successful varieties, wherever bred, may be presumed to offer some economic

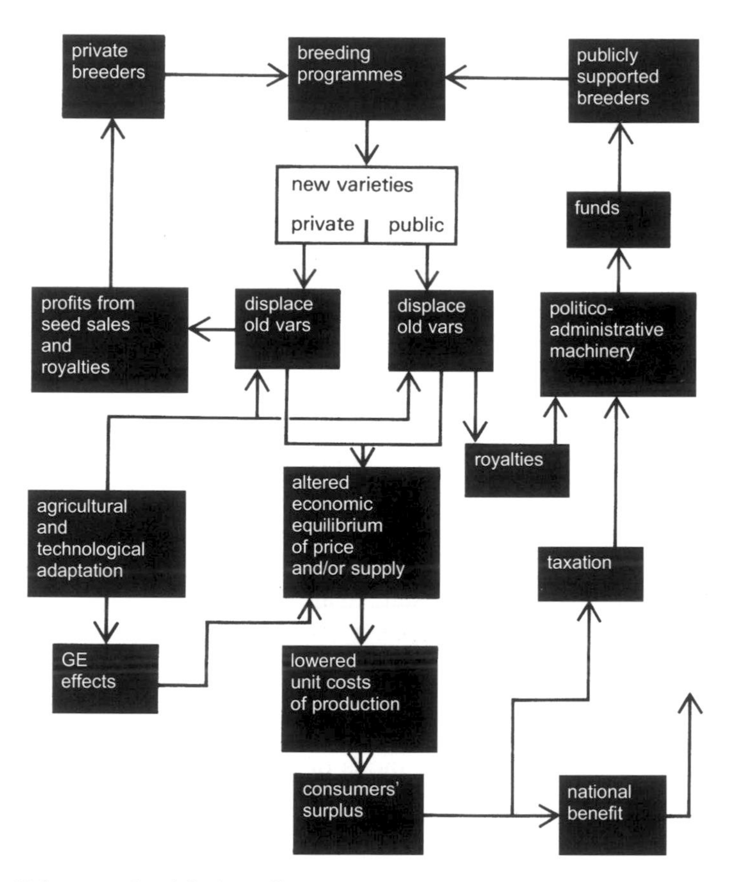

Fig. 11.4 Private and public breeding.
Notes: Both private and public breeding, when successful, generate social benefits (bottom part of diagram, compare Figs 11.2 and 11.3).Both also generate royalties which are the primary objective of the private breeder; the state sector has national benefit as its primary objective and royalties are but indirect evidence of success.

attractions to farmers and so to generate social benefits by the mechanisms that were discussed above. Note, however, that it need be no part of the intention of the private breeder to generate social benefits; he is in business to make a profit and, if he does public good, he does it by doing well for himself. The state-supported institutes are, however, in the social benefit business, supporting the conclusion we reached above, that they could well adopt objectives which were socially favourable though unattractive to the private sector.

Discounted cash flow (DCF)

Nearly all economic activity involves spending money now in the expectation of returns later. Money has a time value, measured by the 'discount rate', r (Table 11.1) which is determined by market forces for the commercial investor but may be a notional figure in the context of government spending (where the choice may be both difficult and critical). However the rate is chosen, money is taken to be 'discounted; or 'accumulated' at compound interest annually. Discounting applies to future costs or receipts and the discounted value is always less than the present cash value; accumulation, by contrast, applies to past items and the present value is greater than the nominal. The word 'discounting' is often taken to cover both processes. By this means uneven flows of cash over time can be brought to 'present values', as shown in Table 11.1. The sum of two flows (costs and receipts, expenditure and revenue, etc.) is a 'net present value' (NPV). When the NPV is zero the project breaks even and would be regarded as economically acceptable; obviously, positive values are favourable, negative ones unfavourable. There is usually (but not always if the flows are erratic) a value of r at which the NPV = 0; this is the 'internal rate of return' which is close to 10 per cent in the example of Table 11.1. An analyst considering this project would be inclined to view it favourably if the chosen r were 10 per cent or less but not if r were greater. Three other points emerge from the table. First, the size of the NPV – but not its sign – depends upon when time is taken to be zero; the magnitudes of different, possibly competing, cash flows can therefore only be compared on identical time scales. Second, the R/C ratio (a measure of economic attractiveness) depends upon r but not upon time scale; obviously, it is unity when NPV = 0. And third, the choice of r is critical, especially for long time scales, so long-delayed returns are likely to be unattractive if r must be large.

There are many other methods of investment appraisal but the discounted cash flow is dominant. The commercial plant breeder, for example, would almost certainly use it in deciding whether or not to undertake a new breeding project. On the cost side he would tabulate staff, new buildings and equipment, machinery, land, maintenance, rent, rates, taxes, power, multiplication, overheads, shareholders' interest and so on; on the receipts side there would be royalties, seed sales perhaps fees from other breeders. The cash flow would certainly be long-continued (twenty years at least) and the risks and uncer-

Table 11.1 Example of discounted cash flow. Adapted from Simmonds (1975, Table 1) by permission of the Scottish Plant Breeding Station.

Years	Cash		r = 0.05			r = 0.10			r = 0.15		
$(t - t_0)$	Costs −	Receipts +	F	C −	R +	F	C −	R +	F	C −	R +
0	50	0	1.00	50	0	1.00	50	0	1.00	50	0
–1	100	0	0.95	95	0	0.91	91	0	0.87	87	0
–2	200	0	0.91	182	0	0.83	166	0	0.76	152	0
–3	80	40	0.86	69	34	0.75	60	30	0.66	53	26
–4	0	120	0.82	0	98	0.68	0	82	0.57	0	68
–5	0	190	0.78	0	148	0.62	0	118	0.50	0	95
–6	0	250	0.75	0	188	0.56	0	140	0.43	0	108
Totals (1)	430	600		396	468		367	370		342	297
	+170			+72			+3			–45	
Totals (2)	430	600		532	628		648	657		789	688
	+170			+97			+9			–101	
R/C ratio	1.40			1.18			1.01			0.87	

Notes: r is the discount rate (0.05 = 5 per cent), F is the discount factor and $F = (1 + r)^{t - t_0}$. The columns headed Cash represent r = 0, F = 1. Totals (1) refer to the discounted figures, as tabulated; Totals (2) are accumulated, with the time scale changed to +6- - -0; net totals are net present values (NPV) and their size depends on the choice of r and when time is set at zero. Note that the receipt–cost (R/C) ratio depends only upon the cash flow and r, not upon the choice of time scale. The interested reader should try plotting totals and net totals, discounted and accumulated, against the discount rate; it is also instructive to plot cumulative net totals against time at different discount rates.

tainties would be very considerable. The choice of discount rate could well be critical for the decision as to whether to start or not; it would probably be determined either by the rate at which the company could borrow or, if the project were to be internally financed, from consideration of alternative uses for its money.

Clearly, DCF arithmetic is simple enough. Getting good estimates on which to do the arithmetic will be far more difficult, especially estimates of long-deferred royalties and sales which will, of their nature, depend upon the success or otherwise of the breeding programme designed to generate them. Nevertheless all business expenditure must rest on just such forecasts – though not always on quite such long time scales. Some projects will succeed, some will fail; the successful firm will achieve an average NPV equal to, or a bit better than, zero.

Cost–benefit analysis

The rationale of cost–benefit (CB) analysis is to apply the financial logic of the discounted cash flow to the economic appraisal of public expenditure. The method has been widely used in the past fifty years, especially in the USA, as a technique for deciding whether a project is acceptable or, alternatively, for making choices between projects competing for limited funds. Before discussing its possible use in analysing the socio-economic consequences of plant breeding, a brief general account is necessary.

The basic requirements are that all costs and benefits shall be identified, that they shall be estimated in cash terms and that their time-flows shall be established. Where the commercial analyst works purely in cash terms, the social analyst must somehow convert social effects into cash. He must first identify and estimate crude benefits, by way of the consumer surplus and/or resource–cost calculations mentioned above; then he must bring in welfare or distributional effects and, somehow, quantify those, too; he must seek possible implementation costs, those that may be necessary to realize the benefit but do not appear on his own prospective list of expenditure; and, finally, he will have to identify potential 'externalities', side-effects of the project, positive or negative, additional to the effects he has already listed under distribution and implementation. The categories are not sharply distinct but they are all, at least sometimes, relevant. Finally, the analyst will take in the direct cash elements of the project, e.g. construction, maintenance, revenue, and do his DCF calculation.

Formally, the results might be used to make a decision on that particular project: if the NPV $\geqslant 0$, go ahead. Or they might be used to rank the project in competition with others; it can be shown that, if projects are ranked in descending order of B/C ratio, then the procedure which maximizes total benefits within a specified total cost is to take projects in order up to the cost limit. However, one would have to have considerable confidence in the accuracy of the calculations to adopt any such procedures. In practice, the accuracy, at least for complex projects, is never good enough to do more than provide an order of magnitude result and the decision-maker will proceed very cautiously, using common sense and experience rather than financial rules to guide him.

The reasons for inaccuracy are many, roughly listed as follows. First, available economic data are often poor, despite the gigantic volume of information generated by contemporary governments; CB analysis is fraught with plausible guesswork. Second, there are many unresolved technical–economic problems in the estimation of benefits. Third, all calculations involving welfare-distributional effects impute costs and prices on the basis of subjective value judgements as to what is socially desirable; no such judgement can command general acceptance. Fourth, the future is uncertain and unforeseen risks abound; 'reasonable' calculations to include uncertainty can usually be done but any possibility of simple rule-of-thumb decision-making then disappears. Fifth, the choice of discount rate (which may well turn out to be critical for the outcome of the DCF) presents a formidable problem; since no objective value presents itself, as it does for the businessman, choice must be subjective or political or merely arbitrary in nature. Sixth and finally, both B minus C (i.e. NPV) and the

B/C ratio are intrinsically unsatisfactory criteria, the first because it depends, as we saw above, upon the time scale chosen, the second because it is not always easy to distinguish costs from negative benefits ('disbenefits'); B/C ratios can be – indeed, have been – quite seriously distorted by 'netting out' costs from benefits.

What use, then, is CB analysis? Regarded as a formal, decision-making procedure for large, complex public projects (new airports, towns, motorways), it is clearly unsatisfactory; regarded, rather, as a method of thinking about the consequences of doing such projects, however, it has the great merit of making complex socio–economic issues explicit; so the ideas may be useful if not the arithmetic. Even at the level of much simpler projects, it seems doubtful whether CB analysis has much to offer. Thus, it has often been argued that it should provide a means of appraising R & D projects, that is, of providing criteria whereby some proposals would be accepted, some rejected. Wise (1978) has, however, shown pretty convincingly, that it can not be a serious candidate for this function. Not only is it methodologically defective, as we have seen above, but it is also laborious and expensive to do and it helps the policy maker and research administrator little or not at all in grappling with two of their central problems: the budget (how to choose the 'best' total expenditure?) and the portfolio (how to achieve the 'best' mixture within a limited budget?). This is not to say that economic arguments have no place in R & D planning; cautious, order-of-magnitude CB calculations may indeed be valuable in indicating how and where research may, or may not, pay off. But this is a long way from making economic analysis the basis for decision; informed good sense will generally be a better guide than ill-founded arithmetic.

There remains one other use for CB analysis that is of interest in our context, namely retrospective analysis of what has already happened. This is at least free of future uncertainties. Evidently any attempt to assess the socio–economic effects of new technology must adopt CB methods, simply because no others are available. Several retrospective studies of agricultural R & D have been published. They nearly all show substantial positive NPV and $B/C > 1$. This is hardly surprising, because the subjects were chosen for their probable success (e.g. the pioneering work of Griliches, 1958, on hybrid maize). The claim made by some authors, on the basis of these and other related studies, that agricultural research in general is economically highly productive, can hardly be regarded as well established until a far better sample of cases has been examined (including evident failures as well as successes). This conclusion applies to plant breeding as well as to the wider field. There are a few established successes, e.g. hybrid maize and sorghum in the USA (Griliches, 1958); rice in Japan (Akino and Hayami, 1975); rubber in Malaysia (Pee, 1977); potatoes in the UK (see Fig. 11.5); see also several chapters in Arndt *et al.*, 1977); but there is no good evidence on plant breeding as a whole.

Probably, it is indeed economically worthwhile but this has yet to be demonstrated. More retrospective studies are wanted; given the experience and methodology that should come from such studies, it may be that some cautious use of CB methods in a prospective (*ex ante*) role will yet be possible. It is likely that we shall find that plant breeding often not only generates benefits (as consumers' surplus or resource savings) but is also attractive in having relatively low implementation costs, good distributional features (cheaper food benefits everyone) and lack of unfavourable externality.

Economic criteria in plant breeding

Plant breeding is an economic activity, yet explicit economic criteria very rarely enter into the definition of either objectives or selection criteria. It is usually held to be self-evident that increased yield, enhanced disease resistance, improved quality and so forth offer economic advantages worthy of pursuit. And so, clearly, they sometimes do, as is evidenced by the many successes of new varieties; if a new variety succeeds it must, presumably, be of economic advantage to some section of society (farmers, breeders, processors) even though the benefits, in a truly competitive situation, will subsequently become much more widely diffused, as we saw above.

However, many new varieties fail, sometimes for not very obvious reasons; do they sometimes, perhaps, represent successes which achieved economically the wrong objectives? Contrariwise, seemingly excellent varieties sometimes fail from unforeseen defects which again must often reflect failure of prior economic apprehension as to what is important and what is not. Unfortunately, this is an area which has not had systematic study and of which we are, therefore, almost totally ignorant. This is a pity because it is seemingly self-evident that instinct or feeling can not be a good substitute for calculation; economic arguments are rarely either obvious or simple. Four generalizations, however, seem to be possible, as follows.

First, assumptions and approach must depend, at least to some extent, upon who does the calculations. Thus the private breeder concerned to maximize the profits of his own seed firm would tend to favour objectives which promoted sales of seed and discouraged the retention of seed by farmers; thus HYB varieties are favoured over OPP or IBL and there is little encouragement to breed resistance to seedborne diseases (e.g. viruses in potatoes). Again, the private breeder subsidiary to a chemical firm might not favour a disease resistance that obviated the use of a spray made by the parent company. Both might breed excellent varieties that generated social benefits (Fig. 11.4) but their economic weights would differ and neither would be the same as those adopted by the state-supported breeder. The last would have to adopt, in principle at least, social benefits as estimates of economic weightings. Again, an industry-supported breeding scheme (sugar, cotton, rubber and other plantation crops come to mind) would have to take a view of the economic interests of the whole industry and adopt objectives that differed from both those of private breeders concerned to sell seed to farmers and state-supported breeders with social

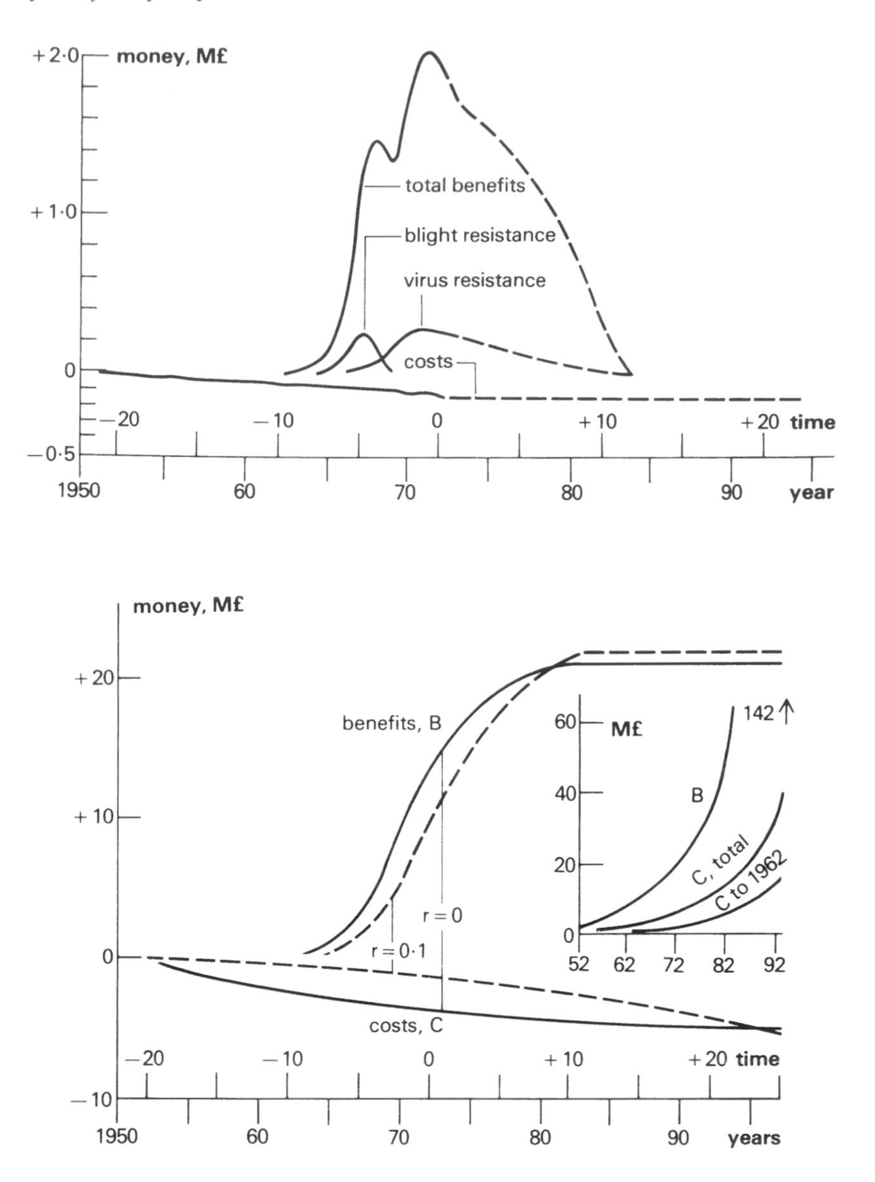

benefits in view (see Fig. 6.6). Clearly, then, there must be differences and, as argued above, the society is probably best served which maintains a state of competition.

Second, instructive calculations will vary very widely in scale from full-blown CB analysis to quite simple sums on individual varietal characters. At one extreme, the state-supported breeder might feel the need for a massive comparison of social benefits of various objectives, trying to weight – probably in terms of resource-cost savings – yield, disease resistances, quality characters and so on. This would be a formidable undertaking. Partial calculations might

Fig. 11.5 Cost-benefit analysis of a plant breeding project: two potato varieties, Pentland Crown and Pentland Dell, bred at the Scottish Plant Breeding Station, Edinburgh. Based on Simmonds (1974).
Notes: The analysis was completed in 1972 (when $t = 0$) and was part retrospective, part prospective (distinguished as solid and dashed lines in the upper part of the figure). Costs of potato breeding at the Station were calculated from accounts and adjusted to 1970 money values. Benefits were estimated on a resource-cost-saving basis: increased yields (about 10 per cent over standard), leading to reduced inputs of some factors of production and hence reduced costs for an approximately constant total crop, were the principal item; a transient (major-gene) blight resistance in one variety saved spraying costs for a few years and virus resistance in the other reduced the resource-cost of seed. The varieties, though successful, were predicted to have a limited life and to be displaced by others in time.

The top part of the figure shows the raw cash-flow, year by year, undiscounted; the lower part shows the cumulative flows at $r = 0$ and $r = 0.10$, the latter figure being the arbitrary 'test-discount-rate' chosen by the government of the time as appropriate to public expenditure. Costs, C, are well estimated but no reasonable calculation as to which portion of them should be allocated to the achievement of the specific benefits, B, is possible; since breeding is a continuous process, some costs before 1950 would be relevant and some costs during the 1950s would probably better be allocated to later products. To take (for illustration only) all costs 1951–62 as being appropriate, we have: at $r = 0$, NPV = 21.4 and B/C = 44; at $r = 0.10$, NPV = 19.0 and B/C = 10. This is quite arbitrary and implies that costs after 1962 should be allocated to later breeding achievements. The only firm conclusion is that, however costs are reckoned and whatever the discount rate (in the range 0–10 per cent), $B > C$ and $B/C > 1$. The inset shows the effect of shifting the time scale so that $t = 0$ at decades 1952–92 (i.e. the whole range from discounting all figures to accumulating all); B can be made any figure we please in the range 3–142 M£, NPV varies accordingly but B/C is unaffected (see Table 11.1).

Other economic consequences of the varieties were also considered. No externalities or implementation costs could be identified. Distributional effects are likely to be broadly favourable because potatoes are the food of poorer rather than richer people; the virus resistance of Pentland Crown, however, meant that growers can keep their own seed for longer and hence buy less seed from the professional seed-raisers, who must thereby suffer some loss of trade. Overall, the effects of the two potatoes are economically attractive and socially benign and this is arguably true of plant breeding as a whole.

---

however be possible and very instructive. Thus, given estimates of the incidences of several diseases, the damage they do, the cost of control measures and other parameters of production, order-of-magnitude calculation of the benefits of various levels of resistance would certainly be feasible: two examples are shown in Fig. 11.5 to which we can add that calculations of the value of a good horizontal resistance to potato blight suggest that it would be roughly equal to a substantial (10 per cent) yield advance. It goes without saying, of course, that any such calculations, however rough, should be on a serious economic basis; the enhanced profits of innovating farmers multiplied up to a whole-crop scale are *not* identical with social benefits (though they must be related to them). The private breeder, by contrast, might well regard the innovating farmer's prospective profits as a reasonable candidate for adoption as an economic criterion *per se.* We conclude that partial economic calculations about objectives, even if fairly rough, would often be feasible and could be informative; they seem to have been unduly neglected.

The third point is this. Any economic weighting of objectives not only affects the strategy of a programme – choice of breeding plans, spectrum of parents chosen, characters to be considered and so on – but also immediately leads on to the tactics of it, specifically to selection procedures. We saw in Section 5.8 that multiple selection may proceed by successive truncations but that, much more generally, some kind of index selection prevails (Figs 5.11, 5.12). The classical index developed by animal breeders applies to the progressive improvement of an outbred population and it takes heritabilities and genetic correlations into account. The plant breeder far more commonly encounters the need to select on several characters simultaneously, within groups of entities (IBL, CLO for example) which are already, in effect, genetically fixed. Any joint selection makes, by implication, some value (i.e. economic) judgement. The judgement may be, and often is, intuitive but is always open to economic definition, as in the example in Fig. 11.6. Whether there would be much advantage in doing so, we cannot know until we shall have tried. The example of Fig. 11.7 (already seen in Chapter 5) vividly amplifies from practical agriculture the reality of economic progress in a major crop.

Fourth and finally, even if we conclude (as we should) that explicit economic arguments have their uses, we recall that a breeding programme is ultimately dominated by the biological facts and possibilities. The breeder will either know

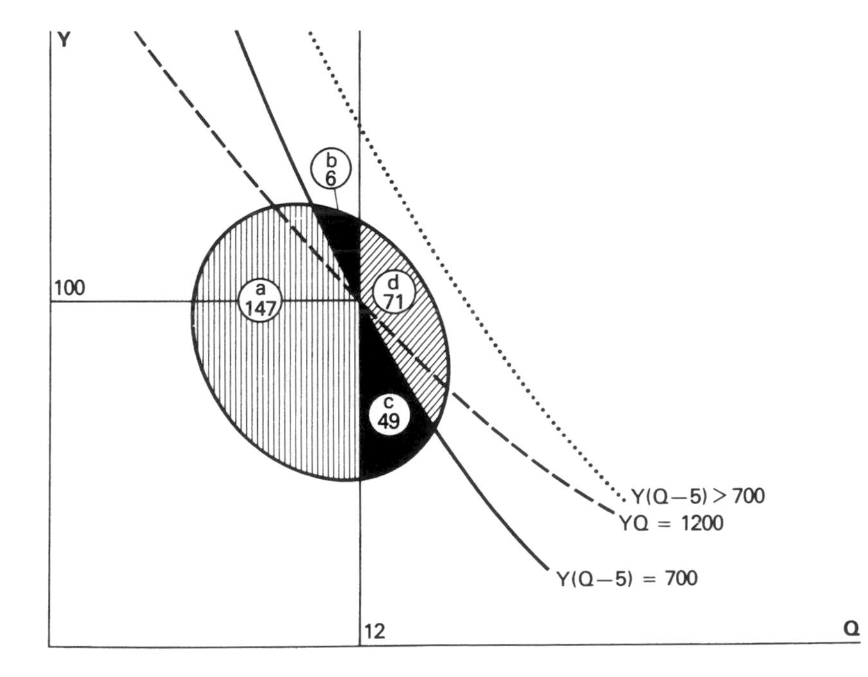

with confidence what is possible or will, at least, have some intuition for it. Economic calculations are no help in this area; they can sharpen choices between objectives and help to define selection procedures but no more.

11.4 Tropical development

The green revolution

The term 'green revolution' was invented in 1958 in Washington D.C. by a then administrator in USAID. It subsequently gained wide circulation and persists today, though now, as we shall see, in some disrepute. Essentially it applies to the transfer of technology-based cereal agriculture from the temperate countries (which developed it) to tropical countries operating at lower levels of both inputs and yields. The background was the realization, which has been developing since the 1940s but which only reached general consciousness about 1960,

---

Fig. 11.6 Economic index selection in sugarcane (Simmonds, 1972).
Notes: Sugarcane clones are selected on yield of cane (Y) and sugar content per cent of cane (Q), among other characters. YQ (sugar per hectare) could serve as an economic index but workers in Mauritius, the West Indies and Mexico have found that the profitability of the whole operation, field and factory together, is related rather to $Y(Q-m)$ where m is an economic parameter that depends upon harvesting and processing costs per tonne of cane. Hence a high sugar content, even at some cost to sugar yield, is economically favourable. The value of m has been variously calculated as about 4–6 and is here taken to be 5. This leads to a family of isoprofit curves of which one can be chosen to fit the standards in any given set of trials. For the Jamaican trials summarized in the figure, $Y(Q-5)=700$ and the equivalent curve on the sugar-per-hectare criterion ($YQ=1200$) is shown for comparison. As breeding progresses and standards improve, the appropriate isoprofit line will move outwards ($Y(Q-5)>700$). The distribution of 273 clones in the trials is shown by the shaded area; no correlation between Y and Q is apparent. Areas (b + d) show that 77 clones appeared to be at least equal to standards on the profitability criterion and the population so selected is different from that which would have been taken on the basis of YQ. A majority of clones would not have been selected on any criterion, suggesting the possibility of using the diagram to promote trials economy. In practice, Q is generally thought to have higher clonal repeatability than Y, so an effective pre-selection for Q, before the trials stage may be possible. Clearly, if the population could be efficiently truncated at $Q=12$, then 92 per cent (71/77) of the best clones would be retained for 44 per cent ((71 + 49)/273) of the trials effort; and the steeper the curve (i.e. large m) the stronger the conclusion.
Similar relations could be developed whenever a crop is grown for extraction of a specific fraction (sugar, starch, drugs, insecticides). They apply also in several animal breeding situations (e.g. weaners per sow and food conversion ratio, eggs per hen and market age); as in plant breeding contexts, the definition of profitability depends upon who writes the equations (Moav, 1973). In the sugarcane example above, the interests of the industry as a whole will determine m; the chosen equation need not reflect social benefit; farmers' interests will depend upon how they are paid for their cane (usually, but not always, on an arbitrary scale that favours high Q). Note that the equation $Y(Q-5)=700$ could be approximated in the relevant range by $I=Y+10Q=220$ and $YQ=1200$ by $I=Y+5Q=160$, i.e. by conventional additive indices (cf. Fig. 5.10).

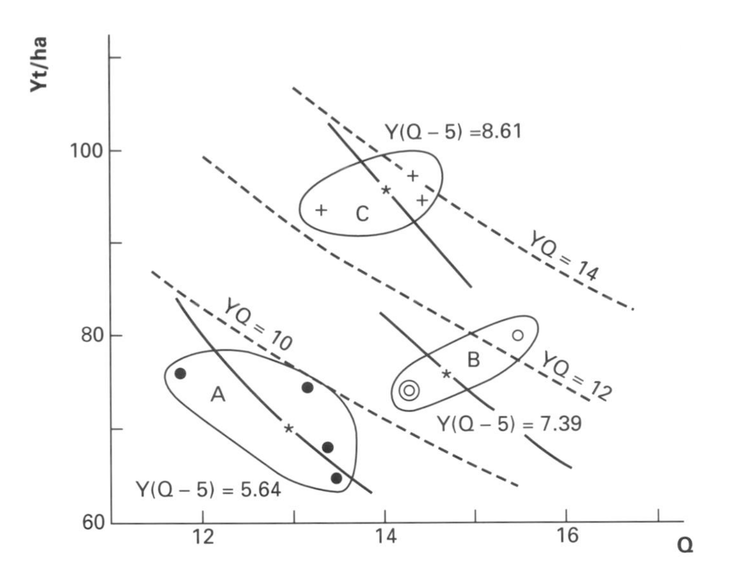

Fig. 11.7 Progress in sugar cane breeding.
Notes: The figure, data from Copersucar (1987, p. 231) summarizes three cycles of selection in sugarcane in Sao Paulo State, Brazil. Four old varieties (A) are compared with three more recent ones (B) and with the three latest (C), dominant at the time of writing. Data are from commercial productions from huge areas of 307 kha from 28 sugar mills in the year 1986–87. Y is cane yield in t/ha and Q is sucrose per cent cane, so that YQ is t sucrose/ha. The equations in Y (Q-5) are economic selection indices (cf. Fig. 10.6). Enormous gains, in both sugar yield and index, are apparent. Note that the two possible criteria of selection are different and that the economic index favours Q. The breeder is always selecting towards the top right hand of the diagram. Note that both Y and Q are high and that this is one of the most productive sugar industries in the world. This is one of the best available examples of the gains due to plant breeding because all data are measured, the samples are from agriculture (not trials) and the samples are huge.

that world population was growing at least as fast as food supplies were increasing. Since, in even the moderately short-term, food was unlikely to emulate the exponential capacity of populations for increase, Malthusian disaster was widely prophesied. And disasters were, indeed, staved off several times by US grain supplies to poor countries during the 1950s and 1960s. These issues, with technological, sociological, economic and political ramifications, are complex and the subject of an enormous literature. They lie beyond our scope. For our present purpose, the following can be taken to be self-evident: the desired outcome, a stable food-population balance at a decent level of nutrition, can only be achieved in the long run by stabilization of populations at levels which can be fed by sustainable technology; in the short run, there is an absolute necessity to increase food production in the countries (all poor and mostly tropical) which are at greatest risk. Acceptance of this necessity was at the heart of one of the most influential documents on the subject, FAO's

Table 11.2 Economic aspects of plant breeding, based on Simmonds (1988, 1989, 1990). See also the sugar cane examples, Figs 11.6 and 11.7.

$W = f(X_1\ X_2\ X_3 \ldots)$	Economic index of worth, W, a function of traits, $X_1\ X_2\ X_3$ etc.
$W_N = W_0 + R_W$	Genetic advance in W, denoted R_W, is necessary.
$S_0 = Y_0P_0 - bY_0 - C_0 = 0$	At equilibrium, farmer surplus (S_0) is zero because returns (Y_0P_0) balance costs ($bY + C$). Y yield, P price, C costs per ha.
$S_N = \alpha Y_0P_0 - \alpha\gamma bY_0 - \beta C_0 > 0$	Surplus now exceeds zero because breeder has changed α, β or γ, $R > 0$ and new variety succeeds.
$S_0 = \alpha Y_0P_N - \alpha\gamma bY_0 - \beta C_0 = 0$	Market adjusts P_N downwards to return S_0 to zero; new equilibrium.
$P_N = \gamma b + \dfrac{\beta C_0}{\alpha Y_0}$	Outcome is lower price to consumers, dependent on breeder's success in adjusting α, β, γ. Inverse relation between P and Y.

Indicative World Plan for Agriculture. By the time the Plan was published (1969), the green revolution, bearing very directly on the leading objective, was already well under way.

The beginning was in Mexico in the 1940s where a joint programme by the Rockefeller Foundation and the Mexican Government set out to enhance crop yields using well-established temperate-zone technology, including plant breeding, agronomy, irrigation, fertilizer use, pest control and so forth. The chosen crops were, primarily, wheat and potatoes; the potato work did much to establish the importance and potential of horizontal resistance (Section 7.2) and the wheat work led on to the green revolution. Dwarf wheats (Norin dwarfs), the first available, were brought to the USA from Japan by Salmon in 1947, were brilliantly exploited by Vogel and his colleagues in the northwestern USA and taken to Mexico by Borlaug in 1954. Borlaug selected dwarf lines out of crosses in two crops per year grown at different latitudes. The product was a series of dwarf wheats which were day-length neutral, widely adapted and high-yielding if adequately watered and fertilized. Some did outstandingly well in the irrigated wheat lands of Mexico and promising lines were vigorously promoted in other wheat-growing tropical countries by CIMMYT; CIMMYT is the programme's successor-organization, initiated by the Rockefeller and Ford Foundations jointly, with a more international remit (see Table 11.3). From small beginnings in the middle 1960s, over 10 Mha were sown to the new wheats by 1970, mostly in India and Pakistan; large areas in other countries followed.

Rice followed a remarkably similar pattern a few years later. IRRI was started by the same Foundations (Table 11.3), with the wheat model in view. A cross (1962) of a dwarf, early-maturing Japonica-type rice from Taiwan (Dee-geo-woo-gen) by the tall Indonesian variety Peta yielded an F_5 selection which was bulked up as IR8 from 1966 (obviously unimpeded by PVR rules). IR8 was,

Table 11.3 The international agricultural research institutes as established by the late 1970s.

Institute	Activities	Initiated
International Rice Research Institute – IRRI *Los Baños, Philippines*	Tropical rices, especially in Asia. Great rice collection (42 000 entries)	1960 by Ford and Rockefeller Foundations
Centro Internacional de Mejoramiento de Maiz y Trigo – CIMMYT *el Batan, Mexico*	Maize and wheat; also other cereals (triticale, barley, sorghum). Maize collection (8000 entries)	1966 by Ford and Rockefeller Foundations, on basis of earlier Rockefeller Foundation Mexican Programme (1943 onwards)
Centro Internacional de Agricultura Tropical – CIAT Cali, Colombia	Cassava and beans; also maize and rice in collaboration with CIMMYT and IRRI. Also beef, pastures and farming systems. Latin American emphasis. Collections of beans (10 000), forage grasses and legumes being built up	1967, by Ford and Rockefeller Foundations
International Institute of Tropical Agriculture – IITA *Ibadan, Nigeria*	Grain legumes, roots and tubers; also maize and rice in collaboration with IRRI and CIMMYT. Also farming systems. African emphasis. Collections of grain legumes (especially cowpeas) and tuber crops being built up	1968, by Ford and Rockefeller Foundations
West African Rice Development Association – WARDA *Monrovia, Liberia*	Regional cooperative rice research in collaboration with IITA and IRRI	1971, by West African Governments
Centro Internacional de Papa – CIP *Lima, Peru*	Potatoes. Collection being built up (11 000 entries). Breeding, especially for tropical latitudes	1971 by CGIAR
International Crops Research Institute for the Semi-Arid Tropics – ICRISAT *Hyderabad, India*	Sorghum, millets, grain legumes (pigeon pea, groundnut, chickpea). Dry land farming systems. Collections being built up (cereals 23 000)	1972 by CGIAR
International Laboratory for Reserch on Animal Diseases – ILRAD *Nairobi, Kenya*	Animal diseases	1973 by CGIAR

Contd

Table 11.3 *Contd.*

Institute	Activities	Initiated
International Board for Plant Genetic Resources – IBPGR *now International Plant Genetic Resources Institute* – IPGRI *Rome, Italy*	Genetic conservation in general; an agency rather than an institute (see Section 10.3)	1974 by CGIAR
International Livestock Centre for Africa – ILCA *Addis Ababa, Ethiopia*	Livestock production systems	1974 by CGIAR
International Centre for Agricultural Research in Dry Areas – ICARDA *Aleppo, Syria*	Barley, wheat, lentils and mixed farming systems. Also sheep. Emphasis on Mediterranean climates	1976 by CGIAR

Note: Sources are various, including CGIAR (1974, 1976), Crawford in Arndt, Dalrymple and Ruttan (1977), miscellaneous annual reports and helpful criticism by Prof. A.H. Bunting. The structure of the system has recently (1990s) been somewhat revised.

like the wheats, dwarf, day-length neutral, fertilizer-responsive and capable of very high yields under good conditions.

It was early-maturing, strongly tillering (an advantage in transplanted rice if not in densely drilled wheat) and had erect leaves thought to contribute to photosynthetic efficiency of the leaf canopy. Like the wheats, it was widely promoted and it and its immediate successors occupied a similar area in 1970 (about 10 Mha, mostly in India, the Philippines, Pakistan and Indonesia).

The two programmes had much in common. Both sought to exploit dwarfness (hence enhanced partition towards grain yield – Section 3.2) and day-length neutrality (hence wide latitudinal and seasonal adaptation); both recognized that no plant breeding could much improve yields if soil fertility were limiting so sought to exploit the inevitable GE component by promoting, not the variety alone, but the appropriate variety-agronomy 'package' (Figs 11.8, 11.9). In short, both programmes sought to do quickly, in one step, what temperate programmes on cereals had done over several decades (Hutchinson, 1971b). The agronomic parts of the packages were complex, containing water control, fertilizers, weed control and disease control as principal elements. This, and the essential GE component, turned out, as we shall see, to be the weakest part of the operation.

The literature of the green revolution is now enormous, much of it is conflicting and it is hard to form a balanced view. The gross over-optimism of the early days to the effect that 'the food-population problem' had been 'solved' (which never was a short-term possibility) has given way to a range of views from, roughly: (1) that the green revolution has had both considerable local successes and failures, the process of adjustment is inevitable and can not be socially painless and lessons have been learned in the only possible way; to (2)

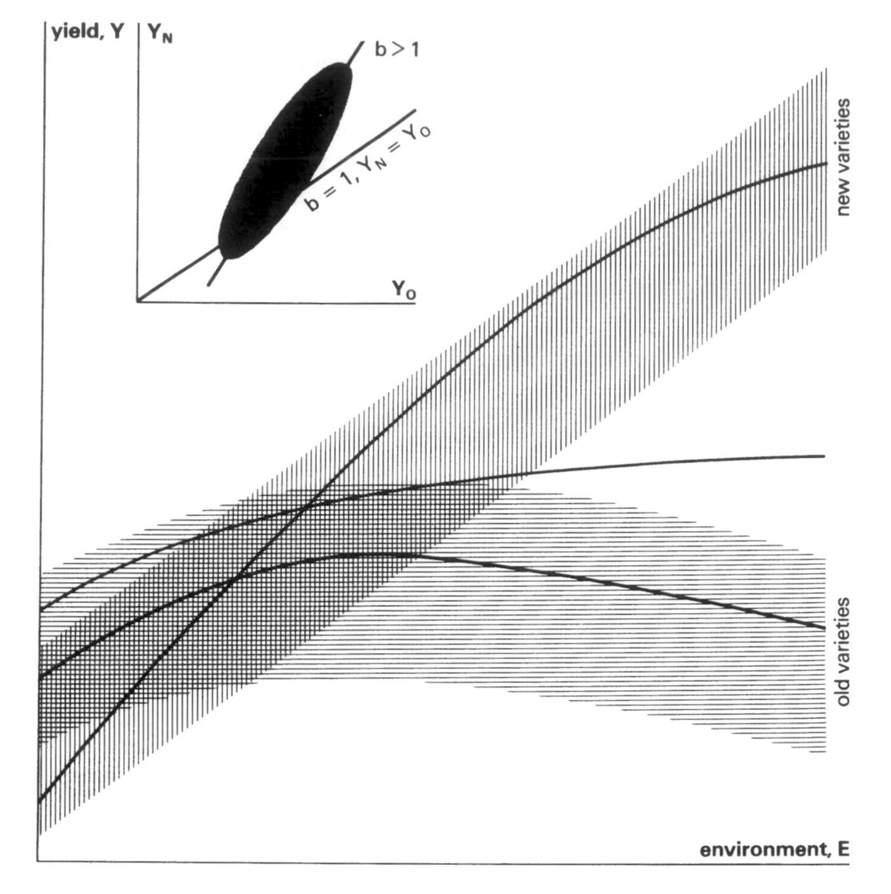

Fig. 11.8 GE interaction and the 'green revolution'.
Notes: the figure is a purely diagrammatic summary of many trials and much experience. Environment (abscissa) is to be taken to mean diverse factors, alone or collectively, e.g. fertility, water supply, freedom from weeds etc. Very broadly (exceptions could no doubt be found) old varieties were selected locally for consistent fair performance at a low level of whatever environmental factors dominated; their response curves are low and may indeed decline towards the right. By contrast, the dwarf wheats and rices upon which the 'green revolution' was based were selected for steep response curves regardless of possible inferiority to old varieties at the lower end of the environmental scale. An analysis of variance of a trial comparing varieties over several specified environments (e.g. N applications) would show a large GE item. Regression analysis of new or old over a sample of randomly varying environments (inset) would show $b>1$ (new less stable or more responsive than old – see Section 6.2). The 'green revolution' sought to make the transition from one response curve to the other in a single step. The intermediate line in the figure shows that there are other possibilities which might accord better with social needs: such should emerge from the more cautious programmes of the international institutes.

that what little it has done to improve food supplies has been done at an unacceptable social cost and that this is yet one more example of a misapplied 'technological fix'.

To generalize, the criticisms of the green revolution fall into three categories: of the varieties themselves, of agronomic/technological/economic weaknesses and of adverse social effects. The first two are criticisms on grounds of partial non-success; the third flows from the opposite criticism, namely too much success (in the purely technological sense). Criticisms of the varieties relate to quality and disease resistance. IR8 was widely disliked for its cooking quality (chalky, brittle grain) and the first wheats in India were disliked because they were red rather than amber-coloured. These defects were corrected in later cycles of breeding and there is no reason to suppose they need to be a permanent feature. Of more substance were fears of potential defects of disease resistance as few pure-line varieties (potentially carrying VR) spread over very large areas previously occupied by a mosaic of many local varieties presumptively carrying much HR. The risks, on all experience (Chapter 7), are obviously real: serious epidemics would be peculiarly damaging to small farmers and to poor societies. So far, at least, serious epidemics do not seem to have followed in the wheats (Saari and Wilcoxon, 1974); it is hard to assess scattered reports of local increases of blast, viruses and leafhoppers in rice. The risk having been recognized, the correct response is clearly to promote diversity among varieties and to favour HR over VR; this is now, of course, no more than conventional wisdom.

Weaknesses of execution lay in: (1) the inability of extension services to persuade farmers to adopt either the whole package or nothing; adoption of bits of it (e.g. the new varieties without fertilizer or variety-plus-fertilizer without good weed control) tends to produce poor yields at high cost; (2) unanticipated socio-economic weaknesses lay in obstacles arising from systems of land tenure and in failure to provide adequate credit facilities, chemical inputs, water supplies, storage and marketing facilities for enhanced crops of grain and so on. The list is a long one and details lie beyond our scope.

If the above were the only just criticisms the practical problems would still be formidable but there would be reasonable hope that, in time, a combination of education and sensible administrative action could solve them. The 'technological fix' could still probably be made to work. The third category of criticism, however, the social, is more weighty. Briefly, the richer farmers are those who can afford to apply the prescribed variety-agronomy package efficiently. They prosper and buy machinery in the interests of timely operations and labour-saving. The displaced labour is unemployed and moves to already dreadfully overcrowded towns. As grain supplies increase, prices fall, small marginal producers fail and join the landless unemployed in the towns; the larger farmers then acquire their land and yet more machines. So, paradoxically, enhanced supplies of grain may do little to benefit a population too poor to buy it even at reduced prices. No doubt there is a net benefit but it is unevenly spread; the

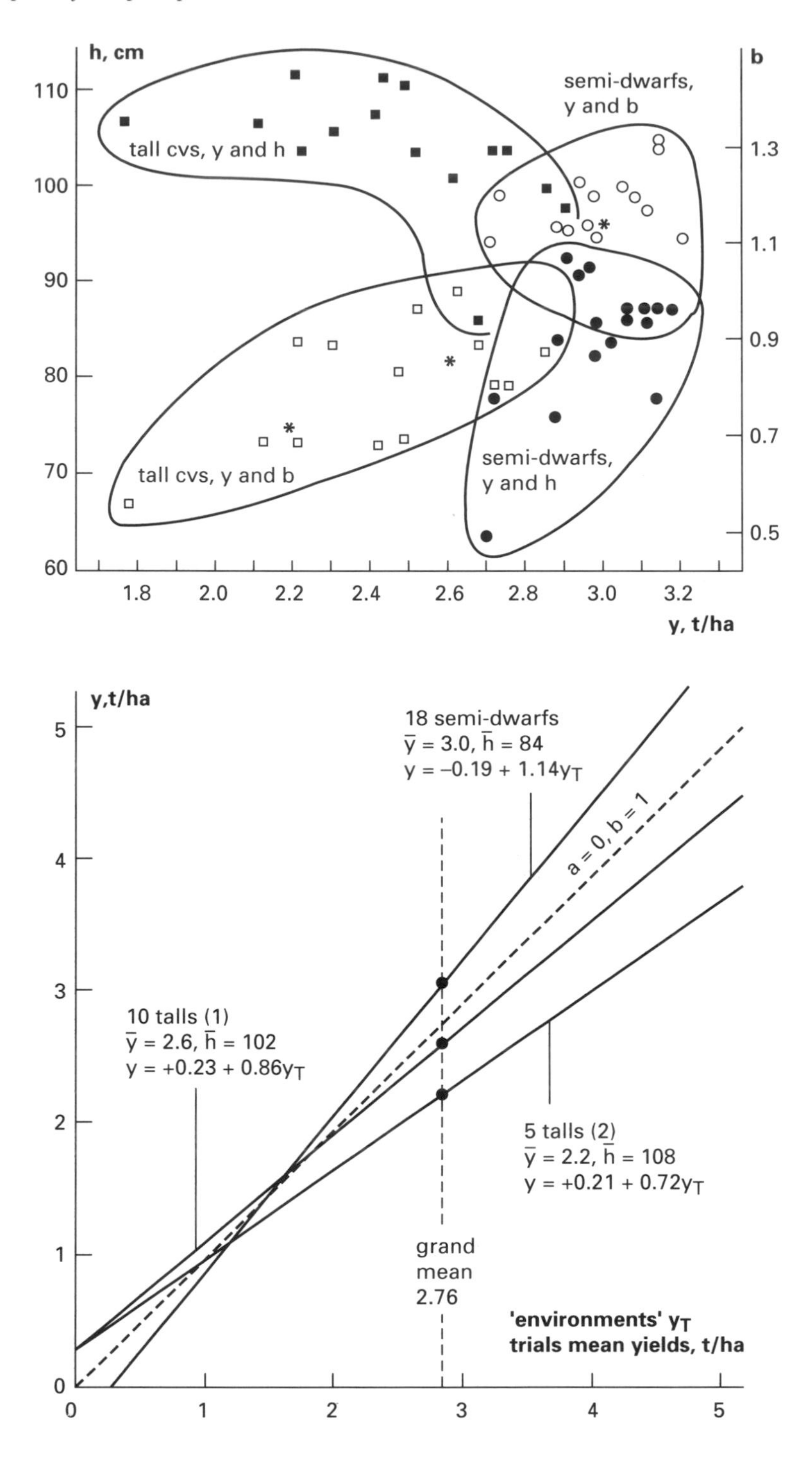
h, cm
b
110
100
90
80
70
60
1.3
1.1
0.9
0.7
0.5
semi-dwarfs, y and b
tall cvs, y and h
tall cvs, y and b
semi-dwarfs, y and h
1.8
2.0
2.2
2.4
2.6
2.8
3.0
3.2
y, t/ha
y,t/ha
18 semi-dwarfs
ȳ = 3.0, h̄ = 84
y = –0.19 + 1.14y_T
a = 0, b = 1
10 talls (1)
ȳ = 2.6, h̄ = 102
y = +0.23 + 0.86y_T
5 talls (2)
ȳ = 2.2, h̄ = 108
y = +0.21 + 0.72y_T
grand mean 2.76
'environments' y_T
trials mean yields, t/ha
0
1
2
3
4
5

Fig. 11.9 Adaptation of wheat varieties in the International Spring Wheat Yield Nurseries, 1969–71. Scatter diagram to show the relations between grain yields and stature and between yields and the stability regression coefficients; below, typical regressions for the three cultivar groups represented. Based on the data of Laing and Fischer (1977).
Notes: There were 33 cvs common to 44 trials (22 sites × 2 years): the 18 semi-dwarfs were nearly all of Mexican (ultimately Norin 10) origin; the 15 talls were heterogeneous, 10 bred for low latitude dryland conditions, 5 (less well adapted) for higher latitudes. The regression coefficients of variety yields on trials means are a measure of adaptability; values of b ≠ 1 indicate GE effects (Section 5.9, 6.2). The semi-dwarfs are high-yielding on average but adapted to good growing conditions (steep regressions); they would be inferior to the better adapted tall cvs below about 2 t/ha. The regressions shown in the lower part of the diagram are median for the three groups of cultivars, as indicated by the stars in the upper part of the diagram.

problem is thus one of distribution (or welfare). This criticism is fundamental and it is not helpful to observe that the process described is just what happened in rich north temperate countries over the past 200 years: our current prosperity was based on agrarian change which displaced workers into industry, not into unemployment.

The green revolution, in any reasonably strict sense of the term, must be said to have failed. No one seems to have tried to count the cost in global CB terms but there seems no doubt that, despite substantial net benefits locally (wheat in Mexico and Pakistan, perhaps), the social disbenefits of maldistribution are high (Farmer, 1977). The need for more food produced locally remains, however. This can only be achieved by joint adjustment of G, E and GE; the biological principle of the green revolution is still perfectly sound; any success in future practice will depend upon cautious, pragmatic application of that principle in socially tolerable ways, a viewpoint which, happily, seems to inform the programmes of the international institutes. In retrospect, it seems a pity that the term, green revolution, ever gained currency: it raised expectations which could never be fulfilled and it were now better abandoned.

The international institutes

The history of the Consultative Group for International Agricultural Research (CGIAR) and the institutes it supports has been mentioned in Section 10.3 in relation to genetic conservation. Conservation is, indeed, one of the leading functions, not only of the IBPGR (and its successor IPGRI) specifically, but also of several of the specialist institutes (Table11.3). The broad remit, however, is much wider. It is to enhance, by way of applied R & D and education, food crop production in the tropics. Industrial and export crops are excluded because they themselves mostly support (indeed have long supported) competent research and, anyway, do not bear directly on the main objective. In order of priority, cereals are basic; then food legumes, tuber crops and cattle; followed by other food crops and other stock. There is much emphasis on farming systems and on how individual crops and stock can be fitted into them; on socio-economic studies intended to define 'appropriate technology' and to facilitate the smooth assimilation of new methods into traditional systems; and on education and extension (CGIAR, 1974; Crawford in Arndt *et al.*, 1977) (Fig. 11.10).

The ten institutes are spread through the tropics (Table 11.2) in places determined by historical, climatic, geographical and agricultural factors. In the eight institutes primarily devoted to crops, a degree of crop specialization (sometimes complete, as with IRRI) is apparent. This means that the breeders and agronomists working with a crop may have virtually pan-tropical responsibility for it (though assisted, to be sure, by other institutes and local government agencies). This is a situation which plant breeders have never before had to face on this sort of scale: of breeding a crop in one place for adaptation elsewhere to a vast range of climates, latitudes, soils, diseases, agricultural systems and consumer preferences. Most plant breeders work for adaptation to one or a few rather well-defined environments within a few hundred miles of home.

In these circumstances, the best that the breeders can do, for a start at least, is to try to define the likely environments (on admittedly imperfect knowledge) and then ensure that large volumes of potentially adapted stocks, on a very wide genetic base, are tried in appropriate places spread around the world (Fig. 11.10). Such programmes call for the exploitation of genetic variability on a scale never before attempted, so the linkage, in several institutes, between breeding and genetic conservation could hardly be more appropriate. CIMMYT, for example, recognizes nine basic maize environments, on the basis of maturity, altitude and latitude, and four grain types, potentially thirty-six combinations in all; widely based composites, built up from the maize collection, are maintained under a PIM routine and improved populations widely distributed for trial and local selection. Much attention is also paid to high–low input comparisons in trials, so that varieties which respond to good conditions as well as others which tolerate poor ones should be identified (CIMMYT, 1976). Essentially similar methods, allowing for differences in breeding systems, are followed in other crops. Promising lines, of whatever crop, when identified, are freely available for local use as varieties and/or as parents in local programmes. Lines of inbreeders are only made as pure as they need be for agricultural purposes since there are no PVR systems to be contended with. In maize, CIMMYT chose OPP rather than HYB (surely the correct decision) on grounds of speed, economy, wide genetic base and cheap seed; locally, maize agriculture can then move, wholly or partly, to HYB varieties if and when circumstances encourage the change (see Sections 5.5 and 5.9 on Kenya maize).

All this bears little resemblance to the green revolution and indeed one has the impression that the use of the phrase itself, in literature about the work of the institutes, is now avoided. Revolutions are not in prospect. Rather,

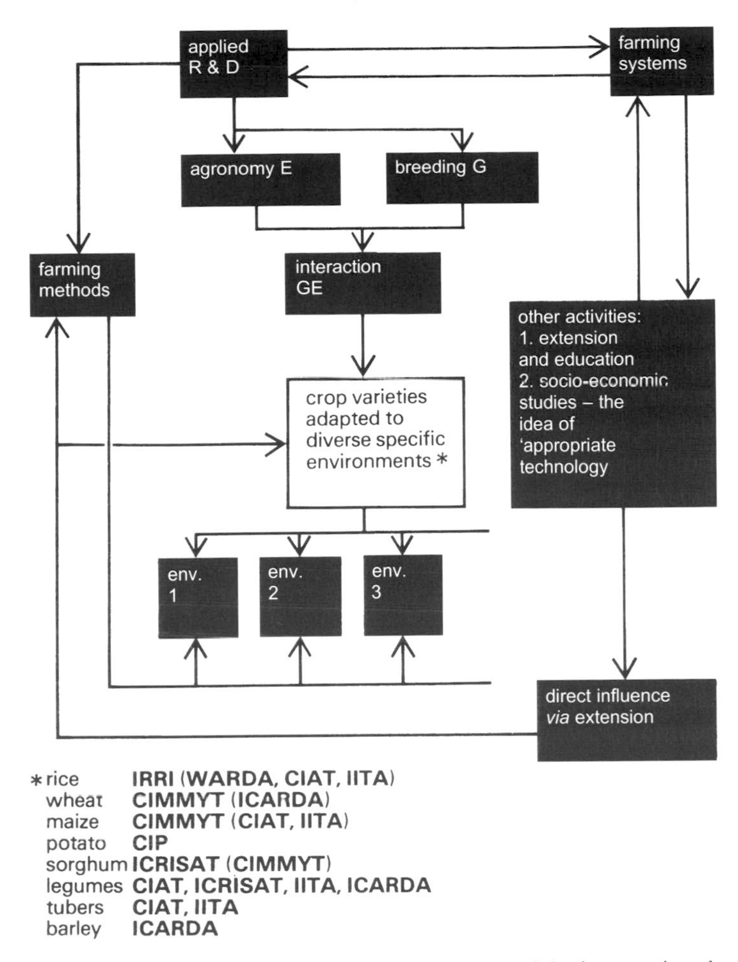

Fig. 11.10 Plant breeding in relation to the programmes of the international agricultural institutes.
Notes: interactions are likely to be complex with research on specific crops, ideas on farming systems, extension experience and socio-economic studies each influencing the others. Plant breeding is the G component of crops research. For abbreviations at the bottom of the diagram see Table 11.3.

adjustment of farming systems to use the results of agricultural R & D in socially acceptable ways is the intention. In the long run the promotion of multiple cropping and of diversification of cropping may be almost as important as increasing the yields of the major crops themselves. Whatever the developments, it is clear that useful advances will always depend here, as elsewhere and in the past, on the three components of crops R & D, breeding, agronomy and the interaction. To the extent that the green revolution demonstrated this in the context of two important food crops, it was valuable, perhaps inevitable. The principle, more cautiously exploited, underlies the work of the international institutes, themselves evoked by the revolution that went wrong.

11.5 Summary

Agricultural research and development (R & D) is undertaken by agencies, private or public, in the expectation of economic advantage. R & D applied to crops has three components: the husbandry, agronomic or environmental (E) component; the breeding or genetic (G) component; and the interaction (GE). Successful plant breeding, with or without an element of interaction, lowers the cost per unit of supply. Innovating farmers will profit transiently but, as the new variety spreads, market forces will tend to produce a new equilibrium in which economic benefits are distributed to consumers collectively in the form of lower prices. In principle, 'social benefits' (the gain to society as a whole) are calculable in terms of lowered prices (the idea of 'consumer surplus') or of resources saved in production, or both; but there are considerable technical–economic problems in doing so. Social benefit must be the objective of state-supported breeding programmes, whereas private breeders must aim at private profit; some (minor) conflict is inherent but, whoever breeds them, successful new varieties generate benefits.

The discounted cash flow (DCF) is a device for financial appraisal of prospective expenditure that takes account of the cost of money over time. It is routinely used for business purposes. Cost–benefit (CB) analysis adopts the financial logic of the DCF in the context of public expenditure. It requires the translation of *all* social costs and benefits inherent in a programme into cash terms defined over time. Because of factual deficiencies, inescapable value-judgements, economic problems and future uncertainty, prospective analyses of large social projects rarely command wide acceptance. Small projects, however, may be less controversial and the methods are as applicable to retrospective as to prospective study. A few analyses of plant breeding programmes agree in indicating that they generate substantial net benefits but the examples were chosen for presumptive success. *Probably*, plant breeding as a whole is indeed economically attractive in CB terms but this has yet to be proved.

Economic criteria are always implicit in the objectives and in the conduct of selection in plant breeding; explicit criteria are rarely defined but would probably often be useful. Some action is needed: the nature of a maximum profit/benefit criterion depends upon who writes the equations.

The 'green revolution' was a response to the food-population crisis which has developed (or at least become apparent) over the past fifty to sixty years. It started around 1960 with wheat in Mexico and rice in the Philippines. Dwarf, fertilizer-responsive, day-length-neutral varieties of each crop were widely disseminated in the poorer tropical countries as one part of a variety-agronomy 'package'. Given good growing conditions (but only then) the varieties were capable of very high yields, so the green revolution was essentially based on exploiting a GE interaction on the grand scale: it sought to transfer a substantial piece of well-established temperate-zone agricultural technology to poor tropical countries in one step. Despite substantial local successes in raising grain production, it failed, for complex reasons, of which the most important was probably adverse distributional effects among small farmers and labourers. The need to increase food production in the poorer tropical countries, however, remains but a more cautious and more widely based approach now prevails, in the form of the international agricultural institutes. Between them, they deal with all the major tropical food crops and seek to disseminate new varieties and new methods integrated into farming systems in a socially acceptable way. The plant breeding programme on any particular crop is aimed at, in effect, pan-tropical adaptation by an array of varieties collectively; it involves the breeders in coping with a diversity of environments and GE effects on a scale never before attempted. The need for genetic diversity and genetic conservation has never been more apparent so it is good that breeding and conservation are closely linked in several of the institutes.

11.6 Literature

CB analysis, general

Layard (1974); Mishan (1971).

Economics of agricultural research

Akino and Hayami (1975); Anon (1996); Arndt *et al.* (1977); Evenson and Kislev (1975); Fishel (1971); Griliches (1958); Lindner and Jarrett (1977); Pee (1977); Simmonds (1974, 1988, 1989, 1990); Wise (1975, 1977, 1978).

Economics, applications

Carlson and Main (1976); Simmonds (1972).

Green revolution, tropical development

Allaby (1977); Brown (1970); Bunting (1970); Falcon (1970); Farmer (1977); Sen (1974); Vyas (1975); Wilkes and Wilkes (1972).

International institutes

Arndt *et al.* (1977); CGIAR (1974, 1976); CIMMYT (1975, 1976); IRRI (1977).

Animal breeding

Lerner and Donald (1966).

Bibliography

Adair, C.R. and Jones, J.W. (1946) Effect of environment on the characteristics of plants surviving in bulk hybrid populations in rice. *J. Amer. Soc. Agron.*, **38**, 708–16.

Åkerberg, A. and Hagberg, A. (1963) *Recent Plant Breeding Research*. Wiley, New York.

Akerman, A., Tedin, O. and Fröier, K. (1948) *Svalöf, 1886–1946. History and Present Problems*. Lund, Sweden.

Akino, M. and Hayami Y. (1975) Efficiency and equity in public research: rice breeding in Japan's economic development. *Amer. J. Agric. Econ.*, **57**, 1–10.

Allaby, M. (1977) *World Food Resources, Actual and Potential*. Applied Science Publishers, London.

Allard, R.W. (1960) *Principles of Plant Breeding*. Wiley, New York and London.

Allard, R.W. (1961) Relationship between genetic diversity and consistency of performance in different environments. *Crop Sci.*, **1**, 127–33.

Allard, R.W. and Adams, J. (1969) Population studies in predominantly self pollinating species, XII. Intergenotypic competition and population structure in barley and wheat. *Amer. Nat.*, **103**, 620–45.

Allard, R.W. and Bradshaw, A.D. (1964) Implications of genotype–environmental interactions in applied plant breeding. *Crop Sci.*, **4**, 503–8.

Allard, R.W. and Hansche, P.E. (1964) Some parameters of population variability and their implications in plant breeding. *Adv. Agron.*, **16**, 281–325.

Allard, R.W. and Jain, S.K. (1962) Population studies in predominantly self pollinated species. II. Analysis of quantitative genetic changes in bulk-hybrid populations of barley. *Evolution*, **16**, 90–101.

Allard, R.W., Jain, S.K. and Workman, P.L. (1968) The genetics of inbreeding populations. *Adv. Genet.*, **14**, 55–131.

Allard, R.W. and Kannenberg, L.W. (1968) Population studies in predominantly self-pollinated species. XI. Genetic divergence among members of the *Festuca microstachya* complex. *Evolution*, **22**, 517–28.

Allard, R.W. and Workman, P.L. (1963) Population studies in predominantly self-pollinated species. IV. Seasonal fluctuations in estimated values of genetic parameters in Lima bean populations. *Evolution*, **17**, 470–80.

Anderson, E. (1952) *Plants, Man and Life*. Little and Brown, Boston, USA.

Anderson, M.K., Taylor, N.L. and Hill, R.R. (1974) Combining ability in I_0 single crosses of red clover. *Crop Sci.*, **14**, 417–19.

Anderson, M.K., Taylor, N.L. and Kirthavip, R. (1972) Development and performance of double cross hybrid red clover. *Crop Sci.*, **12**, 240–2.

Anon (1996) Will the world starve? – Feast and famine. *Economist*, Nov. 16, 25–9.

AOSCA (1971) *AOSCA Certification Handbook.* AOSCA (Association of Official Seed Certifying Agencies), USA. (Publ. 23.)

Arasu, N.T. (1968) Self-incompatibility in Angiosperms: a review. *Genetica*, **39**, 1–24.

Arboleda-Rivera, F. and Compton, W.A. (1974) Differential response of maize to mass selection in diverse selection environments. *Theor. Appl.. Genet.*, **44**, 77–81.

Arndt, T.M., Dalrymple, D.G. and Ruttan, V.W., eds (1977) *Resource Allocation and Productivity in National and International Agricultural Research.* University of Minnesota Press, Minneapolis, USA.

Arnold, M.H., ed. (1976) *Agricultural Research for Development, the Namulonge Contribution.* Cambridge University Press.

Aronson, A.I. (1994) *Bacillus thuringiensis* and its use as a biological insecticide. *Pl Breed. Rev.*, **12**, 19–46.

Arás, P. and Moreno-González, J. (1993) Marker assisted selection, pp. 314–31 in M.D. Hayward, N.O. Bosemark and I. Romagosa (eds) *Plant Breeding – Principles and Prospects*. Chapman & Hall, London.

Atherton, J.G. and Radich, J., eds (1986) *The Tomato Crop – A Scientific Basis for Improvement.* Chapman & Hall, London.

Atkins, A.E. (1953) Effect of selection upon bulk barley populations. *Agron. J.*, **45**, 311–14.

Atkins, R.E. and Murphy, H.C. (1949) Evaluation of yield potentialities of oat crosses from bulk hybrid tests. *Agron. J.*, **41**, 41–5.

Auerbach, C. (1976) *Mutation Research.* Chapman and Hall, London.

Austin, R.B. and Jones, H.G. (1975) The physiology of wheat. *Pl. Br. Inst. Camb. Ann.. Rep.*, 1974, 20–73.

Babcock, E.B. and Clausen, R.E. (1927) *Genetics in Relation to Agriculture.* McGraw-Hill, New York (2nd edn.).

Baihaki, A., Stucker, R.E. and Lambert, J.W. (1976) Association of genotype × environment interactions with performance level of soybean lines in preliminary yield tests. *Crop Sci.*, **16**, 718–21.

Bailey, T.R. and Comstock, R.E. (1976) Linkage and synthesis of better genotypes in self-fertilizing species. *Crop Sci.*, **16**, 363–70.

Baker, H.G. and Stebbins, G.L., eds (1965) *The Genetics of Colonizing Species.* Academic Press, New York.

Baker, L.H. and Curnow, R.N. (1969) Choice of population size and use of variation between replicate populations in plant breeding selection programs. *Crop Sci.*, **9**, 555–60.

Baker, R.J. (1968) Genotype-environment interaction variances in cereal yields in western Canada. *Can. J. Plant Sci.*, **48**, 293–8.

Baker, R.J. (1971) Effects of stem rust and leaf rust of wheat on genotype–environment interaction for yield. *Can. J. Plant Sci.*, **51**, 457–61.

Baker, R.J. and Campbell, A.B. (1971) Evaluation of screening tests for quality of bread wheat. *Can. J. Plant Sci.*, **51**, 449–55.

Baker, R.J. and Leisle, D. (1970) Comparison of hill and rod row plots in common and durum wheats. *Crop Sci.*, **10**, 581–3.

Bakhteyev, F.K. (1968) Reminiscences of N.I. Vavilov (1887–1943) on the eightieth anniversary of his birthday. *Theor. Appl.. Genet.*, **38**, 79–84.

Bal, B.S., Suneson, C.A. and Ramage, R.T. (1959) Genetic shift during 30 generations of natural selection in barley. *Agron. J.*, **51**, 555–7.

Banks, D.J. (1976) Peanuts: germplasm resources. *Crop Sci.*, **16**, 499–502.

Barz, W., Reinhard, E. and Zenk, M.H., eds (1977) *Plant Tissue Culture and its Biotechnological Application.* Springer, Berlin, Heidelberg and New York.

Bateman, A.J. (1947a) Contamination in seed crops. I. Insect pollination. *J. Genet.*, **48**, 257–75.

Bateman, A.J. (1947b) Contamination of seed crops. II. Wind pollination. III. Relation with isolation distance. *Heredity*, **I**, 235–46, 303–36.

Baumann, L.F., Mertz, E.T., Carballo, A. and Sprague, E.W., eds (1975) *High-Quality Protein Maize.* Dowden, Hutchinson & Ross, Stroudsburgh, Penn., USA.

Bechtold, N. and Bouchez, D. (1995) In planta *Agrobacterium*-mediated transformation of adult *Arabidopsis thaliana* plants by vacuum infiltration. In: *Gene Transfer to Plants* (eds I. Potrykus and G. Spangenberg) pp. 19–23. Berlin: Springer.

Becker, J., Vos, P., Kuiper, M., Salamini, F. and Heun, M.J.N. (1995) Combined mapping of AFLP and RFLP markers in barley. *Molecular and General Genetics*, **249**, 65–73.

Becker, W.A. (1975) *Manual of Quantitative Genetics.* Students Book Corporation, Pullman, Washington, USA.

Beg, A., Emery, D.A. and Wynne, J.C. (1975) Estimation and utilization of intercultivar competition in peanuts. *Crop Sci.*, **15**, 633–7.

Bemis, W.P., Berry, J.W., Weber, C.W. and Whitaker, T.W. (1978) The buffalo gourd, a new horticultural crop plant? *Hort. Sci.*, **13**, 235–40.

BFPP (Terminology Sub-committee of the British Federation of Plant Pathologists) (1973) A guide to the use of terms in plant pathology. *Commonw. Mycol. Inst., Phytopath Paper*, **17**, p. 55.

Bhatt, G.M. (1973) Comparison of various methods of selecting parents for hybridization in common bread wheat. *Aust. J. Agric. Res.*, **24**, 457–64.

Bhatt, G.M. (1977) Response to two-way selection for harvest index in two wheat (*Triticum aestivum*) crosses. *Aust. J. Agric. Res.*, **28**, 29–36.

Bhowjani, S.S., Evans, P.K. and Cocking, E.C. (1977) Protoplast technology in relation to crop plants: progress and problems. *Euphytica*, **26**, 343–60.

Bilbro, J.D. and Ray, L.L. (1976) Environmental stability and adaptation of several cotton cultivars. *Crop Sci.*, **16**, 821–4.

Bingefors, S. and Ellerström, S. (1964) Polyploidy breeding in red clover. *Zeitschr. f. Pflz-Zücht.*, **51**, 315–34.

Bingham, E.T. (1971) Isolation of haploids of tetraploid alfalfa. *Crop Sci.*, **11**, 433–5.

Bingham, E.T., Hurley, L.V., Kaatz, D.M. and Saunders, J.W. (1975) Breeding alfalfa which regenerates from callus tissue in culture. *Crop Sci.*, **15**, 719–21.

Bingham, E.T. and Saunders, J.W. (1974) Chromosome manipulations in alfalfa: scaling the cultivated tetraploid to seven ploidy levels. *Crop. Sci.*, **14**, 474–7.

Bingham, J., Riley, R. and Johnson, R., eds (1979) *Wheat. The Scientific Basis for Improvement.* Chapman and Hall, London.

Bird, R. McK. and Neuffer, M.G. (1987) Induced mutations in maize. *Pl. Breed. Rev.*, **5**, 139–80.

Bishop, C.J. (1963) Reviews in genetics and cytology. I. Plant Breeding. *Canad. J. Genet. Cytol.*, **5**, 1–11.

Bjarnason, M. and Vasal, S.K. (1992) Breeding of quality protein maize. (QPM) *Pl. Breed Rev.*, **9**, 181–217.

Bjorling, K. (1966) Virus resistance problems in plant breeding. *Acta Agric. Scand. Suppl.*, **16**, 119–36.

Bliss, F.A. and Brown, J.W.S. (1983) Breeding common bean for improved quantity and quality of seed protein. *Pl. Breed. Rev.*, **1**, 59–102.

Boerma, H.R. and Cooper, R.L. (1975a) Comparison of three selection procedures for yield in soybeans. *Crop Sci.*, **15**, 225–9.

Boerma, H.R. and Cooper, R.L. (1975b) Performance of pure lines obtained from superior-yielding heterogeneous lines in soybeans. *Crop Sci.*, **15**, 300–2.

Boerma, H.R. and Cooper, R.L. (1975c) Effectiveness of early generation yield selection of heterogeneous lines in soybeans. *Crop Sci.*, **15**, 313–15.

Bond, D.A. and Fyfe, J.L. (1961) Breeding field beans. *Pl. Br. Inst. Camb. Ann. Rep., 1960–61*, 4–26.

Borlaug, N.E. (1959) The use of multibreed or composite varieties to control airborne epidemic diseases of self-pollinated crop plants. *Proc I. Intnl. Wheat Genet. Symp.*, **12–26.**

Bos, I. and Caligari, P. (1985) *Selection Methods in Plant Breeding.* Chapman & Hall, London.

Boyce, S., Copp, L. and Frankel, O.H. (1947) The effects of selection for yield in wheat. *Heredity*, **1**, 223–33.

Bradshaw, J.E., Meyer, R.C., Milbourne, D., McNicol, J.W., Phillips, M.S. and Waugh, R. (1998) Identification of AFLP and SSR markers associated with quantitative resistance to *Globodera pallida* (Stone) in tetraploid potato (*Solanum tuberosum subsp. tuberosum*) with a view to marker-assisted selection. *Theoretical and Applied Genetics*, **97**, 202–10.

Braun, A.C. (1958) A physiological basis for autonomous growth of crown gall tumor cell. *Proceedings of the National Academy of Science, USA*, **44**, 344–9.

Bravo, J.E. and Evans, D.A. (1985) Protoplast fusion for crop improvement. *Pl. Breed. Rev.*, **3**, 193–218.

Breese, E.L. (1969) The measurement and significance of genotype-environment interactions in grasses. *Heredity*, **24**, 27–44.

Breese, E.L. and Hayward, M.D. (1972) The genetic basis of present breeding methods in forage crops. *Euphytica*, **21**, 324–36.

Bretting, P.K. and Widrlechner, M.P. (1995) Genetic marker and plant genetic resource management. *Plant Breeding Reviews*, **13**, 11–86.

Brewbaker, J.L. (1964) *Agricultural Genetics.* Prentice-Hall, Englewood Cliffs, N.J., USA.

Brian, P.W. and Garrett, S.D., eds (1972) A discussion on disease resistance in plants. *Proc. Roy. Soc. London, B.*, **181**, 213–351.

Bridge, R.R., Meredith, W.R. and Chism, J.F. (1971) Comparative performance of obsolete varieties and current varieties of upland cotton. *Crop Sci.*, **11**, 29–32.

Briggle, L.W. (1963) Heterosis in wheat – a review. *Crop Sci.*, **3**, 407–12.

Briggs, D.E. (1978) *Barley.* Chapman & Hall, London.

Briggs, F.N. (1959) Backcrossing – its development and present application. *Proc. I. Intnl. Wheat Genet. Symp.*, 8–9.

Briggs, F.N. and Knowles, P.F. (1967) *Introduction to Plant Breeding.* Reinhold, New York.

Briggs, K.G., Bushuk, W. and Shebeski, L.H. (1969) Variation in breadmaking quality of systematic controls in a wheat breeding nursery and its relationship to plant breeding procedures. *Can. J. Plant Sci.*, **49**, 21–8.

Briggs, K.G. and Shebeski, L.H. (1968) Implications concerning the frequency of control plots in wheat-breeding nurseries. *Can. J. Plant Sci.*, **48**, 149–53.

Briggs, K.G. and Shebeski, L.H. (1970) Visual selection for yielding ability of F_3 lines in a hard red spring wheat breeding program. *Crop Sci.*, **10**, 400–2.

Briggs, K.G. and Shebeski (1971) Early generation selection for yield and breadmaking quality of hard red spring wheat'. *Euphytica*, **20**, 453–63.

Brim, C.A. (1966) A modified pedigree method of selection in soybeans. *Crop Sci.*, **6**, 200.

Brim, C.A., Johnson, H.W. and Cockerham, C.C. (1959) Multiple selection criteria in soybeans. *Agron. J.*, **51**, 42–6.

Brim, C.A. and Schutz, W.M. (1968) Inter-genotypic competition in soybeans. II. Predicted and observed performance of multiline mixtures. *Crop Sci.*, **8**, 735–9.

Brim, C.A. and Stuber, C.W. (1973) Application of genetic male sterility to recurrent selection schemes in soybeans. *Crop Sci.*, **13**, 528–30.

Broekhuizen, S. *et al.*, eds (1964) *Barley Genetics. I.* Wageningen, The Netherlands (*Proc. I. Intnl. Barley Genetics Symp*, 1963).

Brookhaven (1956) *Genetics in Plant Breeding, Brookhaven Natnl. Lab. Symp. Biol.*, 9.

Brousseau, R. (1997) Transgenic insect-resistant plants: cents and sense. *PBI Bulletin*, National Research Council of Canada, May 1997, 1–3.

Brown, A.H.D. and Allard, R.W. (1971) Effect of reciprocal recurrent selection for yield on isozyme polymorphisms in maize. *Crop Sci.*, **11**, 888–93.

Brown, A.H.D. and Daniels J. (1973) Mass selection of commercial hybrid sugarcane populations in Fiji. *Expl. Agric.*, **9**, 321–8.

Brown, A.H.D., Daniels J. and Latter, B.D.H. (1968, 1969) Quantitative genetics of sugarcane. I. Analysis of variation in a commercial hybrid sugarcane population. II. Correlation analysis of continuous characters in relation to hybrid sugarcane breeding. *Theor. Appl.. Genet.*, **38**, 361–9; **39**, 1–10.

Brown, A.H.D., Daniels, J., Latter, B.D.H. and Krishnamurthi, M. (1969) Quantitative genetics of sugarcane. III. Potential for sucrose selection in *Saccharum spontaneum. Theor. Appl.. Genet.*, **39**, 79–87.

Brown, A.H.D., Daniels, J. and Stevenson, N.D. (1971) The mass selection reservoir and sugarcane selection. *Theor. Appl.. Genet.*, **41**, 174–80.

Brown, A.H.D., Frankel, O.H., Marshall, D.R. and Williams, J.T. (1989) *The Use of Plant Genetic Resources.* Cambridge University Press.

Brown, D.C.W. and Thorpe, T.A. (1995) Crop improvement through tissue culture. *World Journal of Microbiology and Biotechnology*, **11**, 409–415.

Brown, L.R. (1970) *Seeds of Change: the Green Revolution and Development in the 1970s.* Praeger, New York.

Brown, T.A. (1995) *Gene Cloning*, Chapman & Hall, London, 3rd edn.

Browning, J.A. and Frey, K.H. (1969) Multiline cultivars as a means of disease control. *Ann. Rev. Phytopathol.*, **7**, 355–82.

Bunting, A.H., ed. (1970) *Change in Agriculture.* Duckworth, London.

Bunting, E.S. and Gunn, R.E. (1974) Maize in Britain – a survey of research and breeding. *Pl. Br. Inst. Camb. Ann.. Rep., 1973*, 32–74.

Burley, J. and Styles, B.T. (eds) (1976) *Tropical Trees: Variation, Breeding and Conservation.* Academic Press for Linnean Society of London, London.

Burton, G.W. (1976) Gene loss in pearl millet germplasm pools. *Crop Sci.*, **16**, 251–5.

Burton, G.W. and Powell, J.B. (1968) Pearl millet breeding and cytogenetics. *Adv. Agron.*, **20**, 50–89.

Busbice, T.H. (1968) Effect of inbreeding on fertility in *Medicago sativa. Crop Sci.*, **8**, 231–4.

Busbice, T.H. (1969) Inbreeding in synthetic varieties. *Crop Sci.*, **9**, 601–4.

Busbice T.H. (1970) Predicting yield of synthetic varieties. *Crop Sci.*, **20**, 265–9.

Busbice, T.H., Hunt, O.J., Elgin, J.H. and Peaden, R.N. (1974) Evaluation of effectiveness of polycross and self-progeny tests in increasing the yield of alfalfa synthetic varieties. *Crop Sci.*, **14**, 8–11.

Busch, R.H., Hammond, J. and Frohberg, R.C. (1976) Stability and performance of hard red spring wheat bulks for grain yield. *Crop Sci.*, **16**, 256–9.

Busch, R.H., Janke, J.C. and Frohberg, R.C. (1974) Evaluation of crosses among high and low yielding parents of spring wheat and bulk prediction of line performance. *Crop Sci.*, **14**, 47–50.

Buzzell, R.I. and Haas, J.H. (1972) Natural and mass selection estimate of relative fitness for the soybean *rps* gene. *Crop Sci.*, **12**, 75–80.

Byrne, N. (1992) Patents for plants and genes under the European Patent Convention. *Proc. Royal Soc. Edinburgh.*, **99B**, 141–52.

Byrne, N. (1993) Patents for biological inventions in the European Communities. *World Patent Information*, **15**, 77–80.

Byrne, N. (1993) Patents for human genes, ownership of biological materials and other issues in patent law. *World Patent Information*, **15**, 199–202.

Byth, D.E., Caldwell, B.E. and Weber, C.R. (1969) Specific and non-specific index selection in soybeans, *Glycine max. Crop Sci.*, **9**, 702–5.

Byth, D.E., Weber, C.R. and Caldwell, B.E. (1969) Correlated truncation selection for yield in soybeans. *Crop. Sci.*, **9**, 699–702.

Caldwell, B.E. and Weber, C.R. (1965) General, average and specific selection indices for yield in F_4 and F_5 soybean populations. *Crop Sci.*, **5**, 223–6.

Campbell, G.K.G. and Russell, G.E. (1964) Breeding sugar beet. *Pl. Br. Inst. Camb. Ann.. Rep., 1963–64*, 6 31.

Campbell, L.G. and Lafever, H.N. (1977) Cultivar × environment interactions in soft red winter wheat yield tests. *Crop Sci.*, **17**, 604–8.

Cannell, M.G.R. (1978) Improving per hectare forest productivity. *Proc. V.N. Amer. Forest Biology Wkshop*, Gainesville, Florida, USA.

Cannell, M.G.R. and Last, F.T., eds (1976) *Tree Physiology and Yield Improvement.* Academic Press, London, New York and San Francisco.

Carlson, G.A. and Main C.E. (1976) Economics of disease-loss management. *Ann. Rev. Phytopathol.*, **14**, 381–403.

Carlson, I.T. (1971) Randomness of mating in a polycross of orchard grass. *Dactylis glomerata. Crop Sci.*, **11**, 499–502.

Carlson, P.S., Smuth, H.H. and Dearing, R.D. (1972) Parasexual interspecific plant hybridisation. *Proc. Nat. Acad. Sci., USA*, **69**, 2292.

Carnahan, H.L. and Hill, H.D. (1961) Cytology and genetics of forage grasses. *Bot. Rev.*, **27**, 1–162.

Casas, A.M., Kononowicz, A.K., Bressan, R.A. and Hasegawa, P.M. (1995) Cereal transformation through particle bombardment. *Pl. Breed. Rev.*, **13**, 235–64.

Ceccarelli, S., Grando, S. and Hamblin, J. (1992) Relationship between barley grain yield measured in low-and high-yielding environments. *Euphytica*, **61**, 49–58.

CGIAR (1974) *International Research in Agriculture*. CGIAR, New York.

CGIAR (1976) *Consultative Group on International Agricultural Research.* CGIAR, New York.

Chandraratna, M.F. (1964) *Genetics and Breeding of Rice*. Longman, London.

Chang, T.T. (1964) Present knowledge of rice genetics and cytogenetics. *Intl. Rice Res. Inst. Tech. Bull.*, **1**, p. 96.

Chang, T.T. and Tagumpay, O. (1970) Genotype association between grain yield and six agronomic traits in a cross between rice varieties of contrasting plant type. *Euphytica*, **19**, 356–63.

Chase, S.S. (1964) Analytic breeding of amphipolyploid plant varieties. *Crop Sci.*, **4**, 334–7.

Chase, S.S. (1969) Monoploids and monoploid derivatives of maize. *Bot. Rev.*, **35**, 117–67.

Chebib, F.S., Helgason, S.B. and Kaltsikes, P.J. (1973) Effect of variation in plant spacing, seed size and genotype on plant-to-plant variability in wheat. *Z. f. PflzZüchtung*, **69**, 301–32.

Chongbiao You and Zhangliang Chen, eds (1983) *Biotechnology in Agriculture.* Kluwer Academic Publishers, Dordrecht.

Choo, T.M., Reinbergs, E. and Kasha, K.J. (1985) Use of haploids in breeding barley. *Pl. Breed. Rev.*, **3**, 219–52.

Christey, M.C., Makaroff, C.A. and Earle, E.D. (1991) Atrazine-resistant cytoplasmic male-sterile-nigra broccoli obtained by protoplast fusion between cytoplasmic male-sterile *Brassica oleracea* and atrazine-resistant *Brassica campestris. Theor. Appl. Gen.*, **83**, 201–8.

Christian, C.S. and Grey, S.G. (1941) Interplant competition in mixed wheat populations and its relation to single plant selection. *J. Coun. Sci. Ind. Res., Australia*, **14**, 59–68.

CIMMYT (1975) *World-wide Maize Improvement in the 1970s and the Role for CIMMYT*, El Batán, Mexico.

CIMMYT (1976) *CIMMYT Review 1976. Tenth Anniversary, 1966–76.* El Batán, Mexico.

Clay, R.E. and Allard, R.W. (1971) A comparison of performance of homogeneous and heterogeneous barley populations. *Crop Sci.*, **9**, 407–12.

Clifford, M. and Willson, K., eds (1993) *Coffee: Botany, Biochemistry and Production of Beans and Beverage.* Chapman & Hall, London.

Cochran, W.G. and Cox, G.M. (1957) *Experimental Designs.* Wiley, New York.

Cockerham, G. (1970) Genetical studies on resistance to potato viruses X and Y. *Heredity*, **25**, 309–48.

Coffman, F.A., ed. (1961) *Oats and Oat Improvement.* American Society of Agronomy, Madison, USA.

Collins, G.B. (1977) Production and utilization of anther-derived haploids in crop plants. *Crop Sci.*, **17**, 583–6.

Collins, G.B., Legg, P.D. and Kasperbauer, M.J. (1974) Use of anther-derived haploids in *Nicotiana.* I. Isolation of breeding lines differing in total alkaloid content. *Crop Sci.*, **14**, 77–80.

Committee on Genetic Vulnerability of Major Crops (1972) *Genetic Vulnerability of Major Crops*. National Academy of Sciences, Washington, D.C., USA.

Comstock, R.E. and Robinson, H.F. (1948) The components of genetic variation in populations of biparental progenies and their use in estimating average degree of dominance. *Biometrics*, **4**, 254–66.

Comstock, R.E. and Robinson, H.F. and Harvey, P.H. (1949) A breeding procedure designed to make maximum use of both general and specific combining ability. *Agron. J.*, **41**, 360–7.

Cook, A.H., ed. (1962) *Barley and Malt.* Academic Press, New York and London.

Cooke, D.A. and Scott, R.K., eds (1993) *The Sugar Beet Crop.* Chapman & Hall, London.

Cooke, G.W. *et al.*, eds (1977) *Agricultural Efficiency.* The Royal Society, London.

Cooper, J.P., ed. (1975) *Photosynthesis and Productivity in Different Environments.* Cambridge University Press (International Biological Programme, 3).

Cooper, R.L. (1985) Breeding semi-dwarf soybean. *Pl. Breed. Rev.*, **3**, 289–311.

Copersucar (1987) *International Sugarcane Breeding Workshop*, São Paulo, Brazil.

Corbin, D.R., Greenplate, J.T. and Purcell, J.P. (1998) The identification and development of proteins for control of insects in genetically modified crops. *Hortscience*, **33**, 614–17.

Cornelius, P.L. and Dudley, J.W. (1974) Effects of inbreeding by selfing and full-sib mating in a maize population. *Crop Sci.*, **14**, 815–19.

Cornelius, P.L. and Dudley, J.W. (1976) Genetic variance components and predicted response to selection under selfing and full-sib mating in a maize population. *Crop Sci.*, **16**, 333–9.

Cox, D.R. (1958) *Planning of Experiments.* Wiley, New York, London, Sydney.

Crane, M.B. and Lawrence, W.J.C. (1938) *The Genetics of Garden Plants.* Macmillan, London.

Creech, J.L. and Reitz, L.P. (1971) Plant germ-plasm now and for tomorrow. *Adv. Agron.*, **23**, 1–49.

Crill, P. (1977) An assessment of stabilizing selection in crop variety development. *Ann. Rev. Phytopath.*, **15**, 185–202.

Crow, J.F. and Kimura, M. (1970) *An Introduction to Population Genetics Theory.* Harper and Row, New York.

Culbertson, J.O. (1954) Seed flax improvement. *Adv. Agron.*, **6**, 143–82.

Culp, T.W. and Harrell, D.C. (1973) Breeding methods for improving yield and fiber quality of Upland Cotton (*Gossypium hirsutum*). *Crop Sci.*, **13**, 686–9.

Curnow, I. (1961) Optimal programmes for varietal selection. *J. Roy. Stat. Soc. B*, **23**, 282–318.

Curtis, G.J. and Hornsey, K.G. (1972) Competitive and yield compensation in relation to breeding sugar beet. *J. Agric. Sci. Camb.*, **79**, 115–19.

Dale, P.J. (1976) Tissue culture in plant breeding. *Welsh Pl. Br. Sta. Ann. Rep.*, **1976**, 101–13.

Dale, P.J. (1995) R&D regulation and field trialling of transgenic crops. *Trends in Biotechnology*, **13**, 398–403.

Darlington, C.D. (1973) *Chromosome Botany and Origins of Cultivated Plants.* Allen and Unwin, London, 3rd edn.

Darlington, C.D. and Wylie, A.P. (1956) *Chromosome Atlas of Flowering Plants.* Allen and Unwin, London, 2nd edn.

Darrah, L.L., Eberhart, S.A. and Penny, L.H. (1972) A maize breeding methods study in Kenya. *Crop Sci.*, **12**, 605–8.

Darrah, L.L. and Hallauer, A.R. (1972) Genetic effects estimated from generation means in four diallel sets of maize inbreds. *Crop Sci.*, **12**, 615–21.

Darwin, C.R. (1876) *The Effects of Cross-and Self-Fertilisation in the Vegetable Kingdom.* Murray, London.

Darwin, C.R. (1882) *The Variation of Animals and Plants under Domestication.* Murray, London, 2nd edn.

Davey, V.M. (1959) Cultivated Brassicae: information available to the breeder. *Ann. Rep. Scott. Pl. Br. Sta.*, **39**, 23–62.

Davidson, B.R. (1962) Crop yields in experiments and on farms. *Nature, Lond.*, **194**, 458–9.

Davis, D.N. (1974) Synthesis of commercial F_1 hybrids in cotton. I. *Crop Sci.*, **14**, 745–9.

Day, P.R. (1973) Genetic variability of crops. *Ann. Rev. Phytopathol.*, **11**, 293–312.

Day, P.R. (1974) *Genetics of Host-Parasite Interaction.* Freeman, San Francisco.

De Candolle, A. (1886) *Origin of Cultivated Plants.* Hafner, New York (reprint of 2nd edn., 1959).

Demarly, Y. (1977) *Génétique et Amélioration des Plantes.* Masson, Paris.

De Pauw, R.M. and Shebeski, L.H. (1973) An evaluation of an early generation yield testing procedure in *Triticum aestivum. Can. J. Plant Sci.*, **53**, 465–70.

Dermen, H. (1960) Nature of plant sports. *Amer. Hort. Mag.*, **39**, 123–73.

Dixon, G.E. (1960) A review of wheat breeding in Kenya. *Euphytica*, **9**, 209–21.

Dobzhansky, T. (1949) *Genetics and the Origin of Species.* Columbia University Press, New York, 2nd edn.

Dobzhansky, T., Ayala, F.J., Stebbins, G.L. and Valentine, J.W. (1977) *Evolution.* Freeman, San Francisco.

Doggett, H. (1970) *Sorghum.* Longman, London.

Doggett, H. (1972) Recurrent selection in sorghum populations. *Heredity*, **28**, 9–29.

Doggett, H. and Majisu, B.N. (1968) Disruptive selection in crop development. *Heredity*, **23**, 1–23.

Doggett, H. and Majisu, B.N. (1972) Fertility improvement in autotetraploid sorghum 3. Yields of cultivated tetraploids. *Euthytica*, **21**, 86–9.

Donald, C.M. (1963) Competition among crops and pasture plants. *Adv. Agron.*, **15**, 1–118.

Donald, C.M. (1968) The breeding of crop ideotypes. *Eupthytica*, **17**, 385–403.

Donald, C.M. and Hamblin, J. (1976) The biological yield and harvest index of cereals as agronomic and plant breeding criteria. *Adv. Agron.*, **28**, 361–405.

Driscoll, C.J. (1972) Genetic suppression of homoeologous chromosome pairing in hexaploid wheat. *Canad. J. Genet. Cytol.*, **14**, 39–42.

Dubois, J.B. and Fossati, A. (1981) Influence of nitrogen uptake and nitrogen partitioning efficiency on grain yield and grain protein concentration of twelve winter wheat genotypes. *Zeitschr. f. Pflanzen*, **86**, 41–9.

Dudley, J.W., ed. (1974) *Seventy Generations of Selection for Oil and Protein in Maize.* Crop Science Society of America, Madison, Wis., USA.

Dudley, J.W. and Alexander, D.E. (1969) Performance of advanced generations of autotetraploid maize synthetics. *Crop Sci.*, **9**, 613–15.

Dudley, J.W. and Moll, R.H. (1969) Interpretation and use of estimates of heritability and genetic variances in plant breeding. *Crop Science*, **9**, 257–62.

Dumet, D.J., Benson, E.E., Reed, B., Harding, K., Millam, S. and Brennen, R. (1998) Developing Cryopreservation strategies for genebanks of vegetatively propagated crop plants. *In Vitro Cellular and Developmental Biology*, **34** (2), 49A.

Dunbier, M.W. and Bingham, E.T. (1975) Maximum heterozygosity in alfalfa: results using haploid-derived autotetraploids. *Crop Sci.*, **15**, 527–31.

Dunbier, M.W., Eskew, D.L., Bingham, E.T. and Schrader, L.E. (1975) Performance of genetically comparable diploid and tetraploid alfalfa: agronomic and physiological parameters. *Crop Sci.*, **15**, 211–14.

Duncan, D.R. and Widholm, J.M. (1986) Cell selection for crop improvement. *Pl. Breed. Rev.*, **4**, 153–74.

Duncan, E.N., Pate, J.B. and Turner, J.H. (1963) Performance of AHA derivative synthetic varieties of cotton. *Crop Sci.*, **3**, 233–4.

Dunder, E., Dawson, J., Suttie, J. and Pace, G. (1995) Maize transformation by microprojectile bombardment of immature embryos. In: *Gene Transfer to Plants* (eds I. Potrykus and G. Spangenberg) pp. 127–38. Berlin, Springer.

Duvick, D.N. (1965) Cytoplasmic pollen sterility in corn. *Adv. Genet.*, **13**, 1–56.

Duvick, D.N. (1974) Continuous backcrossing to transfer prolificacy to a single-eared inbred line of maize. *Crop Sci.*, **14**, 69–71.

Eagles, H.A. and Frey, K.J. (1974) Expected and actual gains in economic value of oat lines from five selection methods. *Crop Sci.*, **14**, 861–4.

Eagles, H.A., Hinz, P.N. and Frey, K.J (1977) Selection of superior cultivars of oats using regression coefficients. *Crop Sci.*, **17**, 101–5.

Early, H.L. and Qualset, C.O. (1971) Complementary competition in cultivated barley. *Euphytica*, **20**, 400–9.

Eastin, J.D., Haskins, F.A., Sullivan, C.Y. and Bavel, C.H.M. Van, eds. (1969) *Physiological Aspects of Crop Yield.* American Society of Agronomy, Madison, USA.

Eastin, J.D. and Munson, R.D., eds (1971) *Moving off the Yield Plateau.* American Society of Agronomy, Madison (ASA Special Publication, 20).

Eberhart, S.A., Debela, S. and Hallauer, A.R. (1973) Reciprocal recurrent selection in the BSSS and BSCBI maize populations and half-sib selection in BSSS. *Crop Sci.*, **13**, 451–6.

Eberhart, S.A., Harrison, M.N. and Ogada, F. (1967) A comprehensive breeding system. *Züchter*, **37**, 169–74.

Eberhart, S.A. and Russell, W.A. (1966) Stability parameters for comparing varieties. *Crop Sci.*, **6**, 36–40.

Eberhart, S.A. and Russell, W.A. (1969) Yield and stability for a 10-line diallel of single cross and double cross hybrids. *Crop Sci.*, **9**, 357–61.

Edwardson, J.R. (1956, 1970) Cytoplasmic male sterility. *Bot. Rev.*, **22**, 696–738; **36**, 341–420.

Eenink, A.H. (1974) Matromorphy in *Brassica oleracea*, 1–5. *Euphytica*, **23**, 429–33, 435–45, 711–18, 719–24, 725–36.

Eenink, A.H. (1976) Genetics of host-parasite relationships and uniform differential resistance. *Neth. J. Pl. Path.*, **82**, 133–45.

Efron, Y. and Everett, H.L. (1969) Evaluation of exotic germ plasm for improving corn yields in Northern United States. *Crop Sci.*, **9**, 44–7.

Ehdaie, B. and Cress, C.E. (1973) Simulation of cyclic single cross selection. *Theor. Appl. Genet.*, **43**, 374–80.

Ehrenberg, L. (1960) Induced mutations in plants: mechanisms and principles. *Genet. Agraria*, **12**, 364–89.

Eigsti, C.J. and Dustin, P. (1955) *Colchicine in Agriculture, Medicine, Biology and Chemistry.* Iowa State College Press, Ames, Iowa, USA.

Elgin, J.H., Hill, R.R. and Zeiders, K.E. (1970) Comparison of four methods of multiple trait selection for five traits in alfalfa. *Crop Sci.*, **10**, 190–3.

Ellerström, S. and Zagorcheva, L. (1977) Sterility and apomictic embryo-sac formation in *Raphanobrassica. Hereditas*, **87**, 107–20.

Elliott, F.C. (1958) *Plant Breeding and Cytogenetics.* McGraw-Hill, New York, Toronto and London.

Empig, L.T. and Fehr, W.R. (1971) Evaluation of methods for generation advance in bulk hybrid soybean populations. *Crop Sci.*, **11**, 51–4.

England, F.J.W. (1967, 1968) Non-sward densities for the assessment of yield in Italian ryegrass, I, II. *J. Agric. Sci. Camb.*, **68**, 235–41; **70**, 105–8.

England, F.J.W. (1974) A general approximate method for fitting additive and specific combining abilities to the diallel cross with unequal numbers of observations in the cells. *Theor. Appl. Genet.*, **44**, 378–80.

England, F.J.W. (1977) Response to family selection based on replicated trials. *J. Agric. Sci. Camb.*, **88**, 127–34.

Erskine, W. (1977) Adaptation and competition in mixtures of cowpea (*Vigna unguiculata*). *Euphytica*, **26**, 193–202.

Evans, L.T., ed. (1975) *Crop Physiology, some Case Histories.* Cambridge University Press.

Evans, L.T. (1993) *Crop Evolution, Adaptation and Yield.* Cambridge University Press.

Evenson, R.E. and Kislev, Y. (1975) *Agricultural Research and Productivity.* Yale University Press, New York and London.

Ewans, W.J. (1969) *Population Genetics.* Methuen, London.

Faegri, K. and Pijl, L. van der (1971) *The Principles of Pollination Ecology.* Pergamon, Oxford, 2nd edn.

Falcon, W.P. (1970) The green revolution: generations of problems. *Amer. J. Agric. Econ.*, **52**, 698–710.

Falconer, D.S. (1989) *Introduction to Quantitative Genetics*. Longman, Harlow, 3rd edn.

FAO (1961) *Agricultural and Horticultural Seeds.* FAO, Rome (FAO Agricultural Studies, 55).

FAO (1967) *FAO/IBP Technical Conference on the Exploration, Utilisation and Conservation of Plant Genetic Resources.* FAO, Rome.

FAO (1986) *Breeding for durable resistance in perennial crops.* (FAO, Plant Production and Protection Paper, 70). FAO, Rome.

FAO (1988) *Breeding for durable disease and pest resistance.* (FAO Plant Production and Protection Paper, 55 (2nd edn)). FAO, Rome.

Farmer, B.H., ed. (1977) *Green Revolution?* Macmillan, London.

Feaster, C.V. and Turcotte, E.L. (1973) Yield stability in doubled haploids of American Pima cotton. *Crop Sci.*, **13**, 232–3.

Fedak, G. and Fejer, S.O. (1975) Yield advantage in F_1 hybrids between spring and winter barley. *Can. J. Plant Sci.*, **55**, 547–53.

Federer, W.T. (1963) Procedures and designs useful for screening material in selection and allocation. *Biometrics*, **19**, 553–87.

Federer, W.T. and Sprague, G.F. (1947) A comparison of variance components in corn yield trials I. *J. Amer. Soc. Agron.*, **39**, 453–63.

Fehr, W.R. and Rodriguez, S.R. (1974) Effect of row spacing and genotype frequency on the yield of soybean blends. *Crop Sci.*, **14**, 521–5.

Feistritzer, W.P., ed. (1975) *Cereal Seed Technology*. FAO–UNO, Rome (FAO Agric.. Paper, 98, p. 238).

Ferweda, F.P. and Wit, F., eds (1969) *Outlines of Perennial Crop Breeding in the Tropics.* Veenman and Zonen, Wageningen, The Netherlands.

Fiddian, W.E.H. *et al.* (1973) Prospects for the health of cereal crops. *Ann. Appl.. Biol.*, **75**, 123–49.

Finlay, K.W. and Shepherd, K.W. (1968) *Proceedings of the Third International Wheat Genetics Symposium.* Australian Academy of Science, Canberra.

Finlay, K.W. and Wilkinson, G.N. (1963) The analysis of adaptation in a plant breeding programme. *Aust. J. Agric.. Res.*, **14**, 742–54.

Finney, D.J. (1958) Plant selection for yield improvement. *Euphytica*, **7**, 83–106.

Finney, D.J. (1988, 1989) Was this in your statistics textbook?
1. Agricultural Scientist and Statistician *Exper. Agric.*, **24**, 153–61.
2. Data handling, ibid, **24**, 343–53.
3. Design and Analysis, ibid., **24**, 421–32.
4. Frequency Data, ibid, **25**, 11–25.
5. Transformation of Data, ibid, **25**, 165–75.
6. Regression and Covariance, ibid., **25**, 291–311.

Finney, D.J. *et al.* (1964) In memoriam Ronald Aylmer Fisher, 1890–1962. *Biometrics*, 237–373.

Finzat, Y. and Atkins, R.E. (1953) Genetic and environmental variability in segregating barley populations. *Agron. J.*, **45**, 414–20.

Fischer, R.A. and Kertesz, Z. (1976) Harvest Index in spaced populations and grain weight in microplots as indicators of yielding ability in spring wheat. *Crop Sci.*, **16**, 55–9.

Fishel, W.L., ed. (1971) *Resource Allocation in Agricultural Research.* University of Minnesota Press, Minneapolis, USA.

Fleming, A.A. (1971) Performance of stocks within long time inbred lines of maize in test crosses. *Crop Sci.*, **11**, 620–2.

Flor, H.H. (1971) Current status of the gene-for-gene concept. *Ann. Rev. Phytopath*, **9**, 275–96.

Forster, B.P. *et al.* (1997) Genetic engineering of crop plants; from genome to gene. *Exper. Agricul.*, **33**, 15–33.

Foster, C.A. (1971a) Interpopulational and intervarietal hybridisation in *Lolium perenne* breeding: heterosis under non-competitive conditions. *J. Agric. Sci. Camb.*, **76**, 107–30.

Foster, C.A. (1971b) A study of the theoretical expectation of F_1 hybridity resulting from bulk interpopulational hybridisation in herbage grasses. *J. Agric. Sci. Camb.*, **76**, 295–300.

Foster, C.A. (1971c) Interpopulational and intervarietal hybrids in *Lolium perenne*: heterosis under simulated sward conditions. *J. Agric. Sci. Camb.*, **76**, 401–10.

Foster, C.A. (1973a) Interpopulational and intervarietal F_1 hybrids in *Lolium perenne*: performance in field sward conditions. *J. Agric. Sci. Camb.*, **80**, 463–77.

Foster, C.A. (1973b) Prospects for the development of F_1 hybrid varieties of perennial ryegrass and other herbage grasses. *J. Agric. Sci. Camb.*, **81**, 33–7.

Fowler, D.B. and Roche, L.A. de la (1975) Wheat quality evaluation 1, 2, 3. *Can. J. Plant Sci.*, **55**, 241–9, 251–62, 263–9.

Frankel, O.H. (1939) Analytical yield investigations on New Zealand wheat. *J. Agric. Sci. Camb.*, **29**, 249–61.

Frankel, O.H. (1947) The theory of plant breeding for yield. *Heredity*, **1**, 109–20.

Frankel, O.H. (1950) The development and maintenance of superior genetic stocks. *Heredity*, **4**, 89–102.

Frankel, O.H. (1958) The dynamics of plant breeding. *J. Aust. Inst. Agric. Sci.*, **24**, 112–23.

Frankel, O.H. and Bennett, E., eds (1970) *Genetic Resources in Plants, their Exploration and Conservation.* Blackwell, Oxford (Intnl Biological Programme, Handbook 11).

Frankel, O.H. and Hawkes, J.G., eds (1975) *Crop Genetic Resources for today and tomorrow.* Cambridge University Press. (International Biological Programme, 2.)

Frankel, R. and Galun, E. (1977) *Pollination Mechanisms, Reproduction and Plant Breeding.* Springer, Berlin, Heidelberg and New York.

Freeman, G.H. (1973) Statistical methods for the analysis of genotype–environment interactions. *Heredity*, **31**, 339–54.

Frey, K.J. (1954) The use of F_2 lines in predicting the performance of F_3 selections in two barley crosses. *Agron. J.*, **46**, 541–4.

Frey, K. (1962) Effectiveness of visual selection upon yield in oat crosses. *Crop Sci.*, **2**, 102–5.

Frey, K.J., ed. (1966) *Plant Breeding.* Iowa State University Press, Ames, Iowa, USA.

Frey, K.J. (1967) Mass selection for seed width in oat populations. *Euphytica*, **16**, 341–9.

Frey, K.J. (1968) Expected genetic advances from three simulated selection schemes. *Crop Sci.*, **8**, 235–8.

Frey, K.J. (1972) Stability indexes for isolines of oats (*Avena sativa*). *Crop Sci.*, **12**, 809–12.

Frey, K.J., Browning, J.A. and Simons, M.D. (1973) Management of host resistance genes to control diseases. *Z. PflKrankh. PflSchutz*, **80**, 160–80.

Frey, K.J. and Horner, T. (1955) Comparison of actual and predicted gains in barley selection experiments. *Agron. J.*, **47**, 186–8.

Frey, K.J. and Huang, T.F. (1969) Relation of seed weight to grain yield in oats. *Euphytica*, **18**, 417–24.

Frey, K.J. and Maldonado, V. (1967) Relative productivity of homogeneous and heterogeneous oat cultivars in optimum and suboptimum environments. *Crop Sci.*, **7**, 532–5.

Fry, W.E. (1975) Integrated effects of polygenic resistance and a protective fungicide on development of potato late blight. *Phytopath.*, **65**, 908–11.

Funk, C.R. and Anderson, J.C. (1964) Performances of mixtures of field corn (*Zea mays*) hybrids. *Crop Sci.*, **4**, 353–6.

Fyfe, J.L. and Gilbert, N.E. (1963) Partial diallel crosses. *Biometrics*, **19**, 278–86.

Gale, M.D. and Law, C.N. (1977) The identification and exploitation of Norin 10 semi-dwarfing genes. *Pl. Br. Inst. Camb. Ann. Rep., 1976*, 21–35.

Galinat, W.C. (1965) The evolution of corn and corn culture in North America. *Econ. Bot.*, **19**, 350–7.

Galinat, W.C. (1973) Intergenomic mapping of maize, teosinte and *Tripsacum. Evolution*, **27**, 644–55.

Gallegly, M.E. (1968) Genetics of pathogenicity of *Phytophthora infestans. Ann. Rev. Phytopath.*, **6**, 375–96.

Gallun, R.L., Starks, K.J. and Guthrie, W.D. Plant resistance to insects attacking cereals. *Ann. Rev. Entomol.*, **20**, 337–57.

Gama, E.E.G. and Hallauer, A.R. (1977) Relations between inbred and hybrid traits in maize. *Crop Sci.*, **17**, 703–6.

Gardner, C.O. (1961) An evaluation of effects of mass selection and seed irradiation with thermal neutrons on yield of corn. *Cop Sci.*, **1**, 241–5.

Gardner, C.O., Harvey, P.H., Comstock, R.E. and Robinson, H.F. (1953) Dominance of genes controlling quantitative characters in maize. *Agron. J.*, **43**, 186–91.

Gardner, C.O. and Lonnquist, J.H. (1959) Linkage and the degree of dominance of genes controlling quantitative characters in maize. *Agron. J.*, **51**, 524–8.

Gatehouse, A.M.R. and Gatehouse, J.A. (1998) Identifying proteins with insecticidal activity: Use of encoding genes to produce insect-resistant crops. *Pesticide Science*, **52**, 156–75.

Gaul, H., ed. (1976) *Barley Genetics III Münich (Proc. III. Intnl. Barley Genet. Symp.*, 1975).

Geadelmann, J.L. and Frey, K.J. (1975) Direct and indirect mass selection for grain yield in bulk oat populations. *Crop Sci.*, **15**, 490–4.

Geadelmann, J.L. and Peterson, R.H. (1976) Effect of yield component selection on the general combining ability of maize inbred lines. *Crop Sci.*, **16**, 807–11.

Gedge, D.L., Fehr, W.R. and Walker, A.K. (1977) Intergenotypic competition between rows and within blends of soybeans. *Crop Sci.*, **17**, 787–90.

Genter, C.F. (1971) Yields of S_1 lines from original and advanced synthetic varieties of maize. *Crop. Sci.*, **11**, 821–4.

Genter, C.F. (1976a) Recurrent selection for yield in the F_2 of a maize single cross. *Crop Sci.*, **16**, 350–2.

Genter, C.F. (1976b) Mass selection of a composite of intercrosses of Mexican races of maize. *Crop Sci.*, **16**, 556–8.

Genter, C.F. and Eberhart, S.A. (1974) Performance of original and advanced maize populations and their diallel crosses. *Crop Sci.*, **14**, 881–5.

Gerstel, D.U. (1945) Inheritance in *Nicotiana tabacum*. XX. The addition of *Nicotiana glutinosa* chromosomes to tobacco. *J. Hered.*, **36**, 197–206.

Gerstel, D.U. (1966) Evolutionary problems in some polyploid crop plants. *Proc. 2nd Wheat Genet. Symp., Lund*, **2**, 481–504.

Gilbert, N.E. (1958) Diallel cross in plant breeding. *Heredity*, **12**, 477–92.

Gilbert, N.E. (1961) Polygene analysis, I, II. *Genet. Res. Camb.*, **2**, 96–105, 456–60.

Gilbert, N.E. (1967) Additive combining abilities fitted to plant breeding data. *Biometrics*, **23**, 45–9.

Gilbert, N.E. (1973) *Biometrical Interpretation.* Clarendon Press, Oxford.

Gilbert, N.E., Dodds, K.S. and Subramaniam, S. (1973) Progress of breeding investigations with *Hevea brasiliensis*. V. Analysis of data from earlier crosses. *J. Rubb. Res. Inst. Malaya*, **23**, 365–80.

Giles, R.J., McConnell, G. and Fyfe, J.L. (1974) The frequency of natural cross-fertilization in a composite cross of barley grown in Scotland. *J. Agric. Sci. Camb.*, **84**, 447–50.

Gladstones, J.S. (1970) Lupins as crop plants. *Field Crop Abstr.*, **23**, 123–48.

Gladstones, J.S. (1975) Lupin breeding in Western Australia. *J. Agric. West. Austr.*, **16**, 44–9.

Glover, D.M. and Hames, B.D., eds (1995) *DNA Cloning I Core Techniques, A Practical Approach.* Oxford University Press, 2nd edn.

Glover, D.M. and Hames, B.D., eds (1995) *DNA Cloning 2 A Practical Approach.* Oxford University Press, 2nd edn.

Good, R.L. and Horner, E.S. (1974) Effect of normal cytoplasms on resistance to southern corn leaf blight and on other traits in maize. *Crop Sci.*, **14**, 368–70.

Gorz, H.J. and Haskins, F.A. (1971) Evaluation of cross-fertilization in forage crops. *Crop Sci.*, **11**, 731–3.

Goulas, E.K. and Lonnquist, J.H. (1976) Combined half-sib and S_1 family selection in a maize composite population. *Crop Sci.*, **16**, 461–4.

Gowen, J.W., ed. (1952) *Heterosis.* Iowa State College, Ames, Iowa, USA.

Gowen, S., ed. (1995) *Bananas and Plantains*. Chapman & Hall, London.
Gowers, S. (1973) The production of F_1 hybrid swedes (*Brassica napus* ssp *rapifera*) by the utilization of self-incompatibility. *Euphytica*, **23**, 205–8.
Grafius, J.E. (1972) Competition for environmental resources by component characters. *Crop Sci.*, **12**, 364–7.
Grafius, J.E., Nelson, W.L. and Dirks, V.A. (1952) The heritability of yields in barley as measured by early generation bulked progenies. *Agron. J.*, **44**, 253–7.
Grafius, J.E., Thomas, R.L. and Barnard, J. (1976) Effect of parental component complementation on yield and components of yield in barley. *Crop Sci.*, **16**, 673–7.
Grant, V. (1963) *The Origin of Adaptations.* Columbia University Press, New York.
Grant, V. (1971) *Plant Speciation.* Columbia University Press, New York.
Grant, V. (1975) *Genetics of Flowering Plants.* Columbia University Press, New York.
Grant, V. (1977) *Organismic Evolution.* Freeman, San Francisco.
Gressel, J. (1993) Advances in achieving the needs for biotechnologically derived herbicides. *Pl. Breed. Rev.*, **11**, 155–98.
Grierson, D. and Schuch, W. (1993) Control of ripening. *Philosophical Transactions of the Royal Society of London* B342: 241–50.
Griffing, B. (1956a) A generalized treatment of the use of diallel crosses in quantitative inheritance. *Heredity*, **10**, 31–50.
Griffing, B. (1956b) Concept of general and specific combining ability in relation to diallel crossing systems. *Aust. J. Biol. Sci.*, **9**, 465–93.
Griffing, B. (1975) Efficiency changes due to use of doubled haploids in recurrent selection methods. *Theor. Appl. Genet.*, **46**, 367–86.
Griliches, Z. (1958) Research costs and social returns: hybrid corn and related innovations. *J. Polit. Econ.*, **66**, 419–31.
Grosser, J.W. and Gmitter, F.G. Jr. (1990) Protoplasm fusion and citrus improvement. *Pl. Breed. Rev.*, **8**, 339–74.
Grubben, G.J.H. (1977) *Tropical Vegetables and Their Genetic Resources.* IBPGR, Rome.
Grumet, R. (1994) Development of virus resistant plants via engineering. *Pl. Breed. Rev.*, **12**, 47–80.
Grümmer, G. and Roy, S.K. (1966) Intervarietal mixtures of rice and incidence of brown spot disease. *Nature, Lond.*, **209**, 1265–7.
Guitard, A.A., Newman, J.A. and Hoyt, P.B. (1961) The influence of seeding rate on the yield and yield components of wheat, oats and barley. *Can. J. Pl. Sci.*, **41**, 751–8.
Gustafsson, A. (1953) The cooperation of genotypes in barley. *Hereditas, Lund*, **39**, 1–18.
Gustafsson, A. (1954) Mutations, viability and population structure. *Acta Agric. Scand.*, **4**, 601–32.
Gustafsson, A. (1968) Reproduction mode and crop development. *Theor. Appl. Genet.*, **38**, 109–17.
Gustafsson, A., Hagberg, A., Persson, G. and Wiklund, K. (1971) Induced mutations and barley improvement. *Theor. Appl. Genet.*, **41**, 239–48.
Haag, W.L. and Hill, R.R. (1974) Comparison of selection methods for autotetraploids. II. Selection for disease resistance in alfalfa. *Crop Sci.*, **14**, 591–3.
Hagedoorn, A.L. (1950) *Plant Breeding.* Crosby Lockwood, London.
Hallauer, A.R. (1970) Genetic variability for yield after four cycles of reciprocal recurrent selection in maize. *Crop Sci.*, **10**, 482–5.
Hallauer, A.R. (1972) Third phase in the yield evaluation of synthetic varieties of maize. *Crop Sci.*, **12**, 16–18.

Hallauer, A.R. and Sears, J.H. (1969) Mass selection for yield in two varieties of maize. *Crop Sci.*, **9**, 47–50.

Hallauer, A.R. and Sears, J.H. (1973) Changes in quantitative traits associated with inbreeding in a synthetic variety of maize. *Crop Sci.*, **13**, 327–30.

Hamblin, J. (1977) Plant breeding interpretation of the effects of bulk breeding on four populations of beans (*Phaseolus vulgaris*). *Euphytica*, **26**, 157–68.

Hamblin, J. and Donald, C.M. (1973) The relationship between plant form, competitive ability and grain yield in a barley cross. *Euphytica*, **23**, 535–42.

Hamblin, J. and Morton, J.R. (1977) Genetic interpretation of the effects of bulk breeding on four populations of beans (*Phaseolus vulgaris*). *Euphytica*, **26**, 75–83.

Hamblin, J. and Rowell, R.G. (1975) Breeding implications of the relationship between competitive ability and grain yield in a barley cross. *Euphytica*, **24**, 221–8.

Hammat, N. (1992) Progress in the biotechnology of trees. *World Journal of Microbiology and Biotechnology*, **8**, 369–77.

Hammond, J.J. and Gardner, C.O. (1974a) Modification of the variety-cross diallel model for evaluating cycles of selection. *Crop Sci.*, **14**, 6–8.

Hammond, J.J. and Gardner, C.O. (1974b) Effect of genetic sampling technique on variation within populations derived by crossing, selfing or random-mating other populations. *Crop Sci.*, **14**, 63–6.

Hammons, R.O. (1976) Peanuts: genetic vulnerability and breeding strategy. *Crop Sci.*, **16**, 527–30.

Hanson, A.A. and Carnahan, H.L. (1956) Breeding perennial forage grasses. *USDA Tech. Bull.*, **1145**, p. 116.

Hanson, A.A. and Weiss, M.G., eds (1964) *Plant Breeders' Rights.* Crop Science Society of America, Madison, USA (*ASA Spec. Publ. 3*, p. 78).

Hanson, C.H., Busbice, T.H., Hill, R.R., Hunt, O.J. and Oakes, A.J. (1972) Directed mass selection for developing multiple pest resistance and conserving germ plasm in alfalfa. *J. Environm. Qual.*, **1**, 106–11.

Hanson, W.D. (1970) Genotypic stability. *Theor. Appl. Genet.*, **40**, 226–31.

Hanson, W.D. and Johnson, H.W. (1956) Methods for calculating and evaluating a general selection index by pooling information for two or more experiments. *Genetics*, **42**, 421–32.

Hanson, W.D. and Robinson, H.F., eds (1963) *Statistical Genetics and Plant Breeding.* National Academy of Sciences and National Research Council, Washington, USA.

Harborne, J.B., Boulter, D. and Turner, B.L. (eds) (1971) *Chemotaxonomy of the Leguminosae.* Academic Press, London.

Harding, J. and Allard, R.W. (1968) Population studies in predominantly self-pollinated species. XII. Interactions between loci affecting fitness in a population of *Phaseolus lunatus. Genetics*, **61**, 721–36.

Harding, J., Allard, R.W. and Smeltzer, D.G. (1966) Population studies in predominantly self-pollinated species. IX. Frequency-dependent selection in *Phaseolus lunatus. Proc. Natl. Acad. Sci. Washington*, **56**, 99–104.

Hardwick, R.C. and Wood, J.T. (1972) Regression methods for studying genotype-environment interactions. *Heredity*, **28**, 209–22.

Harlan, H.V. and Martini, M.L. (1929) A composite hybrid mixture. *J. Amer. Soc. Agron.*, **21**, 487–90.

Harlan, H.V. and Martini, M.L. (1938) The effect of natural selection in a mixture of barley varieties. *J. Agric. Res.*, **57**, 189–99.

Harlan, H.V., Martini, M.L. and Stevens, H. (1940) A study of methods in barley breeding. *USDA Tech. Bull.*, **720**, p. 26.

Harlan, J.R. (1956) Distribution and utilization of natural variability in cultivated plants. *Brookhaven Symp. Biol.*, **9**, 191–208.

Harlan, J.R. (1971) Agricultural origins: centers and non-centers. *Science*, **174**, 468–74.

Harlan, J.R. (1972) Genetics of disaster. *J. Environm. Qual.*, **1**, 212–15.

Harlan, J.R. (1975a) *Crops and Man.* Madison, Wis., USA.

Harlan, J.R. (1975b) Geographic patterns of variation in some cultivated plants. *J. Hered.*, **66**, 184–91.

Harlan, J.R. (1976a) Diseases as a factor in plant evolution. *Ann. Rev. Phytopathol.*, **14**, 31–51.

Harlan, J.R. (1976b) Genetic resources in wild relatives of crops. *Crop Sci.*, **16**, 329–33.

Harlan, J.R., De Wet, J.M.J. and Price, E.G. (1973) Comparative evolution of cereals. *Evolution*, **271**, 311–25.

Harlan, J.R. and De Wet, M.J. (1975) On O. Winge and a prayer: the origins of polyploidy. *Bot. Rev.*, **41**, 361–90.

Harland, S.C. (1944) The selection experiment with the Peruvian Tangüis cotton. *Inst. Cotton. Genet., Lima, Peru, Bull.*, **1**, p. 98.

Harrington, J.B. (1932) Predicting the value of a cross from an F_2 analysis. *Can. J. Res.*, **6**, 21–37.

Harrington, J.B. (1937) The mass-pedigree method in the hybridization improvement of cereals. *J. Amer. Soc. Agron.*, **29**, 379–84.

Harrington, J.B. (1940) Yielding capacity of wheat crosses as indicated by bulk hybrid tests. *Can. J. Res.*, **57**, 189–99.

Harris, P.M., ed. (1978) *The Potato Crop. The Scientific Basis for Improvement.* Chapman and Hall, London.

Harris, D.R. (1972) The origins of agriculture in the tropics. *Amer. Sci.*, **60**, 180–93.

Harris, P., ed. (1991) *The Potato Crop.* Chapman & Hall, London.

Harris, R.E., Gardner, C.O. and Compton, W.A. (1972) Effects of mass selection and irradiation in corn measured by random S_1 lines and their testcrosses. *Crop Sci.*, **12**, 594–8.

Harris, R.W., Moll, R.H. and Stuber, C.W. (1976) Control and inheritance of prolificacy in maize. *Crop Sci.*, **16**, 843–50.

Hartwig, E.E. (1972) Utilization of soybean germplasm strains in a soybean improvement program. *Crop Sci.*, **12**, 856–9.

Harvey, P.H., Levings, C.S. and Wernsman, E.A. (1972) The role of extrachromosomal inheritance in plant breeding. *Adv. Agron.*, **24**, 1–28.

Haunold, A. (1975) Use of triploid males for increasing hop yield. *Crop Sci.*, **15**, 833–40.

Hawkes, J.G., ed. (1968) *Chemotaxonomy and Serotaxonomy.* Academic Press, London.

Hawkes, J.G. (1983) *The Diversity of Crop Plants.* Harvard University Press, Cambridge, Mass.

Hawkes, J.G. and Lange, W., eds (1973) *European and Regional Gene-Banks.* Eucarpia, Wageningen, The Netherlands.

Hawkes, J.G., Williams, J.T. and Hanson, J. (1976) *A Bibliography of Plant Genetic Resources.* IBPGR, Rome (with supplement by J.T. Williams).

Hayes, J.D. and Paroda, R.S. (1974) Parental generation in relation to combining ability analysis in spring barley. *Theor. Appl. Genet.*, **44**, 373–7.

Hayes, H.K. (1963) *A Professor's Story of Hybrid Corn.* Burgess, Minneapolis, USA.

Hayes, H.K., Immer, F.R. and Smith, D.C. (1955) *Methods of Plant Breeding.* McGraw-Hill, New York (2nd edn.).

Hayman, B.I. (1954) The theory and analysis of diallel crosses. *Genetics*, **39**, 789–809.

Hayter, A.M. and Allison, M.J. (1976) High diastase barley. *ARC Res. Rev.*, **2**, 42–4.

Hayward, M.D. *et al.*, eds (1993) *Plant Breeding Principles and Prospects.* Chapman & Hall, London.

Hebblethwaite, P.D., ed. (1983) *The Faba Bean.* Butterworths, London.

Hecker, R.J. (1972) Inbreeding depression in diploid and autotetraploid sugar beet, *Beta vulgaris. Euphytica*, **21**, 106–11.

Hehn, E.R. and Barmore, M.A. (1965) Breeding wheat for quality. *Adv. Agron.*, **17**, 85–114.

Heiser, C.B. (1973) *Seed to Civilization.* Freeman, San Francisco.

Hermsen, J.G.T. (1969) Frequencies of reciprocally compatible single, three-way and double crosses. *Euphytica*, **18**, 170–7.

Hernandez Bermejo, J.E. and Léon, J. (1994) *Neglected Crops: 1492 from a different perspective.* FAO, Rome.

Herrera-Estrella, L., Depicker, A., Van Montagu, M. and Schell, J. (1983) Expression of chimaeric genes into plant cells using a Ti-plasmid derived vector. *Nature (London)*, **303**, 209–13.

Heslop-Harrison, J. (1974) Genetic resource conservation: the end and the means. *J. Roy. Soc. Arts*, **122**, 157–69.

Heslop-Harrison, J. and Lewis, D., eds (1975) A discussion on incompatibility in flowering plants. *Proc. Roy. Soc. London. B.*, **188**, 233–375.

Hill, J. (1975) Genotype-environment interactions – a challenge for plant breeding. *J. Agric. Sci. Camb.*, **85**, 477–94.

Hill, R.R. (1971a) Selection in autotetraploids. *Theor. Appl. Genet.*, **41**, 181–6.

Hill, R.R. (1971b) Effect of the number of parents on the mean and variance of synthetic varieties. *Crop Sci.*, **11**, 283–6.

Hill, R.R. and Haag, W.L. (1974) Comparison of selection methods for autotetraploids I. Theoretical. *Crop Sci.*, **14**, 587–90.

Hill, R.R., Leath, K.T. and Zeiders, K.E. (1972) Combining ability among four-clone alfalfa synthetics. *Crop Sci.*, **12**, 627–30.

Hinson, K. and Hanson, W.D. (1962) Competition studies in soybeans. *Crop Sci.*, **2**, 117–23.

Hodgson, R.E., ed. (1961) *Germ Plasm Resources.* American Association for the Advancement of Science, Washington, Publ. 66.

Hoegemeyer, T.C. and Hallauer, A.R. (1976) Selection among and within full-sib families to develop single crosses of maize. *Crop Sci.*, **16**, 76–81.

Hogarth, D.M. (1968) A review of quantitative genetics in plant breeding with particular reference to sugar cane. *J. Aust. Inst. Agric. Sci.*, **34**, 108–20.

Hogarth, D.M. (1971, 1977) Quantitative inheritance studies in sugarcane. I, II, III. *Aust. J. Agric. Res.*, **22**, 92–102, 103–10; **28**, 257–68.

Holcomb, J., Tolbert, D.M. and Jain, S.K. (1977) A diversity analysis of genetic resources in rice. *Euphytica*, **26**, 441–50.

Holden, J.H.W. (1977) Potato breeding at Pentlandfield. *Scott. Pl. Br. Sta. Ann. Rep.*, **56**, 66–97.

Hollaender, A. *et al.*, eds (1977) *Genetic Engineering for Nitrogen Fixation.* Plenum, New York.

Holmes, F.O. (1954) Inheritance of resistance to viral diseases in plants. *Adv. Virus Res.*, **2**, 1–30.

Holmes, F.O. (1965) Genetics of pathogenicity in viruses and of resistance in host plants. *Adv. Virus Res.*, **11**, 139–61.

Holton, C.S. *et al.*, eds (1959) *Plant Pathology, Problems and Progress, 1908–58.* University of Wisconsin Press, Madison, Wis., USA.

Hooker, A.L. (1977) A plant pathologist's view of germplasm evaluation and utilization. *Crop Sci.*, **17**, 689–94.

Hooks, J.A., Williams, J.H. and Gardner, C.O. (1971) Estimates of heterosis from a diallel cross of inbred lines of castors. *Ricinus communis*, *Crop Sci.*, **11**, 651–5.

Horner, E.S. (1968) Effect of a generation of inbreeding on genetic variation in corn as related to recurrent selection procedures. *Crop Sci.*, **8**, 32–5.

Horner, E.S., Chapman, W.H., Lundy, H.W. and Lutrick, M.C. (1972) Commercial utilization of the products of recurrent selection for specific combining ability in maize. *Crop Sci.*, **12**, 602–4.

Horner, E.S., Lundy, H.W., Lutrick, M.C. and Chapman, W.H. (1973) Comparison of three methods of recurrent selection in maize. *Crop Sci.*, **13**, 485–9.

Horner, T.W. and Frey, K.J. (1957) Methods for determining natural areas for oat varietal recommendations. *Agron. J.*, **49**, 313–15.

Horner, T.W. and Weber, C.R. (1956) Theoretical and experimental study of self-fertilized populations. *Biometrics*, **12**, 404–14.

Hornsey, K.G. (1975) The exploitation of polyploidy in sugar-beet breeding. *J. Agric. Sci. Camb.*, **84**, 543–57.

Horsch, R.B., Fry, J.E., Hoffman, N.L., Eichholtz, D., Rogers, S.G. and Fraley, R.T. (1985) A simple and general method for transferring genes into plants. *Science*, **227**, 1229–31.

Horsfall, J.G. and Dimond, A.E. (1959, 1960) *Plant Pathology*, *An Advanced Treatise*, *I*, *II*, *III*. Academic Press, New York and London.

Howard, H.W. (1961) Potato cytology and genetics. *Bibl. Genet.*, **19**, 87–216.

Howard, H.W. (1962) Some potato breeding problems. *Pl. Br. Inst. Camb. Ann. Rep*, *1961–62*, 5–21.

Howard, H.W. (1970) *Genetics of the Potato.* Logos, London.

Howard, H.W., Johnson, R., Russell, G.E. and Wolfe, M.S. (1969) Problems in breeding for resistance to pests and diseases. *Pl. Br. Inst. Camb. Ann. Rep.*, *1968–69*, 6–36.

Hoyt, E. (1988) *Conserving the Wild Relatives of Crops.* IBPGR, Rome.

Hughes, H.D., Heath, M.E. and Metcalfe, D.S., eds (1962) *Forages* Iowa State University Press, Ames, Iowa, USA, 2nd edn.

Hunt, O.J., Peaden, R.N., Nielson, M.W. and Hanson, C.H. (1971) Development of two alfalfa populations with resistance to insect pests, nematodes and diseases. I Aphid resistance. *Crop Sci.*, **11**, 73–5.

Hunter, H. and Leake, H.M. (1933) *Recent Advances in Agricultural Plant Breeding.* Churchill, London.

Hutchinson, J.B. (1959) *The Application of Genetics to Cotton Improvement.* Cambridge University Press.

Hutchinson, J.B., ed. (1965) *Essays in Crop Plant Evolution.* Cambridge University Press.

Hutchinson, J.B. (1971a) Changing concepts in crop plant evolution. *Expl. Agric.*, **7**, 273–80.

Hutchinson, J.B. (1971b) High cereal yields. *J. Roy. Soc. Arts*, **119**, 104–14.

Hutchinson, J.B., ed. (1974) *Evolutionary Studies in World Crops*. Cambridge University Press.

Hutchinson, J.B., Clark, G., Jope, E.M. and Riley, R., eds (1976) The early history of agriculture. *Phil. Trans. R. Soc. Lond. B*, **275**, 1–213.

Hutton, E.M. (1970) Tropical pastures. *Adv. Agron.*, **22**, 2–74.

Hutton, E.M. (1976) Selecting and breeding tropical pasture plants. *SPAN*, **19**, 21–4.

Huxley, J., ed. (1940) *The New Systematics*. Oxford University Press.

Huxley, J. (1942) *Evolution, the Modern Synthesis*. Allen and Unwin, London.

Hyams, E. (1971) *Plants in the Service of Man*. Dent, London.

IAEA (1964) *The Use of Induced Mutations in Plant Breeding*. International Atomic Energy Agency, Vienna.

IAEA (1969) *Induced Mutations in Plants*. International Atomic Energy Agency, Vienna.

IAEA (1970a) *Manual on Mutation Breeding*. International Atomic Energy Agency, Vienna.

IAEA (1970b) *Improving Plant Protein by Nuclear Techniques*. International Atomic Energy Agency, Vienna.

IAEA (1972) *Induced Mutations and Plant Improvement*. International Atomic Energy Agency, Vienna.

IAEA (1973) *Induced Mutations in Vegetatively Propagated Plants*. International Atomic Energy Agency, Vienna.

IAEA (1974) *Polyploidy and Induced Mutations in Plant Breeding*. International Atomic Energy Agency, Vienna.

IBPGR (1975) *The Conservation of Crop Genetic Resources*. IBPGR, Rome.

IBPGR (1976) *Priorities among Crops and Regions*. IBPGR, Rome.

Imam, A.G. and Allard, R.W. (1965) Population studies in predominantly self-pollinated species. VI Genetic variability between and within natural populations of wild oats from differing habitats in California. *Genetics*, **51**, 49–62.

Immer, F.R. (1941) Relation between yielding ability and homozygosis in barley crosses. *J. Amer. Soc. Agron.*, **33**, 200–6.

Ingham, J.L. (1972) Phytoalexins and other natural products as factors in plant disease control. *Bot. Rev.*, **38**, 343–424.

Ingham, J. (1973) Disease resistance in higher plants. *Phytopath. Z.*, **78**, 314–35.

Innes, N.L. and Jones, G.B. (1977) Performance of seed mixtures and multilines of Upland cotton in Uganda. *J. Agric. Sci. Camb.*, **88**, 47–54.

IRRI (1972) *Rice Breeding*. International Rice Research Institute, Manila, Philippines.

IRRI (1977) *Research Highlights for 1976*. International Rice Research Institute, Los Baños, Philippines.

Jain, H.K. and Kulshrestha, V.P. (1976) Dwarfing genes and breeding for yield in bread wheat. *Z. f. PflzZüchtung*, **76**, 102–12.

Jain, S.K. (1959) Male sterility in flowering plants. *Bibl. Genet.*, **18**, 101–66.

Jain, S.K. (1961) Studies on the breeding of self-pollinated cereals. The composite cross bulk population method. *Euphytics*, **10**, 315–24.

Jain, S.K. and Allard, R.W. (1960) Population studies in predominantly self-pollinated species. I. Evidence for heterozygote advantage in a closed population of barley. *Proc. Natnl. Acad. Sci. Wash.*, **46**, 1371–7.

Jain, S.K. and Jain, K.B.L. (1962) The progress of inbreeding in a pedigree-bred population of barley. *Euphytica*, **11**, 229–32.

Jain, S.K. and Marshall, D.R. (1967) Population studies in predominantly self-pollinated species. X. Variation in natural populations of *Avena fatua* and *A. barbata*. *Amer. Nat.*, **101**, 19–33.

Jain, S.K., Qualset, C.O., Bhatt, G.M. and Wu, K.K (1975) Geographical patterns of phenotypic diversity in a world collection of durum wheats. *Crop Sci.*, **15**, 700–4.

Jain, S.K. and Suneson, C.A. (1964) Population studies in predominantly self-pollinated species. VII. Survival of a male-sterility gene in relation to heterozygosis in barley populations. *Genetics*, **50**, 905–13.

James, E. (1967) Preservation of seed stocks. *Adv. Agron.*, **19**, 87–106.

James, N.I. and Miller, J.D. (1975) Selection in six crops of sugarcane. II. Efficiency and optimum selection intensities. *Crop Sci.*, **15**, 37–40.

James, W.C. (1974) Assessment of plant diseases and losses. *Ann. Rev. Phytopathol.*, **12**, 27–48.

James, W.C. and Shih, C.S., Hodgson, W.A. and Calbeck, L.C. (1972) The quantitative relationship between late blight of potato and loss in tuber yield. *Phytopath*, **62**, 92–6.

Janick, J., ed. (1996) *Progress in New Crops*. ASHS Press, Alexandria, Virginia.

Janick, J. and Moore, J.N., eds (1975) *Advances in Fruit-Breeding*. Purdue University Press, Lafayette, Indiana, USA.

Janick, J. and Simon, J.E., eds (1990) *Advances in New Crops*. Timber Press, Portland, Oregon.

Janick, J. and Simon, J.E., eds (1993) *New Crops*. J. Wiley, New York.

Jan-orn, J., Gardner, C.O. and Ross, W.M. (1976) Quantitative genetic studies of the NP3R random-mating grain sorghum population. *Crop Sci.*, **16**, 489–96.

Janossy, A. and Lupton, F.G.H., eds (1976) *Heterosis in Plant Breeding*. Elsevier, Amsterdam.

Jenkins, M.T., Robert, A.L. and Findley, W.R. (1954) Recurrent selection as a method for concentrating genes for resistance to *Helminthosporium turcicum* leaf blight in corn. *Agron. J.*, **46**, 89–94.

Jennings, P.R. and de Jesus, J. (1968) Studies on competition in rice I. Competition in mixtures of varieties. *Evolution*, **22**, 119–24.

Jennings, P.R. and Herrera, R.M. (1968) Studies on competition in rice II. Competition in segregating populations. *Evolution*, **22**, 332–6.

Jensen, N.F. (1952) Intra-varietal diversification in oat breeding. *Agron. J.*, **44**, 30–4.

Jensen, N.F. (1965) Multiline superiority in cereals. *Crop Sci.*, **5**, 566–8.

Jensen, N.F. (1966) Broadbase hybrid wheats. *Crop Sci.*, **6**, 376–7.

Jensen, N.F. (1970) A diallel selective mating system for cereal breeding. *Crop Sci.*, **10**, 629–35.

Jensen, N.F. and Federer, W.T. (1964) Adjacent-row competition in wheat. *Crop Sci.*, **4**, 641–5.

Jensen, N.F. and Federer, W.T. (1965) Competing ability in wheat. *Crop Sci.*, **5**, 449–52.

Jinks, J.L. (1955) A survey of the genetical basis of heterosis in a variety of diallel crosses. *Heredity*, **9**, 223–38.

Jinks, J.L., ed. (1970) *Fifty Years of Genetics*. Oliver and Boyd, Edinburgh.

Johnson, H.W. and Bernard, R.L. (1962) Soybean genetics and breeding. *Adv. Agron.*, **14**, 149–222.

Johnson, H.W., Robinson, H.F. and Comstock, R.E. (1955a) Estimates of genetic and environmental variability in soybean. *Agron. J.*, **47**, 314–18.

Johnson, H.W., Robinson, H.F. and Comstock, R.E. (1955b) Genotypic and phenotypic correlations in soybeans and their implications in selection. *Agron. J.*, **47**, 477–82.

Johnson, L.N. and Aksel, R. (1959) Yielding capacity in barley. *Can. J. Genet. Cytol.*, **1**, 208–65.

Johnson, R. and Taylor, A.J. (1976) Spore yields of pathogens in investigations of the race specificity of host resistance. *Ann. Rev. Phytopathol.*, **14**, 97–119.

Johnson, R.R. and Brown, C.M. (1976) Chemical control of pollination in wheat and oats. *Crop Sci.*, **16**, 584–7.

Johnson, V.A. and Schmidt, J.W. (1968) Hybrid wheat. *Adv. Agron.*, **20**, 199–233.

Johnson, V.A., Schmidt, J.W. and Mattern, P.J. (1968) Cereal breeding for better protein impact. *Econ. Bot.*, **22**, 16–25.

Johnston, T.D. (1968) Studies on a diallel cross and double-cross hybrids among various *Brassica oleracea* types. *Euphytica*, **17**, 63–73.

Jones, D.F. (1958) Heterosis and homeostasis in evolution and applied genetics. *Amer. Nat.*, **92**, 321–8.

Jones, M.B. and Lazenby, A., eds (1988) *The Grass Crop – the Physiological Basis of Production.* Chapman & Hall, London.

Joppa, L.R., Lebsock, K.L. and Busch, R.H. (1971) Yield stability of selected spring wheat cultivars in the Uniform Regional Nurseries, 1959 to 1968. *Crop Sci.*, **11**, 238–41.

Jurado-Tovar, A. and Compton, W.A. (1974) Intergenotypic competition studies in corn (*Zea mays*), I. Among experimental hybrids. *Theor. Appl. Genet.*, **45**, 205–10.

Justus, N. (1960) Residual heterozygosity in a variety of Upland cotton. *Agron J.*, **52**, 555–9.

Kaepller, H.F., Somers, D.A., Rines, H.W. & Cockburn, A.F. (1990) Silicon carbide fiber-mediated DNA delivery into plant cells. *Plant Cell Reports*, **9**, 415–18.

Kannenberg, L.W. and Allard, R.W. (1967) Population studies in predominantly self-pollinated species VIII. Genetic variability in the *Festuca microstachys* complex. *Evolution*, **21**, 227–40.

Kannenberg, L.W. and Hunter, R.B. (1972) Yielding ability and competitive influence in hybrid mixtures of maize. *Crop Sci.*, **12**, 274–7.

Kasha, K.J., ed. (1974) *Haploids in Higher Plants, Advances and Potential.* University of Guelph, Canada.

Kasperbauer, M.J. and Collins, G.B. (1974) Anther-derived haploids in tobacco: evaluation of procedures. *Crop Sci.*, **14**, 305–7.

Kaufmann, M.L. and McFadden, A.D. (1960) The competitive interaction between barley plants grown from large and small seeds. *Can. J. Pl. Sci.*, **40**, 623–9.

Kaufmann, M.L. and McFadden, A.D. (1963) The influence of seed size on the result of barley yield trials. *Can. J. Pl. Sci.*, **43**, 51–8.

Kavanagh, T.A. and Spillane, C. (1995) Strategies for engineering virus resistance in transgenic plants. *Euphytica*, **85**, 149–58.

Kearsey, M.J. (1993) Biometrical genetics in plant breeding, pp. 163–83, in M.D. Haywood, N.O. Bosemark and I. Romagosa (eds) *Plant Breeding – Principles and Prospects.* Chapman & Hall, London.

Kehr, W.R. (1976) Cross fertilization in seed production in relation to forage yield of alfalfa. *Crop Sci.*, **16**, 81–6.

Kempthorne, O. (1957) *An Introduction to Genetic Statistics.* Wiley, New York; Chapman and Hall, London.

Kempton, R.A. and Fox, P.N. eds (1996) *Statistical Methods for Plant Variety Evaluation*. Chapman and Hall, London.

Kempton, R.A. and Lockwood, G. (1984) Inter-plot competition in variety trials of field beans (*Vicia faba*). *J.Agric. Sci. Cambr.*, **103**, 293–302.

Keuls, M. and Sieben, J.W. (1955) Two statistical problems in plant selection. *Euphytica*, **4**, 34–44.

Khalifa, M.A. and Qualset, C.O. (1974, 1975) Intergenotypic competition between tall and dwarf wheats. I. In mechanical mixtures, II. In hybrid bulks. *Crop Sci.*, **14**, 795–9; **15**, 640–4.

Khan, T.N (1973) A new approach to the breeding of pigeon pea (*Cajanus cajan*); formation of composites. *Euphytica*, **22**, 373–7.

Khush, G.S. and Coffmann, W.R. (1977) Genetic evaluation and utilization (GEU) program. The rice improvement program of the International Rice Research Institute. *Theor. Appl.Genet.*, **51**, 97–110.

Killick, R.J. and Malcolmson, J. (1973) Inheritance in potatoes of field resistance to late blight. *Phys. Pl. Path.*, **3**, 121–31.

Kim, S.K. and Brewbaker, J.L. (1977) Inheritance of general resistance in maize to *Puccinia sorghi. Crop Sci.*, **17**, 456–61.

Kimber, G. and Riley, R. (1963) Haploid angiosperms. *Bot. Rev.*, **29**, 480–531.

Kingsman, S.M. and Kingsman, A.J. (1988) *Genetic Engineering.* Blackwell, Oxford.

Klein, T.M., Wolf, E.D., Wu, R. and Sanford, J.C. (1987) High-velocity microprojectiles for delivering nucleic acids into living cells. *Nature (London)*, **327**, 70–73.

Kleinhofs, A. and Behki, R. (1977) Prospects for plant genome modification by nonconventional methods. *Ann. Rev. Genet.*, **11**, 79–101.

Klinkowski, M. (1970) Catastrophic plant diseases. *Ann. Rev. Phytopath.*, **8**, 37–60.

Knight, R. (1970) The measurement and interpretation of genotype environment interactions. *Euphytica*, **19**, 225–35.

Knight, R. (1973) The relation between hybrid vigour and genotype-environment interactions. *Theor. Appl. Genet.*, **43**, 311–18.

Knight, R.L. (1945) The theory and application of the backcross technique in cotton breeding. *J. Genet.*, **47**, 78–86.

Knight, R.L. (1956) Blackarm disease of cotton and its control. *Proc. II Intnl. Plant Prot. Conf., Fernhurst, England*, 53–9.

Knott, D.R. (1972) Effects of selection for F_2 plant yield on subsequent generations in wheat. *Can. J. Plant Sci.*, **52**, 721–6.

Knott, D.R. and Dvořak, J. (1976) Alien germ-plasm as a source of resistance to disease. *Ann. Rev. Phytopathol.*, **14**, 211–35.

Knott, D.R. and Kumar, J. (1975) Comparison of early generation yield testing and a single-seed descent procedure in wheat breeding. *Crop Sci.*, **15**, 295–9.

Kojima, K., ed. (1970) *Mathematical Topics in Population Genetics.* Springer, Berlin, Heidelberg, New York.

Komamine, A., Matsumoto, M., Tsukahara, M. Fujiwara, A., Kawahara, R., Ito, M., Smith, J., Nomura, K. and Fujimura, T. (1990) Mechanisms of somatic embryogenesis in cell cultures – physiology, biochemistry and molecular biology. In: *Progress in Plant Cellular and Molecular Biology* (eds H.J.J. Nijkamp, L.H.W. van der Plas and J. van Aartrijk), 307–13. Kluwer Academic Publishers. Dordrecht.

Konzak, C.F. (1977) Genetic control of the content, amino-acid composition and processing properties of proteins in wheat. *Adv. Genet.*, **19**, 408–582.

Konzak, C.F., Kleinhofs, A. and Ullrich, S.E. (1984) Induced mutations in seed-propagated crops. *Pl. Breed. Rev.*, **2**, 13–72.

Kranz, A.R. (1973) Wildarten and primitivformen des Roggens (*Secale*). *Fortschr. PflzZüchtung*, **3**, 1–60.

Kranz, J., ed. (1974) *Epidemic Plant Diseases: Mathematical Analysis and Modelling.* Springer, Berlin, Heidelberg and New York.

Krikorian, A.D. and Berquam, D.L. (1969) Plant cell and tissue cultures: the role of Haberlandt. *Bot. Rev.*, **35**, 59–88.

Krikorian, A.D. and Kann, R.P. (1986) Oil palm improvement via tissue culture. *Pl. Breed. Rev.*, **4**, 175–202.

Kronstad, W.E. and Foote, W.H. (1964) General and specific combining ability estimates in winter wheat. *Crop Sci.*, **4**, 616–19.

Krug, H., Wriedt, G. and Weber, W.E. (1974) Untersuchungen zur Frühselektion in der Kartoffelzüchtung, 1, 2. *Z. f. PflzZüchtung*, **73**, 141–62, 163–93.

Laing, D.R. and Fischer, R.A. (1977) Adaptation of semidwarf wheat cultivars to rainfed conditions. *Euphytica*, **26**, 129–39.

Lamb, C.J., Ryals, J.A., Ward, E.R. and Dixon, R.A. (1992) Emerging strategies for enhancing crop resistance to microbial pathogens. *Bio/Technology*, **10**, 1436–45.

Lamberts, H. *et al.*, eds (1968) Cereal improvement. Hybrid varieties and aspects of culm shortening. *Euphytica*, **17**, suppl. 1, p. 322.

Laubscher, F.X. *et al.* (1967) The significance of genetic variability in wheat cultivars. *S. Afr. J. Agric. Sci.*, **10**, 631–40.

Laude, H.M. and Swanson, A.F. (1943) Natural selection in varietal mixtures of winter wheat. *J. Amer. Soc. Agron.*, **34**, 270–4.

Law, C.N. and Worland, A.J. (1973) Aneuploidy in wheat and its uses in genetic analysis. *Pl. Br. Inst. Camb. Ann. Rep., 1972*, 25–65.

Lawrence, W.J.C. (1951) *Practical Plant Breeding.* Allen and Unwin, London, 3rd edn.

Layard, R., ed. (1974) *Cost–Benefit Analysis.* Penguin Books, Harmondsworth, England.

Leakey, C.L.A., ed. (1970) *Crop Improvement in East Africa.* Farnham Royal, England (*Tech. Comm. 19, Comm. Bur. Pl. Br. Genet.*).

Leppik, E.E. (1970) Gene centers of plants as sources of disease resistance. *Ann. Rev. Phytopath*, **8**, 323–44.

Lerner, I.M. (1950) *Population Genetics and Animal Improvement.* Cambridge University Press.

Lerner, I.M. (1954) *Genetic Homeostasis.* Oliver and Boyd, Edinburgh.

Lerner, I.M. (1958) *The Genetic Basis of Selection.* Wiley, New York.

Lerner, I.M. and Donald, H.P. (1966) *Modern Development in Animal Breeding.* Academic Press, London and New York.

Lewers, K.S. and Palmer, R.G. (1997) Recurrent selection in soybean. *Pl. Breed. Rev.*, **15**, 275–314.

Lewontin, R.C. (1974) *Genetic Basis of Evolutionary Change.* Columbia University Press, New York.

Li, C.C. (1955) *Population Genetics.* University of Chicago Press.

Li, H.L. (1970) The origin of cultivated plants in Southeast Asia. *Econ. Bot.*, **24**, 3–19.

Lim, S.M. (1975) Diallel analysis for reaction of eight corn inbreds to *Helminthosporium maydis* race T. *Phytopathology*, **65**, 10–15.

Lin, C.C. and Torrie, J.H. (1971) Alternate row multi-strain culture in soybeans. *Crop Sci.*, **11**, 331–4.

Lindner, R.K. and Jarrett, F.G. (1977) Measurement of the level and distribution of research benefits. *Austr. Agric. Econ. Soc. Ann. Conf.*, 21.

Litz, R.E. and Gray, D.J. (1995) Somatic embryogenesis for agricultural improvement. *World Journal of Microbiology and Biotechnology*, **11**, 416–25.

Lonnquist, J.H. (1964) A modification of the ear-to-row procedure for the improvement of maize populations. *Crop Sci.*, **4**, 227–8.

Lonnquist, J.H. (1968) Further evidence on testcross versus line performance in maize. *Crop. Sci.*, **8**, 50–3.

Lonnquist, J.H. and Gardner, C.O. (1961) Heterosis in intervarietal crosses in maize and its implication in breeding procedures. *Crop Sci.*, **1**, 179–83.

Lonnquist, J.H. and Lindsey, M.F. (1964) Topcross *vs* S_1 line performance in corn. *Crop Sci.*, **4**, 580–4.

Lonnquist, J.H. and McGill, D.P. (1956) Performance of corn synthetics in advanced generations of synthesis and after two cycles of recurrent selection. *Agron. J.*, **48**, 249–53.

Loomis, R.S., Williams, W.A. and Hall, A.E. (1971) Agricultural productivity. *Ann.Rev. Pl. Physiol.*, **22**, 431–68.

Lowe, C.C., Cleveland, R.W. and Hill, R.R. (1974) Variety synthesis in alfalfa. *Crop Sci.*, **14**, 321–5.

Luedders, V.D., Duclos, L.A. and Matson, A.L. (1973) Bulk pedigree and early generation testing breeding methods compared in soybeans. *Crop Sci.*, **13**, 363–4.

Lupton, F.G.H. (1961) Studies in the breeding of self-pollinating cereals, 3. Further studies in cross prediction. *Euphytica*, **10**, 209–24.

Lupton, F.G.H. (1965) Studies on the breeding of self-pollinating cereals, 5. Use of the incomplete diallel in wheat breeding. *Euphytica*, **124**, 331–52.

Lupton, F.G.H., Jenkins, G. and Johnson, R., eds (1972) *The Way Ahead in Plant Breeding.* Plant Breeding Institute, Cambridge (Proceedings of the Sixth Congress of Eucarpia).

Lupton, F.G.H. and Johnson, R. (1970) Breeding for mature-plant resistance to yellow rust in wheat. *Ann. Appl. Biol.*, **66**, 137–43.

Lupton, F.G.H. and Kirby, E.J.M. (1967) Application of physiological analysis to cereal breeding. *Pl. Br. Inst. Camb. Ann. Rep., 1966–67*, 5–26.

Lupton, F.G.H. and Whitehouse, R.N.H. (1957) Studies on the breeding of self-pollinating cereals I. Selection methods in breeding for yield. *Euphytica*, **6**, 169–84.

Lush, J.L. (1945) *Animal Breeding Plans.* Iowa State College Press, Ames, Iowa, USA, 3rd edn.

Lyrene, P.M. (1977) Heritability of flowering in sugar cane. *Crop. Sci.*, **17**, 462–4.

Macer, R.C.F. (1963) Developments in cereal pathology. *Pl. Br. Inst. Camb. Ann. Rep., 1962–3*, 5–33.

Mackay, G.R. (1973) Interspecific hybrids between forage rape and turnip as alternatives to forage rape. I. An exploratory study with single pair crosses. *Euphytica*, **22**, 495–9.

Mackay, G.R. (1977) The introgression of S-alleles into forage rape from turnip. *Euphytica*, **26**, 511–19.

Mangelsdorf, P.C. (1974) *Corn: Its Origin, Evolution and Improvement.* Harvard University Press, Cambridge, Mass., USA.

Manning, H.L. (1956) Yield improvement from a selection index technique with cotton. *Heredity*, **10**, 303–22.

Mantell, S.H. and Smith, H., eds (1983) *Plant Biotechnology*. Cambridge University Press.

Marani, A. (1975) Simulation of response of small self-fertilizing populations to selection for quantitative traits. I. Effect of number of loci, selection intensity and initial heritability under conditions of no dominance. *Theor. Appl. Genet.*, **46**, 221–31.

Marani, A., Fischler, G. and Amirav, A. (1971) The inheritance of resistance to blue mold (*Peronospora tabacina*) in two cultivars of tobacco. *Euphytica*, **21**, 97–105.

Maris, B. (1966) The modifiability of characters important in potato breeding. *Euphytica*, **15**, 18–31.

Maris, B. (1969) Studies on maturity, yield, under-water weight and some other characters of potato progenies. *Euphytica*, **18**, 297–319.

Markley, K.S. (1971) The babassú oil palm of Brazil. *Econ. Bot.*, **25**, 267–304.

Marshall, D.R. and Allard, R.W. (1974) Performance and stability of mixtures of grain sorghum I. Relationship between level of genetic diversity and performance. *Theor. Appl. Genet.*, **44**, 145–52.

Marshall, D.R. and Brown, A.H.D. (1973) Stability of performance of mixtures and multilines. *Euphytica*, **22**, 405–12.

Marshall, H.G. (1976) Genetic changes in oat bulk populations under winter survival stress. *Crop. Sci.*, **16**, 9–15.

Mather, K. (1971) On biometrical genetics *Heredity*, **26**, 349–64.

Mather, K. and Jinks, J.L. (1971) *Biometrical Genetics.* Chapman and Hall, London, 2nd edn.

Mather, K. and Jinks, J.L. (1977) *Introduction to Biometrical Genetics.* Chapman and Hall, London.

Matsuo, T., ed. (1975a) *Gene Conservation.* University of Tokyo (IBP Synthesis, 5).

Matsuo, T., ed. (1975b) *Adaptability in Plants with Special Reference to Crop Yield.* University of Tokyo (IBP Synthesis, 6).

Matzinger, D.F. and Wernsman, E.A. (1968) Four cycles of mass selection in a synthetic variety of an autogamous species. *Nicotiana tabacum. Crop Sci.*, **8**, 239–43.

Matzinger, D.F., Wernsman, E.A. and Ross, H.F. (1971) Diallel crosses among Burley varieties of *Nicotiana tabacum* in F_1 & F_2 generations. *Crop Sci.*, **11**, 275–9.

Maxwell, F.G., Jenkins, J.N. and Parrott, W.L. (1972) Resistance of plants to insects *Adv. Agron.*, **24**, 187–265.

McCough, S.R. and Doerge, R.W. (1995) QTL mapping in rice. *Trends in Genetics*, **11**, 482–7.

McGill, D.P. and Lonnquist, J.H. (1955) Effects of two cycles of recurrent selection for combining ability in an open-pollinated variety of corn. *Agron. J.*, **47**, 319–23.

McGregor, S.E. (1976) *Insect Pollination of Cultivated Crop Plants.* U.S.D.A., Washington, DC, USA (*U.S.D.A. Agric. Handb.*, **496**, p 411).

McGuire, C.F. and McNeal, F.H. (1974) Quality response of 10 hard red spring wheat cultivars to 25 environments. *Crop Sci.*, **1**, 175–8.

McNaughton, I.M. (1973) Synthesis and sterility of Raphanobrassica. *Euphytica*, **22**, 70–88.

McNaughton, I.H. and Cross, C.L. (1978) Interspecific and intergeneric hybridization in the Brassicae with special emphasis on the improvement of forage crops. *Scott. Pl. Br. Sta. Ann. Rep.*, **56**, 75–110.

McWilliam, J.R. and Latter, B.D.H. (1970) Quantitative genetic analysis in *Phalaris* and its breeding applications. *Theor. Appl. Genet.*, **40**, 63–72.

McWilliam, J.R., Schroeder, H.E., Marshall, D.R. and Oram, R.N. (1971) Genetic stability of Australian Phalaris (*Phalaris tuberosa*) under domestication. *Aust. J. Agric. Res.*, **22**, 895–908.

Meer, Q.P., Van der and Nieuwhof, M. (1968) Production of hybrid seed using male sterility or self-incompatibility. *Euphytica*, **17**, 284–8.

Melchers, G. (1972) Haploid higher plants for plant breeding. *Z. f. PflzZüchtung*, **67**, 19–32.

Mendiburu, A.D. and Peloquin, S.J. (1976) Sexual polyploidization: some terminology and definitions. *Theor. Appl. Genet.*, **48**, 137–43.

Mendiburu, A.D. and Peloquin, S.J. (1977) The significance of 2N gametes in potato breeding. *Theor. Appl. Genet.*, **49**, 53–61.

Mendoza, H.A. and Haynes, F.L. (1974) Genetic basis of heterosis for yield in the autotetraploid potato. *Theor. Appl. Genet.*, **45**, 21–5.

Meredith, W.R. and Bridge, R.R. (1972) Heterosis and gene action in cotton, *Gossypium hirsutum. Crop Sci.*, **12**, 304–10.

Meredith, W.R. and Bridge, R.R. (1973) The relationship between F_2 and selected F_3 progenies in cotton (*Gossypium*). *Crop Sci.*, **13**, 354–6.

Mettler, L.E. and Gregg, T.G. (1969) *Population Genetics and Evolution.* Prentice-Hall, Englewood Cliffs, NJ, USA.

Meyer, V.G. (1973) A study of reciprocal hybrids between upland cotton (*Gossypium hirsutum*) and experimental lines with cytoplasms from seven other species. *Crop Sci.*, **13**, 439–49.

Meyer, V.G. (1974) Interspecific cotton breeding. *Econ. Bot.*, **28**, 56–60.

Miller, J.D. (1977) Combining ability and yield component analyses in a five-parent diallel cross in sugarcane. *Crop Sci.*, **17**, 545–7.

Miller, J.D. and James, N.I. (1975) Selection in six crops of sugarcane. I. Repeatability of three characters. *Crop Sci.*, **15**, 23–5.

Miller, J.F. and Lucken, K.A. (1976) Hybrid wheat seed production methods for North Dakota. *Crop Sci.*, **16**, 217–21.

Miller, P.A. and Lee, J.A. (1964) Heterosis and combining ability in varietal top-crosses of upland cotton. *Crop Sci.*, **4**, 646–9.

Milner, M., ed. (1975) *Nutritional Improvement of Food Legumes by Breeding.* Wiley, New York.

Mishan, E.J. (1971) *Cost-Benefit Analysis.* Unwin, London.

Moav, R., ed. (1973) *Agricultural Genetics: Selected Topics.* Wiley, New York and Toronto.

Mode, C.J. (1958) A mathematical model for the co-evolution of obligate parasites and their hosts. *Evolution*, **12**, 158–65.

Mok, D.W.S. and Peloquin, S.J. (1975) Breeding value of 2n pollen (diplandroids) in tetraploid × diploid crosses in potatoes. *Theor. Appl. Genet.*, **46**, 307–14.

Moll, R.H., Bari, A. and Stuber, C.W. (1977) Frequency distribution of maize yield before and after reciprocal recurrent selection. *Crop Sci.*, **17**, 794–6.

Moll, R.H. and Stuber, C.W. (1971) Comparisons of response to alternate selection procedures initiated with two populations of maize. *Crop Sci.*, **11**, 706–11.

Moll, R.H. and Stuber, C.W. (1974) Quantitative genetics – empirical results relevant to plant breeding. *Adv. Agron.*, **26**, 277–314.

Moll, R.H., Stuber, C.W. and Hanson, W.D. (1975) Correlated responses and responses to index selection involving yield and ear height in maize. *Crop Sci.*, **15**, 243–8.

Moran-Val, C.A. and Miller, P.A. (1975) Inter-row competitive effects among four cotton cultivars. *Crop Sci.*, **15**, 479–82.

Mumaw, C.R. and Weber, C.R. (1957) Competition and natural selection in soybean varietal composites. *Agron. J.*, **49**, 154–60.

Munck, L. (1972) Improvement of nutritional value of cereals. *Hereditas, Lund*, **72**, 1–128.
Murashige, T. (1974) Plant propagation through tissue cultures. *Ann. Rev. Plant Physiol.*, **25**, 135–66.
Murfet, I.C. (1977) Environmental interaction and the genetics of flowering. *Ann. Rev. Plant Physiol.*, **28** 253–78.
Murphy, D.J. (1996) Engineering oil production in rapeseed and other oil crops. *Trends in Biotechnology*, **14**, 206–13.
Narayanaswami, S. and Norstog, K. (1964) Plant embryo culture *Bot. Rev.*, **30**, 587–628.
National Research Council (1989) *Lost Crops of the Incas: Little Known Plants of the Andes with Promise for Worldwide Cultivation.* National Academy Press, Washington, DC.
Neilson-Jones, W. (1969) *Plant Chimeras.* Methuen, London, 2nd edn.
Nelson, O.E. (1969) Genetic modification of protein quality in plants. *Adv. Agron.*, **21**, 171–94.
Nelson, R.R., ed. (1973) *Breeding Plants for Disease Resistance.* Pennsylvania State University Press, University Park and London.
Nester, M.R. (1996) An applied statistician's creed. *Appl. Stat.*, **45**, 401–10.
Nettancourt, D. de (1977) *Incompatibility in Angiosperms.* Springer, Berlin, Heidelberg and New York.
Netterich, E.D. (1968) The problem of utilizing heterosis in wheat. *Euphytica*, **17**, 54–62.
Nickell, L.G. (1977) Crop improvement in sugarcane: studies using *in vitro* methods. *Crop Sci.*, **17**, 717–19.
Nilan, R.A., ed. (1971) *Barley Genetics II.* Washington State University, USA (*Proc. II Intnl. Barley Genet. Symp.*, 1970).
Norrington-Davies, J. (1972) Diallel analysis of competition between some barley species and their hybrids. *Euphytica*, **21**, 292–308.
Nuding, J. (1936) Leistung und Ertragstruktur von Winterweizensorten in Reinsaat und Mischung in verschiedenen deutschen Anbaugebieten. *Pflanzenbau*, **12**, 382–447.
Obajami, A.O. and Bingham, E.T. (1973) Inbreeding cultivated alfalfa in one tetraploid-haploid-tetraploid cycle: effects on morphology, fertility and cytology. *Crop Sci.*, **13**, 36–9.
Obilana, A.T. and Hallauer, A.R. (1974) Estimation of variability in quantitative traits in BSSS by using unselected maize inbred lines. *Crop Sci.*, **14**, 99–103.
OECD (1969) *OECD Scheme for the Varietal Certification of Cereal Seed moving in International Trade. Guide to the Methods used in Plot Tests and to the Methods of Field Inspection of Cereal Seed Crops.* OECD, Paris.
Olby, R.C. (1966) *Origins of Mendelism.* Constable, London.
Old, R.W. and Primrose, S.B. (1994) *Principles of Gene Manipulation – an Introduction to Genetic Engineering.* Blackwell, Oxford.
Ordish, G. and Dufour, D. (1969) Economic bases for protection against plant diseases. *Ann. Rev. Phytopath*, **7**, 31–50.
Otsuka, Y., Eberhart, S.A. and Russell, W.A. (1972) Comparisons of prediction formulas for maize hybrids. *Crop Sci.*, **12**, 325–31.
Pahlen, A. von der (1968) Rendimiento y estabilidad en mezclas de mutantes de Cebada. *Centr. Invest. Ciencia agron., INTA, Castelar, Argentina, Publ. tecn.*, **446**, p. 11.
Painter, R.H. (1951) *Insect Resistance in Crop Plants.* Macmillan, New York.
Painter, R.H. (1958) Resistance of plants to insects. *Ann. Rev. Entomol.*, **3**, 267–90.

Palmer, T.P. (1952) Population and selection studies in a *Triticum* cross. *Heredity*, **6**, 171–85.

Park, S.J., Walsh, E.J., Reinbergs, E., Song, S.P. and Kasha, K.J. (1976) Field performance of doubled haploid barley lines in comparison with lines developed by the pedigree and single-seed descent methods. *Can. J. Plant Sci.*, **56**, 464–74.

Parlevliet, J.E. (1973) The genetic variability of the yield components in the Kenyan pyrethrum population. *Euphytica*, **23**, 377–84.

Parlevliet, J.E. and Zadocks, J.C. (1977) The integrated concept of disease resistance: a new view including horizontal and vertical resistance in plants. *Euphytica*, **26**, 5–21.

Parra, J.R. and Hallauer, A.R. (1997) Utilization of exotic maize germplasm. *Pl. Breed. Rev.*, **14**, 165–88.

Paterniani, E. and Lonnquist, J.H. (1963) Heterosis in inter-racial crosses of corn. *Crop Sci.*, **3**, 504–7.

Paterson, A.H. *et al.* (1991) Mendelian factors underlying quantitative traits, in tomato: comparison across species, generations and environments. *Genetics*, **127**, 181–97.

Patterson, H.D. (1979) Routine least squares estimation of variety means in incomplete tables. *J. Natnl. Inst. Agric. Bot.*

Patterson, H.D., Silvey, V., Talbot, M. and Weatherup, S.T.C. (1977) Variability of yields of cereal varieties in UK trials. *J. Agric. Sci. Camb.*, **89**, 239–45.

Patterson, H.D. and Simmonds, N.W. (1989) Tables to calculate means in a doubly truncated bivariate normal population. *Euphytica*, **42**, 241–9.

Paschal, E.H. and Wilcox, J.R. (1975) Heterosis and combining ability in exotic soybean germplasm. *Crop Sci.*, **15**, 344–9.

Pearce, S.C. (1983) *The Agricultural Field Experiment.* J. Wiley, Chichester.

Pederson, D.G. (1969) The prediction of selection response in a self-fertilizing species. I. Individual selection. II. Family selection. *Austr. J. Biol. Sci.*, **22**, 117–29, 1245–57.

Pederson, D.G. (1972) A comparison of four experimental designs for the estimation of heritability. *Theor. Appl. Genet.*, **42**, 371–7.

Pederson, D.G. (1974a) Arguments against intermating before selection in a self-fertilizing species. *Theor. Appl. Genet.*, **45**, 157–62.

Pederson, D.G. (1974b) The stability of varietal performance over years, 1, 2. *Heredity*, **32**, 85–94; **33**, 217–28.

Pederson, M.W. and Hill, R.R. (1972) Combining ability in alfalfa hybrids made with cytoplasmic male sterility. *Crop Sci.*, **12**, 500–2.

Pee, T.Y. (1977) *Social Returns from Rubber Research in Peninsular Malaysia.* Ph.D. Thesis, Michigan State University.

Pelletier, G. (1993) Somatic hybridization, pp. 93–106 in M.D. Hayward, N.O. Bosemark and I. Romagosa (eds) *Plant Breeding – Principles and Prospects.* Chapman & Hall, London.

Pelton, J.S. (1964) Genetic and morphogenetic studies of Angiosperm single-gene dwarfs. *Bot. Rev.*, **30**, 479–512.

Penny, L.H. and Eberhart, S.A. (1971) Twenty years of reciprocal recurrent selection with two synthetic varieties of maize. *Crop Sci.*, **11**, 900–3.

Pérez de la Vega, M. (1993) Biochemical characterization of populations, pp. 184–200 in M.D. Hayward, N.O. Bosemark and I. Romagosa (eds). *Plant Breeding – Principles and Prospects.* Chapman & Hall, London.

Person, C. and Groth, J.V. and Mylyk, O.M. (1976) Genetic change in host-parasite populations. *Ann. Rev. Phytopathol.*, **14**, 177–88.

Pesek, J. and Baker, R.J. (1969) Comparison of tandem and index selection in the modified pedigree method of breeding self-pollinated species. *Can. J. Plant Sci.*, **69**, 773–81.

Pesek, J. and Baker, R.J. (1970) Application of index selection to the improvement of self-pollinated species. *Can. J. Plant Sci.*, **50**, 267–76.

Peterson, F.R. (1958) Twenty five years of progress in breeding new varieties of wheat for Canada. *Emp. J. Exp. Agric.*, **26**, 104–22.

Peterson, G.A. and Foster, A.E. (1973) Malting barley in the United States. *Adv. Agron.*, **25**, 328–78.

Pfahler, P.L. (1964) Fitness and variability in fitness of the cultivated species of *Avena. Crop Sci.*, **4**, 29–31.

Pfahler, P.L. (1965) Genetic diversity for environmental variability within the cultivated species of *Avena. Crop Sci.*, **5**, 47–50.

Pfahler, P.L. (1971) Heritability estimates for grain yield in oats (*Avena* spp.). *Crop Sci.*, **11**, 378–81.

Phung, T.K. and Rathjen, A.J. (1976) Frequency-dependent advantage in wheat. *Theor. Appl. Genet.*, **48**, 289–97.

Phung, T.K. and Rathjen, A.J. (1977) Mechanisms of frequency-dependent advantage in wheat. *Aust. J. Agric. Res.*, **28**, 187–202.

Pickersgill, B. (1977) Taxonomy and the origin and evolution of cultivated plants in the New World. *Nature, Lond.*, **268**, 591–5.

Pierik, R.L.M. and Ruibing, M.A. (1997) Developments in the micropropagation industry in The Netherlands. *Plant Tissue Culture and Biotechnology*, **3**, 152–6.

Poehlman, J.M. (1959) *Breeding Field Crops.* Holt, New York.

Plaisted, R.L. and Peterson, L.C. (1959) A technique for evaluating the ability of selections to yield consistently in different locations or seasons. *Amer. Potato J.*, **36**, 381–5.

Plaisted, R.L., Sanford, L., Federer, W.T., Kehr, A.E. and Peterson, L.C. (1962) Specific and general combining ability for yield in potatoes. *Amer. Potato J.*, **39**, 185–97.

Posler, G.L., Wilsie, C.P. and Atkins, R.E. (1972) Inbreeding *Medicago sativa* by selfing, sib-mating and inter-generational crossing. *Crop Sci.*, **12**, 49–52.

Potrykus, I. (1993) Gene transfer to plants: approaches and available techniques, pp. 126–37 in M.D. Hayward, N.O. Bosemark and I. Romagosa (eds) *Plant Breeding – Principles and Prospects.* Chapman & Hall, London.

Prakash, J. and Pierik, R.L.M., eds (1993) *Plant Biotechnology – Commercial Prospects and Problems.* Oxford and IBH Publishing, New Delhi.

Probst, A.H. (1957) Performance of variety blends in soybeans. *Agron. J.*, **49**, 148–51.

Purdy, L.H. *et al.* (1968) A proposed standard method for illustrating pedigrees of small grain varieties. *Crop Sci.*, **8**, 405–6.

Purseglove, J.W. (1957) History and function of botanic gardens, with special reference to Singapore. *Trop. Agriculture, Trin.*, **34**, 165–89.

Purseglove, J.W. (1968, 1972) *Tropical Crops, Dicotyledons 1, 2, Monocotyledons 1, 2.* Longman, London.

Qualset, C.O. and Granger, R.M. (1970) Frequency-dependent stability of performance in oats. *Crop Sci.*, **10**, 386–9.

Quinby, J.R. (1963) Manifestations of hybrid vigour in sorghum. *Crop Sci.*, **3**, 288–91.

Quinby, J.R. (1967) The maturity genes of sorghum. *Adv. Agron.*, **19**, 267–305.

Quinby, J.R. (1973) The genetic control of flowering and growth in sorghum. *Adv. Agron.*, **25**, 126–62.

Quinby, J.R. (1975) The genetics of sorghum improvement. *J. Hered.*, **66**, 56–62.

Quinby, J.R. and Martin, J.H. (1954) Sorghum improvement. *Adv. Agron.*, **6**, 305–59.

Quisenberry, J.E. and Kohel, R.J. (1971) Phenotypic stability in cotton. *Crop Sci.*, **11**, 827–9.

Rachie, K.O. and Roberts, L.M. (1974) Grain legumes in the lowland tropics. *Adv. Agron.*, **26**, 2–132.

Raina, S.K. (1997) Double hoploid breeding in cereals. *Pl. Breed. Rev.*, **15**, 141–86.

Rajhathy, T. (1976) Haploid flax revisited. *Z. f. PflzZüchtung*, **76**, 1–10.

Rasmusson, D.C. (1968) Yield and stability of yield in barley populations. *Crop Sci.*, **8**, 600–2.

Rasmusson, D.C. and Byrne, I. (1972) Fitness of heterozygotes in barley. *Crop Sci.*, **12**, 640–43.

Rasmusson, D.C. and Cannell, R.Q. (1970) Selection for grain yield and components of yield in barley. *Crop Sci.*, **10**, 51–4.

Rasmusson, D.C. and Lambert, J.W. (1961) Variety × environment interactions in barley variety tests. *Crop Sci.*, **1**, 261–2.

Rathjen, A.J. and Lamacraft, R.R. (1972) The use of computers for information management in plant breeding. *Euphytica*, **21**, 502–6.

Redden, R.J. and Jensen, N.F. (1974) Mass selection and mating systems in cereals. *Crop Sci.*, **14**, 345–50.

Reddy, B.V.S. and Comstock, R.E. (1976) Simulation of the backcross breeding method. I. Effects of heritability and gene number on fixation of desired alleles. *Crop Sci.*, **16**, 825–30.

Rédei, G.P. (1974) Economy in mutation experiments. *Z. f. PflzZüchtung*, **73**, 87–96.

Reed, H.S. (1942) *A Short History of the Plant Sciences.* Chronica Botanica, Waltham, Mass., USA.

Reich, V.H. and Atkins, R.E. (1970) Yield stability of four population types of grain sorghum in different environments. *Crop Sci.*, **10**, 511–17.

Reinbergs, E., Park, S.J. and Song, L.S.P. (1976) Early identification of superior barley crosses by the doubled haploid technique. *Z. f. PflzZüchtung*, **76**, 215–24.

Reinert, J. and Bajaj, Y.P.S. (1977) *Plant Cell, Tissue and Organ Culture.* Springer, Berlin, Heidelberg and New York.

Reitz, L.P. and Salmon, S.C. (1968) Origin, history and use of Norin 10 wheat. *Crop Sci.*, **8**, 686–9.

Rice, T.B. and Carlson, P.S. (1975) Genetic analysis and plant improvement. *Ann. Rev. Pl. Physiol.*, **26**, 279–308.

Richmond, T.R. (1952) Procedures and methods of cotton breeding with special reference to American cultivated species. *Adv. Genet.*, **4**, 213–45.

Richmond, T.R. and Lewis, C.F. (1951) Evaluation of varietal mixtures in cotton. *Agron. J.*, **43**, 66–70.

Riggs, T.J. (1970) Trials of cotton seed mixtures in Uganda. *Cott. Gr. Rev.*, **47**, 100–11.

Riggs, T.J. and Snape, J.W. (1977) Effects of linkage and interaction in a comparison of theoretical populations derived by diploidized haploid and single seed descent methods. *Theor. Appl. Genet.*, **49**, 111–15.

Riley, R. and Kimber, G. (1965) The transfer of alien genetic variation to wheat. *Pl. Br. Inst. Camb. Ann. Rep., 1964–65*, 6–36.

Riley, R. and Lewis, K.R., eds (1966) *Chromosome Manipulations and Plant Genetics.* Oliver and Boyd, Edinburgh.

Rinke, E.M. and Hayes, H.K. (1964) General and specific combining in diallel crosses of 15 inbred lines of corn. *Bot. Bull. Acad. Sinica*, **5**, 36–41.

Roane, C.W. (1973) Trends in breeding for disease resistance in crops. *Ann. Rev. Phytopathol.*, **11**, 463–86.

Roberts, E.H. *et al.* (1976) *Report of IBPGR Working Group on Engineering, Design and Cost Aspects of Long-term Seed Storage Facilities.* IBPGR, Rome (AGPE: IBPGR/76/25).

Roberts, H.F. (1929) *Plant Hybridization before Mendel.* Princeton University Press, Princeton, N.J., USA.

Robertson, A. (1960) A theory of limits in artificial selection. *Proc. R. Soc. London, B*, **153**, 234–49.

Robinson, H.F., Comstock, R.E. and Harvey, P.H. (1949) Estimates of heritability and degree of dominance in corn. *Agron. J.*, **41**, 353–9.

Robinson, H.F., Comstock, R.E. and Harvey, P.H. (1951) Genotypic and phenotypic correlations in corn and their implications in selection. *Agron. J.*, **43**, 283–7.

Robinson, H.F., Comstock, R.E. and Harvey, P.H. (1955) Genetic variances in open pollinated varieties of corn. *Genetics*, **40**, 45–60.

Robinson, R.A. (1969) Disease resistance terminology. *Rev. Appl. Mycol.*, **48**, 593–606.

Robinson, R.A. (1971) Vertical resistance. *Rev. Pl. Path.*, **50**, 233–9.

Robinson, R.A. (1973) Horizontal resistance. *Rev. Pl. Path.*, **52**, 483–501.

Rodrigues, C.J., Bettencourt, A.J. and Rijo, L. (1975) Races of the pathogen and resistance to coffee rust. *Ann. Rev. Phytopath.*, **13**, 49–70.

Roemer, T. and Rudorf, W., eds (1955–62) *Handbuch der Pflanzenzüchtung.* Parey, Berlin and Hamburg (Six volumes, 2nd edn.' 1st edn., five volumes, 1938–50).

Rogers, H.H. (1976) Forage legumes. *Pl.Br. Inst. Camb. Ann. Rep., 1975*, 22–57.

Rhode, R.A. (1972) Expression of resistance in plants to nematodes. *Ann. Rev. Phytopath.*, **10**, 233–52.

Rosales, F.E. and Davis, D.D. (1976) Performance of cytoplasmic male-sterile cotton under natural crossing in New Mexico. *Crop Sci.*, **16**, 99–102.

Rosielle, A.A., Eagles, H.A. and Frey, K.J. (1976) Application of restricted selection indexes for improvement of economic value in oats. *Crop Sci.*, **17**, 359–61.

Rosielle, A.A. and Frey, K.J. (1975) Application of restricted selection indices for grain yield improvement in oats. *Crop Sci.*, **15**, 544–7.

Rosielle, A.A. and Frey, K.J. (1977) Inheritance of harvest index and related traits in oats. *Crop Sci.*, **17**, 23–8.

Ross, J.M. and Brookson, C.W. (1966) Progress of breeding investigations with *Hevea brasiliensis.* III. *J. Rubb. Res. Inst. Malaya*, **19**, 158–72.

Ross, W.M. (1965) Yield of grain sorghum hybrids alone and in blends. *Crop Sci.*, **5**, 593–4.

Rowe, P.R. and Andrew, R.H. (1964) Phenotypic stability for a systematic series of corn genotypes. *Crop Sci.*, **4**, 563–7.

Rowe, P.R. and Rosales, F.E. (1996) Bananas and plantains, pp. 167–211 in Janick, J.J. and Moore, J.N. (eds) *Plant Breeding I. Trees and Tropical Fruits.* J. Wiley, Chichester.

Rowlands, D.G. (1964) Self incompatibility in sexually propagated cultivated plants. *Euphytica*, **13**, 157–62.

Roy, N.N. and Murty, B.R. (1970) Effect of environment on the efficiency of selection in bread wheat. *Z. f. PflzZüchtung*, **63**, 56–60.

Roy, S.K. (1960) Interaction between rice varieties. *J. Genet.*, **57**, 137–52.

Russell, G.E. (1978) *Plant Breeding for Pest and Disease Resistance.* Butterworth, London.

Russell, W.A. (1969) Hybrid performance of maize inbred lines selected by testcross performance in low and high plant densities. *Crop. Sci.*, **9**, 185–8.

Russell, W.A. (1972) Effect of leaf angle on hybrid performance in maize (*Zea mays*). *Crop Sci.*, **12**, 90–2.

Russell, W.A. (1974) Comparative performance for maize hybrids representing different eras of maize breeding. *Proc. Ann. Corn and Sorghum Res. Conf.*, **29**, 81–101.

Russell, W.A. and Eberhart, S.A. (1975) Hybrid performance of selected maize lines from reciprocal recurrent selection and testcross selection programmes. *Crop Sci.*, **15**, 1–4.

Russell, W.A., Eberhart, S.A. and Vega, U.A. (1973) Recurrent selection for specific combining ability for yield in two maize populations. *Crop Sci.*, **13**, 257–61.

Russell, W.A. and Vega, U.A. (1973) Genetic stability of quantitative characters in successive generations in maize inbred lines. *Euphytica*, **22**, 172–80.

Saari, E.E. and Wilcoxon, R.D. (1974) Plant disease situation of high yielding dwarf wheats in Asia and Africa. *Ann. Rev. Phytopathol.*, **12**, 49–68.

Sage, G.C.M. (1971) Inter-varietal competition and its possible consequences for the production of F_1 hybrid wheat. *J. Agric. Sci. Camb.*, **77**, 491–8.

Sage, G.C.M. (1976) Nucleo-cytoplasmic relationships in wheat *Adv. Agron.*, **28**, 267–300.

Sakai, K., Takahashi, R. and Akemine, H., eds (1958) *Studies in the Bulk Method of Plant Breeding.*

Salamini, F. and Motto, M. (1993) The role of gene technology in plant breeding, pp. 138–60 in M.D. Hayward, N.O. Bosemark and I. Romagosa (eds) *Plant Breeding – Principles and Prospects.* Chapman & Hall, London.

Salmon, S.C., Mathews, O.R. and Leukel, R.W. (1953) A half-century of wheat improvement in the United States. *Adv. Agron.*, **5**, 3–141.

Sampson, D.R. (1972) Evaluation of nine oat varieties as parents in breeding for short, stout straw with high grain yield. *Can. J. Plant Sci.*, **52**, 21–8.

Sandfaer, J. (1954) The influence of natural selection on breeding work with self-fertilized cereal species. *Tidsskr. Planteavl.*, **58**, 333–54.

Sandfaer J. (1970) An analysis of the competition between some barley varieties. *Danish Atomic Energy Comm., Risö Rep.*, **230**, p. 114.

Sandfaer, J. and Haahr, V. (1975) Barley stripe mosaic virus and the yield of old and new barley varieties. *Z. f. PflzZüchtung*, **74**, 211–22.

Sandford, J.C. and Johnson, S.A. (1985) The concept of parasite-derived resistance-deriving resistance genes from the parasites own genome. *Journal of Theoretical Biology*, **113**, 395–405.

Sauer, C.O. (1952) *Agricultural Origins and Dispersals.* American Geographical Society, New York.

Schnell, F.W. (1975) Type of variety and average performance in hybrid maize. *Z. f. PflzZüchtung*, **74**, 177–88.

Schutz, W.M. and Brim, C.A. (1967) Intergenotypic competition in soybeans I. Evaluation of effects and proposed field plot design. *Crop Sci.*, **7**, 371–6.

Schutz, W.M. and Brim, C.A. (1971) Intergenotypic competition in soybeans. III. An evaluation of stability in multiline mixtures. *Crop Sci.*, **11**, 684–9.

Schutz, W.M., Brim, C.A. and Usanis, S.A. (1968) Intergenotypic competition in plant populations. I. Feedback systems with stable equilibria in populations of autogamous homozygous lines. *Crop Sci.*, **8**, 61–6.

Schutz, W.M. and Usanis, S.A. (1969) Intergenotypic competition in plant populations II. Maintenance of allelic polymorphisms with frequency dependent selection and mixed selfing and random mating. *Genetics*, **61**, 875–91.

Schwanitz, F. (1955) *Die Entstehung der Nutpflanzen als Modell für Evolution der gesamten Pflanzenwelt.* Stuttgart, 2nd edn.

Schwanitz, F. (1966) *The Origin of Cultivated Plants.* Harvard University Press, Cambridge, Mass., USA.

Sears, E.R. (1972) Chromosome engineering in wheat. *Stadler Symp. Univ. Missouri*, **4**, 23–38.

Sen, B. (1974) *The Green Revolution in India, a Perspective.* Wiley, New Delhi.

Shank, D.B. and Adams, M.W. (1960) Environmental variability within inbred lines and single crosses of maize. *J. Genet.*, **57**, 119–26.

Sharma, D., Tandon, J.P. and Batia, J.N. (1967) Effect of tester on combining ability estimates of maize germplasm complexes. *Euphytica*, **16**, 370–76.

Shaw, N.H. and Bryan, W.W., eds (1976) *Tropical Pasture Research, Principles and Methods.* Commonwealth Agricultural Bureaux, Farnham Royal, England (*Comm. Bur. Pastures and Field Crops, Bull.*, 51).

Shebeski, L.M. (1966) Quality and yield studies in hybrid wheat. *Can. J. Genet. Cytol.*, **8**, 375–86.

Shewry, P.R., Napier, J.A., Sayonova, O., Smith, M., Cooke, D.T., Stoker, K.G., Hill, D.J., Stobart, A.K. and Lapinskas, P. (1997) The use of biotechnology to develop new crops and products, pp. 76–87 in J. Smartt and N. Haq (eds). *Domestication, Production and Utilization of New Crops.* International Centre for Underutilized Crops, Southampton.

Shillito, R.D., Saul, M.W., Paszowkski, J., Muller, M. and Potrykus, I. (1985) High frequency direct gene transfer to plants. *Bio/Technology*, **3**, 1099–1103.

Siddig, M.A. (1967) Evaluation of several selection procedures for predicting yield performance of progenies in two populations of cotton. *Euphytica*, **16**, 377–84.

Sidhu, G. (1975) Gene for gene relationships in plant parasitic systems. *Sci. Progr. Oxf.*, **62**, 467–85.

Sidwell, R.J., Smith, E.L. and McNew, R.W. (1976) Inheritance and interrelationships of grain yield and selected yield-related traits in a hard red winter wheat cross. *Crop Sci.*, **16**, 650–4.

Simmonds, N.W. (1961) Mating systems and the breeding of perennial crops. *Advancem. Sci. London*, **18**, 183–6.

Simmonds, N.W. (1962a) Variability in crop plants its use and conservation. *Biol. Rev.*, **37**, 422–65.

Simmonds, N.W. (1962b) *The Evolution of the Bananas.* Longman, London.

Simmonds, N.W. (1966) *Bananas.* Longman, London, 2nd edn.

Simmonds, N.W. (1969a) Genetical bases of plant breeding. *J. Rubb. Res. Inst. Malaya*, **21**, 1–10.

Simmons, N.W. (1969b) Prospects of potato improvement. *Scott. Pl. Br. Sta. Ann. Rep.*, **48**, 18–38.

Simmonds, N.W. (1972) Profitability selection in relation to trials economy. *ISSCT Sugarcane Br. Newsl.*, **29**, 20–23.

Simmonds, N.W. (1973) Plant breeding. *Phil. Trans. R. Soc. Lond. B.*, **267**, 145–56.

Simmonds, N.W. (1974) Costs and benefits of an agricultural research institute. *R. and D. Management*, **5**, 23–8.

Simmonds, N.W. (1975) The place of economics. *Scott. Pl. Br. Sta. Ann. Rep.*, **54**, 44–53.

Simmonds, N.W., ed. (1976a) *Evolution of Crop Plants.* Longman, London and New York.

Simmonds, N.W. (1976b) Neotuberosum and the genetic base in potato breeding. *ARC Res. Rev.*, **2**, 9–11.

Simmonds, N.W. (1976c) Progress in sugarcane breeding. *ISSCT Sugarcane Br. Newsl.*, **38**, 75–8.

Simmonds, N.W. (1983) Plant breeding. The state of the art. In Kosuge *et al.*, eds. *Genetic Engineering of Plants.* Plenum Press, New York and London, 5–25.

Simmonds, N.W. (1985) Two-stage selection strategy in plant breeding. *Heredity*, **55**, 393–9.

Simmonds, N.W. (1986) Theoretical aspects of synthetic/polycross populations of rubber seedlings. *J. Nat. Rubber Res.*, **1**, 1–15.

Simmonds, N.W. (1988) Environmental features of plant breeding. In *Environmental Features of Applied Biology.* Association of Applied Biologists Wellesbourne, 3–10.

Simmonds, N.W. (1989) Economic aspects of plant breeding with especial reference to economic index selection. *Research and Development in Agriculture*, **6**, 53–62.

Simmonds, N.W. (1990) The social context of plant breeding. *Pl. Breed. Abstr.*, **60**, 337–41.

Simmonds, N.W. (1991a) Genetics of horizontal resistance to diseases of crops. *Biol. Rev.*, **66**, 189–241.

Simmonds, N.W. (1991b) Selection for local adaptation in a plant breeding programme. *Theor. Appl. Genet.*, **82**, 363–7.

Simmonds, N.W. (1993) Introgression and incorporation: strategies for the use of crop genetic resources. *Biol. Rev.*, **68**, 539–62.

Simmonds, N.W. (1994a) Yield and sugar content in sugarbeet. *Internat. Sugar J.*, **96**, 413–16.

Simmonds, N.W. (1994b) The breeding of perennial crops. In Cocoa Research Unit ed. *Conservation, Characterization and Utilization of Crop Genetic Resources in the 21st Century.* CRM/UWI, Trinidad, West Indies.

Simmonds, N.W. (1995a) The relation between yield and protein in cereal grain. *J. Sci. Fd. Agric.*, **67**, 309–315.

Simmonds, N.W. (1995b) *Food Crops: 500 Years of Travels. International Germplasm. Transfer – Past and Present.* CSSA Special Publication 23, pp. 32–45. Crop Science Society of America. Madison, Wisconsin 53711, USA.

Simmonds, N.W. (1996) Family selection in plant breeding. *Euphytica*, **90**, 201–208.

Simmonds, N.W. (1997) A review of potato propagation by means of seed as distinct from clonal propagation by tubers. *Potato Res.*, **40**, 191–214.

Simon, U., Spanakakis, A. and Scheller, H. (1975) Combining effects in a complete reciprocal diallel cross of nine non-inbred lucerne clones. *Z. f. PflzZüchtung*, **74**, 322–30.

Simons, M.D. (1972) Polygenic resistance to plant disease and its use in breeding resistant cultivars. *J. Environm. Qual.*, **1**, 232–40.

Singh, R.K. and Bellmann, K. (1974) Evaluation of selection indices under various parameter combinations in simulated genetic populations. *Theor. Appl. Gen.*, **44**, 63–8.

Sinha, S.K. and Khanna, R. (1975) Physiological, biochemical and genetic basis of heterosis. *Adv. Agron.*, **27**, 123–74.

Skoog, F. and Miller, C.O. (1957) Chemical regulation of growth and organ formation in plant tissues cultured *in vitro*. *Symposia of the Society for Experimental Biology*, **11**, 118–30.

Slootmaker, Z.A.J. and Essen, A. van (1969) Yield losses in barley caused by mildew attack. *Neth. J. Agric. Sci.*, **17**, 279–82.

Smartt, J. (1976) *Tropical Pulses*. Longman, London.

Smartt, J. (1990) *Grain Legumes – Evolution and Genetic Resources*. Cambridge University Press.

Smartt, J., ed. (1994) *The Groundnut Crop – A Scientific Basis for Improvement*. Chapman & Hall, London.

Smartt, J. and Haq, N. (1972) Fertility and segregation of the amphidiploid *Phaseolus vulgaris* × *P. coccineus* and its behaviour in backcrosses. *Euphytica*, **21**, 496–501.

Smartt, J. and Haq, N., eds (1997) *Domestication, Production and Utilization of New Crops*. International Centre for Underutilized Crops, Southampton.

Smartt, J. and Simmonds, N.W., eds (1995) *Evolution of Crop Plants*. Longman, Harlow, 2nd edn.

Smith, C.E. (1968) The New World centers of origin of cultivated plants and the archaeological evidence. *Econ. Bot.*, **22**, 253–66.

Smith, C.E. (1969) From Vavilov to the present – a review. *Econ. Bot.*, **23**, 2–19.

Smith, E.L. and Lambert, J.W. (1968) Evaluation of early-generation testing in spring barley. *Crop Sci.*, **8**, 490–3.

Smith, H.F. (1936) A discriminant function for plant selection. *Ann. Eugen.*, **7**, 240–50.

Smith, H.H. (1968) Recent cytogenetic studies in the genus *Nicotiana Adv. Genet.*, **14**, 1–54.

Snape, J.W. (1976) A theoretical comparison of diploidised haploid and single seed descent populations. *Heredity*, **36**, 275–7.

Snape, J.W. and Riggs, T.J. (1975) Genetical consequences of single seed descent in the breeding of self-pollinating crops. *Heredity*, **35**, 211–19.

Sneath, P.H.A. (1976) Some applications of numerical taxonomy to plant breeding. *Z. f. PflzZüchtung*, **76**, 19–46.

Snedecor, G.W. and Cochran, W.G. (1989) *Statistical Methods* 8th edn, Iowa State University Press, Ames Iowa, USA.

Snoad, B. (1974) A preliminary assessment of 'leafless' peas. *Euphytica*, **23**, 257–65.

Somaroo, B.H. and Grant, W.F. (1972) Crossing relationships between synthetic *Lotus* amphidiploids and *L. corniculatus*. *Crop Sci.*, **12**, 103–5.

Sowunini, M.A. (1985) The beginnings of agriculture in West Africa: botanical evidence. *Curr. Anthrop.*, **26**, 127–9.

Spangelo, L.P.S., Hsu, C.S., Fejer, S.O. and Watkins, R. (1971) Inbred line × tester analysis and the potential of inbreeding in strawberry breeding. *Can. J. Genet. Cytol.*, **13**, 460–9.

Spoor, W. and Simmonds, N.W. (1993) Pot trials as an adjunct to cereal breeding and evaluation of genetic resources. *Field Crops Research*, **35**, 205–213.

Sprague, G.F., ed. (1955) *Corn and Corn Improvement*. Academic Press, New York.

Sprague, G.F. (1967) Plant breeding. *Ann. Rev. Genet.*, **1**, 269–94.

Sprague, G.F. and Federer, W.T. (1951) A comparison of variance components in corn yield trials. *Agron. J.*, **43**, 535–41.

Sprague, G.F. and Schuler, J.F. (1961) The frequencies of seed and seedling abnormalities in maize. *Genetics*, **46**, 1713–20.

Sprague, G.F. and Tatum, L.A. (1942) General *vs* specific combining ability in single crosses of corn. *J. Amer. Soc. Agron.*, **34**, 923–32.

Srb, A.M., ed. (1973) *Genes, Enzymes and Populations.* Plenum Press, New York.

Stam, P. (1977) Selection response under random mating and under selfing in the progeny of a cross of homozygous parents. *Euphytica*, **26**, 169–84.

Stebbins, G.L. (1950) *Variation and Evolution in Plants.* Columbia University Press, New York.

Stebbins, G.L. (1971a) *Chromosomal Evolution in Higher Plants.* Arnold, London.

Stebbins, G.L. (1971b) *Processes of Organic Evolution.* Prentice-Hall, Englewood Cliffs, N.J., USA.

Stern, K. and Roche, L. (1974) *Genetics of Forest Ecosystems.* Springer, Berlin.

Stevens, N.E. (1948) Disease damage in clonal and self pollinated crops. *J. Amer. Soc. Agron.*, **40**, 841–4.

Storey, H.H. *et al.* (1958) East African work on breeding maize resistant to the tropical American rust. *Puccinia polysora. Emp. J. Exp. Agric.*, **26**, 1–17.

Stover, R.H. and Simmonds, N.W. (1987) *Bananas.* Longman, London, 3rd edn.

Streber, W.R. and Willmitzer, L. (1989) Transgenic tobacco plants expressing a bacterial detoxifying enzyme are resistant to 2, 4-D. *Bio/Technology* **7**, 811–16.

Street, H.E., ed. (1977) *Plant Tissue and Cell Culture.* Blackwell, Oxford and London, 2nd edn.

Stringfield, G.H. (1964) Objectives in corn improvement. *Adv. Agron.*, **16**, 101–37.

Stuber, C.W. (1992) Biochemical and molecular markers in plant breeding. *Pl. Breed. Rev.*, **9**, 37–61.

Stuber, C.W. (1994) Heterosis in plant breeding. *Pl. Breed. Res.*, **12**, 227–52.

Stuber, C.W. (1995) Mapping and manipulating quantitative traits in maize. *Trends in Genetics*, **11**, 477–82.

Stuber, C.W., Edwards, M.D. and Wenzel, J.F. (1987) Molecular marker facilitated investigations of Quantitative Trait Loci in Maize II Factors influencing yield and its component traits. *Crop Sci.*, **27**, 639–48.

Stuber, C.W., Williams, W.P. and Moll, R.H. (1973) Epistasis in maize (*Zea mays*): III. significance in predictions of hybrid performances. *Crop Sci.*, **13**, 195–200.

Struthman, D.D. and Steidl, R.P. (1976) Observed gain from visual selection for yield in diverse oat populations. *Crop Sci.*, **16**, 262–4.

Subandi and Compton, W.A. (1974) Genetic studies in an exotic population of corn (*Zea mays*) grown under two plant densities I, II. *Theor. Appl. Genet.*, **44**, 153–9, 193–8.

Subandi, Compton, W.A. and Empig, L.T. (1973) Comparison of the efficiencies of selection indices for three traits in two variety crosses of corn. *Crop Sci.*, **13**, 184–6.

Suneson, C.A. (1949) Survival of four barley varieties in a mixture. *Agron. J.*, **41**, 459–61.

Suneson, C.A. (1951) Male-sterile facilitated synthetic hybrid barley. *Agron. J.*, **43**, 234–6.

Suneson, C.A. (1956) An evolutionary plant breeding method. *Agron. J.*, **48**, 188–90.

Suneson, C.A. (1960) Genetic diversity – a protection against plant diseases and insects. *Agron. J.*, **52**, 319–21.

Suneson, C.A. (1969) Evolutionary plant breeding. *Crop Sci.*, **9**, 119–21.

Suneson, C.A. and Ramage, R.T. (1962) Competition between near-isogenic genotypes. *Crop Sci.*, 249–50.

Suneson, C.A. and Stevens, H. (1953) Studies with bulked hybrid populations of barley. *USDA Tech. Bull.*, 1067, p. 14.

Suneson, C.A. and Weibe, G.A. (1942) Survival of barley and wheat varieties in mixtures. *J. Amer. Soc. Agron.*, **34**, 1052–6.

Suwantaradon, K. and Eberhart, S.A. (1974) Developing hybrids from two improved maize populations. *Theor. Appl. Genet.*, **44**, 206–10.

Suwantaradon, K., Eberhart, S.A., Mock. J.J., Owens, J.C. and Guthrie, W.D. (1975) Index selection for several agronomic traits in the BSSS2 maize population. *Crop Sci.*, **15**, 827–33.

Swaminathan, M.S. and Howard, H.W. (1953) The cytology and genetics of the potato (*Solanum tuberosum*) and related species. *Bibl. Genet.*, **16**, 1–192.

Swanson, M.R., Dudley, J.W. and Carmer, S.G. (1974) Simulated selection in autotetraploid populations. I. Computer methods and effects of linkage. II. Effects of double reduction, population size and selection intensity. *Crop Sci.*, **14**, 625–30, 630–6.

Sybenga, J. (1969, 1973) Allopolyploidization of autopolyploids 1. Possibilities and limitations. 2. Manipulation of the chromosome pairing system. *Euphytica*, **18**, 355–71; **22**, 433–44.

Syme, J.R. (1972) Single plant characters as a measure of field plot performance of wheat cultivars. *Aust. J. Agric. Res.*, **23**, 753–60.

Tai, G.C.C. (1974) A method for quantitative genetic analysis of early clonal generation seedlings of an asexual crop with special application to a breeding population of the potato (*Solanum tuberosum*). *Theor. Appl. Genet.*, **45**, 150–6.

Tai, G.C.C. (1975) Effectiveness of visual selection for early clonal generations seedlings of potatoes. *Crop Sci.*, **15**, 15–18.

Tai, G.C.C. (1976) Estimates of general and specific combining abilities in potato. *Can. J. Genet. Cytol.*, **18**, 463–70.

Takebe, I., Labib, G. and Melchers, G. (1971) Regeneration of whole plants from isolated mesophyll protoplasts of tobacco. *Naturwissenschaften*, **58**, 318–20.

Tanksley, S.D. (1993) Mapping polygenes. *Ann. Rev. Genet.*, **27**, 205–33.

Tanksley, S.D., Medina-Filho, H. and Rick, C.M. (1982) Use of naturally occurring enzyme variation to detect and map genes controlling quantitative traits in an interspecific backcross of tomato. *Heredity*, **49**, 11–25.

Tanksley, S.D., Young, N.D., Paterson, A.H. and Bonierbale, M.W. (1989) RFLP mapping in plant breeding: new tools for an old science. *Bio/Technology*, **7**, 257–64.

Tarn, J.R. and Tai, G.C.C. (1977) Heterosis and variation of yield components in F_1 hybrids between Group Tuberosum and Group Andigena potatoes. *Crop Sci.*, **17**, 517–21.

Tatum, L.A. (1971) The southern corn leaf blight epidemic. *Science*, **171**, 1113–16.

Tavladoraki, P.E., Benvenuto, S., Trinca, D., De Martinis, D., Cattaneo, A. and Galeffi, P. (1993) Transgenic plants expressing a functional single chain Fv antibody are specifically protected from virus attack. *Nature*, **366**, 469–72.

Thomas, H.L. (1969) Breeding potential for forage yield and seed yield in tetraploid versus diploid strains of red clover. *Crop Sci.*, **9**, 365–6.

Thomas, M. (1952) *Back Crossing.* Commonwealth Bureau of Plant Breeding and Genetics, Cambridge (*Tech. Comm.* 16).

Thompson, D.L. (1977) Corn hybrid mixtures in a southern environment. *Crop Sci.*, **17**, 645–6.

Thompson, J.B. and Whitehouse, R.N.H. (1962) Studies on the breeding of self-pollinating cereals. 4. Environment and the inheritance of quality in spring wheats. *Euphytica*, **11**, 181–96.

Thompson, K.F. (1964) Triple cross hybrid kale. *Euphytica*, **13**, 173–7.

Thompson, K.F. (1966) Breeding problems in kale with particular reference to marrowstem kale. *Pl. Br. Inst. Camb. Ann. Rep., 1965–66*, 7–34.

Thompson, T.E. (1977) Haploid breeding technique for flax. *Crop Sci.*, **17**, 757–60.

Thomson, A.J. and Rogers, H.H. (1971) The interrelationship of some components of forage quality. *J. Agric. Sci. Camb.*, **76**, 283–93.

Thomson, A.J. and Wright, A.J. (1972) Principles and problems of grass breeding. *Pl. Br.Inst. Cambridge, Ann. Rep., 1971*, 31–67.

Thomson, N.J. (1972) Effects of the superokra leaf gene on cotton growth yield and quality. *Aust. J. Agric. Res.*, **23**, 285–93.

Thomson, N.J. (1973) Intra-varietal variability and response to single plant selection in *Gossypium hirsutum. J. Agric. Sci. Camb.*, **80**, 135–46, 147–60, 161–70.

Thorne, J.C. and Fehr, W.R. (1970) Exotic germ plasm for yield improvement in 2-way and 3-way soybean crosses. *Crop Sci.*, **10**, 677–8.

Thottappilly, G., Monti, D.R., Mohan Raj and Moore, A.W., eds (1992) *Biotechnology: Enhancing Research on Tropical Crops in Africa.* Technical Centre for Agricultural and Rural Cooperation (CTA) Wageningen.

Thurling, N. (1974) An evaluation of an index method of selection for high yield in turnip rape, *Brassica campestris* spp *oleifera. Euphytica*, **23**, 321–31.

Thurston, H.D., Heidrick, L.E. and Guzman, J.N. (1962) Partial resistance to *Phytophthora infestans* within the Coleccion Central Colombiana. *Amer. Potato J.*, **39**, 63–9.

Townley-Smith, T.F. and Hurd, E.A. (1973) Use of moving means in wheat yield trials. *Can. J. Plant Sci.*, **53**, 447–50.

Trenbath, B.R. (1974) Biomass productivity of mixtures. *Adv. Agron*, **26**, 177–210.

Trick, H.N. and Finer, J.J. (1997) SAAT: Sonication assisted *Agrobacterium*-mediated transformation. *Transgenic Research*, **6**, 329–37.

Troyer, A.F. and Brown, W.L. (1972) selection for early flowering in corn. *Crop Sci.*, **12**, 301–3.

Tucker, C.L. and Harding, J. (1974) Effect of the environment on seed yield in bulk population of Lima beans. *Euphytica*, **23**, 135–9.

Tysdal, H.M. and Crandall, B.H. (1948) The polycross progeny performance as an index of the combining ability of alfalfa clones. *J. Amer. Soc. Agron.*, **40**, 293–306.

Tysdal, H.M. Kiesselbach, T.A. and Westover, H.L. (1942) Alfalfa breeding. *Univ. Nebraska Agric. Exp. Sta. Res. Bull.*, 124.

Ucko, P.J. and Dimbleby, G.W., eds (1969) *The Domestication and Exploitation of Plants and Animals.* Aldine, Chicago.

Ullstrup, A.J. (1972) The impact of the southern corn leaf blight epidemics of 1970–1971. *Ann Rev. Phytopath.*, **10**, 37–50.

Umaerus, V. (1969, 1970) Studies on field resistance to *Phytophthora infestans. Z. f. PflzZüchtung*, **61**, 29–45, 167–94; **62**, 6–15, 357–69; **63**, 1–23.

UPOV (1991) *International Convention for the Protection of New Varieties of Plants, of December 1, 1961 as revised at Geneva on November 10, 1972, on October 23, 1978 and on March 19, 1991.* UPOV Publication no. 644 (E), Geneva.

USDA (1961) *Seeds.* USDA Yearbook of Agriculture, Washington D.C.

Van der Plank, J.E. (1963) *Plant Diseases: Epidemics and Control.* Academic Press, London and New York.

Van der Plank, J.E. (1968) *Disease Resistance in Plants.* Academic Press, New York and London.

Van der Plank, J.E. (1975) *Principles of Plant Infection*: Academic Press, New York, San Francisco and London.

Vasil, I.K., Scowcroft, W.R. and Frey, K.J., eds (1982) *Plant Improvement and Somatic Cell Genetics.* Academic Press, New York.

Vaughan, J.G., MacLeod, A.J. and Jones, B.M.G., eds (1976) *The Biology and Chemistry of the Cruciferae.* Academic Press, London, New York and San Francisco.

Vavilov, N.I. (1951) *The Origin, Variation, Immunity and Breeding of Cultivated Plants.* Chronica Botanica, Waltham, Mass., USA. (Translation from the Russian of selected writings.)

Verhalen, L.M., Baker, J.L. and McNew, R.W. (1975) Gardner's grid system and plant selection efficiency in cotton. *Crop Sci.*, **15**, 588–91.

Vogel, O.A., Allen, R.E. and Paterson, C.J. (1963) Plant and performance characteristic of semi-dwarf winter wheats producing more efficiently in eastern Washington. *Agron. J.*, **55**, 397–8.

Vogel, O.A., Graddoch, J.C., Muir, C.E. Everson, E.H. and Rhode, C.R. (1956) Semi-dwarf growth habit in winter wheat improvement in the Pacific north-west. *Agron. J.*, **48**, 76–8.

Vyas, V.S. (1975) India's high yielding varieties programme in wheat, 1966–67 to 1971–72. CIMMYT, el Batán, Mexico.

Walden, R. (1988) *Genetic Transformation in Plants.* Open University Press, Milton Keynes.

Walejko, R.N. and Russell, W.A. (1977) Evaluation of recurrent selection for specific combining ability in two open-pollinated maize cultivars. *Crop Sci.*, **17**, 647–51.

Walker, J.C. (1953) Disease resistance in the vegetable crops. *Bot. Rev.*, **19**, 606–44.

Walker, J.T. (1960) The use of selection index technique in the analysis of progeny row data. *Emp. Cott. Gr. Rev.*, **37**, 81–107.

Walker, J.T. (1963) Multiline concept and intra-varietal heterosis. *Cott. Gr. Rev.*, **40**, 190–215.

Walker, J.T. (1969) Selection and quantitative characters in field crops. *Biol. Rev.*, **44**, 207–43.

Wall, J.S. and Ross, W.M., eds (1970) *Sorghum Production and Utilization.* Avi, Westport, Conn., USA.

Wallace, D.H. (1985) Physiological genetics of plant maturity adaptation and yield. *Pl. Breed. Rev.*, **3**, 21–168.

Wallace, D.H., Ozbun, J.L. and Munger, H.M. (1972) Physiological genetics of crop yield. *Adv. Agron.*, **24**, 97–146.

Wallace, H.A. and Brown, W.L. (1956) *Corn and its Early Fathers.* Michigan State University, East Lansing, USA.

Walsh, E.J., Park, S.J. and Reinbergs, E. (1976) Hill plots for a preliminary yield evaluation of doubled haploids in barley. *Crop Sci.*, **16**, 862–6.

Wan, Y. and Lemaux, P.G. (1995) Biolistic transformation of microspore-derived and immature zygotic embryos and regeneration of fertile transgenic barley plants. In: *Gene Transfer to Plants* (eds I. Potrykus and G. Spangenberg), pp. 139–46, Berlin: Springer.

Warner, J.W. (1953) The evolution of a philosophy on sugar cane breeding in Hawaii. *Hawaiian Pl. Rec.*, **54**, 139–62.

Watts, L.E. (1963, 1965) Investigations into the breeding system of cauliflower, I, II. *Euphytica*, **12**, 323–40; **14**, 67–77.

Watts, L.E. (1970) Comparative responses of botanical varieties of *Brassica oleracea* to inbreeding and hybridization. *Euphytica*, **19**, 78–90.

Weatherspoon, J.H. (1970) Comparative yields of single, three-way and double crosses of maize. *Crop Sci.*, **10**, 157–9.

Webster, O.J. (1976) Sorghum vulnerability and germ plasm resources. *Crop Sci.*, **16**, 553–6.

Weiss, M.G. (1949) Soybeans. *Adv. Agron.*, **1**, 75–158.

Welch, W.W., ed. (1995) *The Oat Crop – Production and Utilization.* Chapman and Hall, London.

Wellensiek, S.J. (1947) Rational methods for breeding cross fertilizers. *Meded. Landbouwhogeschool, Wageningen*, **48**, 227–62.

Wellensiek, S.J. (1952) The theoretical basis of the polycross test. *Euphytica*, **1**, 15–19.

Wellensiek, S.J., ed. (1968) *Agricultural Sciences and World Food Supply.* Landbouwhogeschool, Wageningen, The Netherlands (*Misc. Papers*, 3).

Wellington, P.S. (1974) Crop varieties: their testing, commercial exploitation and statutory control. *J.R. Agric. Soc. Engl.*, **135**, 84–106.

Wellington, P.S. and Silvey, V. (1997) *A History of the National Institute of Agricultural Botany*. NIAB, Cambridge.

Weltzien, E. and Fischbeck, G. (1990) Performance and variability of local barley landraces in Near-Eastern environments. *Plant Breeding*, **104**, 58–67.

Wenzel, G. and Foroughi-Wehr, B. (1993) *In vitro* selection, pp. 353–70 in M.D. Hayward, N.O. Bosemark and I. Romagosa (eds) *Plant Breeding – Principles and Prospects.* Chapman & Hall, London.

Wernsman, E.A., Matzinger, D.F. and Mann, T.J. (1976) Use of progenitor species germ plasm for the improvement of a cultivated allotetraploid. *Crop Sci.*, **16**, 800–3.

de Wet, J.M.J., ed. (1988) *Biotechnology in Tropical Crop Improvement.* Proceedings of the International Biotechnology Workshop 12–15 January, 1987. ICRISAT Center, Hyderabad, India.

Whan, B.R. and Buzza, G.C. (1977) A flexible computer system for constructing plant breeding trials. *J. Natnl. Inst. Agric. Bot.*, **14**, 262–71.

White, G.A. *et al.* (1971) Agronomic evaluation of prospective new crop species. *Econ. Bot.*, **25**, 22–43.

White, T.G. and Richmond, T.R. (1963) Heterosis and combining ability in top and diallel crosses among primitive foreign and cultivated American upland cottons. *Crop Sci.*, **3**, 58–63.

Whitehouse, R.N.H. (1968) Barley breeding at Cambridge. *Pl. Br. Inst. Camb. Ann. Rep., 1967–68*, 6–29.

Whitehouse, R.N.H., Thompson, J.B. and Ribeiro, M.A.M.V. (1958) Studies on the breeding of self-pollinating cereals. 2. The use of diallel cross analysis in yield prediction. *Euphytica*, **7**, 147–69.

Whyte, R.O., Moir, T.R.G. and Cooper, J.P. (1959) Grasses in agriculture. *FAO Agric. Studies*, **42**, p. 416.

Wiberg, A. (1974) Sources of resistance to powdery mildew in barley. *Hereditas*, **78**, 1–40.

Widstrom, N.W. (1974) Selection indexes for resistance to corn earworm based on realized grains in corn. *Crop Sci.*, **14**, 673–5.

Wilcox, J.R. (1983) Breeding soybeans resistant to diseases. *Pl. Breed. Rev.*, **1**, 182–235.

Wilhem, S. (1974) The garden strawberry: a study of its origin. *American Scientist*, **62**, 264–71.

Wilkes, H.G. and Wilkes, S. (1972) The green revolution. *Environment*, **14**, 32–9.
Williams, E.G., Maheswaran, G. and Hutchinson, J.F. (1987) Embryo and ovule culture in crop improvement. *Pl. Breed. Rev.*, **5**, 181–236.
Williams, J.S. (1962) The evaluation of a selection index. *Biometrics*, **18**, 375–93.
Williams, J.T., ed. (1983) *Pulses and Vegetables.* Chapman & Hall, London.
Williams, J.T., ed. (1995) *Cereals and Pseudocereals.* Chapman & Hall, London.
Williams, P.H. (1975) Genetics of resistance in plants. *Genetics*, **79**, 409–19.
Williams, W. (1960) Relative variability of inbred lines and F_1 hybrids in *Lycopersicum esculentum*'. *Genetics*, **45**, 1457–65.
Williams, W. (1964) *Genetical Principles and Plant Breeding*. Blackwell, Oxford.
Williamson, J. (1975) *Useful plants of Malawi.* University of Malawi, Montfort, Press, Limbe, revised edition.
Willson, K.C. and Clifford, M.N. eds (1985) *Tea – Cultivation to Consumption.* Chapman & Hall, London.
Wise, W.S. (1975) The role of cost-benefit analysis in planning agricultural R. and D. programmes. *Res. Policy*, **4**, 246–61.
Wise, W.S. (1977) Cost benefit analysis of agricultural research: hybrid maize reconsidered. *R. and D. Management*, **8**, 29–32.
Wise, W.S. (1978) The economic analysis of agricultural research. *R. and D. Management*, **9**, 1–5.
Wittwer, S.H. (1974) Maximum production capacity of food crops. *Bio Science*, **24**, 216–24.
Wolfe, M.S. (1977) Yield stability in barley using varietal mixtures and disease control. *Cereals Res. Comm.*, **5**, 119–24.
Workman, P.L. and Allard, R.W. (1962) Population studies in predominantly self-pollinated species. III. A matrix model for mixed selfing and random outcrossing. *Proc. Natnl. Acad. Sci. Washington*, **48**, 1318–25.
Workman, P.L. and Allard, R.W. (1964) Population studies in predominantly self-pollinated species. V. Analysis of differential and random viabilities in mixtures of competing pure lines. *Heredity*, **19**, 181–9.
Wricke, G. (1976) Comparison of selection based on yield of half-sib progenies and of I_1 lines *per se* in rye. *Theor. Appl. Genet.*, **47**, 265–9.
Wricke, E. and Weber, W.E. (1986) *Quantitative Genetics and Selection in Plant Breeding*. De Gruyter, Berlin and New York.
Wright, A.J. (1973) The selection of parents for synthetic varieties of outbreeding diploid crops. *Theor. Appl. Genet.*, **43**, 79–82.
Wright, A.J. (1974) A genetic theory of general varietal ability for diploid crops. *Theor. Appl. Genet.*, **45**, 163–9.
Wright, A.J. (1976) The significance for breeding of linear regression analysis of genotype-environment interactions. *Heredity*, **37**, 83–93.
Wright, A.J. (1977) Inbreeding in synthetic varieties of field beans (*Vicia faba*). *J. Agric. Sci. Camb.*, **89**, 495–501.
Wright, J.W. (1976) *Introduction to Forest Genetics.* Academic Press, London.
Wright, S. (1968, 1969, 1977) *Evolution and the Genetics of Populations, I, II, III.* University of Chicago Press, Chicago and London.
Xoconostle-Cázares, B., Lozoya-Gloria, E. and Herrero-Estrella, L. (1993) Gene cloning and identification, pp. 107–25 in M.D. Hayward, N.O. Bosemark and I. Romagosa (eds) *Plant Breeding – Principles and Prospects.* Chapman & Hall, London.

Yarnell, S.H. (1954, 1962, 1965) Cytogenetics of the vegetable crops. *Bot. Rev.*, **20**, 277–359; **28**, 465–537; **31**, 247–330.

Yates, F. and Cochran, W.G. (1938) The analysis of groups of experiments. *J. Agric. Sci. Camb.*, **28**, 556–80.

Yonezawa, K. and Yamagata, H. (1977) On the optimum mutation rate and optimum dose for practical mutation breeding. *Euphytica*, **26**, 413–26.

Young, S.S.Y. (1961) A further examination of the relative efficiency of three methods of selection for genetic gains under less restricted conditions. *Genet. Res.*, **2**, 106–21.

Young, S.S.Y. and Weiler, H. (1961) Selection for two correlated traits by independent culling levels. *J. Genet.*, **57**, 329–38.

Yoshida, S. (1972) Physiological aspects of grain yield. *Ann. Rev. Pl. Physiol.*, **23**, 437–64.

Yoshida, Y. (1962, 1964) Theoretical studies on the methodological procedures of radiation breeding, I, II. *Euphytica*, **11**, 95–111; **13**, 65–74.

Yunbi Xu (1997) Quantitative Trait Loci; separating pyramiding and cloning. *Pl. Breed. Rev.*, **15**, 85–128.

Zadoks, J.C. (1972) Methodology of epidemiological research. *Ann. Rev. Phytopath.*, **10**, 253–76.

Zambryski, P., Joos, H., Genetello, C., Leemans, J., Van Montagu, M., Schell, J. (1983) Ti plasmid vector for the introduction of DNA into plant cells without alteration of their normal regeneration capacity. *EMBO Journal*, **2**, 2143–50.

Zeven, A.C. (1972) 'Plant density effect on expression of heterosis for yield and its components in wheat and F_1 versus F_3 yields'. *Euphytica*, **21**, 468–88.

Zhao, J., Fan, X., Shi, X. & Fan, Y. (1997) Gene pyramiding: An effective strategy of resistance management for *Helicoverpa armigera and Bacillus thuringiensis. Resistance Pest Management*, **9**, 19–21.

Zhukovsky, P.M. (1962) *Cultivated Plants and their Wild Relatives.* Commonwealth Agricultural Bureaux, Farnham Royal, England (abridged translation from the Russian (1950) by P.S. Hudson).

Zhukovsky, P.M. (1975) *World Gene Pool of Plants for Breeding.* U.S.S.R. Academy of Sciences, Leningrad.

Zillinsky, F.J. (1974) The development of Triticale. *Adv. Agron.*, **26**, 315–48.

Zohary, D. and Hopf, M. (1988) *Domestication of Plants in the Old World.* Oxford University Press.

Zuber, M.S. *et al.* (1971) Evaluation of ten generations of mass selection for corn earworm resistance. *Crop Sci.*, **11**, 16–18.

Index of Crops

General Index